# DNA-targeting Molecules as Therapeutic Agents

**Chemical Biology**

*Editor-in-Chief:*
Tom Brown, *University of Oxford, UK*

*Series Editors:*
Kira J. Weissman, *Lorraine University, France*
Sabine Flitsch, *University of Manchester, UK*
Nick J. Westwood, *University of St Andrews, UK*

*Titles in the Series:*
1: High Throughput Screening Methods: Evolution and Refinement
2: Chemical Biology of Glycoproteins
3: Computational Tools for Chemical Biology
4: Mass Spectrometry in Chemical Biology: Evolving Applications
5: Mechanisms of Primary Energy Transduction in Biology
6: Cyclic Peptides: From Bioorganic Synthesis to Applications
7: DNA-targeting Molecules as Therapeutic Agents

*How to obtain future titles on publication:*
A standing order plan is available for this series. A standing order will bring delivery of each new volume immediately on publication.

*For further information please contact:*
Book Sales Department, Royal Society of Chemistry, Thomas Graham House, Science Park, Milton Road, Cambridge, CB4 0WF, UK
Telephone: +44 (0)1223 420066, Fax: +44 (0)1223 420247,
Email: booksales@rsc.org
Visit our website at www.rsc.org/books

# *DNA-targeting Molecules as Therapeutic Agents*

Edited by

**Michael J. Waring**
*University of Cambridge, UK*
*Email: mjw11@cam.ac.uk*

ROYAL SOCIETY OF CHEMISTRY

THE QUEEN'S AWARDS FOR ENTERPRISE: INTERNATIONAL TRADE 2013

Chemical Biology No. 7

Print ISBN: 978-1-78262-992-4
PDF ISBN: 978-1-78801-292-8
EPUB ISBN: 978-1-78801-428-1
ISSN: 2055-1975

A catalogue record for this book is available from the British Library



The Royal Society of Chemistry is a charity, registered in England and Wales, Number 207890, and a company incorporated in England by Royal Charter (Registered No. RC000524), registered office: Burlington House, Piccadilly, London W1J 0BA, UK, Telephone: +44 (0) 207 4378 6556.

For further information see our web site at www.rsc.org

Printed in the United Kingdom by CPI Group (UK) Ltd, Croydon, CR0 4YY, UK

# *Preface to Sequence-specific DNA Binding Agents*

Twenty or thirty years ago the concept of gene targeting as a means of alleviating disease began to gain currency, whereas it had previously been little more than a pipe dream. Thanks to huge strides in our understanding of basic biology and medicine the concept is no longer a dream but an attainable goal. The talents of traditional biologists and geneticists, but perhaps above all chemists and biochemists, have conspired to create a climate in which the fruits of current research on DNA and gene function are clearly pointing the way to new therapies that will continue the revolutionary progress of chemotherapy as a prime modality for treating cancer. For much of cancer treatment to date has had to rely on drug interactions with DNA, and that is the principal area of medicine that has stimulated research on DNA binders (though it has to be acknowledged that such drugs can find useful employ in other areas of biotechnology and medical science as well). This book represents an attempt to summarise and illustrate some key aspects of the remarkable progress that has been made towards understanding how drugs can bind specifically to nucleic acids, and thus underpin the endeavour to make gene targeting a reality. The brief was to assemble a set of chapters written by senior scientists, who are acknowledged experts in the field, dealing with diverse aspects of the binding of antibiotics and drugs to DNA. Because of the importance of such substances in medicine, perhaps particularly the treatment of cancer, there are chapters that deal with established agents like actinomycin D, the first antibiotic found to be useful for cancer treatment, which is still in use today and forms the subject of chapter 6. After 60 years we are still learning new and surprising things about this

Chemical Biology No. 7
DNA-targeting Molecules as Therapeutic Agents
Edited by Michael J. Waring

Published by the Royal Society of Chemistry, www.rsc.org

remarkable antibiotic. The necessarily historical emphasis of this contribution is complemented by a rare and reflective chapter that describes the coming of age of theoretical and computational studies devoted to understanding how drugs interact specifically with DNA. Then there are contributions focussing on novel agents that show fairly immediate promise for the future of chemotherapy, notably topoisomerase inhibitors, telomerase inhibitors, peptide nucleic acids and triple helix-forming oligonucleotides. Research success is critically dependent upon advances in experimental methodology, so there is an important place for descriptions of new approaches that originate from the study of slow kinetics, melting curve analysis and improvements in classical medicinal chemistry brought about by discoveries originating from sophisticated forays into structural chemistry.

The book concludes with a thoughtful chapter on nucleic acid structures, mostly RNA, that might in due course become important targets for drug action.

The topics chosen by several authors are unique, or almost so, in cutting across the standard divisions of the discipline to provide a novel perspective. As a result, some areas, such as topoisomerase inhibitors and telomerase inhibitors, establish a framework that allows treatment by several different authors in a complementary manner. This may lead to occasional 'overlap' in information content, but has the inestimable virtue of furnishing a varied overview of progress in research from diverse points of view. The reader who progresses systematically through the text will be rewarded with some prime examples of how science works, through glimpses of the story of topoisomerase or telomerase inhibitors all the way from theory to drug development.

This wide compass of subject matter is a feature that may commend the book to students, established research workers, teachers and even historians of science. It is hoped that everyone will find something new and stimulating to read, set within the context of a coherent and multi-faceted attack upon some of the most pressing medical problems of the day. Unifying it all is a clear message of the role played by good chemistry in solving those problems.

# Preface

In the eleven years that have elapsed since the publication by the Royal Society of Chemistry of *Sequence-specific DNA Binding Agents* there have been notable advances in all areas of science pertaining to the chemistry behind the binding of small molecules to DNA. At the same time our understanding of the elements of molecular recognition at the level of nucleotide sequences has become increasingly important for drug discovery and development, not least in the treatment of cancer as well as other diseases that are of genetic origin. Accordingly, the present sequel represents an attempt to bring the reader up to speed with regard to advances in topics dealing with now well-established DNA binders and thus is a legitimate follow-up to the first edition, but now with its horizons greatly expanded through inclusion of an approximately equal number of wholly original chapters devoted to exciting new areas. Thus among the chapter updates that relate fairly directly to drug discovery we hear from David Graves about one of the earliest antibiotics actinomycin D which has saved great numbers of lives, Federico Gago on more antibiotics and drugs studied by computer simulations, Stephen Neidle on quadruplex-binding compounds, Bengt Norden on slow DNA binders, Keith Fox on triplex-forming oligonucleotides, Peter Nielsen on peptide nucleic acids, Yves Pommier on interfacial and topoisomerase inhibitors, and from Peter Dervan on synthetic polyamides designed specifically to target critical nucleotide sequences, such as those associated with prostate cancer. There is also an updated chapter from Brad Chaires expanded to describe new fluorescence methodology for studying drug–nucleic acid interactions, upon which he is a renowned expert. In response to reviewers of the first book as well as the remarks of many colleagues and keen advisors we have beefed up and coordinated topics like metal-based agents by including

Chemical Biology No. 7
DNA-targeting Molecules as Therapeutic Agents
Edited by Michael J. Waring

Published by the Royal Society of Chemistry, www.rsc.org

authoritative contributions from Jackie Barton and Christine Cardin. The field of peptide–DNA recognition, which has seen major advances in the past eleven years and offers equally great promise, is now represented by the inclusion of chapters from Annemieke Madder and Leung Sheh. Finally, the growing importance of finding treatments for dementia among our ageing populations receives mention from Ming-Hon Ho in his chapter on binding of small molecules to trinucleotide DNA repeats associated with neurodegenerative diseases, and the remarkable promise of the new CRISPR/Cas9 technology for genome editing in a brave new world is expounded by Thomas Bentin (a colleague of Peter Nielsen) who was kind enough to put his shoulder to the wheel at a late stage of writing. To all these stalwart authors as well as their collaborators duly acknowledged in the list of contributors, the editor and publishers offer their heartfelt gratitude and echo the sentiments expressed in the preface to the first edition reprinted below, namely that the wide compass of subject matter should commend the book to students, established research workers, teachers and even historians of science. It is hoped that everyone will find something new and stimulating to read, set within the context of a coherent and multi-faceted attack upon some of the most pressing medical problems of our day. Unifying it all is a clear message of the role played by good chemistry in solving those problems.

Michael J. Waring

# Contents



Chemical Biology No. 7
DNA-targeting Molecules as Therapeutic Agents
Edited by Michael J. Waring

Published by the Royal Society of Chemistry, www.rsc.org

CHAPTER 1

# DNA Recognition by Parallel Triplex Formation

KEITH R. FOX,*[a] TOM BROWN*[b] AND DAVID A. RUSLING*[a]

[a] Biological Sciences, Life Sciences Building 85, University of Southampton, Highfield, Southampton SO17 1BJ, UK;
[b] Chemistry Research Laboratory, Department of Chemistry, University of Oxford, Oxford OX1 3TA, UK
*Email: K.R.Fox@soton.ac.uk; Tom.brown@chem.ox.ac.uk; D.A.Rusling@soton.ac.uk

## 1.1 Why Triplexes?

Triplex-forming oligonucleotides (TFOs) bind in the duplex major groove by forming hydrogen bonds with exposed groups on the Watson–Crick (W–C) base pairs, generating a triple-helical structure (*e.g.*, Figure 1.1). The unique base–base recognition properties of these molecules can be exploited as a means to target duplex sequences present or embedded within natural or synthetic DNA.[1,2] Unlike most DNA-recognition agents, such as polyamides, TFOs are capable of targeting extended sequences, with a relatively low propensity to bind to non-target sites. In this way TFOs have been exploited as gene-targeting agents for modulating gene expression,[3,4] as a means to detect and/or isolate plasmid and genomic DNA for molecular biology or diagnostics,[5,6] and as a tool to introduce functionality into DNA nanostructures engineered for bionanotechnology or synthetic biology.[7] Despite this, the applications of TFOs that contain natural nucleotides are often restricted by their low binding affinity and slow association kinetics at neutral pH, as well as a requirement for oligopurine–oligopyrimidine duplex

Chemical Biology No. 7
DNA-targeting Molecules as Therapeutic Agents
Edited by Michael J. Waring

Published by the Royal Society of Chemistry, www.rsc.org

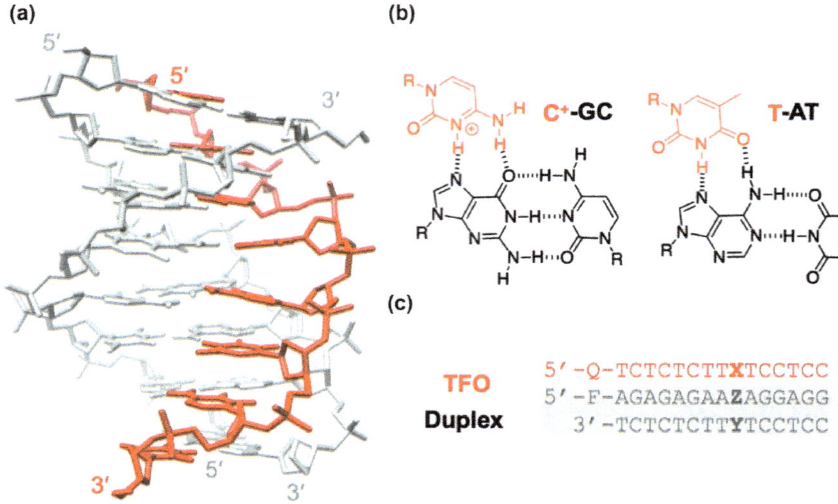


**Figure 1.1** Triplex-directed DNA recognition. (a) Structure of a parallel DNA triplex (PDB code: 1D3X); (b) Chemical structures of $C^+$–GC and T–AT base triplets (R is deoxyribose); (c) Triplex sequence used to characterise the triplex-forming properties of an oligonucleotide containing a single nucleotide analogue (at position X) against each of the four base pairs (at position ZY) by fluorescence melting using molecular beacons (F is a fluorophore: Q is a quencher). In each case the third strand is shown in red and the duplex in black.

target sequences. To overcome these limitations a variety of base, sugar and phosphate modifications have been developed to allow triplex formation at mixed-sequence targets with high affinity at neutral pH. This chapter will review the developments and current state-of-the-art of nucleotide modifications used to improve the triplex-forming properties of oligonucleotides.

### 1.1.1 Triplets and Triplex Motifs

Triplexes were first observed experimentally over 60 years ago by Rich and co-workers after mixing the polyribonucleotides poly-U and poly-A in a 2 : 1 ratio.[8] Additional studies demonstrated that poly-C and poly-G could generate a similar structure under low-pH conditions,[9] and since then a variety of DNA and RNA triplexes have been identified.[10–13] The binding of an oligonucleotide within the major groove is asymmetric and can occur in either a parallel or antiparallel orientation relative to the oligopurine-containing strand of the target duplex. Pyrimidine-rich oligonucleotides bind in a parallel orientation under slightly acidic conditions ($pH < 6.0$), with T and protonated C forming Hoogsteen hydrogen bonds with AT and GC base pairs, generating the base triplets T–AT and $C^+$–GC, respectively (Figure 1.1b).[1,2] (In this chapter the notation X–ZY refers to a triplet, in which the third strand base X interacts with the duplex base pair ZY, forming hydrogen bonds to base Z.) In contrast, purine-rich oligonucleotides bind in an antiparallel orientation, with A and G forming reverse-Hoogsteen hydrogen bonds with AT and GC base pairs respectively, generating A–AT and G–GC triplets.[14,15]

In theory, both triplex motifs could be usefully exploited for the recognition of unique duplex sequences but the greater stability of the parallel motif has meant it has been more widely adopted. Parallel triplexes are intrinsically more stable than their antiparallel counterparts because T–AT and $C^+$–GC triplets are structurally isomorphic; that is, if the C-1′ atoms of their W–C base pairs are superimposed, the positions of the C-1′ atoms of the third strand are almost identical.[16] This minimises backbone distortions of both the third strand and duplex between adjacent triplets. In contrast, antiparallel triplets are not isomorphic and lead to structural distortions at the junctions between consecutive triplets. The use of the antiparallel motif is also hampered by the tendency of purine-containing oligonucleotides to self-associate into structures such as G-quadruplexes and GA-duplexes, which compete with triplex formation and reduce the effective TFO concentration. It should also be noted that both G–GC and T–AT triplets can be generated in both binding motifs, and GT-containing oligonucleotides can therefore be designed to bind in either orientation. However, the non-isomorphic nature of these two triplets means that the most stable orientation is dependent on the number of GpT and TpG steps.[17] This chapter will therefore focus on triplexes generated through the parallel binding motif using pyrimidine-rich oligonucleotides.

### 1.1.2 Base, Sugar and/or Phosphate Modifications

We and others have characterised a variety of novel base, sugar and phosphate modifications designed to improve the triplex-forming properties of oligonucleotides and a wealth of data has been generated on the affinity, kinetics and selectivity of triplexes containing these modifications. However, it is often hard to compare the effectiveness of a given modification between studies, since its properties will depend on its positioning within the third strand, the sequence context, length of third strand and/or duplex, as well as the pH and other solution conditions, such as the presence of divalent cations. Most experiments have involved the characterisation of a single substitution within a third strand by examining its interaction with duplexes containing each of the four base pairs at the same position. In this way an X–ZY triplet is generated, where X is the analogue under study and ZY is either an AT, TA, GC or CG base pair in turn (*e.g.*, Figure 1.1c). Often the modification is compared with the most effective natural nucleotide at the same position, *i.e.*, with T, $C^+$, G or T opposite an AT, GC, TA or CG base pair respectively. The formation of G–TA and T–CG triplets for recognising pyrimidine–purine base pairs will be discussed in Section 1.4.2.

The simplest and most common means for characterising triplex stability is by ultraviolet melting, in which the triplex thermal stability is determined from the temperature-dependent change in absorbance at 260 nm, generating a melting curve, from which the melting temperature ($T_m$) is estimated. However, the analysis of such melting curves is not always straightforward, since the triplex–duplex and duplex–single-strand transitions often overlap. We find that a better approach is to use synthetic oligonucleotides that contain molecular beacons and to measure the fluorescence melting curves.[18] This works best when the fluorescence quencher (*e.g.*, dabcyl) is attached to the TFO and the fluorescent group (*e.g.*, fluorescein) is attached to one of the duplex strands (as shown in Figure 1.1c). In this way, the concentration of the TFO can be varied without altering the background fluorescence. We have used this strategy to characterise 15 of the nucleotide analogues described in this chapter using the same model triplex (Figure 1.1c), with experiments undertaken using the same buffer conditions. (Experiments were performed in 50 mM sodium acetate buffer (pH 6.0) containing 200 mM sodium chloride using a temperature gradient of 0.2 °C $min^{-1}$ and no hysteresis between melting annealing curvès was observed.) To allow the reader to make a useful, and unbiased, comparison between these modifications we have included a table of $T_m$ values later in this chapter (Table 1.1).

**Table 1.1** Comparing nucleotide modifications. Melting temperatures ($T_m$) determined for various parallel triplexes containing a central X–ZY triplet, where X is a natural or modified nucleotide, and ZY is each base pair in turn (sequences shown in Figure 1.1c). Experiments were undertaken as described in the main text and (–) indicates that the melting temperature was lower than 25 °C.

| ZY (base pair) | | | | | |
|---|---|---|---|---|---|
| | X (TFO) | AT | GC | CG | TA |
| Natural nucleotides | A | 28.2 | 30.0 | – | – |
| | G | – | 28 | – | **30.6** |
| | C | – | **40.5** | 28.4 | – |
| | T | **39.4** | 27.7 | **28.3** | – |
| AT recognition | APdU | 41.8 | 30.9 | 29.5 | – |
| | GPdU | 43.7 | 32.3 | 29.9 | – |
| | DMAPdU | 43.8 | 30.7 | 28.4 | – |
| | BAU | 47.5 | 36.5 | – | – |
| | BGU | 47.4 | 35.0 | – | – |
| GC recognition | APdC | 28.8 | 41.1 | 33.7 | – |
| | AP | – | 44.6 | – | – |
| | AE-N7G | – | 40.3 | – | – |
| CG recognition | $^{M}$PP | – | – | 31.5 | – |
| | $^{A}$PP | – | – | 32.1 | – |
| | AE-$^{A}$P | 39.9 | 31.1 | 32.4 | – |
| TA recognition | APdG | – | 30.2 | – | 31.0 |
| | AE-G | – | 36.5 | – | – |
| | S | 30.6 | 30.4 | 33 | 34.6 |
| | AE-S | 31.3 | 31.2 | 33.6 | 36.5 |

## 1.2 Stabilising Triplexes

Triplex stability stems from the formation of two Hoogsteen hydrogen bonds between each base in the third strand and its duplex partner as well as favourable stacking interactions between consecutive bases. Under low-pH conditions the stability of a parallel triplex can be greater than that of its underlying duplex, *i.e.*, the affinity of a third strand for its duplex target is greater than the affinity of a duplex strand for its W–C partner.[19] However, the majority of applications proposed for TFOs require that they bind with high affinity at neutral pH. Although a variety of base analogues have been used to alleviate, at least in part, the pH dependence of the $C^+$–GC triplet (see Section 1.3), the affinity of the third strand can also be improved by increasing the stability of the canonical T–AT and $C^+$–GC triplets.

### 1.2.1 Enhancing Stacking and Hydrophobic Interactions

Base stacking and hydrophobic interactions are important factors that influence the structure and stability of both duplex and triplex DNA. Consequently, several thymine analogues have been prepared with additional aromatic rings across the 4–5 or 5–6 positions, which should increase the aromatic surface area of the base without affecting the hydrogen bonding groups.[20–22] However, and somewhat surprisingly, triplexes containing these analogues did not demonstrate any enhanced stability. The best of these, a non-natural pyrido[2,3-*d*] pyrimidine nucleoside (F; Figure 1.2a), was shown to recognise AT base pairs with an affinity that was similar to, but not greater than, that of unmodified T.[20] However, these studies were undertaken with isolated substitutions and it is likely that only multiple adjacent substitutions will improve stability through stacking interactions. A more successful strategy has been to introduce hydrophobic substituents at the 5-position of the base to increase hydrophobic interactions within the major groove. The simplest addition is a methyl group and probably explains why T–AT triplets are more stable than U–AT and $^{Me}C^+$–GC triplets are more stable than $C^+$–GC.[23,24] The addition of a propynyl group (PdU; Figure 1.2b) further extends the hydrophobic surface and it has been shown that each PdU substitution increases the $T_m$ of the triplex by *ca.* 2.5 °C relative to an

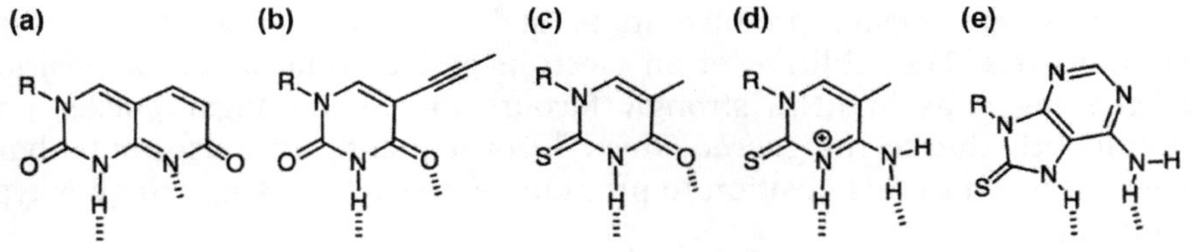


**Figure 1.2** Chemical structures of base modifications that increase stacking/hydrophobic interactions. (a) F; (b) PdU; (c) $s^2T$; (d) $m^5s^2C$; (e) $s^8A$. For each structure R is deoxyribose unless otherwise stated in the main text.

unmodified third strand.[25–29] Subsequent NMR studies revealed that, as expected, the extended aromatic electron cloud of the PdU nucleotide stacks well over the 5′-neighboring nucleotides, and is most probably the cause of the increased stabilisation.[26] A further study examined the properties of four different C5-amino modified deoxyuridines and showed that the order of stability produced by 5-substitutions is alkyne > *E*-alkene > alkane > *Z*-alkene.[29] This order must result from steric factors as well as stacking interactions. The same strategy cannot be applied for increasing the affinity of C for a GC base pair, since the addition of a propynyl group to the 5-position lowers the $pK_a$ of the base and increases the pH dependence of the triplet.[28] Indeed the attachment of a propargylamino chain to the 5-position (APdC), a similar modification that is discussed further below, generated triplexes of equivalent stability to those formed by cytosine at pH 6.0 (Table 1.1).

More recently the introduction of thiocarbonyl groups to the 2-position of thymine ($s^2T$; Figure 1.2c) and 5-methylcytosine ($m^5s^2C$; Figure 1.2d), as well as the 8-position of adenine ($s^8A$; Figure 1.2e), has proved to be a useful strategy for increasing triplex stability.[30,31] Molecular modelling of a parallel triplex containing $s^2T$ in the third strand indicated that the 2-thiocarbonyl group of the 5′-upstream base could interact with the nitrogen atom at the 1 position of the 3′-downstream pyrimidine ring and result in strong stacking effects.[28] Indeed triplexes containing multiple substitutions of $s^2T$ led to a $T_m$ increase of around 5 °C per modification at pH 7.0. A further enhancement in affinity was seen when combined with $m^5s^2C$ and particularly evident for TFOs containing multiple, adjacent substitutions. The $m^5s^2C$ and $s^8A$ analogue were developed for the pH-independent recognition of GC base pairs and are discussed further below (see Sections 1.3.1 and 1.3.2, respectively).

### 1.2.2 Locking the Sugar Pucker

The affinity of a TFO for its duplex target is affected by its ability to adopt N- or S-type sugar pucker conformations and it has been proposed that the former require less distortion of the duplex purine strand upon triplex formation.[32] This is thought to explain why TFOs composed of ribonucleotides exhibit a higher affinity for duplex DNA than those composed of deoxyribonucleotides.[33–37] In general, oligonucleotide modifications that favour N-type sugars produce more stable triplexes than their S-type counterparts. The addition of an electronegative group at the 2′-position of the sugar, as in RNA, strongly favours the N-type sugar pucker predominantly due to the *gauche* effect.[38] Consequently various groups have been attached to this position to promote or restrict the sugar to an N-type configuration.

The first chemical moiety to be added to 2′-position that resulted in improved TFO binding was a methoxy group (2′-OMe; Figure 1.3a).[34,36] Subsequent NMR studies confirmed that this resulted from a reduced distortion

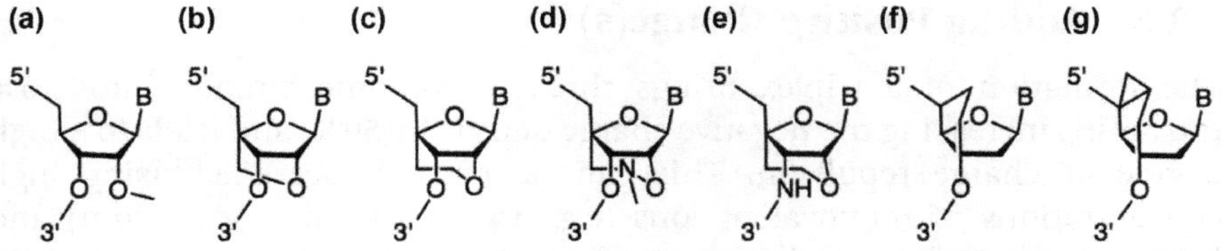


**Figure 1.3** Chemical structures of sugar modifications that restrict the sugar pucker. (a) 2′-OMe; (b) BNA/LNA; (c) ENA; (d) BNA$^{NC}$; (e) 3′-amino-2′,4′-BNA; (f) bicyclo-DNA; and (g) tricyclo-DNA. For each structure B is the DNA base or base analogue.

of the duplex purine strand, enhancing the rigidity of the triplex.[32] A better modification that locks the sugar pucker in an N-type configuration and reduces the rotational freedom of the sugar phosphate backbone is bridged/locked nucleic acid (BNA or LNA; Figure 1.3b).[39,40] This modification exploits a 2′-O,4′-C methylene bridge to constrain the sugar to N-type and was developed independently by the Wengel and Imanishi groups for use in antisense or antigene applications, respectively. TFOs that contain BNA/LNA residues are markedly more stable than their unmodified counterparts but only when substituted every 2–3 nucleotides.[40,41] Further BNA/LNA derivatives have been developed to overcome this sequence restriction. Substitution of the bridge with an ethylene moiety (ENA; Figure 1.3c), which contains an additional carbon, allows triplex formation with fully modified TFOs.[42] Whilst the introduction of an O–N bond (BNA$^{NC}$; Figure 1.3d) further improved the nuclease resistance of the oligonucleotide.[43] TFOs containing the N-methyl derivative of BNA$^{NC}$ were stable in serum for over 90 minutes compared with an unmodified oligonucleotide, which was completely degraded in 5 minutes. Such modifications are likely to be useful for any applications that require the use of TFOs within a physiological setting. Further thermodynamic and kinetic studies revealed that the enhancement in affinity stemmed from a decrease in the dissociation constant of the TFO.[44] An additional advantage of the sugar analogue is that the presence of the nitrogen atom allows functionalisation with other chemical groups, such as fluorophores. Lastly, 3′-amino-2′,4′-BNA has been developed in which the BNA/LNA modification is combined with the N3′-P5′ modification considered below (Figure 1.3e, see Section 1.2.4).[45,46] Although triplexes with this analogue were more stable than their duplex equivalents, they were no more stable than those formed with BNA/LNA alone.[46] Further attempts to constrain the sugar pucker to either S-type or N-type include the bicyclo and tricyclo furanose modifications developed by Leumann and co-workers (Figure 1.3f and g).[47–49] Bicyclo-DNA contains a 3′-O,5′-C ethylene bridge that locks the sugar in an S-type conformation, while the tricyclo derivative contains an additional cyclopropane unit locking the sugar in an N-type pucker. Studies with TFOs composed of tricyclo-modified thymidine showed an increase in $T_m$ of 2 °C per modification at pH 7.0.[49]

## 1.2.3 Adding Positive Charge(s)

The formation of a triplex brings three polyanionic strands into close proximity, increasing the negative charge density by 50%, and leads to a high degree of charge repulsion. This can be partially screened using high concentrations of monovalent ions (*e.g.*, up to 200 mM of sodium) and lower concentrations of divalent or polycationic ions (*e.g.*, up to 10 mM magnesium or spermine).[1,50] Consequently, the incorporation of positively charged moieties into the TFO by their addition to the phosphate, sugar or base has helped increase triplex stability by alleviating in part, some of this charge repulsion.

### *1.2.3.1 Phosphate*

One means to incorporate charges into a TFO is by their appendage to the phosphodiester backbone. For example, Bruice and co-workers have shown that the addition of positively charged guanidinium linkages (DNG; Figure 1.4a) causes a dramatic increase in TFO affinity.[51,52] The synthesis of the ribose derivative has also been reported but to our knowledge this has not yet been studied for its triplex-forming properties.[53] Two further modifications that replace the phosphate residues with either cationic dimethylaminopropyl phosphoramidate linkages (PNHDMAP) or *N,N*-diethyl-ethylenediamine linkages (DEED; Figure 1.4b) have also been characterised.[54–56] TFOs with these modifications generated triplexes which were

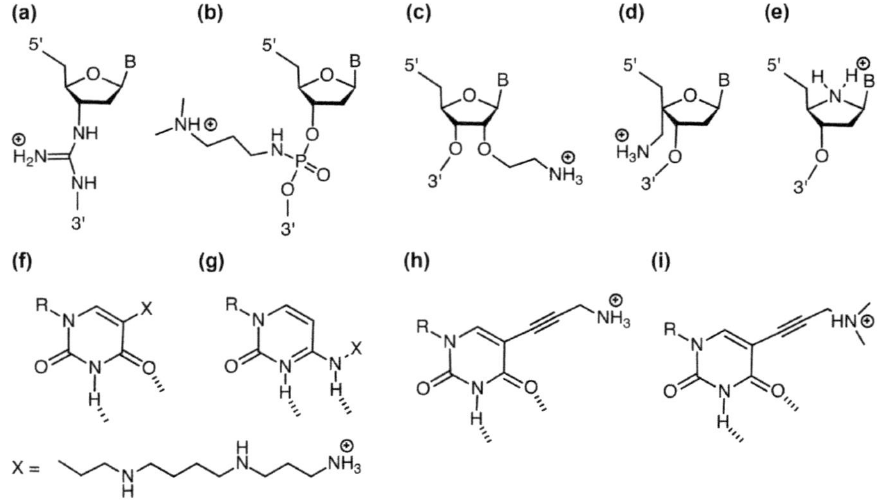


**Figure 1.4** Chemical structures of phosphate, sugar and base modifications that introduce positive charge. (a) DNG; (b) DEED; (c) 2′-AE; (d) 4′-AE; (e) pyrrolidine-DNA; (f) U$^S$; (g) $^{Me}C^S$; (h) APdU; (i) DMAPdU. For each structure B is a base or base analogue; R is deoxyribose unless otherwise stated in the main text.

more stable than the underlying duplex at pH 7.0.[56] More recently, oligonucleotides containing non-nucleosidic monomers composed of partially protonated amines have been prepared. When incorporated at the TFO termini such modifications lead to a significant enhancement of triplex stability, particularly when positioned at the 5′-end of the TFO.[57]

#### 1.2.3.2 Sugar

Perhaps the most exploited approach for the introduction of positive charges into a triplex has relied on their addition to the 2′ (Figure 1.4c)[58] and 4′-positions (Figure 1.4d)[59] of the sugar unit. In both cases, the most stable triplexes were formed by addition of an aminoethoxy side chain with a $T_m$ increase of 3.5 °C and 1 °C per modification at pH 7.0 for the 2′ and 4′ derivatives, respectively. The greater stabilisation afforded by the 2′-derivative has been attributed to the formation of a salt bridge between the positive charge and a pro-R oxygen of the negatively charged phosphate of the purine strand and a favourable N-type sugar pucker as described above.[60] For experiments with psoralen-linked oligonucleotides it was suggested that the 2′-aminoethoxy modification is more effective when the positively charged derivatives are clustered together.[61] In addition, introduction of these cationic modifications in the 5′-region of the TFOs significantly increased the $k_{on}$ values compared with that of natural TFO, while no enhancement in the rate of triplex DNA formation was observed when the modifications were in the middle and at the 3′-region.[62] It is likely that this effect is due to the nucleation zipper mechanism proposed for triplex formation.[63] Replacement of the amine with a guanidinium group, which positions three amines in a plane, has also proved useful.[64] Guanidinylation can be achieved post oligonucleotide synthesis and offers the advantage that the group is protonated over a greater pH-range than the amine; the $pK_a$ of a primary amine is around 9, whilst the $pK_a$ of the guanidinium group is 12.5. It is also offers the potential of forming up to five hydrogen bonds. This modification typically gives the same increase in stability as with the primary amine at neutral pH but in principle should give greater triplex stabilisation at higher pH values. Lastly, substitution of the furanose oxygen with nitrogen, generating pyrrolidine oligonucleotides (Figure 1.4e) can be exploited to position a positive charge next to the pro-R non-bridging phosphate oxygen in the purine strand. However, the degree of stabilisation depends on the attached base; the presence of the cytosine mimic pseudoisocytosine resulted in a $T_m$ increase of 2 °C per modification relative to C (see Section 1.3.1),[65] whilst uracil was destabilising relative to T.[66,67]

#### 1.2.3.3 Base

Since it has been observed that the $C^+$–GC triplet is more stable than T–AT, due to the presence of the positive charge, various polyamines have been appended to different positions of the pyrimidines. Attachment of spermine to the 5-position of uracil[68] ($U^S$; Figure 1.4f) and the N4 position of methylcytosine[69] ($^{Me}C^S$; Figure 1.4g) both led to an increase in triplex stability under physiological pH conditions, though the complexes exhibited decreased sequence selectivity. We have also improved the affinity of T for AT base pairs by preparing the base analogue 5-propargylamino dU (APdU; Figure 1.4h).[70] This analogue bears a positive charge attached to the 5-position of U rather than in the stacked ring system (as seen with protonated C). The presence of the alkyne moiety is also expected to contribute to triplex stability by enhancing stacking interactions in the major groove (as discussed previously). TFOs containing multiple substitutions of APdU are markedly more stable than unmodified TFOs, though the complexes are still pH-dependent on account of the requirement for protonation. However, runs of adjacent substitutions are not destabilising, in contrast to protonated C. This demonstrates that removing the charge from the $\pi$-stack and placing it in the major groove is a useful approach for stabilising triplexes. Each APdU substitution leads to a typical increase in $T_m$ of *ca.* 2 °C relative to T whilst retaining sequence selectivity (Table 1.1). The guanidinylated version of APdU (GPdU) also led to a slight enhancement in affinity with a typical increase of *ca.* 4 °C per modification (Table 1.1).

The synthesis of nucleoside analogues that combine a stabilising base modification with a suitable sugar modification has proved very useful. Bis-amino U (BAU) is a nucleoside analogue that contains both a 5-propargylamino modification on the base and a 2′-aminoethoxy side chain on the sugar.[71,72] At physiological pH both modifications are protonated and substantially increase the $T_m$ of the complexes by *ca.* 8 °C per modification (Table 1.1). The guanidylated version of BAU (BGU) also leads to a similar enhancement in affinity (Table 1.1). The two positive charges act in different ways to enhance triplex stability: the 2′-aminoethoxy group interacts with a phosphate on the duplex purine strand, while the 5-propargylamino group interacts with a third strand phosphate.[71] Positioning this analogue opposite a duplex mismatch decreased the stability of the complex and demonstrated the requirement for positioning the charges at precise locations within the triplex structure.[73] Interestingly, BAU also exhibits a greater sequence selectivity than thymidine, with enhanced discrimination against pyrimidine inversions, and removes the requirement for magnesium ions.[72] Further analysis revealed that triplexes containing BAU exhibit very slow binding kinetics, stemming from a decreased rate of dissociation as the modification had little effect on the association reaction. The sequence selectivity is also due to the slower dissociation of BAU from AT than other base pairs.[74] The 2′-methoxyethyl derivative of APdU has also been prepared and whilst it generates more stable triplexes relative to those that lack the modification it was not as stabilising as BAU.[75] The addition of alkynyl modifications to the 5-position of T has been investigated in the context of BNA/LNA and both ethynyl and propargylamine modifications increased triplex stability by >13.5 °C per modification at pH 7.0.[76]

One of the drawbacks of using such charged nucleoside analogues is the occurrence of side-reactions at the amines during oligonucleotide synthesis and deprotection, which limits their compatibility with other chemical groups. To overcome this problem we have prepared both 5-dimethylaminopropargyl-dU (DMAPdU; Figure 1.4i) and 2′-dimethylaminoethoxy-U, which contain dimethylamines in place of the amine groups.[77,78] Triplexes generated with these analogues are more stable than those containing T but are less stable than those containing the equivalent amine modification. We suggest two possible explanations for their slightly lower affinity: firstly, the addition of the methyl groups could sterically hinder the interaction of the TFO, secondly, the amine group in the parent compound may contribute to a hydrogen bond donor interactions. Analysis of the kinetics of the DMAPdU modification again revealed that the increase in stability stems from a slower dissociation rate of the modified TFO.[78] Another problem that stems from using TFOs containing multiple charges is that they can suffer from off-site binding. To examine this further we used a restriction enzyme protection, selection and amplification assay (REPSA) to isolate sequences that are bound by a heavily modified 9-mer TFO containing six adjacent BAU modifications.[79] The TFO was capable of interacting with a variety of different sequences that contained $A_n$ tracts ($n=6$) even though the surrounding sequence did not match the remainder of the TFO sequence.

## 1.2.4 Removing Negative Charge(s)

An alternative strategy for decreasing the charge repulsion between the three polyanionic strands is to use TFOs that contain non-charged backbones. Replacement of the phosphate linkage with a 3′-5′-methylphosphonate group (Figure 1.5a) was successfully used for triplex formation using short oligonucleotides containing alternating methylphosphonate and phosphodiester linkers.[80] However, subsequent studies with longer fully substituted

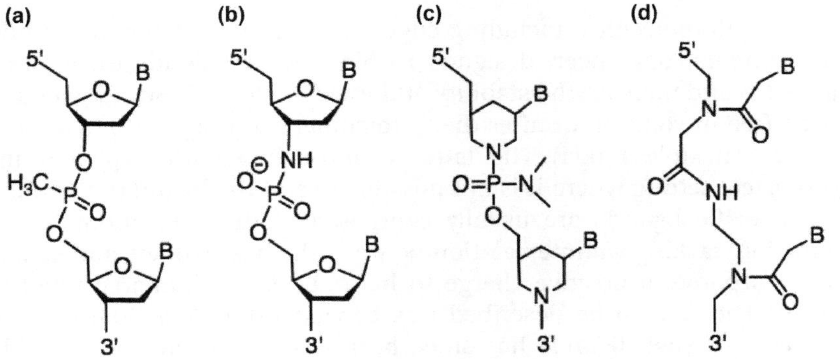


**Figure 1.5** Chemical structures of backbone modifications. (a) 3′–5′-methylphosphonate; (b) N3′–P5′ amidate; (c) morpholino; (d) PNA. For each structure B is a base or base analogue.

TFOs showed that this modification was destabilising.[81,82] The N3′–P5′ amidate modification, where O3′ of the internucleoside phosphate is replaced by NH[83] (Figure 1.5b) increases the binding constant at neutral pH by nearly two orders of magnitude. Triplex binding is probably improved as this modification favours the N-type sugar conformation as discussed above. This modification has also been combined with the addition of a cationic copolymer, which cooperatively stabilises triplex formation and increases association rates by four orders of magnitude.[84] Morpholino oligonucleotides are another interesting class of analogues in which the ribose sugar is replaced with a six-membered morpholino ring and the phosphodiester linkage is replaced by a phosphorodiamidate (Figure 1.5c). TFOs containing this modification are less stable than those containing the N3′–N5′ modification at high concentrations of cations but are more stabilising at low ionic strength.[85–87]

Perhaps the most extensively employed uncharged backbone modification is peptide nucleic acid (PNA; Figure 1.5d). PNA is composed of repeating (2-aminoethyl)glycine units to which nucleobases are linked by methylene bridges.[88,89] PNA usually interacts with duplex DNA *via* a mechanism of strand displacement and P-loop formation, requiring two molecules of PNA,[90] generating a 2 : 1 PNA : DNA triplex. Two pyrimidine-containing PNA molecules form a local triplex with the purine-containing DNA strand. This leaves the pyrimidine DNA strand looped out as a single strand. The resulting triplex is more stable than the equivalent DNA triplex since there is much lower charge repulsion between the three strands. In a few instances PNA can form a 1 : 2 PNA : DNA triplex by simple binding of a PNA third strand to a DNA duplex, though this is usually restricted to cytosine-rich PNAs.[91] Although PNAs demonstrates excellent hybridisation properties the lack of charge often makes such oligomers insoluble unless other cationic groups are attached.

### 1.2.5 Triplex-binding and Cross-linking Agents

Various small-molecules, including edge binders, intercalators and minor-groove binders, have been designed to bind non-covalently to duplex or triplex DNA and increase the stability of these complexes.[92] Such ligands can be used free in solution or after their attachment to the 3′ or 5′-end of the TFO *via* a flexible linker. The latter is most frequently exploited in a physiological setting where it is not possible to add the ligand to the buffer. Triplex-specific ligands are usually composed of extended aromatic ring systems for stacking *via* intercalation between the base triplets and some of these incorporate a positive charge to help alleviate the charge repulsion problem. The first to be described was benzopyridoindole (BePI)[93] and a wide range of such ligands has since been developed and reviewed elsewhere.[94] These compounds have also been used to enhance the affinity of weaker triplexes that are formed at oligopurine sequences that contain pyrimidine interruptions.[95] Another approach that has received much attention is the attachment of DNA cross-linking agents to the TFO. The most frequently employed photo-cross-linking agent is 4,5,8-trimethylpsoralen (psoralen) which preferentially intercalates at TpA steps, and upon photoactivation with long-wavelength UV light leads to a 2 + 2 cycloaddition with the adjacent thymidines, cross-linking the TFO to one or both strands of the duplex.[96] Initially, cross-linking reactions were restricted to TpA steps located at the 5′-end of an oligopurine tract but we have developed phosphoramidite modifications that allow incorporation anywhere along the TFO, as well as at both ends of the oligonucleotide, generating 'triplex staples'.[97,98]

## 1.3 Decreasing pH Dependence

Triplex formation in the parallel motif suffers from a requirement for low-pH conditions necessary for the protonation of cytosine at N3. Without protonation the C–GC triplet contains a single hydrogen bond between the exocyclic N4 of C and 6-keto group of G. The p$K_a$ of cytosine is around 4.5 for the free base but this is elevated within an oligonucleotide and further increased upon triplex formation, particularly in the centre of the triplex.[23,32] Runs of contiguous cytosine residues are destabilising as they reduce the p$K_a$ of each residue due to competition effects.[99–105] Nevertheless, it has been suggested in several reports that under conditions of low pH $C^+$–GC is more stable than T–AT; an effect that has been attributed to electrostatic interactions between the positive charge and the negatively charged phosphodiester backbone and/or favourable stacking interactions between the charged base and the $\pi$-stack.[30,105,107–109] Interestingly, it has also been shown that a silver ion can displace the N3 proton of C to form the base triplet C($Ag^+$)–GC which allowed triplex formation at pH 7.0; in a silver-containing buffer a triplex containing five cytosines in the third strand was stabilised by as much as 30 °C.[106] In our hands the difference in stability of T–AT and $C^+$–GC at pH 5.0 is around 4 °C, which decreases to about 1 °C at pH 6.0 (Table 1.1). To address the pH-dependence of the $C^+$–GC triplet a variety of pyrimidine or purine analogues have been developed.

### 1.3.1 Pyrimidine Analogues

Several cytosine analogues have been prepared that exhibit higher p$K_a$ values than cytosine. Triplets generated with these analogues have the advantage that they are structurally isomorphic with T–AT and the presence of the charge contributes additional stability. The simplest modification to the pyrimidine nucleus is the addition of a methyl group at the 5-position of C generating 5-methylcytosine ($^{Me}$C : Figure 1.6a). The p$K_a$ of the ring nitrogen of $^{Me}$C is increased by 0.1–0.2 pH units.[99,107] Indeed triplexes generated with $^{Me}$C exhibit a lower pH dependence and higher affinity relative to cytosine. Although this was first attributed to the increase in p$K_a$ it has since been suggested that stabilisation might be entropic in origin, resulting from

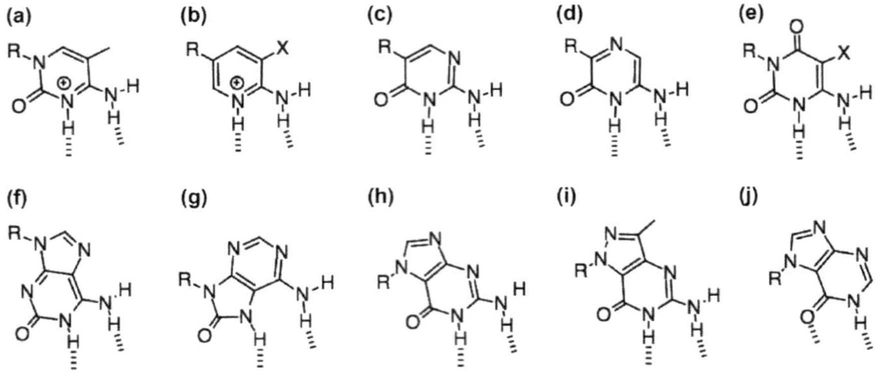


**Figure 1.6** Chemical structures of base analogues that reduce pH dependence. (a) $^{Me}$C; (b) AP (X is H or $CH_3$); (c) $^{\Psi}$C; (d) PyDDA; (e) $^{oxo}$C (X is H or $CH_3$); (f) isoG; (g) $^{oxo}$A; (h) N7G; (i) P1; (j) N7I. For each structure R is deoxyribose unless otherwise stated in the main text.

disruption of the surrounding water structure, greater base stacking, and/or hydrophobic interactions within the major groove.[23,24] Alternatively the improved stacking may increase the residence time of the non-protonated base in the uncharged C–GC triplet, thereby increasing its stability.[108] The 2-thiolyated version of 5-methylcytosine ($m^5s^2C$, Figure 1.2c) gives a further enhancement in affinity with the base exhibiting a much higher p$K_a$ of around 6.3–6.7 depending on sequence context.[30]

Another useful analogue to be developed is the *C*-nucleoside 2-aminopyridine (AP; Figure 1.6b) which was first synthesised independently by the Neidle, Reese and Leumann groups.[109–111] AP differs from cytosine by substitution of a carbon at N1 and removal of the 2-carbonyl. Both β and α-anomers were evaluated since the α-anomer is slightly more basic than the β-anomer.[109,111] Triplexes containing the β-anomer exhibited a lower pH dependency that was attributed to the increased p$K_a$ of around 6.5 for the base, and β-AP generated triplexes that were stable at pH 6.5 even at target sites that contained multiple adjacent GC base pairs.[109] A single AP substitution increased the $T_m$ of a triplex by *ca.* 4 °C at pH 6.0 relative to C (Table 1.1). The 3-methyl and 6-methyl derivatives of AP have also been prepared but did not produce a dramatic improvement in stability.[112,113] We have also shown that AP acts cooperatively with the doubly charged thymine analogue BAU to produce triplexes that have nanomolar binding affinities at pH 7.5 in the absence of any divalent metal cations.[114] The arrangement of the substitutions is important and oligonucleotides in which these analogues are evenly distributed throughout the third strand bind more tightly than those in which they are clustered together. An even greater increase in affinity can be achieved by using the 2′-aminoethoxy derivative of 3-methyl-AP in combination with 2′-aminoethoxy-T, which generated triplexes with nanomolar binding affinities at pH values as high as 9.0.[115] In this instance, the third strand was fully modified with both analogues.

To remove the pH dependency of targeting GC altogether various uncharged analogues of cytosine have been synthesised. The first was pseudoisocytosine ($^{\Psi}$C; Figure 1.6c) and its 2′-*O*-methyl derivative, which formed stable triplexes at pH 7.0 under conditions where deoxycytidine and 2′-*O*-methylcytidine did not.[116,117] As with 2-aminopyridine this analogue can be successfully employed for targeting contiguous GC base pairs. Several derivatives of this base have also been characterised and its complicated synthesis has been streamlined.[118] The deoxyribose derivative exhibits a lower affinity for GC than the 2′-*O*-methyl analogue, presumably because the former adopts the less favourable S-type sugar conformation.[119] The pyrrolidino derivative produced a 2.5–3 °C increase in $T_m$ per modification and can be used to target contiguous guanines.[65,119] Pseudoisocytosine has more frequently been employed for the pH independent recognition of DNA by PNA.[120] A similar analogue is pyrazine (PyDDA; Figure 1.6d), which possesses a nitrogen at the 6-position (instead of the usual 1-position), and it too can be used to produce stable triplexes at pH 7.0.[121]

Lastly, 6-oxo cytosine ($^{oxo}$C; Figure 1.6e) and its 5-methyl derivative have been studied as potential cytosine mimics.[122,123] At low pH these analogues produce triplexes with lower stability than protonated cytosine, though binding is much less pH-dependent. Indeed, at physiological pH it is superior to cytosine. Surprisingly, contiguous substitutions of 6-oxo cytosine are also destabilising.[124] The lower stability of the $^{oxo}$C–GC triplet relative to $C^+$–GC is attributed to unfavourable stacking interactions and/or steric hindrance due to the 6-carbonyl group, which lies close to the furanose oxygen in the *anti*-conformation that is required in triplexes. This has been partially overcome by attaching the base to the backbone *via* an acyclic linker which gives greater flexibility.[123] 2′-*O*-Methyl and ribo derivatives of this base have been synthesised though these produce less stable complexes.[122,125] We have since synthesised the 2′-aminoethoxy derivative of this nucleoside but, surprisingly, only a moderate enhancement in affinity was observed.

### 1.3.2 Purine Analogues

In order to bind to a GC base pair an analogue must present two hydrogen bond donor groups, which can be achieved using a purine nucleus. For example, isoguanine (isoG; Figure 1.6f), which switches the positioning of the amino group to the 2-position of the base and generates stable triplets, presumably through the formation of two hydrogen bonds.[126] Moreover, its 8-aza derivative can also be used to bind to G and exhibits fluorescent properties that can be exploited to monitor third strand binding.[127] Another useful strategy has been to exploit purine analogues that are designed to present the Hoogsteen face of the base by adopting a *syn*-conformation. The first purine analogues to demonstrate successful binding were 8-oxo-adenine[128] ($^{oxo}$A; Figure 1.6g) and its N6-methyl derivative.[129] The presence of the 8-oxo-group forces a *syn*-conformation, which presents the 6-amino and N7 protons in a suitable orientation for recognition of G.

The 7,8-dihydro derivative exhibits similar properties.[130] The same strategy has been exploited using 8-thioadenosine ($s^8A$; Figure 1.2e),[31] which offers an additional advantage that the thiocarbonyl can provide stacking interactions in the major groove. In all cases, these analogues recognise GC in a pH-independent fashion and generate triplexes that have the same, or similar, stability as those containing $^{Me}C$ at low pH. However, the triplets formed by these bases are not structurally isomorphic with T–AT triplets and consequently are better for targeting contiguous rather than isolated guanines, leading to a lower distortion of the TFO backbone.

N7-purine derivatives have also been developed for GC recognition and the first characterised was N7-guanine (N7G; Figure 1.6h).[131] Essentially these analogues alter the antiparallel G–GC triplet so that it can be incorporated within the parallel binding motif. Experiments revealed that the base offers pH insensitivity but suffers from sequence constraints; triplexes were three orders of magnitude less stable with alternating, compared with contiguous substitutions.[132] Similar characteristics were also exhibited by other N7 analogues, (P1; Figure 1.6i)[133,134] and N7-inosine (N7I; Figure 1.6j).[135–137] N7I lacks the amino function of N7G but surprisingly shows stable recognition of guanine. This was attributed to the formation of an unconventional CH–O bond between the carbonyl group of inosine and the CH of guanine. It was postulated that this interaction gives a small, positive, direct electrostatic contribution to stability.[136]

Several strategies have been employed to overcome the sequence constraints imposed by the lack of isomorphism of the triplets formed by these bases. An acyclic glycerol derivative was employed to attach N7G to the oligonucleotide backbone, though the increase in flexibility did not alleviate this constraint.[138] An alternative method to compensate for this loss in binding energy is to add positive charges, as described for some of the analogues described above. We have shown that a single substitution with the 2′ aminoethoxy derivative of N7G is as stable as cytosine at pH 6.0 (Table 1.1), but when this nucleotide is employed at alternate positions it produces less stable complexes than $^{Me}C$ at pH 7.0. Optimum triplex formation may therefore require the use of a combination of such bases; N7-purines for binding to contiguous GC base pairs and pyrimidine analogues for binding to isolated guanines. An alternative strategy would be to develop an isomorphous N7 derivative (such as N7-adenine) for targeting A, though the propensity for purines to bind in an antiparallel orientation may create problems.

## 1.4 Recognising Pyrimidine–Purine Base Pairs

Triplex formation requires oligopurine–oligopyrimidine target sequences and the recognition of sites containing 'inverted' pyrimidine–purine base pairs is much harder to achieve. Although this restriction may seem to limit the application of TFOs for gene-targeting purposes, oligopurine–oligopyrimidine target sequences are surprisingly abundant within the human genome.[139–141] The targeting of pyrimidine bases is hampered since the Hoogsteen face of both C and T offers just a single conventional hydrogen bonding contact within the major groove. Nevertheless, a variety of strategies have been exploited for generating stable triplexes at oligopurine tracts containing pyrimidine inversions.[142]

### 1.4.1 Null Bases and Abasic Linkers

Since the binding of a third strand within the duplex major groove is highly asymmetric it is not possible to switch across the groove to recognise the partner base on the adjacent strand, as this would result in a loss of base stacking, and impose conformational strain on the backbone of the TFO. The simplest means for targeting oligopurine duplex sequences containing pyrimidine interruptions would be to bypass the 'offending' base by placing a null or universal base analogue opposite the inversion site. Such analogues are usually aromatic rings that lack the capacity for hydrogen bonding and stabilise the helical structure through stacking interactions alone.[143] This can also be achieved using an abasic linker, such as 1,2-dideoxy-D-ribose (φ)[144] but results in a loss of binding affinity due to the lack of stacking interactions. Neither of these approaches has yielded stable triple helical structures and both cause a loss of specificity at the skipped base, as any base pair can be tolerated at this position.

### 1.4.2 Natural Bases

Various studies have investigated the stability of all possible triplet combinations composed of natural bases (Table 1.1).[145–147] These have demonstrated that the least destabilising combinations for recognising TA and CG base pairs are G–TA and T–CG (or C–CG). For each of these triplets the third strand forms a single hydrogen bond to the target, resulting in complexes that are less stable and selective than the canonical T–AT and $C^+$–GC triplets. It has been determined that a single base mismatch results in a typical free energy penalty of ~3 kcal $mol^{-1}$.[148,149] The destabilisation is dependent on the nature and position of a mismatch. Central mismatches are more destabilising than terminal ones since they disrupt the cooperative interactions between neighbouring triplets.[149,150]

The G–TA triplet contains a single hydrogen bond between the exocyclic amino group of G and the 4-carbonyl of T.[151] This hydrogen bonding arrangement has been confirmed by comparison with various guanine analogues. Removal of the 2-amino group or the 6-oxo group, generating inosine or 2-aminopurine respectively, produces triplets which are less stable than guanine.[152] The latter is more surprising as the 6-oxo group is not thought to be involved in binding, although 2-aminopurine also differs from guanine in lacking a hydrogen atom on N1. The stability of the G–TA triplet is affected by the sequence context and flanking T–AT triplets produce more stable complexes than flanking $C^+$–GC triplets (especially on the

3′-side). This is thought to be due to the formation of a second (weaker) hydrogen bond with the T of an adjacent T–AT triplet.[151] Stable complexes can be formed when this triplet is present at every fourth position, so long as the triplex contains some $C^+$–GC triplets and T–AT is located on the 3′-side of each G–TA.[153] The interaction is further stabilised by the appropriate use of charged base analogues such as 5-propargylamino-dU. Duplex regions of $(AT)_n$ can be targeted with GT-containing oligonucleotides, forming alternating G–TA and T–AT triplets, though this interaction is only observed if this is anchored by a more stable triplex.[154,155]

The parallel T–CG triplet was first proposed by Yoon *et al.*[146] and has been shown to involve a single hydrogen bond between $O_2$ of the third strand thymine and the free C4-amino proton on the duplex cytosine.[156] This hydrogen bonding pattern can also be generated with a third strand cytosine forming the C–CG triplet. Up to three consecutive T–CG or G–TA triplets can be tolerated in the centre of a triplex, if the interaction is stabilised by a triplex-binding ligand.[157]

## 1.4.3 Analogues for CG Recognition

### *1.4.3.1 Pyrimidine Analogues*

A number of heterocycles based on pyrimidines have been used as a means to recognise inverted CG base pairs within oligopurine tracts. The efficacy of this approach was first demonstrated within the antiparallel motif using 2-pyridone (P; Figure 1.7a).[158] 2-pyridone utilises a carbonyl oxygen at the 2-position for hydrogen bonding with the 4-amino hydrogen of C in a similar manner to C or T. However, it lacks the 3-nitrogen atom and the 4-carbonyl

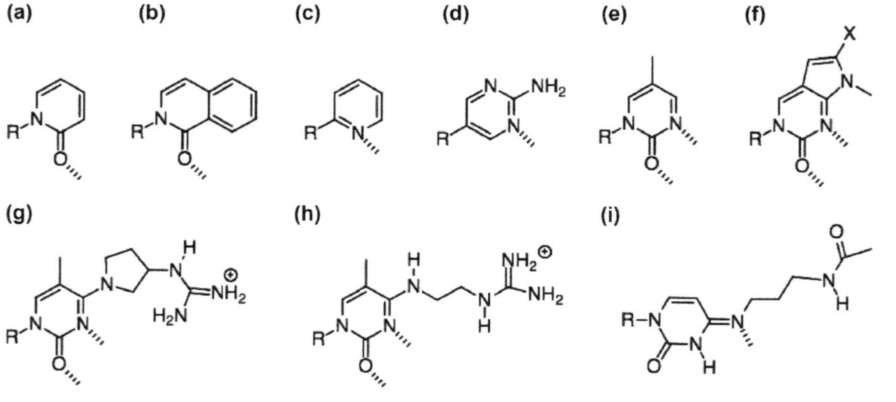


**Figure 1.7** Chemical structures of base analogues for CG recognition. (a) P; (b) $Q^B$; (c) Py; (d) 2APm; (e) $^{4H}T$; (f) $^{A}PP$ (X is aminopropyl or other group); (g) $QP^B$; (h) $^{g}C$; (i) $^{4P}C$. For each structure R is deoxyribose unless otherwise stated in the main text.

or amino group of the pyrimidines which should decrease its ability to bind to either AT or GC base pairs. Imanishi and co-workers were the first to examine the use of P in the parallel motif and prepared its deoxyribose and its 2′,4′-BNA monomer ($P^B$). TFOs positioning a single substitution of $P^B$ opposite CG exhibited a $T_m$ that was 9 °C higher than when it was attached to deoxyribose.[159] However, the triplet was still less stable than either of the canonical base triplets. Nevertheless, this was the first study to demonstrate that combining a base analogue with a stabilising sugar modification compensates, at least in part, for the loss in binding energy at pyrimidine-purine inversions. $P^B$ can also interact with an AT base pair though the binding affinity is much weaker. To improve selectivity the bicyclic analogue 1-isoquinolone ($Q^B$; Figure 1.7b) was examined.[160] It was reasoned that binding to AT base pairs would be sterically hindered by the close proximity of a 4-hydrogen of Q and the 5-methyl group of T on the opposite side of the major groove. Binding to AT was indeed reduced but so too was the desired interaction with CG. We have synthesised the 2′-aminoethoxy derivative of Q but find that in this context it is not effective for recognising CG base pairs.

Two further cytosine analogues that have been utilised for the recognition of CG base pairs are 2-pyridine (Py; Figure 1.7c)[161] and 2-aminopyrimidine (2APm; Figure 1.7d).[162] Since the 2 and 3-nitrogens of these analogues are basic (the $pK_a$ of 2APm is about 3.3) they are unprotonated at all practical pH values. Not only does this prevent their interaction with GC base pairs it also allows these nitrogen atoms to act as strong hydrogen bond acceptors for binding to the exocyclic amino group of C. 2APm also generates a triplet that is more isomorphous with the canonical triplets, in contrast to those formed by either T or C. Consistent with this, Py and 2APm produced triplexes that were equally stable and 4 °C more stable than T at pH 7.0, respectively. In addition, conversion of Py to a BNA derivative ($Py^B$) led to a substantial enhancement in stability without altering its selectivity.

One of the first thymine analogues to be prepared for recognition of CG was 5-methyl-pyrimidine-2-one ($^{4H}T$; Figure 1.7e), which lacks both the 4-carbonyl and 3-NH groups of T.[163] As above $^{4H}T$ exploits the 3-nitrogen as a hydrogen bond acceptor for bonding to the N4 amino group of C. Interestingly, it also positions the 2-carbonyl to form an unconventional C–H–O bond with the 5-hydrogen of C, a similar interaction previously observed for the N7I–GC triplet.[140] Studies showed that $^{4H}T$ had a decreased affinity for AT base pairs, while generating triplexes with a similar stability to those with T–CG at the same position. A further increase in affinity was observed with the 2′-aminoethoxy derivative of $^{4H}T$, producing a melting temperature $T_m$ increase of *ca.* 1.5 °C per substitution relative to T.[164] Within the context of fully modified 2′-AE-RNA it has been used to recognise up to five separated CG interruptions at pH 6.5 with a 33% pyrimidine content in the target strand.[165] It has also been used to bind to targets containing multiple contiguous CG inversions.[166] However, the affinity of the analogue is dependent on sequence context, with the most stable triplexes generated when placed between thymine residues.

We have also developed a series of nucleobases for recognising CG inversions based on methylated 3*H*-pyrrolo[2,3-*d*]pyrimidin-2(7*H*)-one nucleosides (*e.g.*, Figure 1.7f). The core of these structures maintains the hydrogen bonding motif of $^{4H}$T, whilst contributing extra base stacking *via* an additional aromatic ring between the 4 and 5 positions. The simplest analogue, containing a methyl group at the 6-position of the pyrrolo ring ($^{M}$PP), generated stable triplexes when positioned opposite CG, with an increase of *ca.* 3 °C relative to a control triplex (Table 1.1).[167] We have attempted to improve the affinity of this analogue further by attaching different groups to the 6-position that might make additional contacts across the major groove with the guanine of the CG base pair. The addition of either aminoethyl or aminopropyl groups ($^{A}$PP; Figure 1.7f) at this position resulted in moderate increases in $T_m$, which are probably caused by protonation of the pendant amine group and charge-stabilisation interactions, rather than hydrogen bonding (Table 1.1). We reasoned that this may be due to the flexibility of the chain and prepared various phenyl-modified derivatives that included attached amino, acetamido, ureido and guanidino groups, but again no dramatic enhancement in affinity was seen.[168,169] Lastly, we have synthesised the 2′-aminoethoxy variant of the $^{A}$PP analogue, which resulted in a $T_m$ increase of around 4.5 °C per substitution, though it has not yet been assessed in the context of a fully-modified 2′-AE-RNA, which may improve affinity further (Table 1.1).

Several other groups have designed analogues that can bind to cytosine while simultaneously interacting with the O6 and/or N7 acceptors of guanine. In two ambitious studies by Hari and co-workers more than twenty *N*,*N*-disubstituted cytosine derivatives were prepared for CG recognition.[170,171] Unlike C, these derivatives were not expected to bind to GC base pairs, due to steric repulsion of substituents on the amino group, while retaining the use of the 3-nitrogen for recognition of cytosine. The 4-[(3S)-3-guanidinopyrrolidino]-5-methylpyrmidin-2-one variant (GP$^{B}$; Figure 1.7g) exhibited the highest affinity for CG, while discriminating against GC, and its attachment to BNA increased its affinity to that of T for an AT base pair. In a further study the 3-deazacytosine equivalent of GP$^{B}$ showed improved selectivity, through the loss of the 3-nitrogen.[172] In a similar fashion the Seidman group have investigated various *N*$^{4}$-alkyl-5-methylcytosine derivatives and the *N*$^{4}$-(2-guanidoethyl)-5-methylcytosine analogue was the best of these ($^{g}$C; Figure 1.7h).[173] Within the context of a 2′-O-methyl modification, this analogue generated a much more stable triplet than T–CG but was not as stable as T–AT. Various other N4 cytosine derivatives have been developed that utilise the exocyclic nitrogen for recognition of CG.[174,175] The best of these is N$^{4}$-(3-acetamidopropyl)cytosine ($^{4P}$C; Figure 1.7i) which positions a side chain across the major groove allowing the 3-amino group to form a hydrogen bond to the O6 carbonyl group of guanine. UV melting showed this base to be more stable than C–CG but again less stable than the canonical triplets.

#### 1.4.3.2 Other Heterocycles

Some of the first attempts to recognise CG base pairs exploited extended heteroaromatic nucleobases designed to make contact with both partners of the base pair. The first of these was 4-(3-benzamidophenyl)imidazole ($D_3$; Figure 1.8a) which was designed to match the edges of a CG base pair.[176] It was anticipated that the ring nitrogen of the imidazole moiety would form a single hydrogen bond to cytosine, while additional stacking interactions would be possible due to the presence of two aromatic rings positioned across the major groove. Rotational freedom between these two rings could maximise non-bonding interactions. Affinity cleavage experiments showed that $D_3$ bound to CG and TA base pairs with greater affinity than to GC or AT. However, it was later shown that this nucleobase formed triplets that are less stable when they are flanked by $C^+$–GC on the 3′-side. A subsequent NMR study showed that $D_3$ lacked selectivity and intercalated into the adjacent YpR step, thereby skipping the inversion site.[177] Two similar carbocyclic ribofuranose analogues, L1 and L2 were also developed, which exhibited a preference for binding at pyrimidine inversions, and are also thought to bind by an intercalative mechanism.[178] More recently, a variety of imidazole and triazole heterocycles have been attached to BNA sugars and assessed for their ability to target pyrimidine interruptions. Oxazole ($O^B$; Figure 1.8b) recognised CG slightly better than TA but generated a triplex that was less stable than that with T–AT in the same position,[179] most likely through an interaction between the ring oxygen and the exocyclic nitrogen of cytosine. Among the triazole nucleobases examined, a 1-(4-ureidophenyl)triazole ($Tz^B$; Figure 1.8c) provided the best enhancement in affinity but again the triplexes were less stable than their unmodified counterparts.[180]

### 1.4.4 Analogues for TA Recognition

The development of base analogues capable of recognising TA inversions is hampered by the presence of a methyl group at the 5-position of T. One strategy for overcoming this problem is to use a short linker that projects the analogue past the methyl group to allow recognition of the 4-carbonyl group of T. To date this has only been attempted within the context of PNA by the attachment of 3-oxo-2,3-dihydrdopyridazine (E; Figure 1.9a) *via* a β-alanine linker to the backbone. Triplexes containing this analogue exhibited a $T_m$

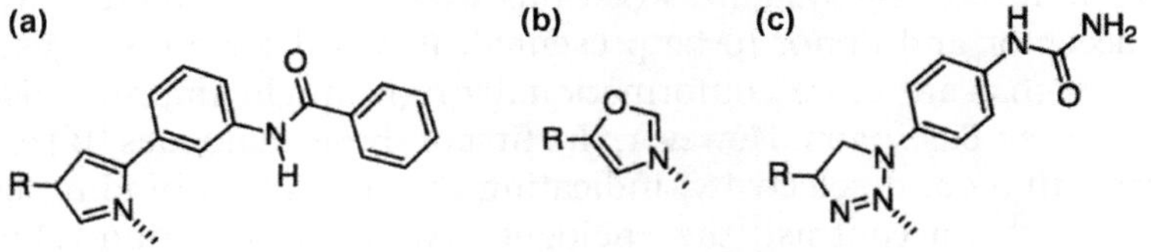


**Figure 1.8** Chemical structures of various heterocycles for CG recognition. (a) $D_3$; (b) $O^B$; (c) $Tz^B$. For each structure R is deoxyribose unless otherwise stated in the main text.

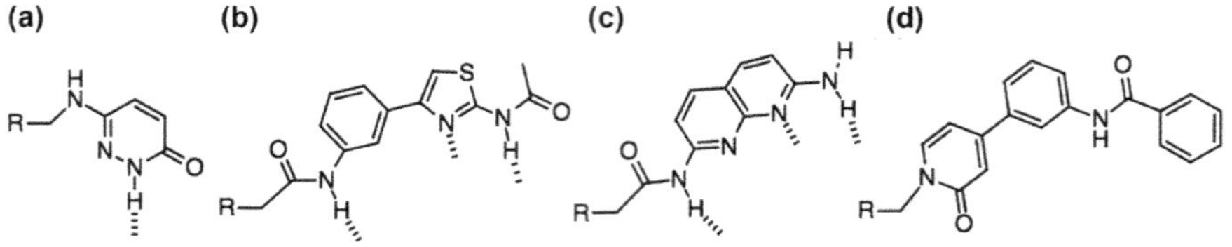


**Figure 1.9** Chemical structures of base analogues for TA recognition. (a) E; (b) S; (c) $^{DA}N_{ac}$; (d) $bP^{B}$. For each structure R is deoxyribose unless otherwise stated in the main text.

increase of 5 °C relative to G when positioned opposite TA.[181] An alternative strategy is to increase the stability of the G–TA triplet using guanine analogues but this has not been successful. For example, the addition of 2′-aminoethoxy groups to the nucleosides of guanine (AE-G) or 2-aminopurine, which is also capable of binding to guanine, decreased both the affinity and selectivity of the resultant triplets.[166] We find that AE-G showed an enhanced affinity for GC base pairs, while having decreased affinity for TA base pairs (Table 1.1). We have also examined the influence of adding a propargylamino group to the 7-position of G (GPdG) but again this resulted in an increase in affinity for GC and not for TA base pairs (Table 1.1).

Greater success has been achieved with the use of extended heteroaromatic nucleobases capable of binding to both partners of the TA base pair. The first to be developed was the unnatural thiazolyl aniline monomer S (Figure 1.9b) which has the capacity for recognising both the 4-carbonyl of T and the 4- and 6-position nitrogen atoms of A.[182] Experiments revealed that the S–TA triplet produced a similar stability to that of the T–AT triplet but also generated an S–CG triplet that was only moderately less stable. We have also shown that the analogue S recognises CG as well as TA at low pH, with little or no discrimination between them, though it binds better to TA at higher pHs. The interaction and selectivity is improved slightly by the addition of a 2′-aminoethoxy group (AE-S) to the sugar; one 2AE-S–TA triplet increases the $T_m$ by 4–6 °C relative to G–TA, and it is 1–2 °C more stable than S–TA (Table 1.1).[183] Although the selectivity problem could be attributed to an intercalative mode of binding it has been suggested that this is not the case since the S–TA triplet was less stable when flanked by $C^+$–GC triplets on either side.[182] Altered specificity is more likely to originate from the rotational freedom of the linker attaching the base to the sugar and/or linking the two unfused ring systems. Rotations would allow a different hydrogen bond acceptor and donor to be presented. It was therefore suggested that analogues that are more conformationally rigid might improve discrimination between base pairs. However, the first of these analogues ($B^t$) exhibited a lower affinity and selectivity, indicating that it favours binding by intercalation.[184,185] In contrast, the analogue N-acetyl-2,7-diamino-1,8-naphtyridine ($^{DA}N_{ac}$; Figure 1.9c), which contains a fused ring system, was shown to recognise TA and CG with different affinities, albeit with a 2–3 °C difference in $T_m$.[186] Lastly, the analogue 4-(3-benzamidophenyl)-2-pyrididone was prepared and shown to exhibit a good affinity for TA, with a difference of 4 °C between its interaction of TA and CG at physiological pH (bP$^B$; Figure 1.9d).[187] Since the nucleobase does not present the appropriate hydrogen bonding pattern, it is likely that it binds through intercalation.

### 1.4.5 Other Approaches

A different approach that allows recognition of both partners of a base pair has arisen from the design and synthesis of a series of 2-aminoquinolone and 2-aminoquinazoline C-glycoside bases (Figure 1.10). These molecules, designated TRIPsides, are designed to bind symmetrically within the major groove, unlike other triplex-forming oligonucleotides, positioning the oligonucleotide backbone in the centre of the groove.[188–190] In this strategy only the purine strand of the target is read, but because the backbone is located in the centre of the major groove, either strand can be recognised by choosing the appropriate TRIPside. The antiCG, antiTA and antiGC monomers have been used in combination and allowed the recognition of a 19-mer target site in which the purines switched from one strand to the other four times.[189] The antiAT monomer has also been prepared but yet to be characterised in combination with the others.[191]

## 1.5 Towards Mixed Sequence Recognition at Neutral pH

It is clear that a large number of base, nucleoside and nucleotide analogues have been developed for improving the triplex-forming properties of TFOs but there are few examples in which these have been used in combination to target mixed DNA sequences under physiological pH and ionic conditions. The majority of studies have focused on triplex formation with oligopurine–oligopyrimidine sequences containing a single pyrimidine–purine inversion, or the use of a single cytosine or thymine analogues within the third strand. The first study to demonstrate the selective recognition of a duplex target containing three of the four base pairs was undertaken in the Leumann

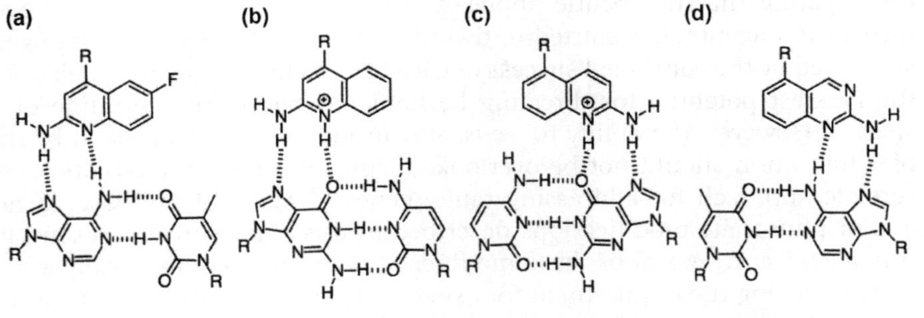


**Figure 1.10** Chemical structures of oligoTRIPsides. (a) AntiAT; (b) AntiGC; (c) AntiCG; (d) AntiTA. For each structure R is deoxyribose.

laboratory.[165] Fully modified 2′-aminoethoxy RNA strands were prepared with T, $^{Me}$C (Figure 1.6a) and $^{4H}$T (Figure 1.7e) for the recognition of AT, GC and CG base pairs, respectively. This combination was able to recognise up to five CG inversions in a 15-mer duplex target with high selectivity and good affinity under near physiological conditions. We have also prepared oligonucleotides that contain combinations of modified nucleosides for the recognition of a duplex target containing all four base pairs.[192] For this we used BAU to target AT with high affinity, AP (Figure 1.6b) for recognition of GC base pairs at elevated pHs, and $^{A}$PP (Figure 1.7f) and S (Figure 1.9b) for recognising CG and TA base pairs, respectively. With this combination we demonstrated triplex recognition at a 19-mer duplex target that contained four pyrimidine interruptions at neutral pH. Moreover, both footprinting and melting experiments demonstrated that this heavily modified oligonucleotide retained its sequence specificity and that changing a single base pair opposite any one of the modified nucleosides led to a large decrease in affinity. The only exception was with the S analogue, which formed stable complexes opposite both TA and CG base pairs, however, this might be improved by using its 2′-aminoethoxy derivative, which shows better discrimination. More recently, the Sekine laboratory has targeted the same duplex sequence used in our study using an oligonucleotide prepared with different nucleoside modifications.[186] 2′-O-methyl modified s$^2$T (Figure 1.2c) and s$^8$A (Figure 1.2e) were used for recognising AT and GC base pairs, respectively. Whilst $^g$C (Figure 1.7h) and $^{DA}N_{ac}$ (Figure 1.9c) were used for the recognition of CG and TA, respectively. Interestingly, their study demonstrated that the use of $^{DA}N_{ac}$ improved the ability to discriminate between TA and CG base pairs, but the overall stability of the complex was lower than that of the triplex examined in our study, with a 10 °C drop in $T_m$ for the complex under the same experimental conditions.

## 1.6 Outlook

Triplex-directed DNA recognition has been driven primarily by a desire to use TFOs for therapeutic applications, *i.e.*, as gene-targeting agents for modulating gene expression. However, in recent years the number of studies investigating the therapeutic application of TFOs, as well as other DNA-recognition agents, has started to dwindle. Perhaps the field is being overshadowed by the continued success of CRISPR–Cas technologies, which offer the greatest potential for directing heritable change to the germline of a species. However, the ability to *transiently* influence gene expression by triplex formation should not be overlooked, since it offers an alternative, reversible approach for addressing faulty genes.[3,4] Many of the base, sugar and/or phosphate modifications described in this chapter will be useful in this regard and several of these modifications are now commercially available; removing the requirement for expertise in phosphoramidite synthesis. Recent improvements in our understanding of the pharmacodynamic and pharmacokinetic properties of oligonucleotides, due in part to a re-emergence of antisense technologies, will also aid in their application.[193] Triplex-directed DNA recognition is also starting to show promise in other fields, such as bionanotechnology, where it provides a means to introduce functionality into DNA nanostructures by exploiting the sequence addressability of duplex regions assembled by strand exchange.[6,194,195]

## Acknowledgements

Original work in the authors' laboratories was supported by BBSRC grants BB/J001694, BBH019219, and BB/C004531.

## References

1. H. E. Moser and P. B. Dervan, *Science*, 1987, **238**, 645–650.
2. T. Le Doan, L. Perrouault, D. Praseuth, N. Habhoub, J. L. Decout, N. T. Thuong, J. Lhomme and C. Hélène, *Nucleic Acids Res.*, 1987, **15**, 7749–7760.
3. M. M. Seidman and P. M. Glazer, *J. Clin. Invest.*, 2003, **112**, 487–494.
4. D. A. Rusling, V. J. Broughton-Head, T. Brown and K. R. Fox, *Curr. Chem. Biol.*, 2008, **2**, 2–11.
5. V. N. Potaman, *Expert Rev. Mol. Diagn.*, 2003, **3**, 481–496.
6. K. R. Fox, *Nucleic Acids Res.*, 1992, **20**, 1235–1242.
7. D. A. Rusling and K. R. Fox, *Methods*, 2014, **67**, 123–133.
8. G. Felsenfeld, D. R. Davies and A. Rich, *J. Am. Chem. Soc.*, 1957, **79**, 2023–2024.
9. F. B. Howard, J. Frazier, M. N. Lipsett and H. T. Miles, *Biochem. Biophys. Res. Commun.*, 1964, **17**, 93–102.
10. A. R. Morgan and R. D. Wells, *J. Mol. Biol.*, 1968, **37**, 63–80.
11. M. Riley and B. Maling, *J. Mol. Biol.*, 1966, **20**, 359–389.
12. C. Marck, D. Thiele, C. Schneider and W. Guschlbauer, *Nucleic Acids Res.*, 1978, **5**, 1979–1996.
13. S. L. Broitman, D. D. Im and J. R. Fresco, *Proc. Natl. Acad. Sci. U. S. A.*, 1987, **84**, 5120–5124.
14. P. A. Beal and P. B. Dervan, *Science*, 1991, **251**, 1360–1363.
15. R. H. Durland, D. J. Kessler, S. Gunnell, M. Duvic, B. M. Pettitt and M. E. Hogan, *Biochemistry*, 1991, **30**, 9246–9255.
16. R. Thenmalarchelvi and N. Yathindra, *Nucleic Acids Res.*, 2005, **33**, 43–55.
17. C. Giovannangéli, M. Rougée, T. Garestier, N. T. Thuong and C. Hélène, *Proc. Natl. Acad. Sci. U. S. A.*, 1992, **89**, 8631–8635.
18. R. A. J. Darby, M. Sollogoub, C. McKeen, L. Brown, A. Risitano, N. Brown, C. Barton, T. Brown and K. R. Fox, *Nucleic Acids Res.*, 2002, **30**, e39.
19. D. A. Rusling,, P. A. Rachwal, T. Brown and K. R. Fox, *Biophys. Chem.*, 2009, **145**, 105–110.

20. A. B. Staubli and P. B. Dervan, *Nucleic Acids Res.*, 1994, **22**, 2637–2642.
21. J. Michel, J. J. Toulmé, J. Vercauteren and S. Moreau, *Nucleic Acids Res.*, 1996, **24**, 1127–1135.
22. F. Godde, J.-J. Toulmé and S. Moreau, *Biochemistry*, 1998, **37**, 13765–13775.
23. D. Leitner, W. Schröder and K. Weisz, *Biochemistry*, 2000, **39**, 5886–5892.
24. L. E. Xodo, G. Manzini, F. Quadrifoglio, G. A. van der Marel and J. H. van Boom, *Nucleic Acids Res.*, 1991, **19**, 5625–5631.
25. N. Colocci and P. B. Dervan, *J. Am. Chem. Soc.*, 1994, **116**, 785–786.
26. A. K. Phipps, M. Tarköy, P. Schultze and J. Feigon, *Biochemistry*, 1998, **37**, 5820–5830.
27. L. Lacroix, J. Lacoste, J. F. Reddoch, J. L. Mergny, D. D. Levy, M. M. Seidman, M. D. Matteucci and P. M. Glazer, *Biochemistry*, 1999, **38**, 1893–1901.
28. B. C. Froehler, S. Wadwani, T. J. Terhorst and S. R. Gerrard, *Tetrahedron Lett.*, 1992, **33**, 5307–5310.
29. J. A. Brazier, T. Shibata, J. Townsley, B. F. Taylor, E. Frary, N. H. Williams and D. M. Williams, *Nucleic Acids Res.*, 2005, **33**, 1362–1371.
30. I. Okamoto, K. Seio and M. Sekine, *Bioorg. Med. Chem. Lett.*, 2006, **16**, 3334–3336.
31. A. Ohkubo, Y. Nishino, A. Yokouchi, Y. Ito, Y. Noma, Y. Kakishima, Y. Masaki, H. Tsunoda, K. Seio and M. Sekine, *Chem. Commun.*, 2011, **47**, 12556–12558.
32. J. L. Asensio, R. Carr, T. Brown and A. N. Lane, *J. Am. Chem. Soc.*, 1999, **121**, 11063–11070.
33. R. W. Roberts and D. M. Crothers, *Science*, 1992, **258**, 1463–1466.
34. M. Shimizu, A. Konishi, Y. Shimada, H. Inoue and E. Ohtsuka, *FEBS Lett.*, 1992, **302**, 155–158.
35. H. Han and P. B. Dervan, *Nucleic Acids Res.*, 1994, **22**, 2837–2844.
36. C. Escudé, J. S. Sun, M. Rougée, T. Garestier and C. Hélène, *C. R. Acad. Sci. III*, 1992, **315**, 521–525.
37. E. Bernal-Méndez and C. J. Leumann, *J. Biol. Chem.*, 2001, **276**, 35320–35327.
38. C. Thibaudeau, J. Plavec, N. Garg, A. Papchikhin and J. Chattopadhyaya, *J. Am. Chem. Soc.*, 1994, **116**, 4038–4043.
39. S. K. Singh, A. A. Koshkin, J. Wengel and P. Nielsen, *Chem. Commun.*, 1998, **22**, 455–456.
40. H. Torigoe, Y. Hari, M. Sekiguchi, S. Obika and T. Imanishi, *J. Biol. Chem.*, 2001, **276**, 2354–2360.
41. B.-W. Sun, B. R. Babu, M. D. Sørensen, K. Zakrzewska, J. Wengel and J.-S. Sun, *Biochemistry*, 2004, **43**, 4160–4169.
42. M. Koizumi, K. Morita, M. Daigo, S. Tsutsumi, K. Abe, S. Obika and T. Imanishi, *Nucleic Acids Res.*, 2003, **31**, 3267–3273.
43. S. M. A. Rahman, S. Seki, K. Utsuki, S. Obika, K. Miyashita and T. Imanishi, *Nucleosides, Nucleotides Nucleic Acids*, 2007, **26**, 1625–1628.

44. H. Torigoe, S. M. A. Rahman, H. Takuma, N. Sato, T. Imanishi, S. Obika and K. Sasaki, *Chemistry*, 2011, **17**, 2742–2751.
45. H. Torigoe, O. Nakagawa, T. Imanishi, S. Obika and K. Sasaki, *Biochimie*, 2012, **94**, 1032–1040.
46. S. Obika, M. Onoda, K. Morita, J. Andoh, M. Koizumi and T. Imanishi, *Chem. Commun.*, 2001, 1992–1993.
47. M. Tarköy and C. Leumann, *Angew. Chem., Int. Ed. Engl.*, 1993, **32**, 1432–1434.
48. R. Steffens and C. J. Leumann, *J. Am. Chem. Soc.*, 1999, **121**, 3249–3255.
49. D. Renneberg and C. J. Leumann, *ChemBioChem*, 2004, **5**, 1114–1118.
50. Y. Kohwi and T. Kohwi-Shigematsu, *Proc. Natl. Acad. Sci. U. S. A.*, 1988, **85**, 3781–3785.
51. R. O. Dempcy, K. A. Browne and T. C. Bruice, *Proc. Natl. Acad. Sci. U. S. A.*, 1995, **92**, 6097–6101.
52. A. Blaskó, R. O. Dempcy, E. E. Minyat and T. C. Bruice, *J. Am. Chem. Soc.*, 1996, **118**, 7892–7899.
53. M. Park and T. C. Bruice, *Bioorg. Med. Chem. Lett.*, 2005, **15**, 3247–3251.
54. S. Chaturvedi, T. Horn and R. L. Letsinger, *Nucleic Acids Res.*, 1996, **24**, 2318–2323.
55. J. M. Dagle and D. L. Weeks, *Nucleic Acids Res.*, 1996, **24**, 2143–2149.
56. T. Michel, F. Debart, F. Heitz and J.-J. Vasseur, *Chembiochem*, 2005, **6**, 1254–1262.
57. N. Bomholt, V. V. Filichev and E. B. Pedersen, *Org. Biomol. Chem.*, 2011, **9**, 4527–4534.
58. B. Cuenoud, F. Casset, D. Hüsken, F. Natt, R. M. Wolf, K.-H. Altmann, P. Martin and H. E. Moser, *Angew. Chem. Int. Ed.*, 1998, **37**, 1288–1291.
59. N. Atsumi, Y. Ueno, M. Kanazaki, S. Shuto and A. Matsuda, *Bioorg. Med. Chem.*, 2002, **10**, 2933–2939.
60. M. J. J. Blommers, F. Natt, W. Jahnke and B. Cuenoud, *Biochemistry*, 1998, **37**, 17714–17725.
61. N. Puri, A. Majumdar, B. Cuenoud, P. S. Miller and M. M. Seidman, *Biochemistry*, 2004, **43**, 1343–1351.
62. Y. Hari, S. Ijitsu, M. Akabane-Nakata, T. Yoshida and S. Obika, *Bioorg. Med. Chem. Lett.*, 2014, **24**, 3046–3049.
63. P. Alberti, P. B. Arimondo, J.-L. Mergny, T. Garestier, C. Hélène and J.-S. Sun, *Nucleic Acids Res.*, 2002, **30**, 5407–5415.
64. T. P. Prakash, A. Püschl, E. Lesnik, V. Mohan, V. Tereshko, M. Egli and M. Manoharan, *Org. Lett.*, 2004, **6**, 1971–1974.
65. A. Mayer, A. Häberli and C. J. Leumann, *Org. Biomol. Chem.*, 2005, **3**, 1653–1658.
66. A. Häberli and C. J. Leumann, *Org. Lett.*, 2002, **4**, 3275–3278.
67. A. Häberli, A. Mayer and C. J. Leumann, *Nucleosides, Nucleotides Nucleic Acids*, 2003, **22**, 1187–1189.
68. H. Nara, A. Ono and A. Matsuda, *Bioconjug. Chem.*, 1995, **6**, 54–61.
69. K. G. Rajeev, V. R. Jadhav and K. N. Ganesh, *Nucleic Acids Res.*, 1997, **25**, 4187–4193.

70. J. Bijapur, M. D. Keppler, S. Bergqvist, T. Brown and K. R. Fox, *Nucleic Acids Res.*, 1999, **27**, 1802–1809.
71. M. Sollogoub, R. A. J. Darby, B. Cuenoud, T. Brown and K. R. Fox, *Biochemistry*, 2002, **41**, 7224–7231.
72. S. D. Osborne, V. E. C. Powers, D. A. Rusling, O. Lack, K. R. Fox and T. Brown, *Nucleic Acids Res.*, 2004, **32**, 4439–4447.
73. D. A. Rusling, T. Brown and K. R. Fox, *Biophys. Chem.*, 2006, **123**, 134–140.
74. D. A. Rusling, V. J. Broughton-Head, A. Tuck, H. Khairallah, S. D. Osborne, T. Brown and K. R. Fox, *Org. Biomol. Chem.*, 2008, **6**, 122–129.
75. C. Lou, Q. Xiao, L. Brennan, M. E. Light, N. Vergara-Irigaray, E. M. Atkinson, L. M. Holden-Dye, K. R. Fox and T. Brown, *Bioorg. Med. Chem.*, 2010, **18**, 6389–6397.
76. S. P. Sau, P. Kumar, B. A. Anderson, M. E. Østergaard, L. Deobald, A. Paszczynski, P. K. Sharma and P. J. Hrdlicka, *Chem. Commun.*, 2009, 6756–6758.
77. L. Brennan, G. Peng, N. Srinivasan, K. R. Fox and T. Brown, *Nucleosides, Nucleotides Nucleic Acids*, 2007, **26**, 1283–1286.
78. D. A. Rusling, G. Peng, N. Srinivasan, K. R. Fox and T. Brown, *Nucleic Acids Res.*, 2008, **37**, 1288–1296.
79. A. S. Cardew, T. Brown and K. R. Fox, *Nucleic Acids Res.*, 2012, **40**, 3753–3762.
80. P. S. Miller, N. Dreon, S. M. Pulford and K. B. McParland, *J. Biol. Chem.*, 1980, **255**, 9659–9665.
81. L. Kibler-Herzog, B. Kell, G. Zon, K. Shinozuka, S. Mizan and W. D. Wilson, *Nucleic Acids Res.*, 1990, **18**, 3545–3555.
82. F. Debart, A. Meyer, J.-J. Vasseur and B. Rayner, *Nucleic Acids Res.*, 1998, **26**, 4551–4556.
83. S. M. Gryaznov, D. H. Lloyd, J. K. Chen, R. G. Schultz, L. A. DeDionisio, L. Ratmeyer and W. D. Wilson, *Proc. Natl. Acad. Sci. U. S. A.*, 1995, **92**, 5798–5802.
84. H. Torigoe, A. Maruyama, S. Obika, T. Imanishi and T. Katayama, *Biochemistry*, 2009, **48**, 3545–3553.
85. L. Lacroix, P. B. Arimondo, M. Takasugi, C. Hélène and J. L. Mergny, *Biochem. Biophys. Res. Commun.*, 2000, **270**, 363–369.
86. J. Basye, *Nucleic Acids Res.*, 2001, **29**, 4873–4880.
87. T. Michel, F. Debart, J.-J. Vasseur, F. Geinguenaud and E. Taillandier, *J. Biomol. Struct. Dyn.*, 2003, **21**, 435–445.
88. P. E. Nielsen, M. Egholm, R. Berg and O. Buchardt, *Science*, 1991, **254**, 1497–1500.
89. P. E. Nielsen, *Curr. Med. Chem.*, 2001, **8**, 545–550.
90. P. E. Nielsen, M. Egholm and O. Buchardt, *J. Mol. Recognit.*, 1994, **7**, 165–170.
91. P. Wittung, P. Nielsen and B. Nordén, *Biochemistry*, 1997, **36**, 7973–7979.

92. D. A. Rusling and K. R. Fox, in *DNA Conjugates and Sensors*, 2012, pp. 75–95.
93. L. A. Yatsunyk, O. Mendoza and J. L. Mergny, *Acc. Chem. Res.*, 2014, **47**, 1836–1844.
94. R. A. J. Darby and K. R. Fox, in *Interaction of Small Molecules with DNA and RNA: from Synthesis to Nucleic Acid Complexes*, ed. M. Demeunynck, C. Bailly and W. D. Wilson, WILEY-VCH Verlag GmbH, 2002, pp. 360–383.
95. S. Kukreti, J.-S. Sun, T. Garestier and C. Helene, *Nucleic Acids Res.*, 1997, **25**, 4264–4270.
96. M. Takasugi, A. Guendouz, M. Chassignol, J. L. Decout, J. Lhomme, N. T. Thuong and C. Helene, *Proc. Natl. Acad. Sci. U. S. A.*, 1991, **88**, 5602–5606.
97. H. Li, V. J. Broughton-Head, G. Peng, V. E. C. Powers, M. J. Ovens, K. R. Fox and T. Brown, *Bioconjug. Chem.*, 2006, **17**, 1561–1567.
98. H. Li, V. J. Broughton-Head, K. R. Fox and T. Brown, *Nucleosides, Nucleotides Nucleic Acids*, 2007, **26**, 1005–1009.
99. J. S. Lee, M. L. Woodsworth, L. J. P. Latimer and A. R. Morgan, *Nucleic Acids Res.*, 1984, **12**, 6603–6614.
100. L. L. Kiessling, L. C. Griffin and P. B. Dervan, *Biochemistry*, 1992, **31**, 2829–2834.
101. J. Volker and H. H. Klump, *Biochemistry*, 1994, **33**, 13502–13508.
102. N. Sugimoto, P. Wu, H. Hara and Y. Kawamoto, *Biochemistry*, 2001, **40**, 9396–9405.
103. M. D. Keppler and K. R. Fox, *Nucleic Acids Res.*, 1997, **25**, 4644–4649.
104. A. M. Soto, J. Loo and L. A. Marky, *J. Am. Chem. Soc.*, 2002, **124**, 14355–14363.
105. R. W. Roberts and D. M. Crothers, *Proc. Natl. Acad. Sci. U. S. A.*, 1996, **93**, 4320–4325.
106. T. Ihara, T. Ishii, N. Araki, A. W. Wilson and A. Jyo, *J. Am. Chem. Soc.*, 2009, **131**, 3826–3827.
107. T. J. Povsic and P. B. Dervan, *J. Am. Chem. Soc.*, 1989, **111**, 3059–3061.
108. S. F. Singleton and P. B. Dervan, *Biochemistry*, 1992, **31**, 10995–11003.
109. P. J. Bates, C. A. Laughton, T. C. Jenkins, D. C. Capaldi, P. D. Roselt, C. B. Reese and S. Neidle, *Nucleic Acids Res.*, 1996, **24**, 4176–4184.
110. S. Hildbrand and C. Leumann, *Angew. Chem. Int. Ed.*, 1996, **35**, 1968–1970.
111. S. A. Cassidy, P. Slickers, J. O. Trent, D. C. Capaldi, P. D. Roselt, C. B. Reese, S. Neidle and K. R. Fox, *Nucleic Acids Res.*, 1997, **25**, 4891–4898.
112. S. Hildbrand, A. Blaser, S. P. Parel and C. J. Leumann, *J. Am. Chem. Soc.*, 1997, **119**, 5499–5511.
113. Y. Hari, M. Nakahara and S. Obika, *Heterocycles*, 2012, **86**, 1135–1139.
114. D. A. Rusling, L. Le Strat, V. E. C. Powers, V. J. Broughton-Head, J. Booth, O. Lack, T. Brown and K. R. Fox, *FEBS Lett.*, 2005, **579**, 6616–6620.
115. C. Lou, M. Shelbourne, K. R. Fox and T. Brown, *Chem. – A Eur. J.*, 2011, **17**, 14851–14856.

116. A. Ono, P. O. P. Ts'o and L. S. Kan, *J. Am. Chem. Soc.*, 1991, **113**, 4032–4033.
117. A. Ono, P. O. P. Ts'o and L. Kan, *J. Org. Chem.*, 1992, **57**, 3225–3230.
118. A. Mayer and C. J. Leumann, *Nucleosides, Nucleotides Nucleic Acids*, 2003, **22**, 1919–1925.
119. T. M. Chin, S. B. Lin, S. Y. Lee, M. L. Chang, A. Y.-Y. Cheng, F. C. Chang, L. Pasternack, D. H. Huang and L. S. Kan, *Biochemistry*, 2000, **39**, 12457–12464.
120. M. Egholm, L. Christensen, K. L. Deuholm, O. Buchardt, J. Coull and P. E. Nielsen, *Nucleic Acids Res.*, 1995, **23**, 217–222.
121. U. von Krosigk and S. A. Benner, *J. Am. Chem. Soc.*, 1995, **117**, 5361–5362.
122. R. Berressem and J. W. Engels, *Nucleic Acids Res.*, 1995, **23**, 3465–3472.
123. G. Xiang, W. Soussou and W. Mclaughlin, *Tetrahedron*, 1994, **11**, 11155–11156.
124. G. Xiang, R. Bogacki and L. W. McLaughlin, *Nucleic Acids Res.*, 1996, **24**, 1963–1970.
125. U. Parsch and J. W. Engels, *Chemistry*, 2000, **6**, 2409–2424.
126. F. Seela and K. I. Shaikh, *Org. Biomol. Chem.*, 2006, **4**, 3993–4004.
127. F. Seela, D. Jiang and S. Budow, *Chembiochem*, 2010, **11**, 1443–1450.
128. P. S. Miller, P. Bhan, C. D. Cushman and T. L. Trapane, *Biochemistry*, 1992, **31**, 6788–6793.
129. S. H. Krawczyk, J. F. Milligan, S. Wadwani, C. Moulds, B. C. Froehler and M. D. Matteucci, *Proc. Natl. Acad. Sci. U. S. A.*, 1992, **89**, 3761–3764.
130. M. C. Jetter and F. W. Hobbs, *Biochemistry*, 1993, **32**, 3249–3254.
131. J. Hunziker, E. S. Priestley, H. Brunar and P. B. Dervan, *J. Am. Chem. Soc.*, 1995, **117**, 2661–2662.
132. K. M. Koshlap, P. Schultze, H. Brunar, P. B. Dervan and J. Feigon, *Biochemistry*, 1997, **36**, 2659–2668.
133. J. S. Koh and P. B. Dervan, *J. Am. Chem. Soc.*, 1992, **114**, 1470–1478.
134. E. S. Priestley and P. B. Dervan, *J. Am. Chem. Soc.*, 1995, **117**, 4761–4765.
135. J. Marfurt, J. Hunziker and C. Leumann, *Bioorg. Med. Chem. Lett.*, 1996, **6**, 3021–3024.
136. J. Marfurt and C. Leumann, *Angew. Chem., Int. Ed. Engl.*, 1998, **37**, 175–177.
137. J. Marfurt, *Nucleic Acids Res.*, 1997, **25**, 1875–1882.
138. A. Clair, G. Xiang and L. W. McLaughlin, *Nucleosides Nucleotides*, 1998, **17**, 925–937.
139. H. Manor, B. S. Rao and R. G. Martin, *J. Mol. Evol.*, 1988, **27**, 96–101.
140. M. J. Behe, *Nucleic Acids Res.*, 1995, **23**, 689–695.
141. J. R. Goñi, X. de la Cruz and M. Orozco, *Nucleic Acids Res.*, 2004, **32**, 354–360.
142. D. M. Gowers and K. R. Fox, *Nucleic Acids Res.*, 1999, **27**, 1569–1577.
143. D. Loakes, *Nucleic Acids Res.*, 2001, **29**, 2437–2447.
144. D. A. Horne and P. B. Dervan, *Nucleic Acids Res.*, 1991, **19**, 4963–4965.
145. L. C. Griffin and P. B. Dervan, *Science*, 1989, **245**, 967–971.

146. K. Yoon, C. A. Hobbs, J. Koch, M. Sardaro, R. Kutny and A. L. Weis, *Proc. Natl. Acad. Sci. U. S. A.*, 1992, **89**, 3840–3844.
147. S. P. Chandler and K. R. Fox, *FEBS Lett.*, 1993, **332**, 189–192.
148. R. W. Roberts and D. M. Crothers, *Proc. Natl. Acad. Sci. U. S. A.*, 1991, **88**, 9397–9401.
149. M. Rougée, B. Faucon, J. L. Mergny, F. Barcelo, C. Giovannangeli, T. Garestier and C. Hélène, *Biochemistry*, 1992, **31**, 9269–9278.
150. J. L. Mergny, J. S. Sun, M. Rougée, T. Montenay-Garestier, F. Barcelo, J. Chomilier and C. Hélène, *Biochemistry*, 1991, **30**, 9791–9798.
151. I. Radhakrishnan, X. Gao, C. de los Santos, D. Live and D. J. Patel, *Biochemistry*, 1991, **30**, 9022–9030.
152. O. A. Amosova and J. R. Fresco, *Nucleic Acids Res.*, 1999, **27**, 4632–4635.
153. D. M. Gowers, J. Bijapur, T. Brown and K. R. Fox, *Biochemistry*, 1999, **38**, 13747–13758.
154. S. P. Chandler and K. R. Fox, *FEBS Lett.*, 1995, **360**, 21–25.
155. D. M. Gowers and K. R. Fox, *Nucleic Acids Res.*, 1998, **26**, 3626–3633.
156. I. Radhakrishnan and D. J. Patel, *J. Mol. Biol.*, 1994, **241**, 600–619.
157. D. M. Gowers and K. R. Fox, *Nucleic Acids Res.*, 1997, **25**, 3787–3794.
158. R. H. Durland, T. S. Rao, G. R. Revankar, J. H. Tinsley, M. A. Myrick, D. M. Seth, J. Rayford, P. Singh and K. Jayaraman, *Nucleic Acids Res.*, 1994, **22**, 3233–3240.
159. S. Obika, Y. Hari, M. Sekiguchi and T. Imanishi, *Angew. Chem., Int. Ed.*, 2001, 2079–2081.
160. Y. Hari, S. Obika, M. Sekiguchi and T. Imanishi, *Tetrahedron*, 2003, **59**, 5123–5128.
161. Y. Hari, S. Matsugu, H. Inohara, Y. Hatanaka, M. Akabane, T. Imanishi and S. Obika, *Org. Biomol. Chem.*, 2010, **8**, 4176–4180.
162. D. L. Chen and L. W. McLaughlin, *J. Org. Chem.*, 2000, **65**, 7468–7474.
163. I. Prévot-Halter and C. J. Leumann, *Bioorg. Med. Chem. Lett.*, 1999, **9**, 2657–2660.
164. S. Buchini and C. J. Leumann, *Tetrahedron Lett.*, 2003, **44**, 5065–5068.
165. S. Buchini and C. J. Leumann, *Angew. Chem. Int. Ed.*, 2004, **43**, 3925–3928.
166. S. Buchini and C. J. Leumann, *Eur. J. Org. Chem.*, 2006, 3152–3168.
167. R. T. Ranasinghe, D. A. Rusling, V. E. C. Powers, K. R. Fox and T. Brown, *Chem. Commun.*, 2005, **2**, 2555–2557.
168. S. R. Gerrard, N. Srinivasan, K. R. Fox and T. Brown, *Nucleosides, Nucleotides Nucleic Acids*, 2007, **26**, 1363–1367.
169. S. R. Gerrard, M. M. Edrees, I. Bouamaied, K. R. Fox and T. Brown, *Org. Biomol. Chem.*, 2010, **8**, 5087–5096.
170. Y. Hari, M. Akabane, Y. Hatanaka, M. Nakahara and S. Obika, *Chem. Commun.*, 2011, **47**, 4424–4426.
171. Y. Hari, M. Akabane and S. Obika, *Chem. Commun.*, 2013, **49**, 7421–7423.
172. M. Akabane-Nakata, S. Obika and Y. Hari, *Org. Biomol. Chem.*, 2014, **12**, 9011–9015.

173. A. Semenyuk, E. Darian, J. Liu, A. Majumdar, B. Cuenoud, P. S. Miller, A. D. Mackerell and M. M. Seidman, *Biochemistry*, 2010, **49**, 7867–7878.
174. C. Y. Huang and P. S. Miller, *J. Am. Chem. Soc.*, 1993, **115**, 10456–10457.
175. C. Y. Huang and P. S. Miller, *J. Am. Chem. Soc.*, 1993, **58**, 5048–5049.
176. L. C. Griffin, L. L. Kiessling, P. A. Beal, P. Gillespie and P. B. Dervan, *J. Am. Chem. Soc.*, 1992, **114**, 7976–7982.
177. K. M. Koshlap, P. Gillespie, P. B. Dervan and J. Feigon, *J. Am. Chem. Soc.*, 1993, 7908–7909.
178. T. E. Lehmann, W. A. Greenberg, D. A. Liberies, C. K. Wada and P. B. Dervan, *Helv. Chim. Acta*, 1997, **80**, 2002–2022.
179. Y. Hari, S. Obika, H. Inohara, M. Ikejiri, D. Une and T. Imanishi, *Chem. Pharm. Bull.*, 2005, **53**, 843–846.
180. Y. Hari, M. Nakahara, J. Pang, M. Akabane, T. Kuboyama and S. Obika, *Bioorg. Med. Chem.*, 2011, **19**, 1162–1166.
181. A. B. Eldrup, O. Dahl and P. E. Nielsen, *J. Am. Chem. Soc.*, 1997, **119**, 11116–11117.
182. D. Guianvarc'h, R. Benhida, J. L. Fourrey, R. Maurisse and J. S. Sun, *Chem. Commun.*, 2001, 1814–1815.
183. Y. Wang, D. A. Rusling, V. E. C. Powers, O. Lack, S. D. Osborne, K. R. Fox and T. Brown, *Biochemistry*, 2005, **44**, 5884–5892.
184. D. Guianvarc, J. Fourrey, R. Maurisse, J. Sun, R. Benhida, I. De Chimie and D. Terrasse, *Tetrahedron*, 2002, 10–13.
185. D. Guianvarc'h, J.-L. Fourrey, R. Maurisse, J.-S. Sun and R. Benhida, *Bioorg. Med. Chem.*, 2003, **11**, 2751–2759.
186. A. Ohkubo, K. Yamada, Y. Ito, K. Yoshimura, K. Miyauchi, T. Kanamori, Y. Masaki, K. Seio, H. Yuasa and M. Sekine, *Nucleic Acids Res.*, 2015, **43**, 5675–5686.
187. S. Obika, H. Inohara, Y. Hari and T. Imanishi, *Bioorg. Med. Chem.*, 2008, **16**, 2945–2954.
188. J.-S. Li, Y.-H. Fan, Y. Zhang, L. A. Marky and B. Gold, *J. Am. Chem. Soc.*, 2003, **125**, 2084–2093.
189. J.-S. Li, F.-X. Chen, R. Shikiya, L. A. Marky and B. Gold, *J. Am. Chem. Soc.*, 2005, **127**, 12657–12665.
190. J.-S. Li, R. Shikiya, L. A. Marky and B. Gold, *Biochemistry*, 2004, **43**, 1440–1448.
191. J.-S. Li and B. Gold, *J. Org. Chem.*, 2005, **70**, 8764–8771.
192. D. A. Rusling, V. E. C. Powers, R. T. Ranasinghe, Y. Wang, S. D. Osborne, T. Brown and K. R. Fox, *Nucleic Acids Res.*, 2005, **33**, 3025–3032.
193. S. T. Crooke, *Nucleic Acid Ther.*, 2017, **27**, 70–77.
194. D. A. Rusling, I. S. Nandhakumar, T. Brown and K. R. Fox, *ACS Nano*, 2012, **6**, 3604–3613.
195. D. A. Rusling, A. R. Chandrasekaran, Y. P. Ohayon, T. Brown, K. R. Fox, R. Sha, C. Mao and N. C. Seeman, *Angew. Chem. Int. Ed.*, 2014, **53**, 3979–3982.

CHAPTER 2

# *Interfacial Inhibitors*

EVGENY KISELEV AND YVES POMMIER*

Developmental Therapeutics Branch and Laboratory of Molecular Pharmacology, Center for Cancer Research, National Cancer Institute, NIH, Bethesda, MD, USA
*Email: pommier@nih.gov

## 2.1 Introduction

Biological systems rely on a variety of macromolecular complexes to perform essential survival functions and maintain structural and biochemical homeostasis. Cells depend on structural polymers to support their architecture (*e.g.* membranes and filaments), store genetic information (*e.g.* nucleic acids), and transduce signals (poly(ADPribosylation), ubiquitylation, SUMOylation). Such structures are assemblies of macromolecules by themselves (*e.g.* microtubules, DNA duplex, ribosomes) and require other multicomponent systems for their synthesis, maintenance and degradation (*e.g.* 5S, 28S and 18S RNA components of eukaryotic ribosomes, 11S regulatory and 20S core particles of proteasomes and RNA polymerase). Macromolecular complexes, once assembled, form a wide spectrum of molecular interfaces that change while such molecular machines fulfil their enzymatic or structural functions. The mechanical movement within such complexes comes from rearrangement of the components comprising the macromolecular assembly, highlighting the dynamic nature of the macromolecular interfaces. It is foreseeable that small molecules binding to such interfaces can obstruct operations of the entire macromolecular complexes, thus creating unique opportunities to develop and discover small molecules targeting transient interfaces of macromolecular machines.

Chemical Biology No. 7
DNA-targeting Molecules as Therapeutic Agents
Edited by Michael J. Waring

Published by the Royal Society of Chemistry, www.rsc.org

To describe such small molecules the term interfacial inhibitor was introduced.[1] The concept of interfacial inhibition originates from the study of topoisomerase inhibitors,[2–4] which are widely used as anticancer drugs, and produce topoisomerase-linked DNA breaks that correspond to normally transient catalytic intermediates of the topoisomerase reactions.[5] The notion rests on the realisation that covalent topoisomerase–DNA intermediates containing DNA strand breaks could be trapped and subsequently turned into abortive double-strand breaks by binding of the topoisomerase inhibitors at the enzyme–DNA interface at the site of initial cleavage. This hypothesis was proposed in the 1990s[2–4] but proved only 10 years later with the resolution of the co-crystal structure of topotecan, an analogue of natural product camptothecin bound to the topoisomerase I (TOP1) cleavage complex (TOP1cc).[6] And more recently a similar molecular picture was obtained for topoisomerase II cleavage complexes (TOP2cc) (see below and Figures 2.2–2.3). This was first shown for the quinolone antibacterials with TOPO IV[7,8] and gyrase.[9,10] Trapping of eukaryotic topoisomerase II was extended to etoposide with TOP2βcc.[11] In this picture once again it was revealed that the location of the drug binding site is at the interface of the enzyme and DNA. Independently, the co-crystal structure of the fungal macrolide brefeldine A has revealed that the inhibitor is bound at the interface of the small GTP-binding protein Arf and its guanine-nucleotide-exchange factor Sec76.[12] The interfacial inhibition concept was extended beyond protein–nucleic acid interfaces and was found to be equally applicable to protein–protein interfaces. The term 'interfacial inhibitor' was coined[1] to describe this previously unanticipated mode of inhibition.

From the time of introduction of the interfacial inhibitor concept, the number of examples and scope of small molecules acting as interfacial inhibitors has expanded, and now includes those targeting receptors (exemplified by the nicotinic receptor inhibitors), signal transducers (exemplified by the mTOR inhibitors), cellular scaffolding structures (exemplified by tubulin inhibitors), and protein–DNA assemblies (exemplified by topoisomerase, gyrase, polymerase and ribosome inhibitors).

This chapter expands on our recent reviews.[13,14] The next sections describe specific examples (case studies) of protein–nucleic acid complexes (topoisomerase and HIV integrase inhibitors) and recent examples of interfacial protein–protein inhibitors for the viral cofactor stimulator of interferon genes (STING) and the actin-related protein 2/3 (Arp2/3) complex.

## 2.2 Case Studies

### 2.2.1 Topoisomerase Inhibitors

Topoisomerases exist in all forms of life where long DNA and RNA polymers need to be tightly packaged and are susceptible to becoming tangled. They are present in all cellular compartments containing DNA and RNA (nucleus, mitochondria, cytoplasm).[5] Topoisomerases are the targets of widely used anticancer drugs (TOP1, TOP2α and TOP2β) and antibiotics (gyrase and TOPO IV).[15,16] Antibiotics of the quinolone family include oxolinic and nalidixic acid analogues. Anticancer agents include irinotecan and topotecan (TOP1 inhibitors) and etoposide, doxorubicin, aclarubicin, dexrazoxane and mitoxantrone (TOP2 inhibitors).[15] Structural and biochemical studies have established their mechanism of action as interfacial inhibitors.[4,5,16–18] Clinically relevant topoisomerase inhibitors bind at the subunit or protein–DNA interfaces and block the topoisomerase catalytic cycle by preventing the rapid interconversion of enzyme–DNA configurations required for the topoisomerase reactions.[13–15,19,20]

#### 2.2.1.1 TOP1 Inhibitors

Human TOP1 is an essential enzyme targeted by the two FDA-approved anticancer drugs topotecan and irinotecan. Both drugs are water-soluble derivatives of the alkaloid camptothecin.[17] In addition to camptothecins, TOP1 can be targeted by another chemical class, the indenoisoquinolines,[18] which also act as interfacial inhibitors.[19–21] When bound at the interface of the TOP1–DNA complex, TOP1 inhibitors intercalate into the DNA at the site of a TOP1-induced nick, and at the same time form a network of direct and water-mediated hydrogen bonds (Figure 2.1). DNA sequence preference differs across chemical classes of TOP1 inhibitors, demonstrating the influence of $\pi$–$\pi$ stacking interactions between drug and DNA nucleobases on the binding and selectivity of the inhibitors. TOP1 mutations at residues R364, D533 and N722 confer resistance towards camptothecin-based inhibitors.[17] These residues were later found to play key roles in drug binding to the TOP1–DNA complex (Figure 2.1B).[6,19–21]

#### 2.2.1.2 TOP2 Inhibitors

TOP2 inhibitors also stabilise cleavage complexes with different topoisomerases. In the case of eukaryotic TOP2, each monomer of the TOP2β dimer accommodates one inhibitor molecule (Figure 2.2A). A similar mode of binding is observed in the case of the quinolone antibacterials with TOPO IV[7,8] and gyrase.[9,10]

Another mode of TOP2 inhibition involves targeting protein–protein interfaces of TOP2 dimers. In the case of eukaryotic TOP2, a small-molecule inhibitor dexrazoxane (ICRF-187) lodges between the subunits of the antiparallel dimer of the TOP2 ATP-binding domain (Figure 2.2B).[22] By binding to the dimer dexrazoxane locks the closed conformation of the ATP-binding domains, thus compromising the strand passage step and trapping TOP2 homodimers on the DNA.

A recent study revealed that the TOP2-related bacterial type II topoisomerase (DNA gyrase) can be inhibited by a mechanism distinct from that of etoposide inhibition of TOP2 or gyrase and TOPOIV inhibition by quinolones.[13,14] Despite the structural and mechanistic similarity between

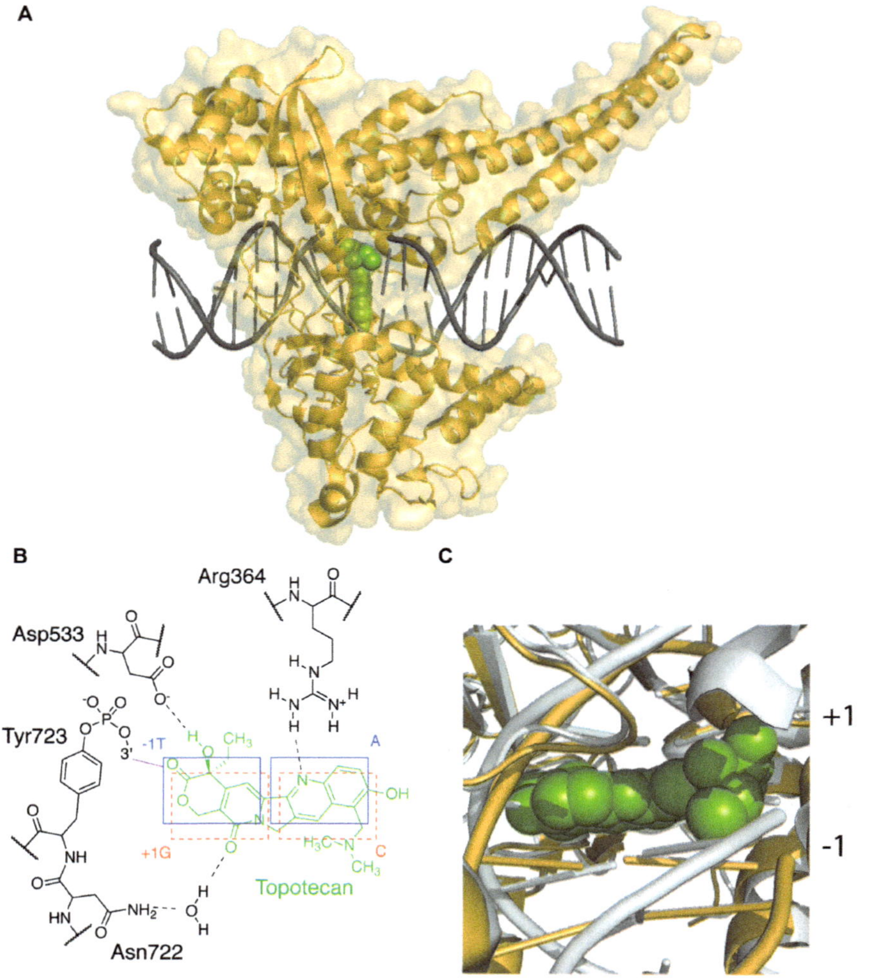


**Figure 2.1** Structure of the topotecan–TOP1–DNA ternary cleavage complex. A. Cartoon and surface representation with TOP1 shown as the gold cartoon and surface, DNA as a dark-grey cartoon and topotecan as green spheres (PDB ID: 1K4T). B. Schematic representation of the molecular interactions of TOP1 (hydrogen bonding) and DNA (stacking, flanking base pairs shown as rectangles) contributing to the binding of topotecan. C. Comparison of the DNA conformation and base pair position at the site of cleavage in the absence (grey, PDB ID: 1K4S) and presence (gold, PDB ID: 1K4T) of topotecan (green spheres). Intercalation of a topotecan molecule results in spatial separation of the base pairs flanking the cleavage site (annotated −1 and +1), which is incompatible with DNA ligation and complex reversal.

DNA gyrase and TOP2, the DNA gyrase is a tetramer (not a dimer as is TOP2) comprised of two GyrA and two GyrB subunits.[15] DNA gyrase is a validated antibacterial drug target, inhibited by fluoroquinone antibiotics such as

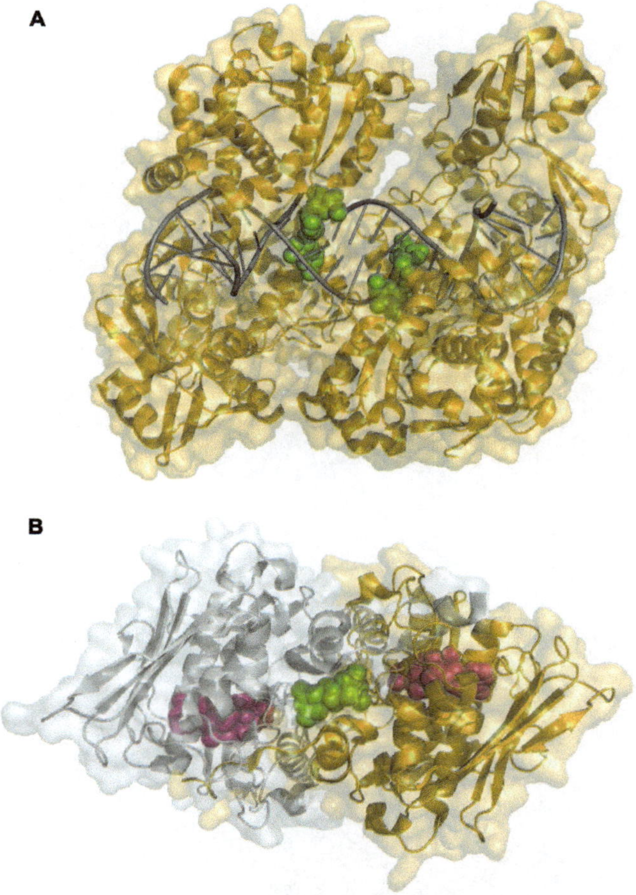


**Figure 2.2** Structure of TOP2 cleavage complexes trapped by the interfacial inhibitors etoposide and dexrazoxane. A. Structure of the etoposide–hTOP2–DNA ternary complex (PDB ID: 3QX3). Etoposide (green spheres) molecules are present in both cleavage sites of the dimeric hTOP2β (gold surface and cartoon) with DNA (black cartoon) in the covalent complex. B. Structure of dexrazoxane (green spheres) bound at the interface of ATP (purple spheres) binding subunits (one grey and one gold cartoon and surface) of TOP2 (PDB ID: 1QZR).

ciprofloxacin. Cyprofloxacin inhibits gyrase by binding to gyrase–DNA cleavage complexes and preventing religation (like etoposide with TOP2; see Figure 2.2A), and ultimately leads to abortive double-strand breaks. The structure of the ternary complex of *Staphylococcus aureus* DNA gyrase bound to novel thiophene-based (Figure 2.3) inhibitors revealed a new drug-binding pocket at the interface of the winged helix domain of GyrA and the TOPRIM (topoisomerase-primase) domain of GyrB without direct contact with DNA.[23] Binding of such small molecules at the GyrA–GyrB interface is likely to

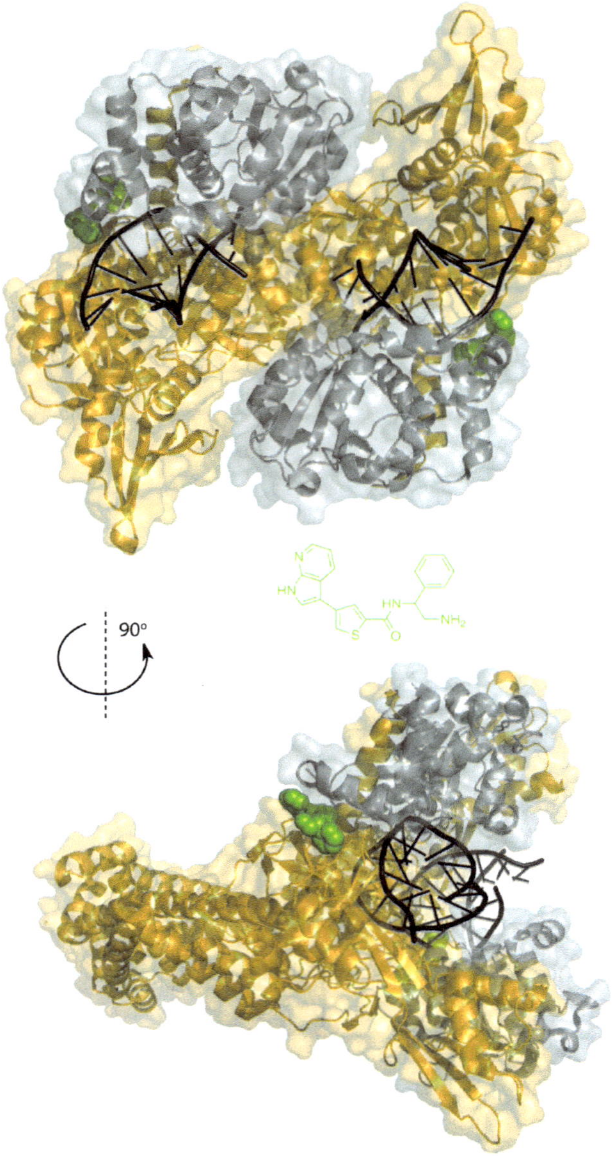


**Figure 2.3** Structure of *Staphylococcus aureus* DNA gyrase (PDB ID: 5NPP). The gyrase tetramer consisting of two GyrA subunits (gold cartoon and surface) and two GyrB subunits (grey cartoon and surface) is shown bound to DNA (black cartoon) with a thiophene-based inhibitor molecule (green spheres with structural formula in the middle) bound at each of two GyrA–GyrB interfaces.

disrupt the relative movement of the gyrase subunits, thereby stalling complexes in cleavage-prone conformations. The discovery of this interfacial binding pocket may lead toward further innovation and improvement in antibacterial treatment as well as overcoming resistance.

### 2.2.2 HIV Integrase Strand Transfer Inhibitors

HIV integrase (IN) is responsible for inserting the retrotranscribed viral DNA into a host chromosome *via* the multistep DNA strand transfer process.[24] The high response of HIV patients has led to the approval of three IN strand transfer inhibitors (INSTIs) by the FDA: raltegravir, elvitegravir and dolutegravir.[25] These inhibitors block integrase-catalysed insertion of viral DNA by stalling the enzyme–DNA system after the initial cleavage (3′-processing) step, which prepares the proviral DNA for the second integrase-catalysed transesterification step (strand transfer). INSTIs such as dolutegravir, elvitegravir and raltegravir bind at the interface of the integrase polypeptide and the processed viral DNA, engaging two catalytic divalent metal ions at the same time (Figure 2.4A).[26] Figure 2.4 shows that dolutegravir coordinates the magnesium cations *via* its co-planar oxygen atoms while the phenyl ring of its benzylamide moiety displaces the terminal adenine of the viral DNA and forms stacking interactions with cytosine (C-2). The displaced adenine stacks with the polycyclic core of dolutegravir.

### 2.2.3 STING Inhibitors

Another example of interfacial inhibition by small-molecule binding at protein–protein interfaces is the action of antiviral drug DMXAA on STING.[27] STING is part of the innate immune response to cytosolic dsDNA, where it acts as a homodimer activating the interferon (IFN) pathway.[28] STING responds to binding of the second messenger cyclic GMP–AMP produced by GMP–AMP synthetase. STING is inhibited by two molecules of DMXAA binding at the subunit interface of the homodimer (Figure 2.5A). Upon binding at the interface each DMXAA molecule contracts a polar interaction with both subunits T262 and T266 of one monomer and R237 of another (Figure 2.5B).

### 2.2.4 Arp2–3 Inhibitors

The multi-subunit macromolecular machinery of the actin-related protein 2–3 (Arp2–3) plays a vital role in regulation of the actin filament network assembly.[29] The internal subunit movement within a conformationally dynamic Arp2–3 complex guides the filament assembly by mimicking the short-pitch actin dimer within filaments, hence executing the spatial control of the overall process. Two chemically distinct small-molecule Arp2–3

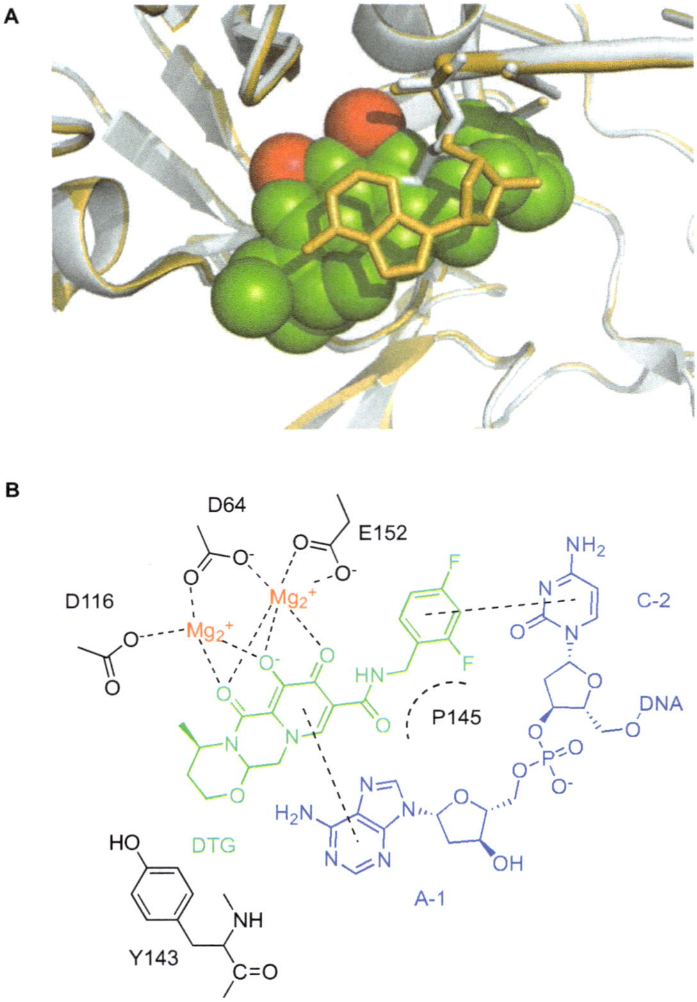


**Figure 2.4** Structure of the dolutegravir retroviral intasome ternary complex. A. Cartoon representation of the complex with integrase shown as grey cartoon representing the IN–DNA complex in the absence of inhibitor (PDB ID: 3OY9), and as gold cartoon when dolutegravir (DTG, green spheres) is present in the ternary complex (PDB ID: 3S3M). Binding of DTG results in deflection of the 3′-terminal nucleobase. (Coordination by magnesium shown as red spheres.) B. Schematic representation of the structure of the ternary complex showing the contributions of integrase (coordination by magnesium shown in red) and DNA (stacking to terminal nucleobases) to the binding of DTG.

inhibitors were discovered to block the activating conformational change of Arp2–3, CK-666 and CK-869. While CK-869 disrupts the protein–protein interface destabilising the active state of Arp2–3, CK-666 utilises the

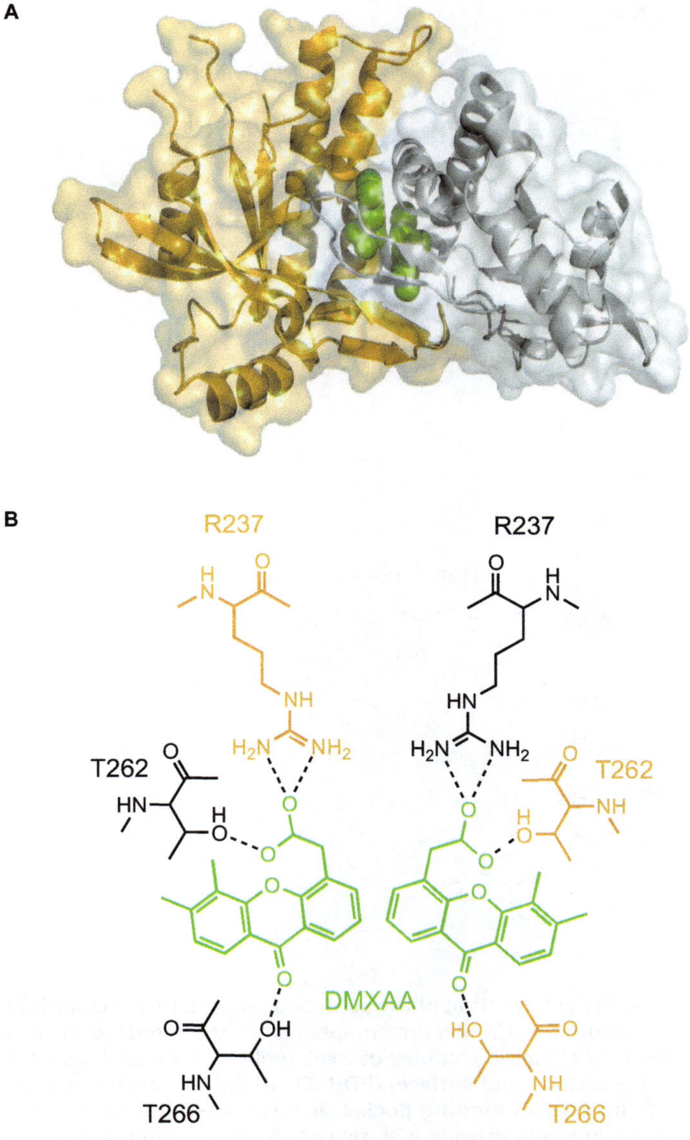


**Figure 2.5** Interfacial inhibition of the signalling protein STING in its homodimeric state with the antiviral DMXAA. A: STING subunits (shown as one grey and one gold cartoon and transparent surface) bind two molecules of DMXAA (green spheres) at their interface (PDB ID: 4LOL). B: Each of two DMXAA molecules bound to the STING homodimer form polar contacts (dashed lines) to both STING subunits.

interface between Arp2 and Arp3 to bind and stabilise the inactive state, preventing the further conformational change into an active state (Figure 2.6).[30]

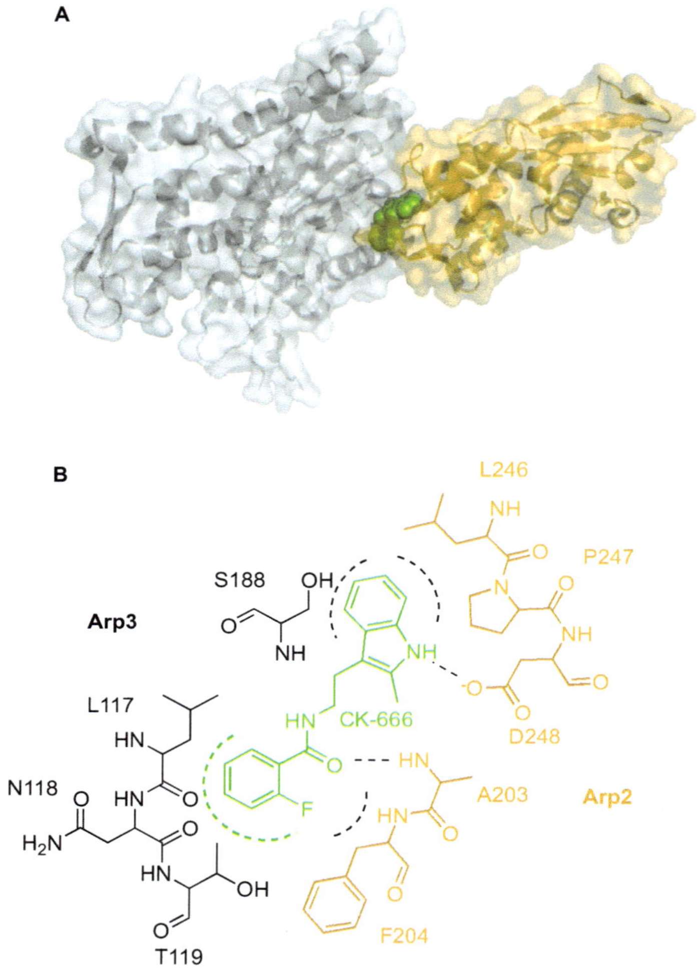


**Figure 2.6** Interfacial inhibition of the actin-related protein 2–3 complex by CK-666. A. Binding of CK-666 (green spheres) at the interface of the conformationally dynamic complex of Arp2 (gold cartoon and surface) and Arp3 (grey cartoon and surface) (PDB ID: 3UKR). B. Schematic representation of the CK-666 binding pocket at the Arp2–3 interface. Both Arp2 and Arp3 subunits provide a steric context to the binding pocket of CK-666 where Arp2 (gold) forms polar contacts (dashed lines) with indole and benzamide moieties of the inhibitor.

## 2.3 Prospects

Interfacial inhibitors open new avenues for drug design and also for elucidating the molecular mechanism of action of medicinal compounds. Rather than disrupting protein–protein or protein–nucleic acid association, interfacial inhibitors exploit the normally transient binding pockets located at the macromolecular subunit interfaces within such assemblies. Interfacial binding pockets are unique to specific complexes, hence providing a valuable basis for inhibitor specificity. The examples of interfacial inhibitor targets described here extend beyond enzymes and range through all major domains of biological macromolecular systems, *e.g.* structural macromolecules, signalling pathway players, macromolecular assembly machinery, *etc.* Depending on the target and particular interface in question, interfacial inhibitors may act as competitive inhibitors by binding at the primary ligand site, or act through allosteric or non-competitive mechanisms to block an intermediate configuration of the macromolecular complex. The interfacial binders of structural complexes (such as the tubulin polymerization inhibitors colchicine and vinblastine) distort the assembled macromolecular architectures, thereby affecting cellular structure. The realisation that molecules act in the form of large protein complexes in a highly dynamic manner makes it likely that the 'undruggable targets', such as chromatin-remodelling complexes, which adjust chromatin to adopt its specific metabolic and structural states, will be within reach once such complexes are properly characterised both structurally and biochemically.

## Acknowledgements

Our studies are supported by the Intramural Program of the National Cancer Institute (Center for Cancer Research), National Institutes of Health, Bethesda, Maryland (Z01-BC006150-19).

## References

1. Y. Pommier and J. Cherfils, *Trends Pharmacol. Sci.*, 2005, **28**, 136–145.
2. G. Capranico, K. W. Kohn and Y. Pommier, *Nucleic Acids Res.*, 1990, **18**, 6611–6619.
3. Y. Pommier, G. Capranico, A. Orr and K. W. Kohn, *Nucleic Acids Res.*, 1991, **19**, 5973–5980.
4. C. Jaxel, G. Capranico, D. Kerrigan, K. W. Kohn and Y. Pommier, *J. Biol. Chem.*, 1991, **266**, 20418–20423.
5. Y. Pommier, Y. Sun, S. N. Huang and J. L. Nitiss, *Nat. Rev. Mol. Cell Biol.*, 2016, **17**, 703–721.
6. B. L. Staker, K. Hjerrild, M. D. Feese, C. A. Behnke, A. B. Burgin, Jr. and L. Stewart, *Proc. Natl. Acad. Sci. U. S. A.*, 2002, **99**, 15387–15392.
7. I. Laponogov, M. K. Sohi, D. A. Veselkov, X. S. Pan, R. Sawhney, A. W. Thompson, K. E. McAuley, L. M. Fisher and M. R. Sanderson, *Nat. Struct. Mol. Biol.*, 2009, **16**, 667–669.
8. I. Laponogov, X. S. Pan, D. A. Veselkov, K. E. McAuley, L. M. Fisher and M. R. Sanderson, *PLoS One*, 2010, **5**, e11338.
9. B. D. Bax, P. F. Chan, D. S. Eggleston, A. Fosberry, D. R. Gentry, F. Gorrec, I. Giordano, M. M. Hann, A. Hennessy, M. Hibbs, J. Huang, E. Jones,

J. Jones, K. K. Brown, C. J. Lewis, E. W. May, M. R. Saunders, O. Singh, C. E. Spitzfaden, C. Shen, A. Shillings, A. J. Theobald, A. Wohlkonig, N. D. Pearson and M. N. Gwynn, *Nature*, 2010, **466**, 935–940.
10. A. Wohlkonig, P. F. Chan, A. P. Fosberry, P. Homes, J. Huang, M. Kranz, V. R. Leydon, T. J. Miles, N. D. Pearson, R. L. Perera, A. J. Shillings, M. N. Gwynn and B. D. Bax, *Nat. Struct. Mol. Biol.*, 2010, **17**, 1152–1153.
11. C. C. Wu, T. K. Li, L. Farh, L. Y. Lin, T. S. Lin, Y. J. Yu, T. J. Yen, C. W. Chiang and N. L. Chan, *Science*, 2011, **333**, 459–462.
12. L. Renault, B. Guibert and J. Cherfils, *Nature*, 2003, **426**, 525–530.
13. Y. Pommier, E. Kiselev and C. Marchand, *Bioorg. Med. Chem. Lett.*, 2015, **25**, 3961–3965.
14. Y. Pommier and C. Marchand, *Nat. Rev. Drug Discovery*, 2012, **11**, 25–36.
15. Y. Pommier, *ACS Chem. Biol.*, 2013, **8**, 82–95.
16. Y. Pommier, E. Leo, H. Zhang and C. Marchand, *Chem. Biol.*, 2010, **17**, 421–433.
17. Y. Pommier, *Nat. Rev. Cancer*, 2006, **6**, 789–802.
18. Y. Pommier and M. Cushman, *Mol. Cancer Ther.*, 2009, **8**, 1008–1014.
19. C. Marchand, S. Antony, K. W. Kohn, M. Cushman, A. Ioanoviciu, B. L. Staker, A. B. Burgin, L. Stewart and Y. Pommier, *Mol. Cancer Ther.*, 2006, **5**, 287–295.
20. A. Ioanoviciu, S. Antony, Y. Pommier, B. L. Staker, L. Stewart and M. Cushman, *J. Med. Chem.*, 2005, **48**, 4803–4814.
21. B. L. Staker, M. D. Feese, M. Cushman, Y. Pommier, D. Zembower, L. Stewart and A. B. Burgin, *J. Med. Chem.*, 2005, **48**, 2336–2345.
22. S. Classen, S. Olland and J. M. Berger, *Proc. Natl. Acad. Sci. U. S. A.*, 2003, **100**, 10629–10634.
23. P. F. Chan, T. Germe, B. D. Bax, J. Huang, R. K. Thalji, E. Bacque, A. Checchia, D. Chen, H. Cui, X. Ding, K. Ingraham, L. McCloskey, K. Raha, V. Srikannathasan, A. Maxwell and R. A. Stavenger, *Proc. Natl. Acad. Sci. U. S. A.*, 2017, **114**, E4492–E4500.
24. Y. Pommier, A. A. Johnson and C. Marchand, *Nat. Rev. Drug Discovery*, 2005, **4**, 236–248.
25. M. Metifiot, C. Marchand and Y. Pommier, *Adv. Pharmacol.*, 2013, **67**, 75–105.
26. S. Hare, S. J. Smith, M. Metifiot, A. Jaxa-Chamiec, Y. Pommier, S. H. Hughes and P. Cherepanov, *Mol. Pharmacol.*, 2011, **80**, 565–572.
27. P. Gao, M. Ascano, T. Zillinger, W. Wang, P. Dai, A. A. Serganov, B. L. Gaffney, S. Shuman, R. A. Jones, L. Deng, G. Hartmann, W. Barchet, T. Tuschl and D. J. Patel, *Cell*, 2013, **154**, 748–762.
28. D. L. Burdette and R. E. Vance, *Nat. Immunol.*, 2013, **14**, 19–26.
29. E. N. Firat-Karalar and M. D. Welch, *Curr. Opin. Cell Biol.*, 2011, **23**, 4–13.
30. B. Hetrick, M. S. Han, L. A. Helgeson and B. J. Nolen, *Chem. Biol.*, 2013, **20**, 701–712.

CHAPTER 3

# *Slow DNA Binding*

PER LINCOLN, L. MARCUS WILHELMSSON AND BENGT NORDÉN*

Chemistry and Biochemistry, Department of Chemistry and Chemical Engineering, Chalmers University of Technology, Kemivägen 10, 41296 Gothenburg, Sweden
*Email: norden@chalmers.se

## 3.1 Introduction—Kinetics *vs.* Thermodynamics of DNA Binding

*In vitro* DNA–drug interaction studies can give information that is important both for therapeutic and diagnostic purposes, such as DNA sequence selectivity and binding geometry, *i.e.*, the thermodynamic equilibrium state of the reaction. However, the path along which the drug reaches this state, the mechanism of binding and the binding kinetics, could be equally important as the rates of association and dissociation might be crucial for the application of the drug. Furthermore, investigating the kinetics of the binding and dissociation reaction could substantially increase understanding of the structure and dynamics of DNA itself and can, for cases where the interaction kinetics is slow, reveal intermediate binding sites that might be significant in the action of a drug.

In the thermodynamic equilibrium state of binding (eqn (3.1)), the number of molecules that are bound to DNA and the number that are free in solution are on average constant (Figure 3.1A), which means that the average

Chemical Biology No. 7
DNA-targeting Molecules as Therapeutic Agents
Edited by Michael J. Waring

Published by the Royal Society of Chemistry, www.rsc.org

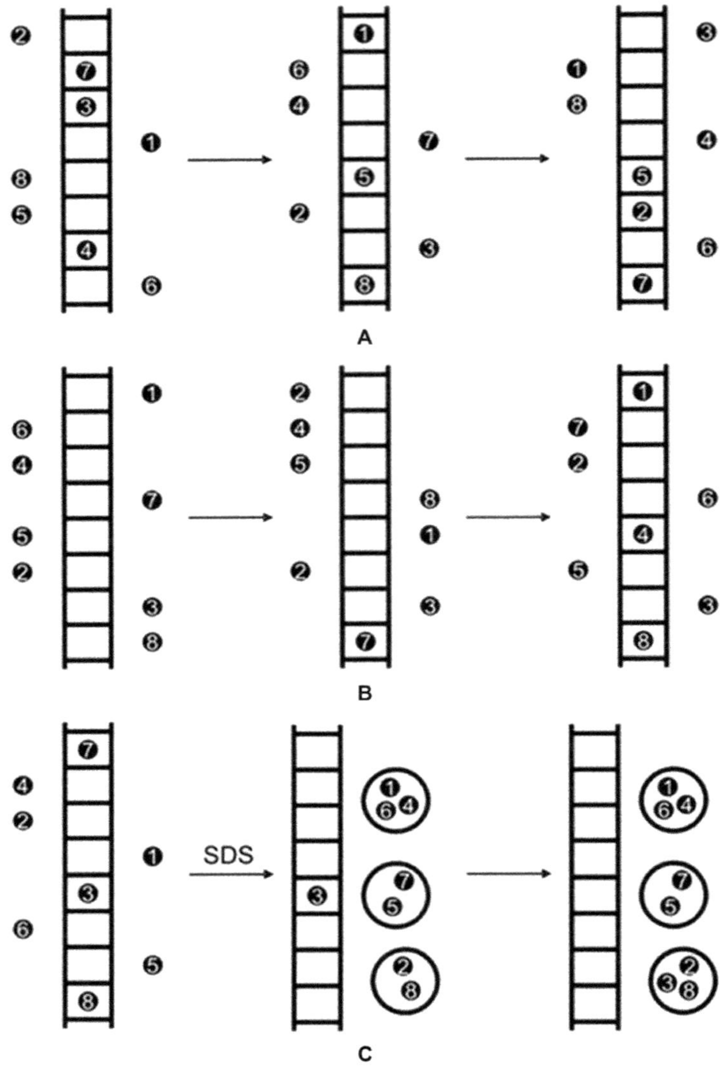


**Figure 3.1** Schematic of DNA-binding processes. (A) Thermodynamic equilibrium state of binding. Filled circles with numbers symbolize DNA-binding molecules. Pictures from left to right are different snapshots after the DNA-binding molecule has reached its equilibrium distribution between free and bound drug. (B) Association process. Pictures from left to right are consecutive snapshots during the DNA association process towards the thermodynamic equilibrium. (C) Dissociation process. Pictures from left to right are consecutive snapshots during the DNA dissociation process towards the thermodynamic equilibrium. SDS (large open circles), forming micelles that sequester free drugs, has been added before the middle snapshot.

number of molecules that dissociates (eqn (3.2)) from DNA at any time is the same as the average number that associates (eqn (3.3)).

$$K = \frac{[\mathrm{DNA} - L]}{[\mathrm{DNA} - \mathrm{site}][L]} \tag{3.1}$$

$$\frac{\mathrm{d}[\mathrm{DNA} - L]}{\mathrm{d}t} = -k_\mathrm{d}[\mathrm{DNA} - L] \tag{3.2}$$

$$\frac{\mathrm{d}[\mathrm{DNA} - L]}{\mathrm{d}t} = k_\mathrm{a}[L][\mathrm{DNA} - \mathrm{site}] \tag{3.3}$$

In contrast, it should be noted that the behaviour of a single molecule may be very dynamic. The rate by which molecules associate to DNA to reach their equilibrium state of binding may vary enormously ($k_\mathrm{a} \sim 10^3$–$10^4$ $\mathrm{M^{-1}\,s^{-1}}$ for actinomycin D[1,2] and $\sim 10^8$ $\mathrm{M^{-1}\,s^{-1}}$ for several groove binders[3–5] at ambient temperature and approximately 0.2 M $\mathrm{Na^+}$; approximately ambient temperature and 0.1–0.2 M $\mathrm{Na^+}$ are the conditions throughout the text unless stated otherwise) as does the rate of dissociation ($k_\mathrm{d} \sim 10^{-3}$ $\mathrm{s^{-1}}$ for actinomycin D[1,2,6–8] and $\sim 10^2$ $\mathrm{s^{-1}}$ for the intercalator proflavine[9] under similar conditions). The equilibrium distribution of molecules between DNA and solution is governed by the relationship between the rate of association and dissociation as shown in eqn (3.4), when the rate of eqn (3.2) equals that of eqn (3.3):

$$K = \frac{k_\mathrm{a}}{k_\mathrm{d}} = \frac{[\mathrm{DNA} - L]}{[\mathrm{DNA} - \mathrm{site}][L]} \tag{3.4}$$

In the literature it has been suggested that not only a high binding constant but also, more importantly, a slow rate of dissociation is crucial for a DNA-binding drug to be able to block binding or movement of a protein along DNA and, as a consequence, interfere with the mechanism of *e.g.* DNA replication or transcription.[8]

On the way towards thermodynamic equilibrium the rate by which the molecules associate to or dissociate from DNA exceeds the other (Figure 3.1B and C) so that in this case the state is really dynamic and shows relaxation kinetics which may be studied, *e.g.*, with the help of stopped-flow spectrophotometric techniques, depending on the rate of the reaction. Association to DNA *in vitro* is normally studied by mixing a solution of drug with a solution of DNA and observing a change in *e.g.* drug fluorescence or absorption intensity with time. Using the same changes in intensity it may also be possible to study the dissociation from DNA by detergent-sequestering, dilution or dialysis methods.

This chapter will focus on slow small-molecule DNA binders. Examples of some established small DNA binders with slow kinetics, actinomycin D and nogalamycin, will be briefly considered, but main focus will be on dimeric ruthenium complexes that exhibit extremely slow binding kinetics and several other interesting properties. As we shall argue, an extremely slow dissociation may be useful both for therapeutic and diagnostic purposes. In addition, slow DNA binders may be used as model systems for getting a 'slow motion' picture of how fast DNA binders may redistribute between solution and DNA as well as between binding sites on DNA. Tables of characteristic rates of association and dissociation for several of the drugs mentioned in the text will be referred to and summarized (Tables 3.1 and 3.2). To begin with we will briefly present the main types of DNA-binding modes, their respective time-scales of interaction and mechanisms of binding.

## 3.2 Different DNA Binding Modes—Different DNA Binding Kinetics

A molecule may interact with DNA either covalently or non-covalently. Although several drugs that bind covalently to DNA are also very interesting we shall here confine ourselves to non-covalent, reversible, types of binding. Such types of binding in the vast majority of strong-binding cases involve positively charged molecules and may, for small molecules, be divided into three different classes of binding: external electrostatic binding (Figure 3.2A), groove binding (Figure 3.2B) and intercalation (Figure 3.2C). A special case of intercalation is threading intercalation (Figure 3.2D) which will be treated separately since it is especially interesting in context of slow DNA binding kinetics.

### 3.2.1 External Electrostatic Binding

Small molecules or ions that bind externally are usually unspecific both with respect to nucleotide sequence and location, their distribution being governed mainly by electrostatic interaction with the negatively charged phosphates located in the DNA backbone. Ions like $Na^+$ (Figure 3.2A), $K^+$, and $Li^+$ all bind in this unspecific way, but with soft transition metal ions there are also examples of ions that can bind specifically to the actual nucleobase (*e.g.* $Ag^+$ and $Hg^{2+}$).[9] Examples of small molecules that in certain studies have been suggested to bind externally to DNA are some of the cationic polyamines.[10–12] External binding being electrostatic in nature and generally requiring insignificant changes of DNA structure, is characterized by a binding kinetics approaching the diffusion limit ($k_a = 10^8$–$10^9$ $M^{-1}\,s^{-1}$).

### 3.2.2 Groove Binding

Groove binding agents bind to DNA in either the major or the minor groove. While a majority of DNA-binding proteins associate into the major groove, most of the smaller molecules bind into the narrow and deep minor groove (Figure 3.2B), which has only a limited number of hydrogen bond donors or acceptors. Despite a smaller number of potential hydrogen bond

**Table 3.1** Rates of association of drugs to DNA. Unless stated otherwise, conditions are around physiological salt concentration and pH and at ambient temperature.

| Drug | Rates of association $k_a$ | | DNA | Ref. |
|---|---|---|---|---|
| DAPI | $6.3\times10^7\ M^{-1}\ s^{-1}$ | | AT | 3 |
| Distamycin | $7\times10^7\ M^{-1}\ s^{-1}$ | Drug : site 1 : 1 | AT-region | 4 |
| | $7\times10^7\ M^{-1}\ s^{-1}$ and $5\ s^{-1}$ | Drug : site 2 : 1 | | |
| Hairpin polyamide | $7\times10^7\ M^{-1}\ s^{-1}$ | Calculated from K | 5′-AGTACT | 5 |
| Daunomycin | $\approx 10^7\ M^{-1}\ s^{-1}$ | | Calf thymus | 47 |
| Daunomycin | $3\times10^6\ M^{-1}\ s^{-1}$, $92\ s^{-1}$, and $4\ s^{-1}$ | | Calf thymus | 48 |
| Adriamycin | $\approx 10^7\ M^{-1}\ s^{-1}$ | | Calf thymus | 47 |
| Ethidium | $\approx 10^7\ M^{-1}\ s^{-1}$ | | AT and GC | 38 |
| Actinomycin D | $\approx 2\times10^4\ M^{-1}\ s^{-1}$ | Various 12-mers | 5′-TGCA | 1 |
| Actinomycin D | $1\times10^3\ M^{-1}\ s^{-1}$ | 10-mer GC center | 5′-GGCC | 2 |
| Actinomycin D | $1\times10^4\ M^{-1}\ s^{-1}$ | 10-mer GC center | 5′-TGCA | 2 |
| Actinomycin D | $1\times10^4\ M^{-1}\ s^{-1}$ | 10-mer GC center | 5′-AGCT | 2 |
| Actinomycin D | $1\times10^4\ M^{-1}\ s^{-1}$ | 10-mer GC center | 5′-CGCG | 2 |
| Nogalamycin | $\approx 10^5\ M^{-1}\ s^{-1}$ (fastest) | Triphasic | Calf thymus | 53 |
| Nogalamycin | $\approx 10^5\ M^{-1}\ s^{-1}$ (fastest) | Biphasic | AT | 53 |
| Nogalamycin | $2.5\times10^3\ M^{-1}\ s^{-1}$ | | GC | 53 |
| ΔΔ-[μ-c4(cpdppz)$_2$(phen)$_4$Ru$_2$]$^{4+}$ | $29\ s^{-1}$ and $4.4\ s^{-1}$ (two fastest) | Multiphasic 45 μM base | AT | 102 |
| ΔΔ-[μ-(11,11′-bidppz)(phen)$_4$Ru$_2$]$^{4+}$ | $1.2\times10^{-3}\ s^{-1}$ and $2\times10^{-4}\ s^{-1}$ [a] (threading) | 160 μM base | Calf thymus | 104 |
| ΔΔ-[μ-(11,11′-bidppz)(phen)$_4$Ru$_2$]$^{4+}$ | $10^{-4}\ s^{-1}$ [b] (shuffling) | 80 μM base | AT | 107 |
| ΛΛ-[μ-(11,11′-bidppz)(phen)$_4$Ru$_2$]$^{4+}$ | $3.4\times10^{-3}\ s^{-1}$ (threading) | 120 μM base | AT | 106 |
| ΛΛ-[μ-(11,11′-bidppz)(bipy)$_4$Ru$_2$]$^{4+}$ | $1.2\times10^{-2}\ s^{-1}$ (first order; threading) | 120 μM base | AT | 106 |
| | $5.0\times10^{-2}\ s^{-1}$ (second order; threading) | | | |

[a] Performed at 50 °C.
[b] Performed at 50 °C, 2 mM $Mg^{2+}$ and 5 mM $K^+$.

**Table 3.2** Rates of dissociation of drugs from DNA. Unless stated otherwise, conditions are around physiological salt concentration and pH and at ambient temperature.

| Drug | Rates of dissociation $k_d$ ($s^{-1}$) | | DNA | Method + ref. |
|---|---|---|---|---|
| DAPI | 8.5 | | AT | SDS; 3 |
| Distamycin | 3 | Drug : site 1 : 1 | AT-region | SDS; 4 |
| | 0.1 | Drug : site 2 : 1 | | |
| Hairpin polyamide | $2\times10^{-3}$ | | 5′-AGTACT | SDS; 5 |
| Daunomycin | 10 | | Calf thymus | T-Jump relaxation; 47 |
| Daunomycin | 170, 10, and 1 | | Calf thymus | SDS; 48 |
| Adriamycin | 3.5 | | Calf thymus | T-Jump relaxation; 47 |
| Ethidium | $\sim10^{1}$ | | AT and GC | P-Jump relaxation; 38 |
| Actinomycin D | $8\times10^{-2}$, $2\times10^{-2}$, and $2\times10^{-3}$ | | Calf thymus | SDS; 8 |
| Actinomycin D | Between $2\times10^{-2}$ and $5\times10^{-4}$ [a] | | Different | Footprinting; 7 |
| Actinomycin D | $\sim5\times10^{-4}$ | Various 5′-TGCA | 5′-TGCA | SDS; 1 |
| Actinomycin D | $6.7\times10^{-4}$ [b] | | *lac* DNA | Transcription assay; 6 |
| Actinomycin D | $3.3\times10^{-3}$ [b] | | Calf thymus | SDS; 6 |
| Actinomycin D | $1.6\times10^{-3}$ | 10-mer GC center | 5′-TGCA | SDS; 2 |
| Actinomycin D | $3.4\times10^{-3}$ | 10-mer GC center | 5′-AGCT | SDS; 2 |
| Actinomycin D | $2.3\times10^{-3}$ | 10-mer GC center | 5′-CGCG | SDS; 2 |
| Nogalamycin | $3.3\times10^{-3}$, $9.8\times10^{-4}$ and $2.3\times10^{-4}$ [c] | | Calf thymus | SDS; 54 |
| Nogalamycin | $4.0\times10^{-2}$ [c] | | AT | SDS; 54 |
| Nogalamycin | $2.7\times10^{-4}$ [c] | | GC | SDS; 54 |
| ΔΔ-[μ-c4(cpdppz)$_2$(phen)$_4$Ru$_2$]$^{4+}$ | $1\times10^{-3}$ [d] | | AT | Calf thymus DNA; 67 |
| ΔΔ-[μ-c4(cpdppz)$_2$(phen)$_4$Ru$_2$]$^{4+}$ | $7.3\times10^{-3}$ | | AT | SDS; 101 |
| ΛΛ-[μ-c4(cpdppz)$_2$(phen)$_4$Ru$_2$]$^{4+}$ | $4.2\times10^{-2}$ | | AT | SDS; 101 |
| Meso-[μ-c4(cpdppz)$_2$(phen)$_4$Ru$_2$]$^{4+}$ | $1\times10^{-2}$ | | AT | SDS; 101 |
| ΔΔ-[μ-c4(cpdppz)$_2$(phen)$_4$Ru$_2$]$^{4+}$ | $4.5\times10^{-3}$ | 95% of process | GC | SDS; 102 |
| Meso-[μ-c4(cpdppz)$_2$(phen)$_4$Ru$_2$]$^{4+}$ | $6.4\times10^{-3}$ and $1.5\times10^{-3}$ | | Calf thymus | SDS; 101 |
| ΔΔ-[μ-(11,11′-bidppz)(phen)$_4$Ru$_2$]$^{4+}$ | $1.9\times10^{-3}$ $s^{-1}$ and $8\times10^{-4}$ $s^{-1}$ [d] | | Calf thymus | SDS; 104 |
| ΔΔ-[μ-(11,11′-bidppz)(phen)$_4$Ru$_2$]$^{4+}$ | $2.4\times10^{-5}$ $s^{-1}$ [d] | | Calf thymus | AT; (Nordell, P., unpublished) |

[a] Salt concentration 10–20 mM $Na^{+}$.
[b] Performed at 37 °C.
[c] Performed at 40 °C.
[d] Performed at 50 °C.

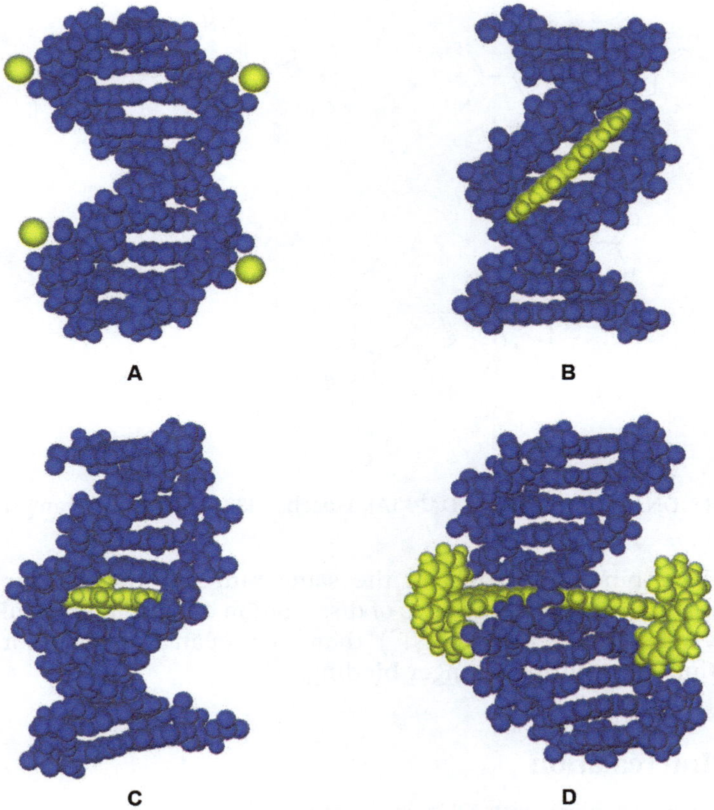


**Figure 3.2** DNA binding modes. (A) External electrostatic binding of positive metal ions. (B) Groove binding of minor groove binder DAPI. (C) Intercalation of ethidium. (D) Threading intercalation of semirigid ruthenium dimer, ΔΔ-$[\mu\text{-}(11,11'\text{-bidppz})(\text{phen})_4\text{Ru}_2]^{4+}$.

interactions in the minor compared with the major groove, binding in the minor groove may in some cases be sequence-selective, mainly as a result of hydrogen bonding but also due to additional effects from van der Waals and hydrophobic contacts as well as more specifically positioned electrostatic interactions. Small minor-groove-binders like DAPI[3,13–16] (Figure 3.3A), Hoechst 33258[17,18] (Figure 3.3B) and distamycin[19–25] (Figure 3.3C) all preferentially bind to AT-rich regions of DNA. From the observation of a sign change in flow linear dichroism (LD) for pure GC contexts, where the exocyclic amino group prevents binding in the minor groove, however, it was concluded that also intercalation may occur for one of the same ligands.[26,27] The groove binders normally have binding kinetics that, as with external binding, approaches the diffusion limit ($k_a \approx 10^8$ M$^{-1}$ s$^{-1}$), while their rate of dissociation could vary considerably ($k_d = 10^{-1}$–$10^3$ s$^{-1}$). In spite of their size, the somewhat larger polyamides,[28–33] which also bind in the minor

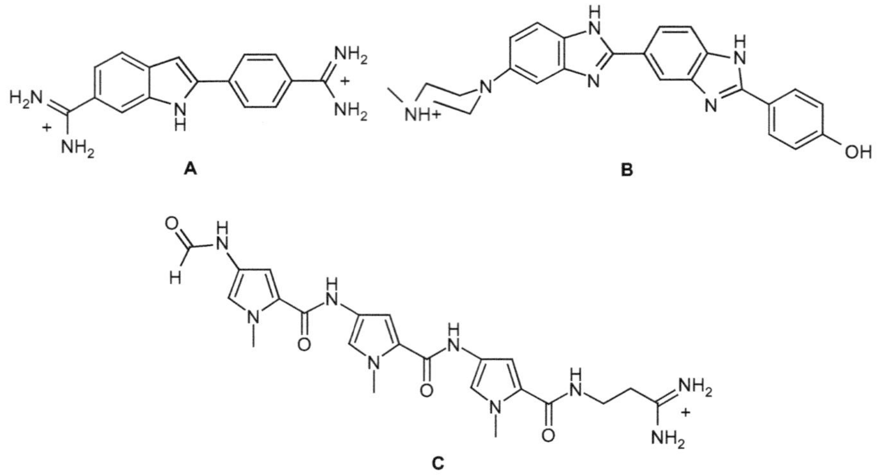


Figure 3.3 DNA groove binders DAPI (A), Hoechst 33258 (B) and distamycin (C).

groove, have binding kinetics in the same range as the smaller minor groove binders.[5] However, the rates of dissociation of hairpin polyamides are generally much slower ($k_d \approx 10^{-3}$ $s^{-1}$)[5] than those of an average minor groove binder, thus constituting stronger binding.

### 3.2.3 Intercalation

Intercalators associate with DNA by inserting themselves between adjacent base pairs (Figure 3.2C). In doing so, they cause elongation of DNA by about the height of one DNA base (*i.e.* 3.4 Å) and a corresponding local helix unwinding. These local distortions are much more pronounced than for groove binders and can extend over several base pairs surrounding the site of intercalation. In addition to the electrostatic attractive contribution to the free energy of binding from normally cationic, classical intercalators like ethidium[34–38] (Figure 3.4A) a substantial part of their binding energy derives from hydrophobic interactions with the surrounding base pairs. As a consequence intercalative binding modes have a lower probability of being sequence-specific. As a result of the requirement for larger and more complex distortions in the DNA structure, and also the need for the intercalator to diffuse along DNA until it finds its binding site, the kinetics of intercalation are generally slower than those for groove-binding. The associated helix distortions often make the mechanism of binding, and therefore also the kinetics, very complex. Consequently, intercalators like daunomycin[39–42] (Figure 3.4B) may show bi- or multiphase association rate laws. Average overall rates of association for daunomycin and the other anthracycline antibiotics adriamycin[43–46] (Figure 3.4C) and iremycin[41] (Figure 3.4D) are generally in the same order of magnitude as that of ethidium

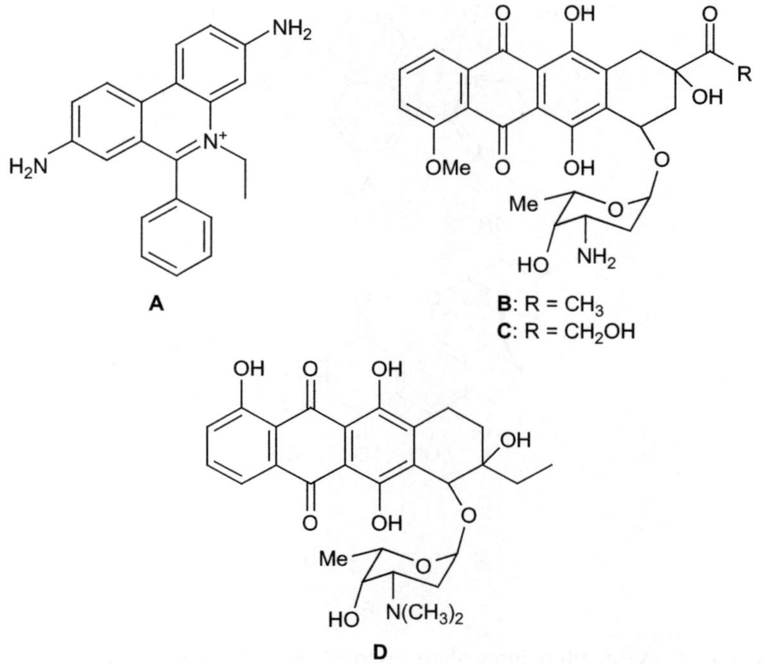


**Figure 3.4** DNA intercalators ethidium (A), daunomycin (B), adriamycin (C) and iremycin (D).

[$k_a$(anthracyclines) = $10^6$–$10^7$ $M^{-1}$ $s^{-1}$ [47,48] and $k_a$(ethidium) = $10^7$ $M^{-1}$ $s^{-1}$ [38]]. Dissociation of these intercalators is approximately as fast ($k_d \approx 1$–$100$ $s^{-1}$)[38,47,48] as for normal groove-binders like distamycin[4] and DAPI.[3]

### 3.2.4 Threading Intercalation

Like their classical counterparts, threading intercalators insert a part of the molecule between base pairs (Figure 3.2D) and interact attractively with DNA through electrostatic binding to the phosphates in the backbone as well as through $\pi$-stacking (hydrophobic and dispersive interactions) with the surrounding base pairs. However, in addition, threading intercalators have bulky or polar substituents tethered to the part of the molecule that gets intercalated, which requires this substituent to be threaded through the DNA-base stack so that the resulting complex has one bulky substituent in each groove (Figure 3.2D). The substituents may or may not contribute to the binding of the threading molecule by hydrophobic interactions and hydrogen bonding to the floor of the grooves. During the process of binding, as the molecule attempts to insert itself between the base pairs, the substituents constitute major obstacles, explaining the slow DNA binding kinetics. In the case of naphthalene diimides[10] (Figure 3.5A) the obstacle to reaching the thermodynamic equilibrium geometry derives from the positively charged

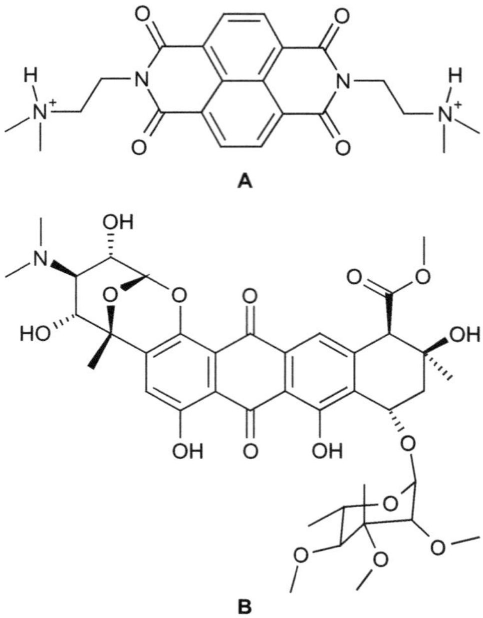


**Figure 3.5** DNA threading intercalators naphthalene diimide (A) and nogalamycin (B).

side chains whereas for nogalamycin[49–52] (Figure 3.5B) it comes from the two bulky sugar moieties attached to the aglycon chromophores. Due to the even more complex distortions of DNA structure than for normal intercalators, and often sequential obstacles for the threading intercalator to find its energetically optimal site of intercalation, the kinetics of association of a threading intercalator are generally multi-exponential.[53] The rate-limitations of reaching, and also leaving, the most stable binding geometry for a threading intercalator can be exemplified by the case of nogalamycin: the average overall rate of association of nogalamycin to DNA is very slow compared with that of the unhindered anthracycline intercalator daunomycin [$k_a$(nogalamycin) ≈ $10^3$–$10^5$ $M^{-1}$ $s^{-1}$,[53] $k_a$(daunomycin) = $10^6$–$10^7$ $M^{-1}$ $s^{-1}$ [47,48]] as is the dissociation from DNA [$k_d$(nogalamycin) ≈ $10^{-3}$ $s^{-1}$ at 0.01–0.5 M $Na^+$ and 40 °C,[54] $k_d$(daunomycin) ≈ 10 $s^{-1}$ [47,48]].

## 3.3 Common Slow DNA Binders

The focus throughout the reminder of the chapter will be on slow DNA binding processes. The rates are slowed down as a result of either the requirement for large conformational change of the DNA and/or of the drug itself or as a result of slow redistribution of the drug between several competing strong DNA binding sites, or both. Examples of both effects have been reported for many DNA–drug interactions in the literature and will be illustrated below for DNA binding of actinomycin D and nogalamycin as representative cases.

### 3.3.1 Actinomycin D

The antitumour antibiotic actinomycin D is composed of two identical cyclic pentadepsipeptides connected to a phenoxazone chromophore (Figure 3.6). It binds to DNA with a relatively high binding constant ($\sim 10^7\ M^{-1}$)[1] by intercalating its aromatic moiety between the base pairs[8] and extending its two pentadepsipeptides in opposite directions from the intercalation pocket along the valley of the minor groove. NMR,[55,56] X-ray crystallography[57–59] and footprinting[23,24] studies have shown that actinomycin D is selective for intercalation at the dinucleotide GpC. The selectivity arises from four specific interactions between N-3 and the C-2 amine of guanine with the threonine amide NH and carbonyl, respectively.[57–59] Besides a preferentiality for the GpC dinucleotide, the binding affinity is also highly dependent on which other bases are flanking the intercalation site. The preferred sequence for actinomycin D binding is 5′-TGCA followed by CGCG, AGCT and GGCC.[60] The difference in binding affinity among these sequences can be correlated to a large variation in dissociation kinetics, where TGCA has a rate that is about four times slower than those from CGCG and AGCT and approximately two orders of magnitude slower than from GGCC.[2,60,61]

Compared with the DNA binding kinetics of normal intercalators mentioned above, both association and dissociation kinetics of actinomycin D

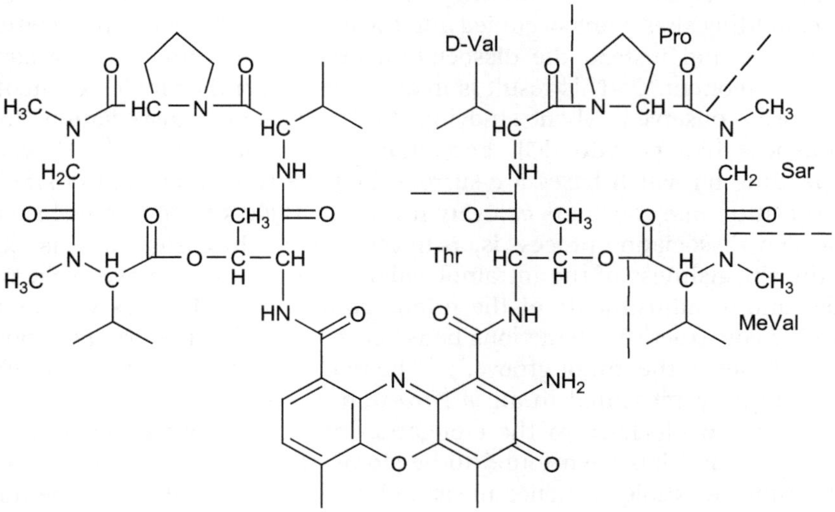


**Figure 3.6** Antitumour antibiotic actinomycin D. Names of the amino acids (Thr = threonine, MeVal = *N*-methylvaline, Sar = sarcosine, Pro = proline and D-Val = D-valine) in cyclic pentadepsipeptides are given on the right hand ring.

are very slow. The association rate constants measured are on the order of $10^3$–$10^4$ $M^{-1}$ $s^{-1}$ [1,2] and the dissociation rate constants measured under the same conditions are roughly $10^{-3}$ $s^{-1}$ [1,2,6–8]. Efforts to find the origins of the slow and multiexponential association kinetics have inspired many studies on the subject and various explanations have been proposed. In one study it has been suggested that as much as five exponentials, of which three are unimolecular, are needed to describe the association and that they originate from slow intramolecular conversion of the two cyclic peptides and/or adjustments of the minor groove of the DNA to optimize the interaction area between the drug and the DNA.[8,62] On the other hand, results from NMR studies indicate that no significant conformational alterations of the cyclic peptides are present upon binding.[56] Another explanation has been that the antibiotic molecule is 'shuffled' between low- and high-binding-constant sites on the DNA.[23,63,64] However, it has been argued that this interpretation is not compatible with the three unimolecular exponentials in the association kinetics.[65] Instead, in an attempt to explain the unimolecular processes it has been suggested that in order to bind to certain sequences of the DNA, actinomycin D has to wait for the thermal occurrence of an 'improbable' distortion of the DNA.[2] This improbable event can be inherent to the DNA but can also result from other actinomycin D molecules binding to DNA sites close to the site of binding of the next drug molecule.[2]

Not unexpectedly, the dissociation kinetics of actinomycin D from a heterogeneous DNA lattice also has to be described by more than a single exponential, where the different exponentials may result from binding sites of different strengths.[7,8] Lowering the drug concentration results in fewer weak binding sites being occupied and the influence of the fast dissociations decreases, and instead the dissociation profile is dominated by a single slow component.[1,66] This result is in agreement with the single-exponential processes observed when studying SDS-sequestered dissociation from sequences like poly(dG–dC)$_2$ and short oligonucleotides.[2,60,61] However, depending on which bases are surrounding the GpC intercalation site the dissociation rate constants can vary from $t_{1/2} = 3000$ s to 100 times faster.[60] The slow dissociation process is, as in the case for the association, ascribed to the sluggishness of the intramolecular conversion of the two cyclic peptides and/or adjustments of the minor groove of the DNA as well as the specific constructive interactions between the drug and the groups exposed by the bases in the minor groove.[8,60] The rate of actinomycin D dissociation has also been measured in an *in vitro* transcription assay that detects the drug-induced blockage to the elongation of an RNA transcript around a AGCT site, and has been found to be slower than the average dissociation rate from the whole sequence measured by the SDS sequestering method.[6] This was not considered surprising since the SDS method gives an average dissociation rate constant whereas the transcriptional assay gives the dissociation from a certain site. Furthermore, recent results have shown that using the detergent sequestering technique can cause overestimations of the dissociation rate constants for slowly dissociating positively charged DNA-binding molecules.[67,68] Also, preliminary results from a similar study with DNA and actinomycin D show that the dissociation rate constants are affected by the concentration of SDS and sodium decyl sulphate (Wilhelmsson, L.M., unpublished results).

### 3.3.2 Nogalamycin

The structurally interesting anthracycline antibiotic nogalamycin is built up of two non-aromatic ring systems, an uncharged nogalose sugar group and a charged (at physiological pH) bicyclic aminoglucose moiety, connected to each end of a planar aglycon chromophore (Figure 3.5B).[69] The various parts of this drug appear to have different functions to make binding robust. The 'dumbbell' shaped molecule intercalates its aglycon group in the DNA base stack placing one bulky sugar moiety in each groove of the DNA double helix.[53,70,71] This threading intercalation is believed to explain how nogalamycin inhibits DNA-directed RNA synthesis *in vivo* as well as *in vitro* and interferes with topoisomerase activity.[72–75] However, unlike the other anthracyclines, daunomycin and adriamycin, a higher cytotoxicity and cardiotoxicity has limited the clinical use of nogalamycin.

Various techniques including viscometry,[70] NMR[76,77] and X-ray crystallography[49–52,78,79] have been used to verify that nogalamycin intercalates its aglycon chromophore, resulting in severe buckling of the surrounding base pairs. Furthermore, NMR and X-ray crystallography have revealed evidence that the nogalose sugar is residing sandwiched between the walls of the minor groove and that the bicyclic aminoglucose moiety is positioned in the major groove of the distorted B-DNA double helix.[49–51,76–79] Numerous studies with different DNA sequences indicate that nogalamycin intercalates selectively at the 5′ side of a purine and the 3′ side of a pyrimidine, *e.g.* CpG and TpG, spanning three base pairs. In structures where nogalamycin binds to CpG sequences,[49–51,53,76–80] it has been proposed that the drug displays two hydrogen bonds in the major groove from the two –OH groups in the aminoglucose to the N7 in guanine and the N4 of the base pairing C as well as one hydrogen bond in the minor groove from the keto oxygen of the methyl ester of the nogalamycin and the N2 hydrogen of guanine.[49–51,76–79] These specific hydrogen bonds, together with several water-mediated hydrogen bonds and van der Waals interactions between the drug and the DNA, constitute the CpG sequence selectivity.[52]

Perhaps the most intriguing characteristic of the nogalamycin–DNA interaction is the sterically hindered threading of the approximately 1 nm wide bulky sugar moieties through the core of the DNA, which results in the slow association [$k_a$(nogalamycin) $\approx 10^3$–$10^5$ $M^{-1}$ $s^{-1}$ at 0.01–0.5 M $Na^+$ and ambient temperature][53] and the very slow dissociation kinetics [$k_d$(nogalamycin) $\approx 10^{-3}$ $s^{-1}$ at 0.01–0.5 M $Na^+$ and 40 °C][54] of nogalamycin. Two mechanism of threading have been proposed in which either a transient local melting of base pairs occurs[71] or there is an extreme elongation of the base stack, without any disruption of hydrogen bonds, combined with severe buckling of surrounding base pairs to create a considerable opening in the duplex, in this way, allowing the nogalamycin to enter its site of intercalation.[78] To be able to describe the kinetics of association to mixed sequence DNA an expression with three exponentials has to be used.[53] The fastest association process is approximately $10^5$ $M^{-1} s^{-1}$ (0.01–0.5 M $Na^+$ and ambient temperature) and the activation energies for the two slower rate constants are both about 90–100 kJ $mol^{-1}$, which would be consistent with a large conformational change of the DNA like, for example, the coherent opening of more than one base pair.[53] Association to poly(dA–dT)$_2$ can be described by the sum of two exponentials, the fastest being on the order of $10^5$ $M^{-1} s^{-1}$ (0.01–0.5 M $Na^+$ and ambient temperature) whereas association to poly(dG–dC)$_2$ is monoexponential with a rate constant of $2.5\times10^3$ $M^{-1} s^{-1}$ (0.01–0.5 M $Na^+$ and ambient temperature).[53] Since AT sequences melt more easily and thus are expected to exhibit more frequent transient openings than GC sequences, the results are in line with the first hypothesis of threading mentioned above. It has also been suggested that the three association constants observed for the mixed sequence DNA most probably originate from heterogeneous binding of nogalamycin to regions with different stability, and even though it is tempting to suggest that the three exponentials are a result of intercalation into AT-rich ($\tau_1$), GC-rich ($\tau_3$) and mixed regions ($\tau_2$) of the DNA this most certainly is an oversimplification.[53] Another proposal is that the complexity of the association to the mixed-sequence DNA may arise from initial binding to easily accessible sites (AT-rich regions) followed by slow redistribution to thermodynamically more favoured sites (*vide infra*).[64]

Compared with other anthracyclines like adriamycin, daunomycin and iremycin ($k_d \approx 1$–$100\ s^{-1}$),[47,48] nogalamycin has an extremely slow rate of dissociation from DNA [$k_d$(nogalamycin) $\approx 10^{-3}\ s^{-1}$ at 0.01–0.5 M $Na^+$ and 40 °C].[54] The slow dissociation kinetics of nogalamycin are suggested, as for the association process, to be a result of the need for threading the bulky sugar moieties through the core of the DNA base stack. In synthetic polynucleotides like poly(dA–dT)$_2$ ($k_d = 4.0\times10^{-2}\ s^{-1}$), poly(dG–dC)$_2$ ($k_d = 2.7\times10^{-4}\ s^{-1}$) and poly(dA–dC) · poly(dG–dT) ($k_d = 3.4\times10^{-4}\ s^{-1}$) the dissociation of nogalamycin can be described by a single exponential.[54] The more than two orders of magnitude slower dissociation from poly(dG–dC)$_2$ than from poly(dA–dT)$_2$ again suggests that nogalamycin has to wait longer for the more stable GC base pairs to transiently melt and enable the drug to escape from the DNA. When studying dissociation of nogalamycin from a mixture of poly(dG–dC)$_2$ and poly(dA–dT)$_2$ a single rate constant, which is close to the one observed for poly(dG–dC)$_2$, was found.[54] This indicates that nogalamycin in its final state of binding has a strong preference for GC compared with AT sequences. For mixed sequence DNA three exponentials ($k_{d1} = 3.3\times10^{-3}\ s^{-1}$, $k_{d2} = 9.8\times10^{-4}\ s^{-1}$ and $k_{d3} = 2.3\times10^{-4}\ s^{-1}$) are needed to get an adequate fit with the experimental data.[54] At higher drug binding ratios it was found that the weight of the two faster processes increases, indicating that these two binding sites are the weaker ones of three different sites. Additional evidence for weak and strong sites came from varying the equilibration time, before starting the SDS sequestering.[54] In these experiments different rate constants were observed, which again indicates a redistribution ('shuffling') of the drug from a site which is easy to thread through to a thermodynamically more favourable site. The conclusion about mixed-sequence DNA, with regions of different stabilities mentioned for the association process above, is further supported here since the three exponentials needed to describe the dissociation in the mixed DNA cannot be explained only by using the single exponentials of the polynucleotides.

## 3.4 Ruthenium Complexes Exhibiting Slow DNA-binding Kinetics

Octahedral ruthenium(II) complexes with aromatic ligands and their DNA-binding properties have been the subject of extensive research during the last couple of decades.[81–93] The positively charged ruthenium centre and the planar aromatic hydrophobic ligands are well suited for interacting with the negatively charged backbone and the hydrophobic core of DNA, respectively. Furthermore, the inherent chirality of the complexes is of interest as the DNA duplex is chiral in itself. The rich photophysical repertoire of the complexes can be exploited in many ways to probe the binding to DNA and the techniques frequently used for this purpose are absorption, emission, circular dichroism (CD) and linear dichroism (LD) measurements.

Early experiments demonstrating binding of ruthenium(II) complexes to DNA focused on $[Ru(phen)_3]^{2+}$ (phen = 1,10-phenanthroline, Figure 3.7A),[94–98] and before that the slowly inverting ferroin complex $[Fe(phen)_3]^{2+}$ was found to show enantiopreferential binding as well as orientation *vs.* DNA.[99] Both groove binding and intercalation have been suggested as the DNA binding mode. From flow-orientation studies it has been inferred that both enantiomers of $[Ru(phen)_3]^{2+}$ bind by partly intercalating one phenanthroline edge into DNA, possibly by 'quasi-intercalation' requiring only indentation of a nucleobase without separation of base pairs.[100] However, when extending one of the phen-ligands to the larger dppz (dipyrido-[3,2-*a*:2′,3′-c]-phenazine) this moiety is found to become deeply intercalated between separated base pairs and the DNA to be unwound. The intercalation of the dppz moiety is associated with an astounding change in photophysical properties first observed for $[Ru(bipy)_2dppz]^{2+}$ (bipy = 2,2-bipyridine) and later also for the more extensively studied $[Ru(phen)_2dppz]^{2+}$ (Figure 3.7B).[81,82] The drastic increase in luminescence quantum yield when these dppz complexes bind to DNA, the so-called 'light-switch'-effect, has been investigated in great detail and is generally believed to originate from the protection of the aza-nitrogens on the dppz moiety from hydrogen bonding to water. When binding to DNA the dppz moiety is intercalated between the base pairs whereas the metal ion and

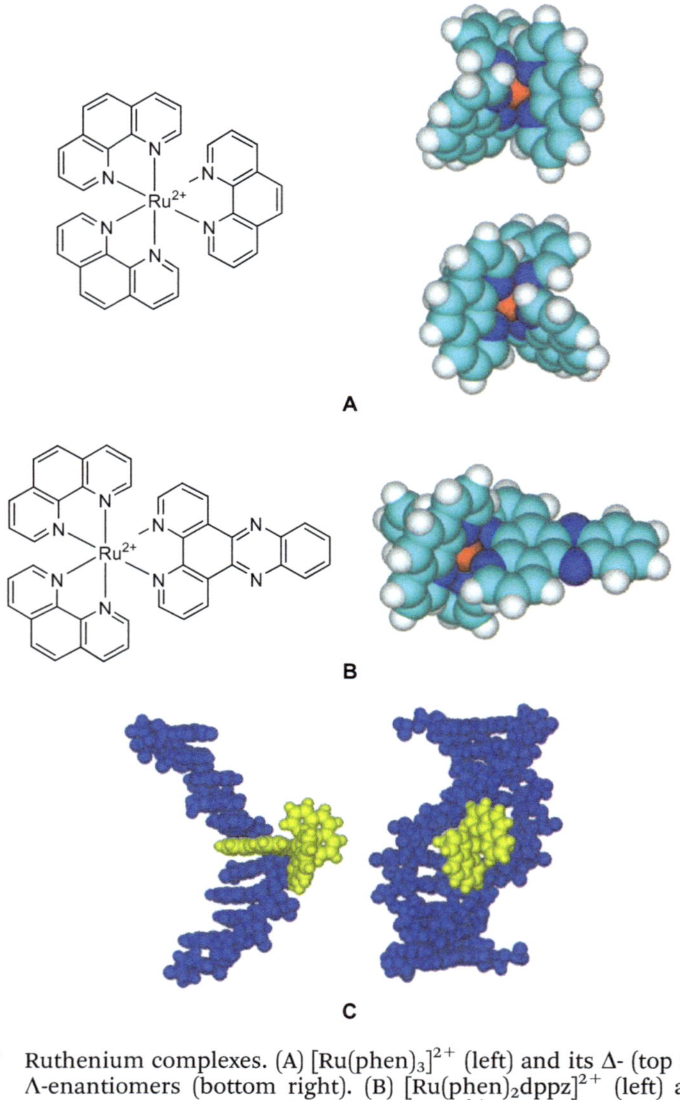


**Figure 3.7** Ruthenium complexes. (A) $[Ru(phen)_3]^{2+}$ (left) and its Δ- (top right) and Λ-enantiomers (bottom right). (B) $[Ru(phen)_2dppz]^{2+}$ (left) and its Δ-enantiomer (right). (C) Δ-$[Ru(phen)_2dppz]^{2+}$ intercalating into DNA with one of the strands omitted for clarity (left) and arrangement of phen ligands of the complex in the DNA minor groove (right).

ancillary ligands are positioned in a groove (Figure 3.7C),[87,89] which groove, however, is still debated.[84,87,92]

### 3.4.1 Bis-intercalating Ru-dimer $[\mu\text{-c4(cpdppz)}_2\text{(phen)}_4\text{Ru}_2]^{4+}$

In an effort to increase DNA binding affinity and sequence-selectivity and enantioselectivity, as well as to decrease the dissociation rates of

Ru(II)polypyridyl compounds compared with the monomer $[Ru(phen)_2dppz]^{2+}$, the three stereoisomers (ΔΔ, ΔΛ = meso, and ΛΛ) of a dimer, $[\mu\text{-c4(cpdppz)}_2(phen)_4Ru_2]^{4+}$ (**1**, Figure 3.8), were synthesized and examined.[101,102] The dimer is built up of two $[Ru(phen)_2dppz]^{2+}$ moieties tethered *via* the dppz ligands and a flexible aliphatic diamide linker, long enough to allow bis-intercalation in two intercalation slots separated by two base pairs.

The initial question to address was whether each monomeric moiety of the dimer was actually interacting with DNA in the same way as $[Ru(phen)_2dppz]^{2+}$, which ought to give rise to a correspondingly higher binding constant, or if the combination of the bulky ancillary phen ligands and the flexible linker of the dimer would prevent it from being intercalated into DNA. The remarkable resemblance in the flow LD spectra of the enantiomerically pure monomeric and dimeric compounds when bound to mixed-sequence calf thymus (ct) DNA confirmed that the binding geometry of each monomeric unit of **1** was indeed very similar to that of $[Ru(phen)_2dppz]^{2+}$ (Figure 3.9, compare with Figure 3.7C).[101] Moreover, the bisintercalative mode of binding was supported by the high luminescence quantum yield of the DNA-bound state.[101] Binding titration experiments indicated that the size of the binding site is four base pairs and that the two dppz ligands of **1** are intercalated separated by two base pairs.[101] Finally, deconvolution of the LD and absorption spectra, into differently polarized spectral components,[89] showed that enantiomers of the dimer **1** both have a clockwise roll of their dppz ligands in the intercalation pocket (7° for ΔΔ-**1** and 19° for ΛΛ-**1**). This small but significant rotation places the two phenanthroline ligands of ΔΔ-**1** directed along, and those of the ΛΛ-**1** roughly perpendicularly to, a groove of the DNA (Figure 3.9).[102] These results are in excellent agreement with the DNA binding of the $[Ru(phen)_2dppz]^{2+}$

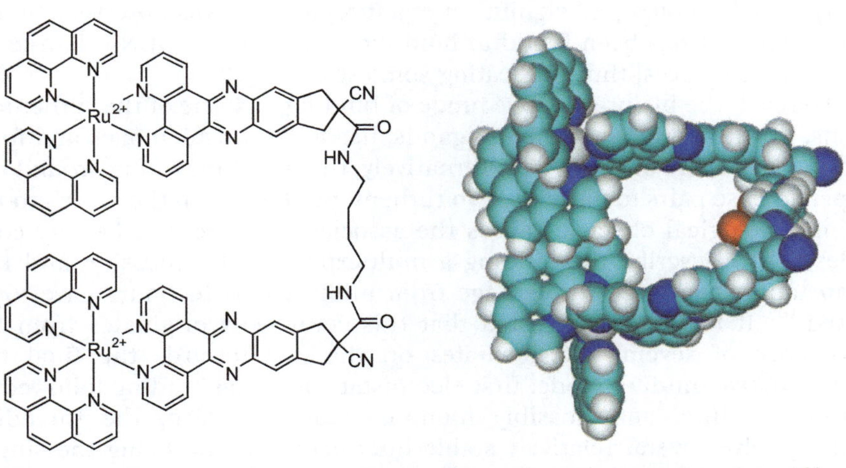


**Figure 3.8** Bis-intercalating ruthenium dimer, $[\mu\text{-c4(cpdppz)}_2(phen)_4Ru_2]^{4+}$ (**1**) (left), and its ΔΔ-enantiomer (right).

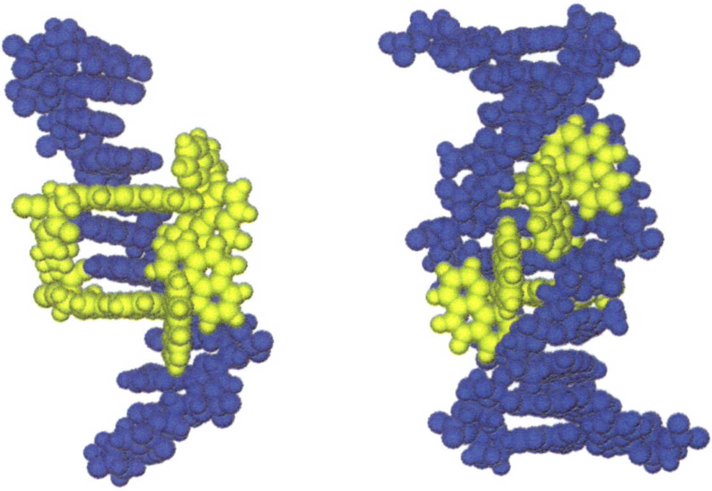

**Figure 3.9** DNA binding of ΔΔ-[μ-c4(cpdppz)$_2$(phen)$_4$Ru$_2$]$^{4+}$. ΔΔ-[μ-c4(cpdppz)$_2$(phen)$_4$Ru$_2$]$^{4+}$ (**1**) bis-intercalating into DNA with one of the strands omitted for clarity (left) and arrangement of phenanthroline ligands of the complex in the DNA minor groove (right).

indicating that the aliphatic diamide tether has no significant influence of the mode of binding for each subunit of the dimer.

It has been observed by calorimetric investigations that the binding of ΔΔ-**1** (Önfelt, unpublished data) and [Ru(phen)$_2$dppz]$^{2+}$ to DNA has a very small reaction enthalpy and is, thus, mainly entropically driven. Furthermore, dilution has been used to estimate the DNA binding constant of **1** to be approximately $10^{10}$ M$^{-1}$ at 50 mM Na$^{+}$ [102] whereas the binding constant of [Ru(phen)$_2$dppz]$^{2+}$ under the same conditions is approximately $10^{6}$ M$^{-1}$. Surprisingly enough, no significant enantioselectivity was observed. On the other hand, **1** has been found to bind more strongly to ct-DNA than to AT and GC sequences, thus, indicating some selectivity.[102]

To reach the bisintercalative mode of binding of **1** one of the ruthenium ions, including the bulky phen ligands, has to be threaded back and forth through the core of the DNA. Alternatively, the tether may sling around two opened base pairs to place the two ruthenium moieties in the same groove. This topological obstacle causes the association process to be very complex.[102] To describe the binding a multiexponential expression and rate constants spanning a time-range from milliseconds to hours have to be used.[102] It has been suggested that this complexity originates from the existence of several intermediates on the way towards the final bis-intercalative binding mode: first electrostatic external binding followed by groove-binding, and possibly mono-intercalation. Also, the threading may involve several relatively stable intermediates, one being the single-threaded geometry, with one ruthenium moiety in each of the grooves.

The association, which has been shown to be slightly enantioselective, is faster to alternating AT than GC sequences.[102] This observation is analogous to what was found for the threading process of nogalamycin (*vide supra*) and indicates that base pair stability is an important factor for determining the rate of association. In a more detailed experiment the two fastest association rate constants for **ΔΔ-1** when binding to 45 μM base of poly(dA–dT)$_2$ have been determined to be $k_{a1} = 29\ s^{-1}$ and $k_{a2} = 4.4\ s^{-1}$, respectively.[102] However, it should be noted that under the same conditions the slowest process, which most probably involves redistribution of **ΔΔ-1**, takes several hours to complete. This slow reorganization is accompanied by a small increase in emission, a large increase in LD amplitude but insignificant changes in the LD spectral profile, thus suggesting that there is a minimal change in binding geometry of **ΔΔ-1** and a local, rather than a global, increase in the orientation of the DNA. It has been proposed that this local increase of orientation comes as a result of differences in binding cooperativity of **1** between early and later stages of the binding process.[102]

Dissociation of **1** from DNA is very slow and it has been suggested that the sluggishness of the process is a result of requirements of both base-pair openings and unstacking as the complex unthreads from the DNA. The slowness of the process can be exemplified with the two rate constants needed to explain the dissociation of meso-**1** from ct-DNA: $k_{d1} = 6.4 \times 10^{-3}\ s^{-1}$ and $k_{d2} = 1.5 \times 10^{-3}\ s^{-1}$.[101] Under the same conditions both dissociation from alternating AT and GC sequences are faster than the corresponding process from ct-DNA. Furthermore, the dissociation of **ΔΔ-1** from AT sequence can be described by a single exponential expression with $k_d = 7.3 \times 10^{-3}\ s^{-1}$ [101] whereas for poly(dG–dC)$_2$ 95% of the dissociation can be assigned to a $k_d$ of $4.5 \times 10^{-3}\ s^{-1}$ [102]. The fact that the dissociation from ct-DNA is slower than those from alternating sequences indicates that it is not only the strength of the base pair hydrogen bonding that dictates the rate of the process but rather the local stability of the DNA duplex as in the case for nogalamycin (*vide supra*). Experiments also demonstrated that the unthreading from poly(dA–dT)$_2$ is fastest for **ΛΛ-1** ($k_d = 4.2 \times 10^{-2}\ s^{-1}$) followed by meso- and **ΔΔ-1** ($k_d = 1 \times 10^{-2}\ s^{-1}$ and $0.73 \times 10^{-2}\ s^{-1}$, respectively).[101] This observation indicates that Λ moieties penetrate, and thus unthread, easier than Δ moieties. The similarity between the rates of dissociation for meso- and **ΔΔ-1** also suggests that the Δ moiety of these stereoisomers serves as an anchor in the DNA during the rate-limiting step from mono-intercalation to free complex in solution.[102] However, this enantioselectivity and also the accuracy of the estimated $k_d$'s for **1** as well as for actinomycin D and nogalamycin should probably be taken with some caution especially in the light of the recent finding that the anionic surfactants employed in the SDS-sequestering method can strongly affect the rate of dissociation depending on the nature of the leaving molecule and its interaction with DNA.[67]

### 3.4.2 Semirigid Ru-dimer [μ-(11,11′-bidppz)(x)$_4$Ru$_2$]$^{4+}$ (x = phen or bipy)

Another dimeric ruthenium complex that has received attention recently is the 'dumbbell'-shaped semirigid [μ-(11,11′-bidppz)(x)$_4$Ru$_2$]$^{4+}$, where x = bipy (**2**) or phen (**3**) (Figure 3.10A, B, and C).[103–107] The complex consists of two [Ru(x)$_2$dppz]$^{2+}$ moieties connected *via* one single bond between the outer rings of the two dppz ligands. The first report on DNA binding of this kind of compound, presented a decade ago, suggested from LD measurements and lack of luminescence that neither the ΔΔ- nor the ΛΛ-enantiomers can intercalate in DNA but instead bind in the grooves with a very high binding constant ($K = 10^{12}$ M$^{-1}$, at 10 mM Na$^+$ and room temperature).[103] Although, the LD data for all four complexes was consistent with groove binding the positive sign of the low-energy part of the spectrum, which was unique for ΔΔ-**3** upon binding to DNA, was intriguing and called for further investigation.

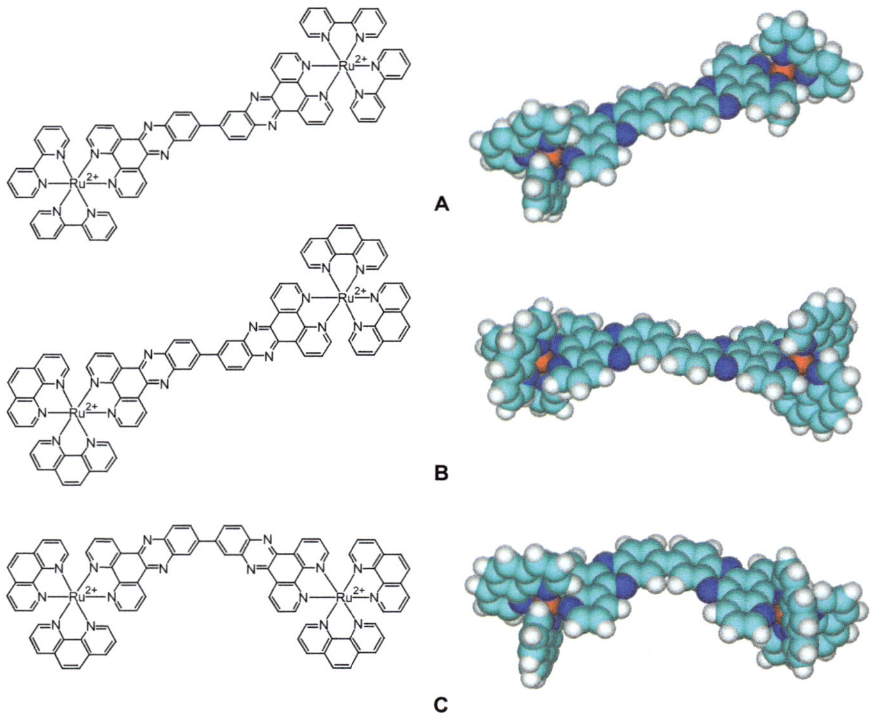


**Figure 3.10** Semirigid Ru-dimer [μ-(11,11′-bidppz)(x)$_4$Ru$_2$]$^{4+}$ (x = phen or bipy). (A) [μ-(11,11′-bidppz)(bipy)$_4$Ru$_2$]$^{4+}$ (**2**) (left) and its ΛΛ-enantiomer (right) with the bidppz ligand in *anti*-conformation. (B) [μ-(11,11′-bidppz)(phen)$_4$Ru$_2$]$^{4+}$ (**3**) (left) and its ΔΔ-enantiomer (right) with the bidppz ligand in *anti*-conformation. (C) [μ-(11,11′-bidppz)(phen)$_4$Ru$_2$]$^{4+}$ (**3**) (left) and its ΛΛ-enantiomer (right) with the bidppz ligand in *syn*-conformation.

### 3.4.2.1 Evidence of Extremely Slow Threading Intercalation of ΔΔ-3

During investigations of the positive low-energy part of the LD spectrum of ΔΔ-**3** upon binding to ct-DNA, a serendipitous observation of a sample that had been left for two weeks (!) at room temperature unexpectedly showed a change from a positive LD amplitude to a negative one.[104] This result indicated a major change in DNA binding geometry of ΔΔ-**3**. To facilitate more efficient studies of the binding process it was accelerated by adding salt and increasing temperature. An isosbestic point in the LD spectra collected at different times, supported by a singular value decomposition (SVD) analysis of the whole spectra, demonstrated that the change involved only two binding geometries: an initial mode of binding corresponding to the positive LD spectrum previously observed and a final mode giving the negative LD spectrum.[104] The final binding mode could be characterized by a binding angle of the bidppz ligand that was roughly perpendicular to the DNA helix axis, a strong luminescence, and luminescence lifetimes that were similar to those of Δ-$[Ru(phen)_2dppz]^{2+}$.[104] Furthermore, it was observed that dissociation from the initial binding mode was effectively instantaneous whereas dissociation from the final binding mode took several days, even at elevated temperatures. It was suggested that the initial binding mode was a groove-bound metastable binding mode from which the ruthenium complex rearranged into the final threading-intercalated mode of binding by threading one of the $Ru(phen)_2$ moieties through the core of the DNA.[104] The kinetics of this rearrangement and threading process, measured at 50 °C, 100 mM $Na^+$ and 160 μM base of ct-DNA required two exponentials ($k_{a1} = 1.2\times10^{-3}$ $s^{-1}$ and $k_{a2} = 0.2\times10^{-3}$ $s^{-1}$) to fit the data. Measurements were also performed at other temperatures and from this threading activation energies of 94 and 83 kJ $mol^{-1}$ were estimated for the processes corresponding to the two exponentials.[104] It should be noted that these values are similar to the threading activation energies of nogalamycin discussed above. The SDS-sequestered dissociation process could, under similar conditions, also be described by a two-exponential expression ($k_{d1} = 1.9\times10^{-3}$ $s^{-1}$ and $k_{d2} = 0.8\times10^{-3}$ $s^{-1}$), but with activation barriers of 65 and 66 kJ $mol^{-1}$, respectively.[104] The surprising result that with SDS the dissociation rate constants were faster than the rearrangement rate constants triggered the discovery of the surfactant-induced dissociation rate enhancement that was mentioned earlier.[67,68] Thus, again the observed values of the rate constants should be treated with caution and only be used for rough comparison of rates of dissociation of DNA drugs. Recently, the non-catalyzed rate of dissociation of ΔΔ-**3** from ct-DNA has been estimated by studying the increase in luminescence as complexes gradually are transferred to the kinetically favoured poly(dA–dT)$_2$. Addition of poly(dA–dT)$_2$ to a pre-equilibrated sample of ΔΔ-**3** and ct-DNA resulted in an emission trace that could be fitted to a single exponential with a rate constant of $2.4\times10^{-5}$ (150 mM NaCl, 50 °C).[108] This rate of dissociation is at least 30 times slower than that determined with the SDS sequestering technique.

### 3.4.2.2 *Final and Metastable DNA Binding Modes of Stereoisomers of 2 and 3*

The initial metastable binding mode of ΔΔ-**3** to ct-DNA characterized by the positive LD, low luminescence quantum yield and instantaneous SDS-driven dissociation was also observed for binding of ΔΔ-**3** to poly(dG–dC)$_2$ and poly(dI–dC)$_2$.[105] The positive LD indicated that this metastable state is groove-bound, but the LD does not reveal if it is major- or minor-groove binding. However, poly(dG–dC)$_2$ and poly(dI–dC)$_2$ display great differences in the minor groove while these polynucleotides are chemically similar in the major groove. Therefore, the similarity of LD of ΔΔ-**3** bound to poly(dG–dC)$_2$ and poly(dI–dC)$_2$ indicated that the primary, metastable binding mode is in the major groove.[105] In contrast, based on the immediate appearance of negative LD and extremely slow dissociation even after short equilibration times it was concluded that ΔΔ-**3** directly threads poly(dA–dT)$_2$ virtually without passing through an intermediate state.[104,105] This substantial 'kinetic recognition' may be utilized for targeting AT-rich DNA sequences among other sequences, both for diagnostic and therapeutic purposes

Both ΛΛ- and meso-**3** interacting with ct-DNA have initial LD spectra that are negative and suggest a binding geometry where their bidppz ligand has an angle relative to the DNA helix axis of approximately 65°. This angle is too large to originate from minor-groove-binding of the bidppz ligand and is larger than for the initial binding mode of ΔΔ-**3** and approaches that of the final intercalated form.[105] However, it was shown that both ΛΛ- and meso-**3** dissociate immediately from this binding mode, thus, suggesting a non-threaded state of binding. It was concluded that the two enantiomers have a metastable binding mode in ct-DNA where the bidppz ligand might be quasi-intercalated in *syn*-conformation (Figure 3.10C) from the major groove side (Figure 3.11A).[105]

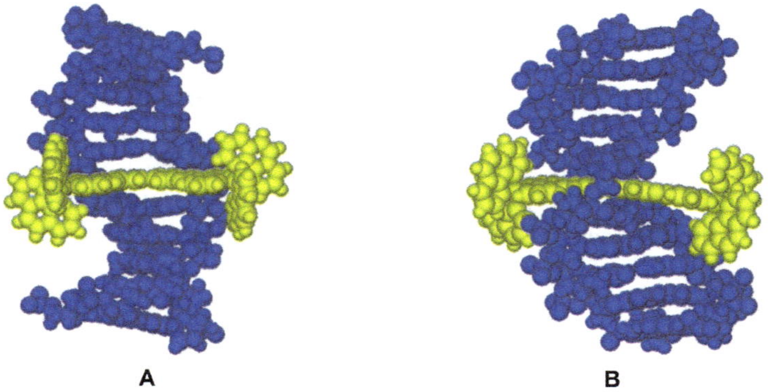


**Figure 3.11** DNA binding of ΔΔ-[μ-(11,11′-bidppz)(phen)$_4$Ru$_2$]$^{4+}$. Initial quasi-intercalative major-groove-bound form (A) and final threading-intercalated form (B).

The general outline of the final binding geometry of all stereoisomers of **2** (Nordell, *et al.*, unpublished results)) and **3** interacting with ct-DNA, poly(dA–dT)$_2$, poly(dG–dC)$_2$, or poly(dI–dC)$_2$ is that the complexes all thread DNA to end up with the bidppz ligand intercalated between the base pairs, which means that one Ru(phen)$_2$ moiety ends up in each groove of the DNA (Figure 3.11B).[105] These conclusions were drawn from the great resemblance of the LD spectra of all stereoisomers of **2** and **3** bound to any of the DNAs. In addition, interaction of the three forms of **3** with all the DNAs except poly(dG–dC)$_2$ resulted in an increase in luminescence and resemblance to the lifetimes of the parent compound [Ru(phen)$_2$dppz]$^{2+}$, which suggested an intercalative mode of binding.[105] Furthermore, since dissociation of ΔΔ-**3** from the final binding mode in poly(dG–dC)$_2$ was found to be very slow, it was concluded that threading intercalation was the equilibrium binding mode also for this system.[105]

In a thorough investigation a detailed final binding geometry of **3** in poly(dA–dT)$_2$ has been suggested (Figure 3.11B).[105] From electrostatic arguments and luminescence properties it was concluded that **3** intercalates asymmetrically with one of the Ru(phen)$_2$ moieties deeply intercalated, leaving the other moiety displaced from the centre of the DNA helix. Furthermore, it was concluded from LD that the central pivot bond of the bidppz ligand of **3** is rotated so that both Ru(phen)$_2$ moieties are clockwise rotated, looking from the corresponding groove side. From CD and differential CD it was suggested that the bridging bidppz ligand is intercalated in its *anti*-conformation (Figure 3.10B). Finally it was proposed from LD and luminescence that the deeply intercalated Ru(phen)$_2$ moiety resides in the minor groove and, with the help of meso-**3**, that there is an enantiopreference for a Λ moiety in the minor groove.

#### *3.4.2.3 DNA Rearrangement Kinetics of ΛΛ-2 and ΛΛ-3*

In the vast majority of studies of the rearrangement of **2** and **3** binding to different DNAs, the kinetics has been found to be multiphasic. Quite unexpectedly, when studying the binding of ΛΛ-**2** to poly(dA–dT)$_2$ by luminescence it was found that the kinetics of the system, basically irrespective of basepair : complex ratio, could be described by a single exponential expression ($k_a = 3.4\times10^{-3}$ s$^{-1}$ at 120 μM of base).[106] This suggested that ΛΛ-**2** was able to rearrange directly from its externally bound state to the threading intercalated state. In contrast, the binding of ΛΛ-**3** to the same kind of alternating DNA at a binding ratio of 8 (basepair:complex) was found to follow a second order rate law: $k\times$[non-threaded complex]$^2$.[106] When the binding density of that system was decreased a gradual transition towards a pseudo-first-order kinetics was observed. To fit this behaviour of both bimolecular and pseudo-first-order kinetics a global fit employing conditional probabilities of a non-cooperative McGhee and von Hippel approach was used. It was concluded that a first order rate constant of $1.2\times10^{-2}$ s$^{-1}$ and second order rate constant of $5.0\times10^{-2}$ s$^{-1}$ (at 120 μM of base, and [ΛΛ-**3**] = 7.5, 3.7,

1.9 and 0.95 μM) gave an excellent fit with the data.[106] It was suggested that the two different pathways originated from direct threading intercalation from a groove-bound form and an indirect (bimolecular) threading intercalation from a groove that is catalyzed by opening of the base stack from the opponent side of DNA by another ΛΛ-**3**.[106]

#### 3.4.2.4 Slow Shuffling Rearrangement of ΔΔ-3

Another intriguing effect was found when instead studying binding of ΔΔ-**3** to poly(dA–dT)$_2$. The association can in this case be described by a bimolecular expression with an additional extremely slow first-order process.[107] The bimolecular processes ($k_{a1} \approx 10^{-3}\,s^{-1}$, at 2 mM $Mg^{2+}$ and 5 mM $K^+$, 50 °C, 80 μM base, and [ΔΔ-**3**] = 6, 4 and 2 μM) which gives rise to a large increase in luminescence quantum yield is, as mentioned above, a result of the complex becoming threading intercalated.[107] What is surprising is that after the complex is threading intercalated another extremely slow first-order process ($k_{a2} \approx 10^{-4}\,s^{-1}$) follows. To investigate this rearrangement process in detail the DNA groove binder DAPI was used as an intrinsic probe. Using observations from CD, emission and energy transfer between DAPI and ΔΔ-**3** it was concluded that the slow reorganization results from a slow 'shuffling' of ΔΔ-**3** from an initial slightly anticooperative to its final thermodynamically most favourable distribution.[107] The slow shuffling process was suggested to be a result of dissociation, lateral diffusion along the DNA, followed by association, like in the case for nogalamycin or, alternatively, it could be explained by a mechanism in which the complex remains threaded and, by sequential base pair opening, diffuses laterally through the DNA.[107]

## 3.5 Addendum to Second Edition

Since the first edition, the field of slow-DNA-binding ruthenium complexes has advanced considerably, and the most significant results are briefly summarized here. Analysis of single-DNA force–extension data recently confirmed the stepwise bis-intercalation of ΔΔ-**1**, and the free-energy landscape of the DNA-extension process strongly suggests that the barriers are due to the threading of the linker chain through the base-pair stack.[109] The semi-rigid binuclear complexes **2** and **3** show an unprecedented selectivity for threading intercalation into long alternating AT-stretches and negatively supercoiled plasmids, compared with linear mixed-sequence DNA.[110–112] Using CD spectroscopy, the threading mechanism has been shown to be a first-order rearrangement also for ΛΛ-**3**,[113] and single-DNA force–extension data supports the notion that one base-pair has to open during the threading process.[114] The initial groove-bound state has been characterized by 2D-NMR to be located in the minor groove for ΛΛ-**2**,[115] and an X-ray crystallography study showed ΔΔ-**3** to be threaded in the *anti*-conformation with one Ru(phen)$_2$ deeply inserted into the minor groove with the bases from an opened A–T base-pair stacked on the ancillary phenanthroline ligands.[116]

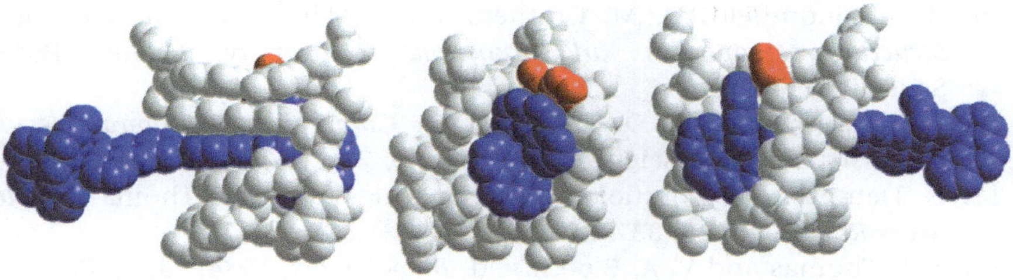

**Figure 3.12** Three views of the X-ray structure of ΔΔ-[μ-(11,11′-bidppz)(phen)$_4$Ru$_2$]$^{4+}$ (blue) intercalating into a hexamer duplex d(CGTACG)$_2$ (white), showing the extruded adenine (red) stacking onto a phenanthroline ligand. Hydrogen atoms have been omitted for clarity.

This interesting result supports the hypothesis that preferred stacking of an opened base-pair on phenanthroline can explain the observed counter-intuitive lower activation energy for threading for the more sterically demanding ΛΛ-**3** complex than for ΛΛ-**2** (Figure 3.12).[110]

However, the *anti*-conformation is not a prerequisite for threading intercalation, since a rigid, planar fluorene analogue of **3** in the *syn*-conformation shows similar threading intercalation characteristics to the parent complex.[117] Elongating the bridging ligand of **3** by introducing an ethyne fragment between the two dppz moieties leads to faster threading kinetics and higher binding affinity, but reduced stereo- and sequence-selectivity compared with the parent complex.[118] In contrast, shortening the bridging ligand by replacing one dppz moiety with imidazo[4,5-*f*][1,10]phenanthroline (ip) leads to binuclear complexes with increased AT-specificity.[119] The four stereoisomers of this unsymmetrical complex further showed remarkably different threading kinetics: the fastest threading rate and highest DNA affinity was found for the hetero-chiral enantiomer with Δ-chirality at the dppz moiety and Λ-chirality at the ip moiety and the slowest rate and lowest affinity for the opposite enantiomer; the rates and affinities for the homochiral ΔΔ- and ΛΛ-enantiomers being in between.[120,121]

## References


1. F. M. Chen, *Biochemistry*, 1988, **27**, 1843.
2. F. M. Chen, *Biochemistry*, 1990, **29**, 7684.
3. F. A. Tanious, J. M. Veal, H. Buczak, L. S. Ratmeyer and W. D. Wilson, *Biochemistry*, 1992, **31**, 3103.
4. R. Baliga and D. M. Crothers, *Proc. Natl. Acad. Sci. U. S. A.*, 2000, **97**, 7814.
5. R. Baliga, E. E. Baird, D. M. Herman, C. Melander, P. B. Dervan and D. M. Crothers, *Biochemistry*, 2001, **40**, 3.
6. D. R. Phillips and D. M. Crothers, *Biochemistry*, 1986, **25**, 7355.
7. M. C. Fletcher and K. R. Fox, *Nucleic Acids Res.*, 1993, **21**, 1339.
8. W. Müller and D. M. Crothers, *J. Mol. Biol.*, 1968, **35**, 251.

9. V. A. Bloomfield, D. M. Crothers, and I. Tinoco Jr., *Nucleic Acids - Structures, Properties, and Functions*, University Science Books, Sausalito, 2000.
10. D. S. Johnson and D. L. Boger, In *Comprehensive Supramolecular Chemistry.* Elsevier Science Ltd., 1996, p. 73.
11. H. Deng, V. A. Bloomfield, J. M. Benevides and G. J. Thomas, *Nucleic Acids Res.*, 2000, **28**, 3379.
12. T. J. Thomas and V. A. Bloomfield, *Biopolymers*, 1984, **23**, 1295.
13. S. Eriksson, S. K. Kim, M. Kubista and B. Nordén, *Biochemistry*, 1993, **32**, 2987.
14. J. Kapuscinski and K. Yanagi, *Nucleic Acids Res.*, 1979, **6**, 3535.
15. M. Kubista, B. Åkerman and B. Nordén, *Biochemistry*, 1987, **26**, 4545.
16. T. A. Larsen, D. S. Goodsell, D. Cascio, K. Grzeskowiak and R. E. Dickerson, *J. Biomol. Struct. Dyn.*, 1989, **7**, 477.
17. C. Bailly, P. Colson, J. P. Henichart and C. Houssier, *Nucleic Acids Res.*, 1993, **21**, 3705.
18. N. Spink, D. G. Brown, J. V. Skelly and S. Neidle, *Nucleic Acids Res.*, 1994, **22**, 1607.
19. M. L. Kopka, C. Yoon, D. Goodsell, P. Pjura and R. E. Dickerson, *Proc. Natl. Acad. Sci. U. S. A.*, 1985, **82**, 1376.
20. A. Abu-Daya, P. M. Brown and K. R. Fox, *Nucleic Acids Res.*, 1995, **23**, 3385.
21. J. G. Pelton and D. E. Wemmer, *Biochemistry*, 1988, **27**, 8088.
22. J. G. Pelton and D. E. Wemmer, *J. Am. Chem. Soc.*, 1990, **112**, 1393.
23. K. R. Fox and M. J. Waring, *Nucleic Acids Res.*, 1984, **12**, 9271.
24. M. W. van Dyke, R. P. Hertzberg and P. B. Dervan, *Proc. Natl. Acad. Sci. U. S. A.*, 1982, **79**, 5470.
25. M. Coll, C. A. Frederick, A. H. J. Wang, A. and Rich, *Proc. Natl. Acad. Sci. U. S. A.*, 1987, **84**, 8385.
26. U. Sehlstedt, S. K. Kim and B. Nordén, *J. Am. Chem. Soc.*, 1993, **115**, 12258.
27. W. D. Wilson, F. A. Tanious, H. J. Barton, R. L. Jones, K. Fox, R. L. Wydra and L. Strekowski, *Biochemistry*, 1990, **29**, 8452.
28. M. Mrksich and P. B. Dervan, *J. Am. Chem. Soc.*, 1993, **115**, 9892.
29. M. Mrksich, M. E. Parks and P. B. Dervan, *J. Am. Chem. Soc.*, 1994, **116**, 7983.
30. M. E. Parks, E. E. Baird and P. B. Dervan, *J. Am. Chem. Soc.*, 1996, **118**, 6147.
31. S. White, J. W. Szewczyk, J. M. Turner, E. E. Baird and P. B. Dervan, *Nature*, 1998, **391**, 468.
32. L. A. Dickinson, R. J. Gulizia, J. W. Trauger, E. E. Baird, D. E. Mosier, J. M. Gottesfeld and P. B. Dervan, *Proc. Natl. Acad. Sci. U. S. A.*, 1998, **95**, 12890.
33. P. B. Dervan, *Bioorg. Med. Chem.*, 2001, **9**, 2215.
34. C. G. Reinhardt and T. R. Krugh, *Biochemistry*, 1978, **17**, 4845.
35. W. Fuller and M. J. Waring, *Ber. Bunsengesellschaft Phys. Chem.*, 1964, **68**, 805.

36. J.-B. LePecq, *Methods Biochem. Anal.*, 1971, **20**, 41.
37. M. J. Waring, *J. Mol. Biol.*, 1965, **13**, 269.
38. R. B. Macgregor, R. M. Clegg and T. M. Jovin, *Biochemistry*, 1985, **24**, 5503.
39. J. B. Chaires, N. Dattagupta and D. M. Crothers, *Biochemistry*, 1982, **21**, 3927.
40. J. B. Chaires, N. Dattagupta and D. M. Crothers, *Biochemistry*, 1982, **21**, 3933.
41. H. Fritzsche, H. Triebel, J. B. Chaires, N. Dattagupta and D. M. Crothers, *Biochemistry*, 1982, **21**, 3940.
42. L. Gianni, B. J. Corden and C. E. Myers, *Rev. Biochem. Toxicol.*, 1983, 1.
43. R. L. Momparler, M. Karon, S. E. Siegel and F. Avila, *Cancer Res.*, 1976, **36**, 2891.
44. M. Eriksson, B. Nordén and S. Eriksson, *Biochemistry*, 1988, **27**, 8144.
45. D. E. Graves and T. R. Krugh, *Biochemistry*, 1983, **22**, 3941.
46. K. M. Tewey, T. C. Rowe, L. Yang, B. D. Halligan and L. F. Liu, *Science*, 1984, **226**, 466.
47. W. Förster and E. Stutter, *Int. J. Biol. Macromol.*, 1984, **6**, 114.
48. J. B. Chaires, N. Dattagupta and D. M. Crothers, *Biochemistry*, 1985, **24**, 260.
49. Y. C. Liaw, Y. G. Gao, H. Robinson, G. A. Vandermarel, J. H. Vanboom and A. H. J. Wang, *Biochemistry*, 1989, **28**, 9913.
50. Y. G. Gao, Y. C. Liaw, H. Robinson and A. H. J. Wang, *Biochemistry*, 1990, **29**, 10307.
51. M. Egli, L. D. Williams, C. A. Frederick and A. Rich, *Biochemistry*, 1991, **30**, 1364.
52. C. K. Smith, G. J. Davies, E. J. Dodson and M. H. Moore, *Biochemistry*, 1995, **34**, 415.
53. K. R. Fox and M. J. Waring, *Biochim. Biophys. Acta*, 1984, **802**, 162.
54. K. R. Fox, C. Brassett and M. J. Waring, *Biochim. Biophys. Acta*, 1985, **840**, 383.
55. T. R. Krugh, E. S. Mooberry and Y. C. C. Chiao, *Biochemistry*, 1977, **16**, 740.
56. S. C. Brown, K. Mullis, C. Levenson and R. H. Shafer, *Biochemistry*, 1984, **23**, 403.
57. H. M. Sobell and S. C. Jain, *J. Mol. Biol.*, 1972, **68**, 21.
58. S. Kamitori and F. Takusagawa, *J. Mol. Biol.*, 1992, **225**, 445.
59. H. Robinson, Y. G. Gao, X. L. Yang, R. Sanishvili, A. Joachimiak and A. H. J. Wang, *Biochemistry*, 2001, **40**, 5587.
60. F. M. Chen, *Biochemistry*, 1988, **27**, 6393.
61. F. M. Chen, *Biochemistry*, 1992, **31**, 6223.
62. R. Bittman and L. Blau, *Biochemistry*, 1975, **14**, 2138.
63. K. R. Fox and M. J. Waring, *Eur. J. Biochem.*, 1984, **145**, 579.
64. K. R. Fox and M. J. Waring, *Nucleic Acids Res.*, 1986, **14**, 2001.
65. S. C. Brown and R. H. Shafer, *Biochemistry*, 1987, **26**, 277.
66. J. J. Duffy and T. J. Lindell, *Biochem. Pharmacol.*, 1985, **34**, 1854.

67. F. Westerlund, L. M. Wilhelmsson, B. Nordén and P. Lincoln, *J. Am. Chem. Soc.*, 2003, **125**, 3773.
68. R. A. Marcus, *J. Phys. Chem. B*, 2005, **109**, 21419.
69. B. K. Bhuyan and C. G. Smith, *Proc. Natl. Acad. Sci. U. S. A.*, 1965, **54**, 566.
70. W. Kersten, H. Kersten and W. Szybalski, *Biochemistry*, 1966, **5**, 236.
71. D. A. Collier, S. Neidle and J. R. Brown, *Biochem. Pharmacol.*, 1984, **33**, 2877.
72. J. Fok and M. J. Waring, *Mol. Pharmacol.*, 1972, **8**, 65.
73. L. H. Li, S. L. Kuentzel, L. L. Murch, L. M. Pschigoda and W. C. Krueger, *Cancer Res.*, 1979, **39**, 4816.
74. H. L. Ennis, *Antimicrob. Agents Chemother.*, 1981, **19**, 657.
75. F. Zunino and G. Capranico, *Anti-Cancer Drug Des.*, 1990, **5**, 307.
76. X. L. Zhang and D. J. Patel, *Biochemistry*, 1990, **29**, 9451.
77. M. S. Searle, J. G. Hall, W. A. Denny and L. P. G. Wakelin, *Biochemistry*, 1988, **27**, 4340.
78. L. D. Williams, M. Egli, G. Qi, P. Bash, G. A. Vandermarel, J. H. Vanboom, A. Rich and C. A. Frederick, *Proc. Natl. Acad. Sci. U. S. A.*, 1990, **87**, 2225.
79. C. K. Smith, J. A. Brannigan and M. H. Moore, *J. Mol. Biol.*, 1996, **263**, 237.
80. K. R. Fox and M. J. Waring, *Biochemistry*, 1986, **25**, 4349.
81. A. E. Friedman, J.-C. Chambron, J.-P. Sauvage, N. J. Turro and J. K. Barton, *J. Am. Chem. Soc.*, 1990, **112**, 4960.
82. A. E. Friedman, C. V. Kumar, N. J. Turro and J. K. Barton, *Nucleic Acids Res.*, 1991, **19**, 2595.
83. A. M. Pyle and J. K. Barton, *Prog. Inorg. Chem.*, 1990, **38**, 413.
84. Y. Jenkins, A. E. Friedman, N. J. Turro and J. K. Barton, *Biochemistry*, 1992, **31**, 10809.
85. C. S. Chow and J. K. Barton, *Methods Enzymol.*, 1992, **212**, 219.
86. I. Haq, P. Lincoln, D. C. Suh, B. Nordén, B. Z. Chowdhry and J. B. Chaires, *J. Am. Chem. Soc.*, 1995, **117**, 4788.
87. C. Hiort, P. Lincoln and B. Nordén, *J. Am. Chem. Soc.*, 1993, **115**, 3448.
88. B. Nordén, P. Lincoln, B. Åkerman, and E. Tuite, in *Metal Ions In Biological Systems*, ed. A. Siegel and H. Siegel, Marcel Dekker, Inc., New York, 1996, vol. 33, p. 177.
89. P. Lincoln, A. Broo and B. Nordén, *J. Am. Chem. Soc.*, 1996, **118**, 2644.
90. K. E. Erikkila, D. T. Odom and J. K. Barton, *Chem. Rev.*, 1999, **99**, 2777.
91. E. Tuite, P. Lincoln and B. Nordén, *J. Am. Chem. Soc.*, 1997, **119**, 239.
92. C. M. Dupureur and J. K. Barton, *J. Am. Chem. Soc.*, 1994, **116**, 10286.
93. C. Moucheron, A. Kirsch DeMesmaeker and S. Choua, *Inorg. Chem.*, 1997, **36**, 584.
94. S. Satyanarayana, J. C. Dabrowiak and J. B. Chaires, *Biochemistry*, 1992, **31**, 9319.
95. S. Satyanarayana, J. C. Dabrowiak and J. B. Chaires, *Biochemistry*, 1993, **32**, 2573.

96. J. P. Rehmann and J. K. Barton, *Biochemistry*, 1990, **29**, 1701.
97. J. P. Rehmann and J. K. Barton, *Biochemistry*, 1990, **29**, 1710.
98. M. Eriksson, M. Leijon, C. Hiort, B. Nordén and A. Gräslund, *J. Am. Chem. Soc.*, 1992, **114**, 4933.
99. B. Nordén and F. Tjerneld, *Febs Lett.*, 1976, **67**, 368.
100. P. Lincoln and B. Nordén, *J. Phys. Chem. B*, 1998, **102**, 9583.
101. B. Önfelt, P. Lincoln and B. Nordén, *J. Am. Chem. Soc.*, 1999, **121**, 10846.
102. B. Önfelt, P. Lincoln and B. Nordén, *J. Am. Chem. Soc.*, 2001, **123**, 3630.
103. P. Lincoln and B. Nordén, *Chem. Commun.*, 1996, 2145.
104. L. M. Wilhelmsson, F. Westerlund, P. Lincoln and B. Nordén, *J. Am. Chem. Soc.*, 2002, **124**, 12092.
105. L. M. Wilhelmsson, E. K. Esbjörner, F. Westerlund, B. Nordén and P. Lincoln, *J. Phys. Chem. B*, 2003, **107**, 11784.
106. P. Nordell and P. Lincoln, *J. Am. Chem. Soc.*, 2005, **127**, 9670.
107. F. Westerlund, L. M. Wilhelmsson, B. Nordén and P. Lincoln, *J. Phys. Chem. B*, 2005.
108. F. Westerlund, P. Nordell, B. Nordén and P. Lincoln, *J. Phys. Chem. B*, 2007, **111**, 9132.
109. M. Bahira, M. J. McCauley, A. A. Almaqwashi, P. Lincoln, F. Westerlund, I. Rouzina and M. C. Williams, *Nucleic Acids Res.*, 2015, **43**, 8856.
110. P. Nordell, F. Westerlund, L. M. Wilhelmsson, B. Nordén and P. Lincoln, *Angew. Chem. Int. Ed.* 2007, **46**, 2203.
111. P. Nordell, F. Westerlund, A. Reymer, A. El-Sagheer, T. Brown, B. Nordén and P. Lincoln, *J. Am. Chem. Soc.*, 2008, **130**, 14651.
112. P. Nordell, E. T. Jansson and P. Lincoln, *Biochemistry*, 2009, **48**, 1442.
113. F. Westerlund, P. Nordell, J. Blechinger, T. Santos, B. Nordén and P. Lincoln, *J. Phys. Chem. B*, 2008, **112**, 6688.
114. A. A. Almaqwashi, T. Paramanathan, P. Lincoln, I. Rouzina, F. Westerlund and M. C. Willams, *Nucleic Acids Res.*, 2014, **42**, 11634.
115. L. Wu, A. Reymer, C. Persson, K. Kazimierczuk, T. Brown, P. Lincoln, B. Nordén and M. Billeter, *Chem. – Eur. J.*, 2013, **19**, 5401.
116. D. R. Boer, L. Wu, P. Lincoln and M. Coll, *Angew. Chem. Int. Ed.* 2014, **53**, 1949.
117. F. Westerlund, M. P. Eng, M. U. Winters and P. Lincoln, *J. Phys. Chem. B*, 2007, **111**, 310.
118. J. R. Johansson, Y. Wang, M. P. Eng, N. Kann, P. Lincoln and J. Andersson, *Chem. – Eur. J.*, 2013, **19**, 6246.
119. J. Andersson, M. Li and P. Lincoln, *Chem. – Eur. J.*, 2010, **16**, 11037.
120. J. Andersson and P. Lincoln, *J. Phys. Chem. B*, 2011, **115**, 14768.
121. A. A. Almaqwashi, J. Andersson, P. Lincoln, I. Rouzina, F. Westerlund and M. C. Williams, *Biophys. J.*, 2016, **110**, 1255.

CHAPTER 4

# *Thermal Denaturation of Drug–DNA Complexes*†

JONATHAN B. CHAIRES

James Graham Brown Cancer Center, University of Louisville, 505 S. Hancock St., Louisville, KY 40202, USA
Email: j.chaires@louisville.edu

## 4.1 Introduction

Only a few months after Watson and Crick presented their model for the DNA double helix,[1] the alkaline and thermal denaturation of DNA, as monitored by changes in UV absorbance, was reported.[2] This seminal work explicitly recognized that the observed hyperchromism resulted from "...the destruction of a secondary molecular structure constituted by labile bonds involving the puric and pyrimidic rings", and specifically referred to the Watson–Crick structure. Soon after, the Doty laboratory published a remarkable and prescient series of studies on the acid, alkaline and thermal denaturation of DNA, employing a wide array of biophysical tools to show that the duplex strands separated upon denaturation.[3–10] That line of research lead directly to the well-known paper by Marmur and Doty[11] that related the DNA melting temperature ($T_m$) to the GC content of the DNA. Concurrently, statistical mechanical theories for the effect of preferential

†This chapter is a revised and updated version of Chapter 7 that appeared in the first edition of *Sequence-specific DNA Binding Agents*, ed. Michael Waring, The Royal Society of Chemistry, 2006, pp. 130–151. Section 3 has been amended to discuss advances in high-throughput thermal denaturation methods.

Chemical Biology No. 7
DNA-targeting Molecules as Therapeutic Agents
Edited by Michael J. Waring

Published by the Royal Society of Chemistry, www.rsc.org

ligand binding on the helix-to-coil transition were developed,[12,13] albeit with explicit reference to protein transitions rather than to DNA melting. These theories were general, however, and were equally applicable to DNA transitions. Only five years after the appearance of the Watson–Crick model, calorimetry was used to study the enthalpy of the acid denaturation of DNA.[14]

Against this background, thermal denaturation became a commonly used tool for the study of drug–DNA complexes from the early 1960s onward. The attraction of melting studies lies in their simplicity and readily available, inexpensive, instrumentation. Melting provides a simple and unambiguous demonstration of drug binding to DNA. Small molecules that bind preferentially to the DNA duplex stabilize the structure and elevate its $T_m$. Intercalators and groove-binders both recognize particular features of duplex DNA, and consequently raise its $T_m$. Apart from a simple qualitative demonstration of binding to DNA, melting studies can be analyzed to obtain quantitative information about the binding interaction. A variety of approaches for such quantitative analysis have appeared over the years,[15–19] and new approaches continue to be developed.[20] One purpose of this chapter is to illustrate the utility of such methods in the quantitative analysis of drug–DNA melting curves to obtain thermodynamic data.

Practical procedures and detailed protocols for properly conducting DNA melting experiments have been published,[21–23] and will not be repeated here. The focus, instead, will be on the analysis and interpretation of such data once the experiment is completed.

Current interest in drug–DNA interactions lies in the specificity of the binding process. Do DNA-binders recognize particular sequences or structures? While the question of binding specificity is usually addressed by chemical and enzymatic footprinting methods, or by competition dialysis, the potential of thermal denaturation methods in this area should not be overlooked. Differential scanning fluorimetry (DSF) will be described as a relatively new high-throughput thermal denaturation method for monitoring DNA binding. Simple experimental designs that allow DSF to be used to explore sequence and structural selectivity of drug binding will be described and illustrated to conclude this chapter.

## 4.2 Thermal Denaturation Tools

### 4.2.1 Analysis of $T_m$ shifts in the Presence of Drug

Figure 4.1 shows sample melting curves for the polydeoxynucleotide poly dA:poly dT in the presence of increasing concentrations of the groove-binder netropsin.[24,25] These experiments show that the melting temperature of the duplex DNA is dramatically increased upon addition of netropsin. At the highest molar ratio of added netropsin shown in this figure, the $T_m$ is elevated from 50.5 to 93.2 °C, a difference of 42.7 °C. In the absence of netropsin, or at saturating concentrations, the melting transition is sharp, with the transition complete over a span of just a few degrees. At lower molar ratios,

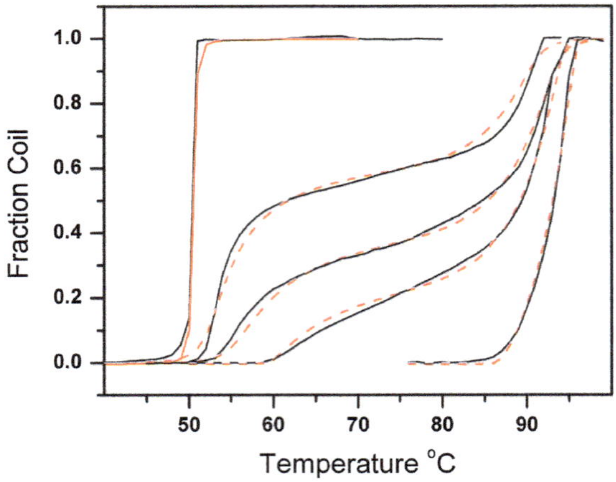


**Figure 4.1** Thermal denaturation of poly dA:poly dT in the presence of netropsin. The polynucleotide concentration was 45 μM bp. The molar ratio of added netropsin (drug $bp^{-1}$) was, from left to right: 0; 0.08; 0.14; 0.18; 0.28.

below saturation, melting curves are seen to be broad and multiphasic. As will be discussed in more detail later, such behavior is a consequence of a complex underlying mechanism that involves ligand redistribution over the course of duplex melting. Qualitatively, the melting curves shown in Figure 4.1 show unambiguously that netropsin binds preferentially to the duplex form, stabilizing it against denaturation. Were netropsin to bind preferentially to single strands, a decrease in the melting temperature would be expected. The experimental behavior in Figure 4.1 is simply an illustration of Le Chatelier's principle at work. Perturbation of the helix-to-coil equilibrium (by drug binding) shifts the reaction in the direction of the favored species, the duplex in this case. Apart from such qualitative conclusions, what quantitative information might be extracted from the data in Figure 4.1?

This is where theories of DNA melting in the presence of ligands are needed. Among the various treatments, those presented by Crothers[15] and by McGhee[17] have proven to be the most accessible and useful. Both McGhee and Crothers provided closed-form equations that permit quantitative interpretation of $T_m$ shifts under certain limiting conditions. Under conditions where the DNA duplex is saturated with ligand, and where there is assumed to be no binding to single-stranded forms, the shift in melting temperature can be described by relatively simple equations. McGhee provided the equation

$$\frac{1}{T_m^0} - \frac{1}{T_m} = \frac{R}{\Delta H_{DNA}} \ln\left[(1 + K_h L)^{1/n}\right] \tag{4.1}$$

where $T_m^0$ and $T_m$ are the melting temperatures in the absence and presence of *saturating* concentrations of drug, respectively, expressed in Kelvin. In eqn (4.1), $R$ is the gas constant, $\Delta H_{DNA}$ is the enthalpy of DNA melting (per bp), $K_h$ is the ligand binding constant to the duplex form, $L$ is the ligand concentration, and $n_h$ is (approximately) the ligand site size in bp. For a full discussion of the origin and limitations of eqn (4.1), the original reference should be studied.[17] Crothers provided an equation with a slightly different algebraic form but which embodies the same underlying mechanism.

By using eqn (4.1), estimates of the binding constant, $K_h$ can be obtained *if* the enthalpy of DNA melting is known, along with a value for the site size ($n_h$). The ligand concentration can normally be assumed to be the total added. Binding constants obtained from eqn (4.1) refer to the temperature for the melting of the complex.

Figures 4.2 and 4.3 map the behavior of eqn (4.1) as a function of the key variables $K_h$ and $n_h$. Figure 4.2 shows that the interplay between the binding constant and the site size results in a complex surface. The contour representation of that surface (Figure 4.3, right) shows that $K_h$ and $n_h$ are tightly coupled, such that several combinations of these variables can elevate $T_m$ to the same extent. $\Delta T_m$ values thus are not direct indications of binding affinity, as is often erroneously asserted, since the shift in $T_m$ is influenced just as strongly by the site size. Figures 4.3A and B show slices of the surface in Figure 4.2 to emphasize the isolated effects of $K_h$ and $n_h$. With a constant binding constant (Figure 4.3A) the magnitude of the elevation of $T_m$ is seen to decrease with increasing site size. That behavior is somewhat counter-intuitive. When the site size is constant (Figure 4.3B), the magnitude of the elevation of $T_m$ increases with increasing binding affinity. For a series of compounds, $\Delta T_m$ values provide a measure of the relative binding affinity *only* if the binding site size is identical for all compounds.

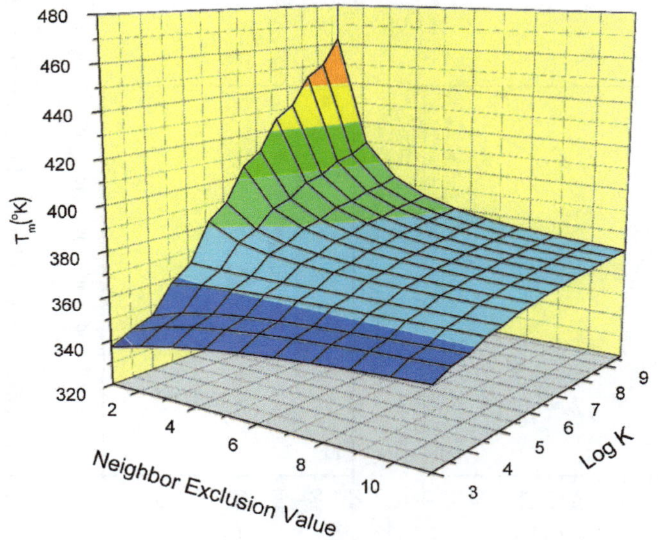


**Figure 4.2** Surface map computed for values of $T_m$ from eqn (4.1). Values for $K$ and $n$ were systematically varied, with the remaining parameters held constant, with $T_0 = 333.15$ K, $\Delta H = 8.0$ kcal mol$^{-1}$ and $L = 20$ μM.

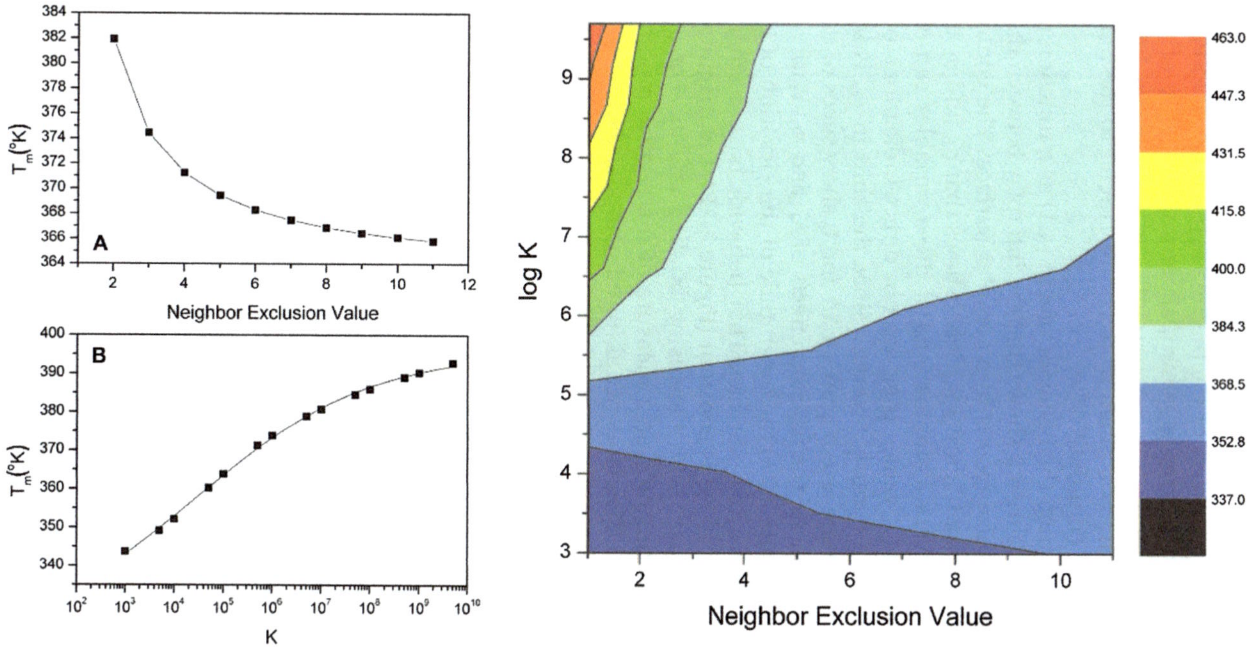


**Figure 4.3** Isolated effects of $K$ and $n$ on the elevation of $T_m$. The right side shows a contour representation of the surface shown in Figure 4.2. Panel A shows a slice across the surface with $K = 5 \times 10^5$ $M^{-1}$ and $n$ varied. Panel B shows a slice with $n = 4$ and $K$ varied.

For netropsin, with $n = 4$, a $\Delta T_m$ value of 53.1 °C was measured for a total ligand concentration of 75 μM for the melting of poly dA:poly dT at 0.3 mM bp. By using eqn (4.1), a binding constant of $1.5 \times 10^8$ $M^{-1}$ was measured at 103.6 °C. Correction of this binding constant to lower temperatures requires knowledge of the binding enthalpy, which can be obtained by differential scanning calorimetry (DSC).

### 4.2.2 Obtaining Binding Enthalpy Values by DSC

Figure 4.4 shows the results of DSC experiments for the melting of poly dA:poly dT in the presence and absence of netropsin. Higher polynucleotide concentrations are required for the DSC experiment compared with the optical melting experiment. With saturating amounts of netropsin, a dramatic elevation in $\Delta T_m$ is again evident, but DSC has the added value of providing enthalpy values directly from the peak areas seen in Figure 4.4.

Analysis of DSC data to obtain ligand binding enthalpies uses Hess's law.[26] The following equilibria are considered:

1. duplex $\rightleftarrows$ 2(single strands) $\Delta H_1$
2. duplex − ligand $\rightleftarrows$ 2(single strands) + ligand $\Delta H_2$
3. duplex − ligand $\rightleftarrows$ duplex + ligand $\Delta H_3 = \Delta H_2 - \Delta H_1$

The derived reaction 3 is the desired binding enthalpy, and can be obtained from the areas of the two peaks shown in Figure 4.4. The enthalpy $DH_3$ needs to be corrected for amount bound ($r_b$), and its sign changed to correspond to binding

$$\Delta H_b = -(\Delta H_3 / r_b)$$

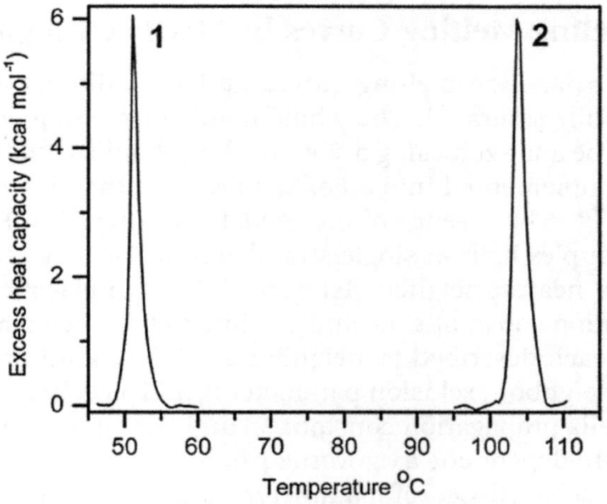


**Figure 4.4** Differential scanning calorimetry experiment with poly dA:poly dT in the absence (1) and presence (2) of netropsin.

For the data shown in Figure 4.4, $DH_1$ was determined to be 10.6 kcal mol$^{-1}$ and $DH_2$ was determined to be 13.6 kcal mol$^{-1}$. The binding ratio $r_b$ was determined to be 0.25, leading to an estimate of $DH_b = -12.0$ kcal mol$^{-1}$.

Once $DH_b$ is known, the binding constant can be calculated for any desired temperature from the van't Hoff equation

$$\ln \frac{K_2}{K_1} = -\frac{\Delta H_b}{R}\left(\frac{1}{T_2} - \frac{1}{T_1}\right)$$

For netropsin, this yields a value for the equilibrium constant of $1.4 \times 10^{10}$ M$^{-1}$ at 20 °C.

The complete thermodynamic profile for netropsin can then be easily derived from standard relationships:

$$\Delta G_{20} = -RT\ln K = -13.6 \text{ kcal mol}^{-1}$$

$$\Delta H_b = -12.0 \text{ kcal mol}^{-1}$$

$$-T\Delta S = \Delta G_{20} - \Delta H_b = -1.6 \text{ kcal mol}^{-1}$$

Netropsin binding to DNA is thus driven by favorable contributions from both enthalpy and entropy. While this approach provides valuable quantitative data, it must be stressed that application of the methods and theory require conditions where the DNA lattice is saturated with ligand. In addition, a rather long extrapolation is required to correct values to lower standard temperatures. Possible heat capacity changes are, of necessity, neglected, perhaps introducing serious error into the estimates of parameters at the lower temperatures. Little can be done to alleviate that drawback, unless the particular drug–DNA system of interest is amenable to study by isothermal titration calorimetry.[27]

### 4.2.3 Modeling Melting Curves by McGhee's Algorithm

It is possible to analyze melting curves under conditions where the DNA lattice is not fully saturated. The whole family of melting curves shown in Figure 4.1 can be analyzed using a statistical mechanical model developed by McGhee and implemented into a Fortran program that contains some 1000 lines of code.[17,18] The essence of the model is portrayed in Figure 4.5. DNA can exist as a duplex helix or single-stranded coil. The helix–coil transition is described by a nearest-neighbor Ising model with nucleation parameter $\sigma$ and a propagation constant $s$. Ligand binding to both helix or coil regions is allowed, with each described by neighbor-exclusion binding with a binding constant $K$, a neighbor-exclusion parameter $n$, and a cooperativity parameter $\omega$. Both the helix propagation constant ($s$) and ligand binding constants ($K$) are temperature-dependent as governed by melting and binding enthalpy values. For a selected set of parameter values, McGhee's program will compute a melting curve that may be compared with a normalized experimental curve. By successive approximation and with judicious constraint of

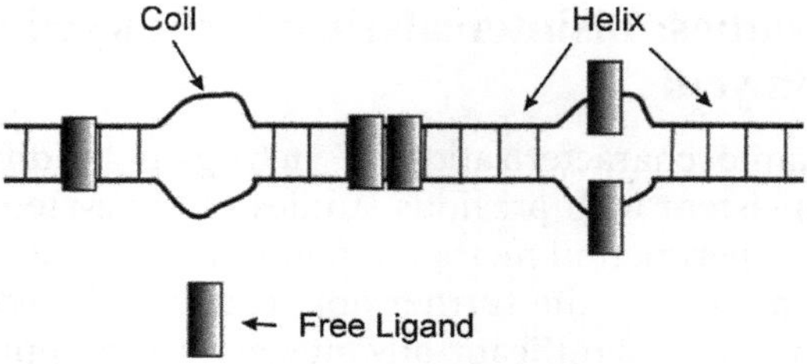


**Figure 4.5** Schematic representation of McGhee's model for the melting of DNA in the presence of ligand.

**Table 4.1** Binding constants obtained from the melting curves shown in Figure 4.1 for the interaction of netropsin with poly dA:poly dT.[a]

| [Netropsin] (μM ) | Molar ratio (Nt : DNA) | $K/10^{10}$ ($M^{-1}$) | $n$ (bp) | $\sigma$ | $\omega$ |
|---|---|---|---|---|---|
| 0 | 0 | 0 | 0 | $5\times10^{-5}$ | 1 |
| 3.7 | 0.08 | 1.3 | 4 | $6\times10^{-4}$ | 1 |
| 6.5 | 0.14 | 1.3 | 4 | $5\times10^{-4}$ | 1 |
| 8.0 | 0.18 | 1.3 | 4 | $1\times10^{-5}$ | 1 |
| 12.5 | 0.28 | 1.3 | 4 | $1\times10^{-5}$ | 1 |

[a]Binding parameters were obtained by successive approximation 'fits' to the experimental data using McGhee's statistical mechanical mode.[17,18] The concentration of poly(dA):poly(dT) was fixed at 45 μM bp. [Netropsin] is the concentration of netropsin used in the experiments. The parameters $K$ (the binding constant at 20 °C), $n$ (the binding site size), $\sigma$ (the nucleation parameter) and $\omega$ (ligand cooperativity constant) were systematically varied to produce the optimal fit to the experimental data. The following parameters were constrained in the simulation: $T_m = 50.5$ °C, $\Delta H_m = -10.6$ kcal mol$^{-1}$, and $\Delta H_b = -12$ kcal mol$^{-1}$.

some parameters, it is possible to estimate a set of parameters that describe an entire family of melting curves. The procedure is described in detail in a useful review by Spink and Wellman.[18] Figure 4.1 shows optimized 'fits' (obtained by successive approximation) to the experimental data for the netropsin system, with parameter values shown in Table 4.1. In this case, binding to single-stranded DNA was found to be negligible. Inspection of Table 4.1 shows, remarkably but as expected, that a single set of binding and melting parameters can (within error) accurately predict the entire family of melting curves. Only the changes in total and free netropsin concentrations dictate the complex shapes and positions of melting curves below saturation.

The complex shapes of melting curves evident in Figure 4.1 deserve comment and explanation. Their complexity, as revealed by McGhee's model, arises from redistribution of ligand over the course of helix denaturation. Binding sites are, by the model, assumed to be homogeneous along the lattice. At less than saturating ligand concentrations, as one region of the helix melts, ligand is transiently released, but then rebinds to any remaining duplex regions. These duplex regions thus become increasingly saturated over the course of the melt, leading to broad, multiphasic transition curves for which there is no well-defined or simple $T_m$ value.

### 4.2.4 Case Studies: Bisintercalating Anthracyclines and Echinomycin

The thermodynamic characterization of netropsin binding as described above is fully consistent with previous studies of the system,[28–30] providing validation of the experimental methods. Additional case studies will now be reviewed that demonstrate the further applicability of these methods, and which also reveal some significant advantages of the approach. These examples are drawn from studies of two types of bisintercalating compounds, first, a new class of rationally designed bisanthracyclines[31–33] and, second, the natural product echinomycin.[34] Each of these cases posed particular difficulties that rendered traditional spectophotometric titration methods useless, making the melting methods described here the only viable alternative available for quantitative studies.

The anthracycline antibiotics daunorubicin and doxorubicin have been mainstays of cancer chemotherapy.[35–37] A new class of bisanthracyclines (Figure 4.6) was rationally designed from considerations of available high-resolution crystal structures of anthracycline–DNA complexes. In high-resolution structures of daunorubicin–DNA complexes, two drug molecules are bound to a DNA hexanucleotide.[38] The two drug molecules are generally intercalated near the ends of the oligonucleotide in a tail-to-tail arrangement, such that their daunosamine moieties lie in the minor groove and are within 7 Å of one another. The proximity of their reactive amine groups suggested a simple design strategy. By simply crosslinking two monomers through their amines using an appropriate linker, a new type of bisanthracycline might be synthesized. Monomers crosslinked through their amine groups would feature an optimal stereochemical fit within the DNA minor groove, with no hindrance to intercalation, and with added favorable minor groove interactions from the linker. The design was realized using a *p*-xylenyl linker to covalently link two daunorubicin molecules to form bisdaunorubicin, WP631 (Figure 4.6).[33]

Bisintercalators ought to possess enormously enhanced binding affinity relative to their monointercalating substituents. The binding constant for a bisintercalator should equal the *square* of the monomer binding constant,[39] $K_B = K_M \times K_M = K_M^2$. Since daunorubicin binds to DNA with $K_M \approx 10^6$ $M^{-1}$, bisdaunorubicin ought to bind to DNA with picomolar affinity, $K_B \approx 10^{12}$ $M^{-1}$. Such a tight binding poses practical problems for the experimental determination of binding constants using traditional spectrophotometric titrations.[40] For accurate determinations of binding isotherms, one must work at concentrations near the reciprocal of the binding constant. At picomolar concentrations, it becomes difficult to accurately monitor absorbance or fluorescence. Fortunately, the melting methods described here do not depend on a signal from the binding ligand, and can be used to measure binding affinity.

For WP631, ultratight binding was in fact observed, and measured by the thermal denaturation methods described in this chapter.[31,33] A binding constant of $3.1 \times 10^{11}$ $M^{-1}$ was determined, with a binding enthalpy of

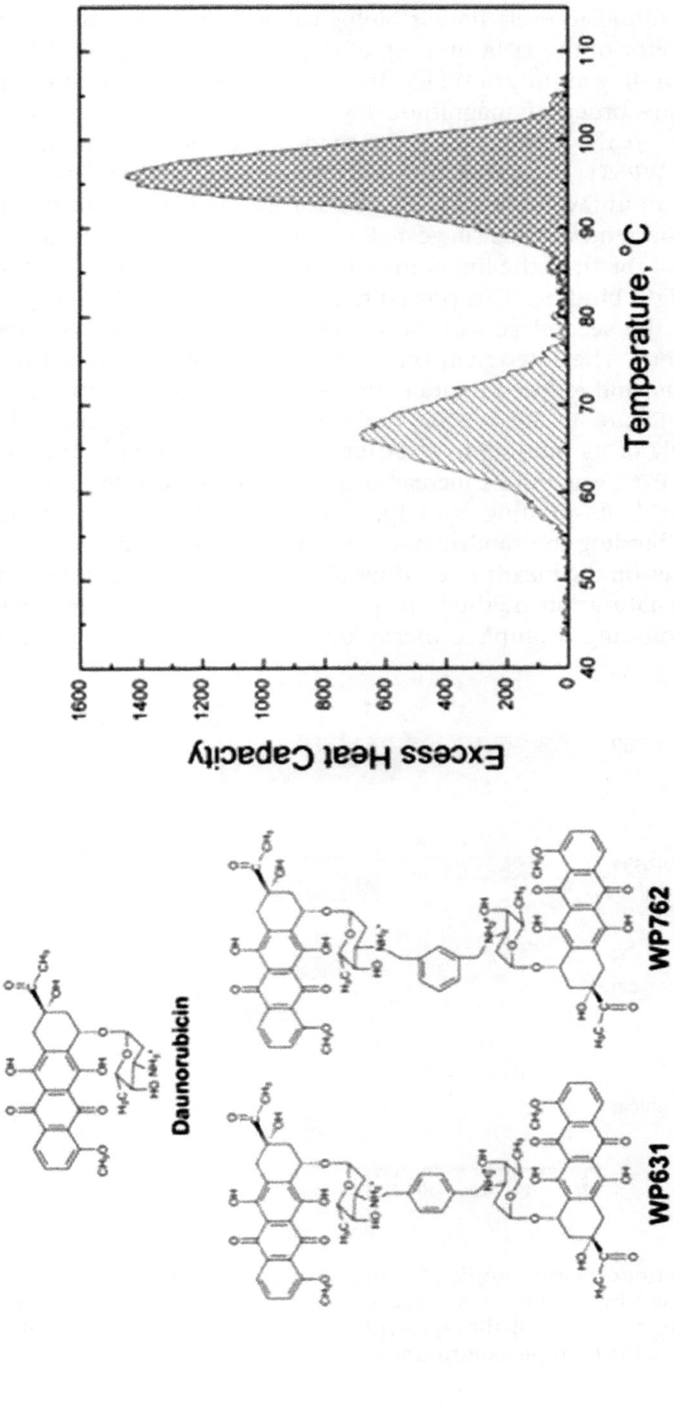


**Figure 4.6** Bisanthracycline compounds. The parent antibiotic daunorubicin (monomer) is shown along with WP631 and WP762, in which monomers are linked by a *p*-xylenyl and *m*-xylenyl linker, respectively. The results of a DSC experiment with herring sperm DNA and WP762 are shown on the right side. Excess Heat Capacity has the units of kcal $mol^{-1}$.

−30.2 kcal mol$^{-1}$. The binding constant is nearly the square of that measured for the monomer daunorubicin under the same conditions. WP631 was subsequently found to exert potent biological activity and proved to be an effective inhibitor of the DNA binding of the transcription factor SP1.[41–43]

While the binding affinity of WP631 to DNA was impressive indeed, the fact that it was one order of magnitude lower than expected suggested that improvements to the design could be made. The thermodynamic profile obtained for WP631 suggested that rotation around linker bonds might contribute to an unfavorable entropic cost for immobilization of the linker. An improved design was thus suggested. Immobilization of linker bonds and optimization of the fit of the linker into the minor groove might improve the already ultratight binding. This proved to be correct. WP762 (Figure 4.6) was a member of the second generation of bisanthracyclines that featured an *m*-xylenyl linker.[44] The different linker slightly adjusted the length between the monomer units, and reduced rotation around linker bonds. Sample DSC data are shown in Figure 4.6 for WP762, and a complete binding and molecular modeling study of its binding to DNA has recently appeared.[44] The binding constant of WP762 was in fact increased in comparison to WP631, to a value of $7.3 \times 10^{12}$ M$^{-1}$. Its binding enthalpy was increased to −33.8 kcal mol$^{-1}$. Anthracycline binding thermodynamics are compared in Figure 4.7.

These studies on the bisanthracyclines illustrate and emphasize the power of thermal denaturation methods to provide quantitative characterization of ultratight binding. Complete thermodynamic profiles may be obtained

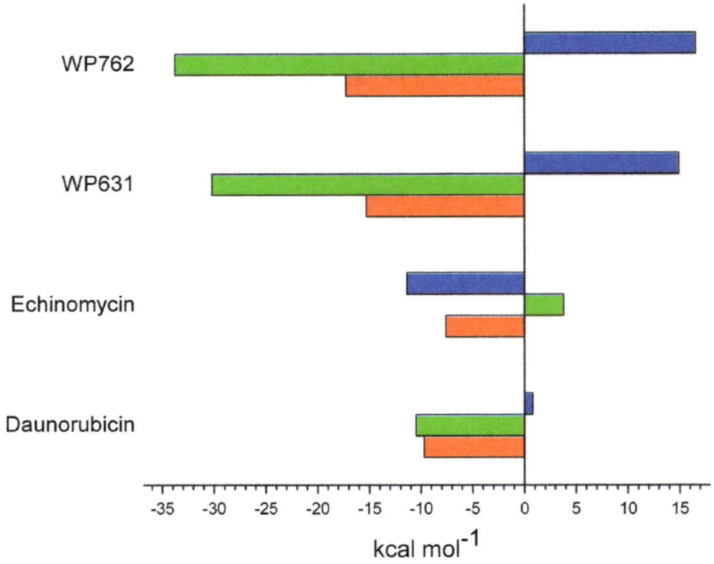


**Figure 4.7** Thermodynamic profiles for the binding of ligands to DNA as determined by thermal denaturation methods. Red fill represents binding free energy, green fill represents binding enthalpy and blue fill represents the entropic contribution, $-T\Delta S$, all in kcal mol$^{-1}$.

in these cases where optical methods cannot be used because of the low concentrations required to properly characterize the equilibrium. At the required low concentrations, the necessary absorbance or fluorescence signals are too weak to be accurately measured.

Another example of the advantage of thermal denaturation approaches for the characterization of binding thermodynamics is provided by the natural product echinomycin (Figure 4.8). Echinomycin was the first bisintercalator to be discovered.[34] Echinomycin is uncharged and hydrophobic. Studies of its DNA binding were hampered by its poor solubility, about 5 μM in aqueous solution. Out of necessity (and perhaps desperation), phase partition methods were devised to study echinomycin binding to DNA.[45] While the phase partition method can reliably determine binding constants, studies of the temperature dependence of binding constants is both difficult and fraught with error because of the simultaneous temperature dependence of the drug partition coefficient. Proper evaluation of the thermodynamic profile was thus difficult, and it was only recently that thermal denaturation methods were successfully applied to alleviate the difficulties.[46] In this case, since thermal denaturation studies are initiated with a preformed drug–DNA complex, the solubility problem is lessened. The requisite complex can be formed by the 'solid-shake' protocol devised by Waring and coworkers, in which a solution containing DNA is equilibrated with solid echinomcyin to bring the lattice to any desired degree of saturation. Figure 4.8 shows sample DSC data. Optical melting studies over a wide range of binding ratios were also done, and may be found in the original publication.[46] These studies provided the first direct calorimetric determination of the enthalpy of echinomycin binding to DNA, and yielded a complete thermodynamic profile for the binding interaction. It was found that $\Delta G^\circ = -7.6$ kcal mol$^{-1}$, $\Delta H = +3.8$ kcal mol$^{-1}$, and $\Delta S = +38.9$ cal mol$^{-1}$ K$^{-1}$ at 20 °C. The binding reaction is clearly entropically driven, a hallmark of a process that is predominantly stabilized by hydrophobic interactions. The binding thermodynamics of echinomcyin provide an interesting contrast to those of the bisanthracylines, as summarized in Figure 4.7. Echinomycin binding is entropically driven, with a positive binding enthalpy, whereas binding of the bisanthracyclines is enthalpically driven, and is opposed by entropy. The difference in binding thermodynamics points to clear differences in the molecular forces that drive the binding of the two types of bisintercalators.

### 4.2.5 Summary: Advantages and Pitfalls

These studies collectively illustrate significant advantages of the thermal denaturation methods. First, they can be used to study ultratight binding interactions, since no optical signal from the binding ligand is required. Second, they can be used for even sparingly soluble ligands, since the drug–DNA complex required for the methods can be formed by equilibrating DNA with solid ligand. Finally, differential scanning calorimetry measures

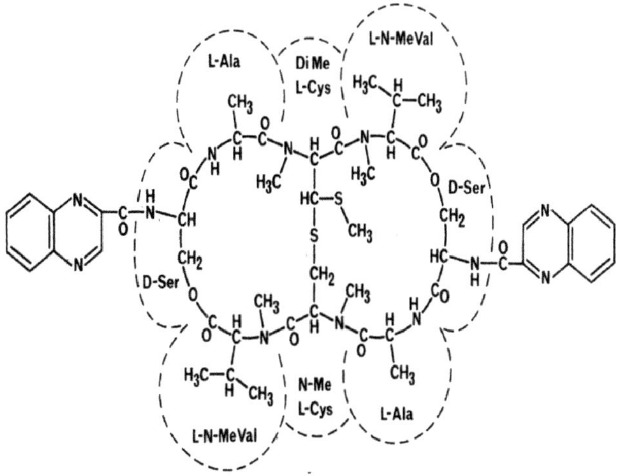


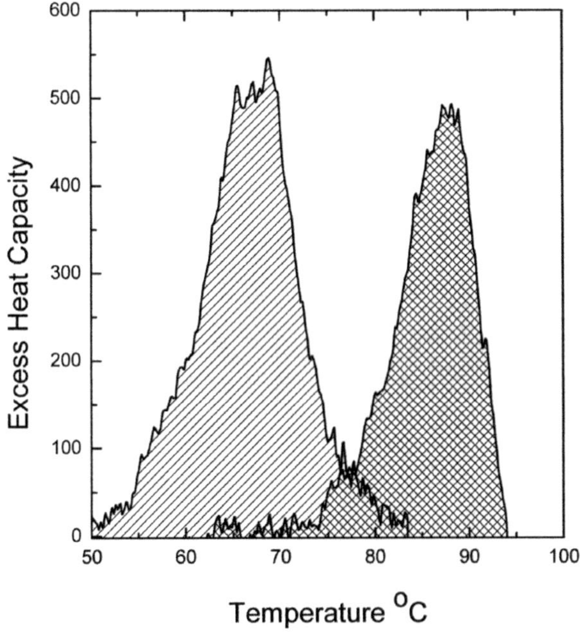


**Figure 4.8** Echinomycin (left) and its effect on the denaturation of herring sperm DNA (right). Excess Heat Capacity has units of $kcal\,mol^{-1}$.

binding enthalpy as directly as possible, yielding complete thermodynamic profiles for complex formation.

There are possible pitfalls. For simple, closed-form equations to be used for analysis of $T_m$ shift, the DNA lattice must be saturated. If the lattice is not saturated, complex multiphasic melting curves will result, which cannot be easily analyzed or interpreted. The magnitude of the increase in melting temperature, even at saturation, is not a simple, unambiguous measure of affinity. $K$ and $n$ values are tightly coupled, and many combinations may combine to produce similar $T_m$ shifts. In order to estimate $K$, some value of $n$ must be known or assumed. Finally, while DSC returns an estimate of binding enthalpy at the melting temperature, it is difficult to obtain values for the heat capacity change for binding, $\Delta C_p = d\Delta H/dT$. Neglect of $\Delta C_p$ may lead to systematic errors in the extrapolation of binding constants to lower temperatures.

Finally, it is necessary to caution that all the discussion presented here refers to polymeric DNA. The theories of McGhee and Crothers assumed a long nucleic acid lattice with homogeneous binding sites. The analysis described here is inappropriate for use with short oligonucleotides.

## 4.3 High-throughput Thermal Denaturation Approaches

As useful and powerful as thermal denaturation methods are, they suffer from being somewhat slow, a decided drawback in efforts to screen even small libraries of compounds for their DNA binding. With typical temperature ramps of 0.5–1.0 °C per minute, a single thermal denaturation experiment will take 1–2 hours for data collection to cover an adequate temperature range, plus the time needed for set up and initial thermal equilibration. Typically only a few experiments can be completed in a single workday. While most spectrophotometers are now routinely equipped with multiple cell holders that might allow 6–8 samples to be run while sampling their absorbance at intervals of a few tenths of a degree, still data collection remains limiting since only 10–20 compounds might be tested per day. Since there are now chemical libraries of potential DNA binders that contain thousands of compounds, such limited throughput compromises the utility of thermal denaturation in drug discovery screening efforts. A method with higher throughput was urgently needed, leading to the implementation of what has come to be known as *differential scanning fluorimetry* (DSF).

### 4.3.1 Differential Scanning Fluorimetry

A high-density miniaturized thermal shift assay as a method for drug discovery was implemented by 3-Dimensional Pharmaceuticals, Inc. (now part of Johnson & Johnson) around the year 2000. The technology was initially dubbed ThermoFluor®, but current variants are generally referred to as differential scanning fluorimetry. The key initial technological development was the design of a microplate assay apparatus that included a temperature-adjusting device for simultaneously heating the samples in the plate along with a detection system for simultaneously recording the fluorescence emission of all samples while the samples are being heated.[47] Small sample volumes, only tens of microliters per well, were an attractive feature of the design. The device provided a rapid and broadly applicable method for obtaining quantitative measurements of drug-binding affinity to virtually any protein or nucleic acid target by measuring changes in melting temperature resulting from ligand binding. The assay for monitoring protein–ligand interactions was fully described in 2001.[48] In that report the thermal denaturation of a number of proteins of interest was monitored by the use of a variety of external fluorescence probes. These probes possessed low fluorescence emission in aqueous solution but enhanced emission upon binding to the hydrophobic surfaces exposed upon protein denaturation. Ligands that bound to the native form were found to stabilize proteins and elevate their $T_m$. The results obtained by ThermoFluor® were validated by rigorous correlation with results obtained by differential scanning calorimetry and by enzymatic kinetic activity measurements. Upon the realization that real-time PCR instruments, once available, could be readily adapted for DSF measurements, the use of high-throughput thermal shift assays for studies of protein structure and function exploded, with several hundred papers using the approach and several thousand citations made to the method. Recent reviews best document that DSF is now widely used for protein drug discovery efforts and as an efficient tool for optimizing protein stability for structural determinations and formulation efforts.[49,50]

The use of DSF in the discovery of nucleic acid binders quickly followed. By the mid-1990s, fluorescent resonance energy transfer (FRET) measurements were routinely used in DNA structural studies. This was possible because the synthesis of well-defined sequences of DNA oligomers became cost effective as well as routine, and fluorescent labels could be conjugated at well-defined positions on a DNA oligomer.[51] A FRET thermal shift assay was used in 2001 in the discovery of quadruplex-targeted telomerase inhibitors, albeit using conventional cuvette spectrofluorimetry.[52] The Fox laboratory first demonstrated that high-throughput measurement of duplex, triplex and quadruplex melting curves using FRET and a commercially available Roche LightCycler instrument could be made, essentially implementing DSF with all of its advantages, especially small volumes and consumption of only modest amounts of material.[53] Several reviews track the use and further development of DSF methods for the discovery of quadruplex binding ligands.[54–56]

### 4.3.2 DSC Compared with DSF: Slow and Expensive but Definitive

Since it measures thermal denaturation heats directly without recourse to an assumed reaction model, DSC remains the gold standard for thermodynamic studies of biological molecules. Compared with DSF, DSC has the advantage of being label-free and can be used to study unmodified receptors. However, DSC does suffer from some disadvantages relative to DSF. First, DSC requires much more material and larger sample volumes, so sample consumption is much greater. Second, data acquisition for DSC remains somewhat slow, with thermal scans taking several hours, sample loading being more involved than for a typical spectrophotometric experiment and, finally, instrument cleaning after each run is more critical and more tedious. An attempt to alleviate these latter issues was made by the design and construction of a commercially available automated DSC instrument that features an autosampling liquid handling system.[57] While this approach relieves the tedium of sample loading and instrument cleaning, DSC measurements still must be run serially, limiting data collection to tens of samples per day at most. At their best these DSC instruments still cannot match the high-throughput of the DSF approach and their sample consumption remains significantly higher. This is not to say that DSC has no place in drug discovery. Its utility lies in furnishing a definitive secondary screening method to validate the results obtained by high-throughput primary screening accomplished by DSF.

### 4.3.3 Illustrations of Differential Scanning Fluorimetry Data and Utility

Figure 4.9 shows examples of DSF and DSC data that illustrate the strengths and weaknesses of the methods. The thermal denaturation profiles of two G-quadruplex conformational forms adopted by the human telomere repeat sequence 5′-AGGG(TTAGGG)$_3$ are shown. For DSF, the sequence must be labeled, in this case by using a FAM–TAMRA donor–acceptor FRET pair. Figure 4.9A shows the normalized, baseline-corrected melting transitions for the folded antiparallel 'basket' form adopted in NaCl (red line) and the 'hybrid' form adopted in KCl (black line). Figure 4.9B shows the corresponding derivative melting curves for these structures. The basket form is seen to melt at 51.6 °C, while the more stable hybrid form melts at 69.6 °C. These melting curves were obtained using a 20 μl sample at 0.2 μM strand concentration, revealing the significant advantage of DSF in minimizing sample consumption. Figure 4.9C shows DSC thermograms of the basket and hybrid forms. These experiments required samples of 1.4 ml of 101 μM strand concentration, over 500 times the concentration needed for DSF. Transition temperatures of 63.7 and 72.4 °C are seen for the basket and hybrid forms, respectively. Notably, these transition temperatures are higher than those observed by DSF. This reveals a possible weakness in DSF, namely that the requisite labeling of the DNA strand may destabilize the structure being studied. In this case, labeling destabilizes the antiparallel basket form (12.1 °C decrease) to a greater extent than the hybrid form (2.8 °C decrease). Despite this pitfall, DSF can still monitor binding interactions, but caution is needed to ensure the integrity of the binding receptor. These data show that

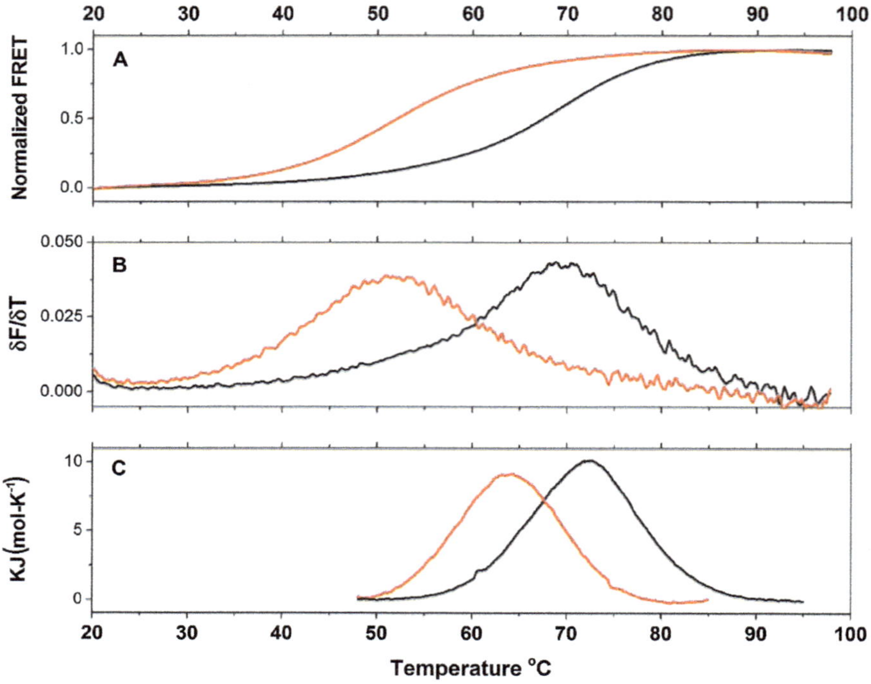


**Figure 4.9** Thermal denaturation of human telomere G-quadruplex forms as measured by DSF and DSC. The DNA sequence used is 5′-FAM-AGGG(TTAGGG)$_3$-TAMRA-3′ in a pH 7.0 solution containing 6 mM $Na_2HPO_4$, 2 mM $NaH_2PO_4$, 1 mM EDTA, and 0.185 M NaCl or KCl. In NaCl, the sequence folds into an antiparallel 'basket' conformation, while in KCl it folds into a 3 + 1 'hybrid' conformation. In all panels, the red line refers to data obtained in NaCl solutions while the black line refers to data obtained in KCl solutions. (A) Normalized primary data obtained by DSF. (B) First derivatives of the data shown in A. (C) Data obtained by DSC.

the hybrid form is less affected by labeling, and perhaps represents a more reliable target structure for DSF.

Figure 4.10 summarizes a screening exercise using DSF. The NCI Natural Products Set I, a library containing 120 compounds, was screened for binding to the human telomere hybrid G-quadruplex form. Figure 4.10A shows the primary, unprocessed, data from DSF scans. The 120 melting transitions were obtained in triplicate efficiently over the course of two working days. Figure 4.10B shows the distribution of measured $T_m$ values, cast into the form of a box plot in Figure 4.10C. The $T_m$ distributions show that most of the compounds in the library do not bind to the quadruplex, and do not shift its melting transition temperature. Seven compounds interact most strongly with the quadruplex and shift its $T_m$ by 6 °C or more. The main point of these data is to illustrate the use of DSF to identify compounds that interact with a target of interest, rather than to discuss in detail the compounds that were

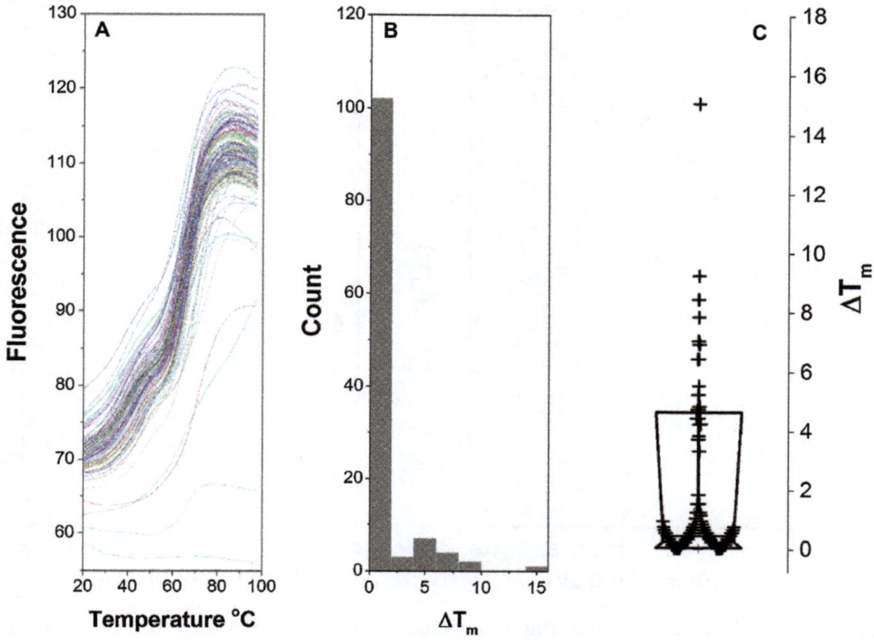


**Figure 4.10** Results obtained from the screening of the NCI Natural Products Set I collection (containing 120 compounds) by DSF. Compound binding to the 'hybrid' human telomere G-quadruplex was measured. (A) Raw primary data from DSF. (B) Distribution of $\Delta T_m$ values. (C) Box plot of the $\Delta T_m$ data from panel B, with the data overlaid. The end lines of the box represent the 10th and 90th percentiles.

discovered, so only a few will be mentioned. The data show that the classic DNA binders daunomycin and echinomycin show significant binding to the hybrid quadruplex, with $T_m$ shifts of 15.1 and 6.9 °C, respectively. DSF reveals a novel quadruplex binding activity for these classic intercalators. Many of the remaining avid binding natural products have unusual chemical scaffolds and no previous known DNA binding activity. As one example from these data dihydroergocristine, an ergot alkaloid, causes a 9.2 °C shift in $T_m$, indicative of strong quadruplex binding.

Apart from use as an efficient primary screen for binding compounds, as illustrated above, DSF is easily adapted for additional exploration of binding interactions once a compound of interest is found. First, as shown in Figure 4.11A, a saturation binding curve can be constructed by measuring the $T_m$ shift as a function of total ligand concentration. By fitting such data to a simple binding hyperbola, an estimate of the binding dissociation constant can be made, $K_d = 21 \pm 2.6$ μM for the data shown. A decided advantage of DSF is that the titration curve shown in Figure 4.11A was obtained using a total of only 320 μl of a 0.2 μM quadruplex solution, far less consumption of material than is normal for spectrophotometric titrations.

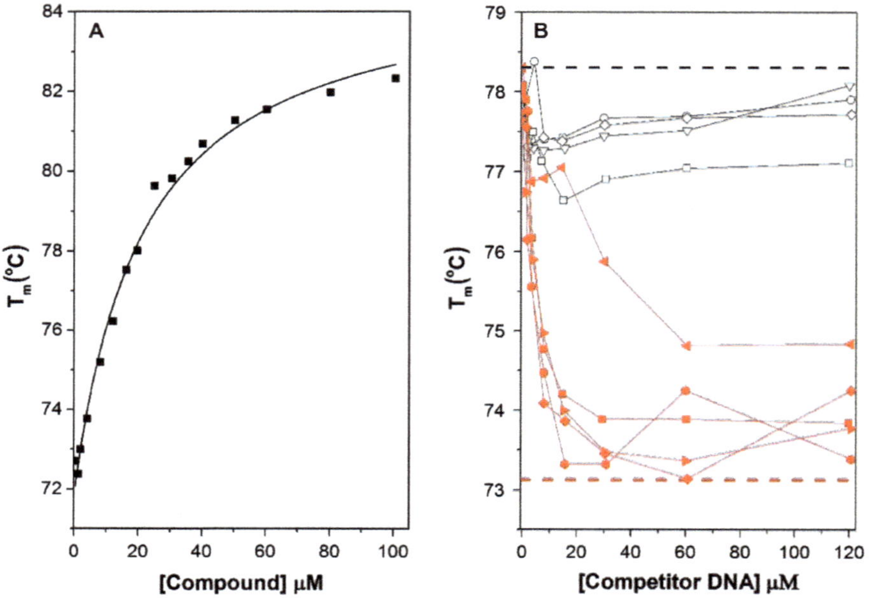


**Figure 4.11** Secondary screening assays using DSF. Compound binding to the FRET-labeled human telomere 'hybrid' quadruplex form. (A) Saturation curve showing the elevation in $T_m$ as a function of ligand concentration. The quadruplex strand concentration is constant at 0.1 μM. The line is the best fit to a simple binding hyperbola with a $K_d = 21 \pm 2.6$ μM and a maximum $\Delta T_m = 13.1 \pm 0.4$ (B) Specificity of ligand binding to the human telomere 'hybrid' quadruplex probed by competition experiments. In these experiments melting of the FRET-labeled quadruplex with bound ligand was measured in the presence of increasing concentrations of unlabeled competitor DNA structures. As ligand is displaced from the labeled quadruplex by binding to the unlabeled competitor structure, the measured $T_m$ will decrease to the value characteristic of the unligated quadruplex.

A second, and perhaps more interesting, application of DSF is to explore the selectivity of the binding interaction, as shown by the competition experiment in Figure 4.11B. In this experiment, ligand (50 μM) is added to a 0.2 μM solution of FRET-labeled hybrid quadruplex, resulting in a $T_m$ shift of 5 °C. If unlabeled competitor DNA is now added, melting of the fluorescent quadruplex can be monitored. The competitor DNA is invisible since it has no fluorescent label. If the added competitor DNA binds the ligand, it sequesters bound ligand from the quadruplex receptor, resulting in a decreased $T_m$ shift. By titrating in increasing amounts of competitor DNA eventually (if the ligand binds to it) all ligand will be displaced from the quadruplex and a $T_m$ equal to that of the unligated quadruplex form will be observed. Almost any unlabeled competitor can be used. In Figure 4.11B a variety of single-stranded, duplex, triplex and alternate quadruplex forms were used to challenge binding to the labeled human telomere hybrid quadruplex.

(The details of the exact sequences and conformations of these competitors do not matter for illustration of this method and will be saved for publication elsewhere.) What the data in Figure 4.11B show is that this particular ligand does not bind appreciably to single-stranded, duplex, or triplex forms (black lines and open symbols) but does bind to alternate quadruplex forms (red lines and filled symbols). This secondary DSF assay thus reveals important information about binding selectivity. This ligand selectively recognizes quadruplex conformations over duplex, triplex and single-stranded forms, but does not distinguish between different types of quadruplexes.

### 4.3.4 Advantages and Prospects

These few examples show that DSF is a valuable tool for the discovery of DNA binders. The primary advantages of DSF are that it allows true high-throughput screening using 96 (and greater) microwell plates and minimizes sample consumption. Disadvantages include the need for sample labeling and the concern that labeling may destabilize the receptor being studied. This latter issue must be addressed in appropriate control experiments. DSF allows all of the powerful thermal denaturation tools described here to be used in an efficient, frugal protocol.

## 4.4 Summary

Thermal denaturation methods utilizing spectrophotometric or calorimetric detection offer fundamentally sound means for quantitatively characterizing drug–DNA binding thermodynamics. The methods are particularly valuable for characterizing the binding interactions of recalcitrant ligands with poor solubility or ultratight binding affinity. In such cases, thermal denaturation may offer the only means for quantitative characterization. Differential scanning fluorimetry provides a platform for using the power of thermal denaturation methods in a high-throughput mode, allowing efficient screening of large libraries of potential DNA interactive compounds.

## Acknowledgements

Supported by grant CA35635 from the National Cancer Institute and GM077422 from the National Institute of General Medical Sciences. Thank to Drs Nichola Garbett and Patricia Raggazon for their helpful comments. Lynn DeLeeuw and Huy Le kindly provided data shown in Figures 4.9–4.11.

## References

1. J. D. Watson and F. H. Crick, *Nature*, 1953, **171**, 737.
2. R. Thomas, *Biochim. Biophys. Acta*, 1954, **14**, 231.
3. P. Doty, *J. Cell. Physiol.*, 1957, **49**, 27.
4. P. Doty and S. A. Rice, *Biochim. Biophys. Acta*, 1955, **16**, 446.

5. J. Marmur and P. Doty, *Nature*, 1959, **183**, 1427.
6. P. Doty, *Harvey Lect.*, 1961, **55**, 103.
7. C. A. Thomas and P. Doty, *J. Am. Chem. Soc.*, 1956, **78**, 1854.
8. P. Doty, *Proc. Natl. Acad. Sci. U. S. A.*, 1956, **42**, 791.
9. S. A. Rice and P. Doty, *J. Am. Chem. Soc.*, 1957, **79**, 3937.
10. P. Ehrlich and P. Doty, *J. Am. Chem. Soc.*, 1958, **80**, 4251.
11. J. Marmur and P. Doty, *J. Mol. Biol.*, 1962, **5**, 109.
12. L. Peller, *J. Phys. Chem.*, 1959, **63**, 1194.
13. J. A. Schellman, *J. Phys. Chem.*, 1958, **62**, 1485.
14. J. M. Sturtevant and P. Geiduschek, *J. Am. Chem. Soc.*, 1958, **80**, 2911.
15. D. M. Crothers, *Biopolymers*, 1971, **10**, 2147.
16. Y. S. Lazurkin, M. D. Frank-Kamenetskii and E. N. Trifonov, *Biopolymers*, 1970, **9**, 1253.
17. J. D. McGhee, *Biopolymers*, 1976, **15**, 1345.
18. C. H. Spink and S. E. Wellman, *Methods Enzymol.*, 2001, **340**, 193.
19. D. Lando, *J. Biomol. Struct. Dyn.*, 1994, **12**, 343.
20. A. S. Benight, *Biopolymers*, 2003, **69**, 406.
21. G. E. Plum, in *Current Protocols in Nucleic Acid Chemistry*, ed. S. L. Beaucage *et al.*, John Wiley & Sons, New York, 2000, pp. 7.3.1–7.3.17.
22. J. D. Puglisi and I. Tinoco, Jr., *Methods Enzymol.*, 1989, **180**, 304.
23. W. D. Wilson, F. Tanious, M. Fernades-Saiz and C. T. Rigl, in *Drug-DNA Interaction Protocols*, ed. K. R. Fox, Humana Press, Totowa, NJ, 1997, vol. 90, pp. 219–240.
24. C. Bailly and J. B. Chaires, *Bioconjug. Chem.*, 1998, **9**, 513.
25. C. Zimmer and U. Wahnert, *Prog. Biophys. Mol. Biol.*, 1986, **47**, 31.
26. S. Glasstone, *Textbook of Physical Chemistry*, 2nd edn. D. Van Nostrand Company, Inc., Princeton, NJ, 1946.
27. I. Haq, T. C. Jenkins, B. Z. Chowdhry, J. Ren and J. B. Chaires, *Methods Enzymol.*, 2000, **323**, 373.
28. K. J. Breslauer, D. P. Remeta, W. Y. Chou, R. Ferrante, J. Curry, D. Zaunczkowski, J. G. Snyder and L. A. Marky, *Proc. Natl. Acad. Sci. U. S. A.*, 1987, **84**, 8922.
29. L. A. Marky and K. J. Breslauer, *Proc. Natl. Acad. Sci. U. S. A.*, 1987, **84**, 4359.
30. L. A. Marky, J. Curry and K. J. Breslauer, *Prog. Clin. Biol. Res.*, 1985, **172B**, 155.
31. F. Leng, W. Priebe and J. B. Chaires, *Biochemistry*, 1998, **37**, 1743.
32. G. G. Hu, X. Shui, F. Leng, W. Priebe, J. B. Chaires and L. D. Williams, *Biochemistry*, 1997, **36**, 5940.
33. J. B. Chaires, F. Leng, T. Przewloka, I. Fokt, Y. H. Ling, R. Perez-Soler and W. Priebe, *J. Med. Chem.*, 1997, **40**, 261.
34. M. J. Waring and L. P. Wakelin, *Nature*, 1974, **252**, 653.
35. F. Arcamone, *Doxorubicin Anticancer Antibiotics*. Academic Press, New York, 1981.
36. *Anthracycline and Anthracenedione-Based Anticancer Agents*, ed. J. W. Lown, Elsevier, Amsterdam, 1988.

37. *Anthracycline Antibiotics: New Analogues, Methods of Delivery and Mechanisms of Action*, ed. W. Priebe, American Chemical Society, Washington, DC, 1995.
38. C. A. Frederick, L. D. Williams, G. Ughetto, G. A. van der Marel, J. H. van Boom, A. Rich and A. H. Wang, *Biochemistry*, 1990, **29**, 2538.
39. D. M. Crothers and H. Metzger, *Immunochemistry*, 1972, **9**, 341.
40. F. G. Loontiens, P. Regenfuss, A. Zechel, L. Dumortier and R. M. Clegg, *Biochemistry*, 1990, **29**, 9029.
41. S. Mansilla, W. Priebe and J. Portugal, *Biochemistry*, 2004, **43**, 7584.
42. J. Portugal, B. Martin, A. Vaquero, N. Ferrer, S. Villamarin and W. Priebe, *Curr. Med. Chem.*, 2001, **8**, 1.
43. B. Martin, A. Vaquero, W. Priebe and J. Portugal, *Nucleic Acids Res.*, 1999, **27**, 3402.
44. J. Portugal, D. J. Cashman, J. O. Trent, N. Ferrer-Miralles, T. Przewloka, I. Fokt, W. Priebe and J. B. Chaires, *J. Med. Chem.*, 2005, **48**, 8209.
45. M. J. Waring, L. P. Wakelin and J. S. Lee, *Biochim. Biophys. Acta*, 1975, **407**, 200.
46. F. Leng, J. B. Chaires and M. J. Waring, *Nucleic Acids Res.*, 2003, **31**, 6191.
47. M. W. Pantoliano, A. W. Rhind and F. R. Salemme, *US Pat.*, 6,020,141, 2000.
48. M. W. Pantoliano, E. C. Petrella, J. D. Kwasnoski, V.S. Lobanov, J. Myslik, E. Graf, T. Carver, E. Asel, B. A. Springer, P. Lane and F.R. Salemme, *J. Biomol. Screening*, 2001, **6**, 429.
49. R. Zhang and F. Monsma, *Curr. Opin. Drug Discovery Dev.*, 2010, **13**, 389.
50. A. Simeonov, *Expert Opin. Drug Discovery*, 2013, **8**, 1071.
51. R. M. Clegg, *Curr. Opin. Biotechnol.*, 1995, **6**, 103.
52. J-L. Mergny, L. Lacroix, M-P. Teulade-Fichou, C. Hounsou, L. Guittat, M. Hoarau, P. B. Arimondo, J-P. Vigneron, J-M. Lehn, J.-F. Riou, T. Garestier and C. Hélène, *Proc. Nat. Acad. Sci. U. S. A.*, 001, **98**, 3062.
53. R. A. Darby, M. Sollogoub, C. McKeen, L. Brown, A. Risitano, N. Brown, C. Barton, T. Brown and K. R. Fox, *Nucleic Acids Res.*, 2002, **30**, e39.
54. A. De Cian, L. Guittat, M. Kaiser and Markus, *et al.*, *Methods*, 2007, **42**, 183.
55. D. Renciuk, J. Zhou and L. Beaurepaire, *et al.*, *Methods*, 2012, **57**, 122.
56. A. De Rache and J-L. Mergny, *Biochimie*, 2015, **115**, 194.
57. V. Plotnikov, A. Rochalski, M. Brandts, J. F. Brandts, S. Williston, V. Frasca and L. N. Lin, *Assay Drug Dev. Technol.*, 2002, **1**, 83.

CHAPTER 5

# Computer Simulations of Drug–DNA Interactions: A Personal Journey

FEDERICO GAGO

Area of Pharmacology, Department of Biomedical Sciences,
University of Alcalá, E-28871 Madrid, Spain
Email: federico.gago@uah.es

## 5.1 Introduction

Deoxyribonucleic acid (DNA), as the cell's repository of genetic information, has been one of the prototypical cancer targets for decades.[1–3] At its lowest level of organization, the three-dimensional structure of this biological macromolecule looks deceptively simple and monotonous: hydrogen-bonded base pairs are stacked along the axis of a right-handed double helix with the sugar–phosphate backbone of each strand on the outside, winding up in antiparallel orientations.[4] As a consequence of this arrangement, the ability of a given stretch of DNA to act as a molecular recognition target appears to be limited, at first sight, to functional group discrimination along the major and minor grooves that lie between the phosphodiester linkages of the two strands.[5] While this is true for many sequence-specific DNA-binding proteins[6,7] and low-molecular-weight ligands (including natural and synthetic information-reading lexitropsins,[8] as well as many clinically used antitumor drugs[9]), it is also accepted nowadays that additional potential for specific recognition is provided by sequence-dependent DNA microheterogeneity.[10,11]

Chemical Biology No. 7
DNA-targeting Molecules as Therapeutic Agents
Edited by Michael J. Waring

Published by the Royal Society of Chemistry, www.rsc.org

The use of models helped Watson and Crick to suggest a structure for the DNA molecule and to hint at a "possible copying mechanism for the genetic material".[12] In the past few years, the way molecular structure is perceived by researchers, educators and students alike has changed dramatically thanks to spectacular advances in structural biology[13,14] and to the availability of high-quality molecular visualization programs from the public domain (*e.g.* RasMol,[15] Swiss-PdbViewer—a.k.a. DeepView—[16], VMD,[17] PyMol,[18] JMol,[19] *etc.*), together with the extremely effective plug-in Chime[20] (largely phased out today and superseded by JmolApplet or JmolViewer[21]), which could be installed on one's favourite internet navigator for free and allowed an easy and highly informative display of both shape and properties of macromolecules. As shown for DNA, the mere visualization of a single molecular structure can help to answer some of the simplest questions regarding function but more complex problems may require the simulation of a process, such as molecular motion or the binding of a ligand. In fact, Sir D. L. Phillips' comment in 1981 that "Brass models of DNA and a variety of proteins dominated the scene and much of the thinking"[22] summarizes the progressive realization that biological molecules are not rigid structures but dynamic systems whose internal motions play important functional roles. The long-range goal of computer-based molecular modelling, as an '*in silico*' approach to biochemistry and pharmacology,[23] is to describe biomolecular interactions in terms of the general laws of chemistry and physics that have been so successfully applied to small molecules. For most practical purposes, the tools of choice to deal with large biological macromolecules are normal mode analysis[24] and classical mechanics in conjunction with molecular dynamics (MD) simulation methods.[25]

By generating hundreds or thousands of time-linked low-energy system configurations of biological macromolecules, MD simulations attempt to capture accurately "the jigglings and wigglings", as Feynman put it,[26] of the atoms that make all living things. MD methods have benefited enormously from the recent convergence of advances in force-field parameters,[27] improved treatment of long-range electrostatic interactions[28] and greatly boosted computer power, not only because of the fulfilment of Moore's law ("CPU speed usually increases by a factor of 2 or so every 18 months") but also due to implementation of software code on graphics processing units (GPUs). Thus, it is now feasible for many research groups to obtain, in a reasonable time, simulated trajectories of solvated macromolecular systems over tens or hundreds of nanoseconds. The concatenation of successively generated 'snapshots', each containing a set of Cartesian coordinates, allows the creation of 'molecular movies' and various analyses of molecular motions in atomic detail. The information derived from these simulations can be used to (i) extract structural and energetic information that is usually beyond current experimental possibilities,[29] (ii) provide independent accounts of experimentally observed behaviour,[30] (iii) help in the interpretation of biochemical or pharmacological results[31] and (iv) open new avenues for research by posing novel relevant questions that can guide the design of new experiments.

In the DNA field, current equilibration and sampling times usually range from a few nanoseconds[32–34] to tens,[35–37] hundreds[38] or thousands of nanoseconds.[39] Although these times are still not sufficiently long[40,41] to monitor large-scale structural changes (in the absence of distance or angular restraints) or to achieve convergence of ion distributions,[34,36,42] relaxation times for helicoidal parameters have been shown to be on the order of 0.5 ns.[36] Thus, trajectories lasting a few nanoseconds can provide adequate dynamical descriptions of DNA and DNA–ligand complexes because the characteristic structural parameters (*e.g.* roll, twist and tilt) settle much more rapidly than ion occupancies.[39]

The initial feeling that theoretical calculations were a waste of time (not least because of the near impossibility of treating such complex systems as macromolecules in living cells with the level of detail that was thought to be necessary to add something of importance to our knowledge[43]) has given way, little by little, to a renewed wave of confidence in simulation methodologies. Ongoing improvements in molecular mechanics (MM) force fields,[44,45] the almost systematic incorporation of solvent molecules and counterions into the models,[46,47] and a reliable treatment of long-range electrostatics by applying particle mesh Ewald methods to cubic and octahedral solvent 'boxes'[48] are making it possible to simulate the dynamics of relatively large systems (including such highly charged ones as nucleic acids[44,49,50]) over considerably longer times, sometimes approaching the millisecond time scale,[51,52] without the instabilities that plagued earlier simulations. Moreover, use of the computationally much faster Generalized Born (GB) model of implicit solvation[53] allows the exclusion of explicit solvent molecules[54] and extends the length of some of these MD simulations even further, as does the use of coarse-grained representations of biomolecules.[55] These methodologies are now well suited to study conformational changes that are coupled to function[56] and, in conjunction with quantum mechanical (QM) methods,[57] the course of reactions in which covalent bonds are broken and formed, such as those taking place in the active sites of enzymes[58] or between some drugs and DNA (see below).

A number of tools also exist that can 'pump' some extra energy into the system to accelerate the crossing of local barriers ('targeted' or 'steered' MD) thus allowing the study of processes that would normally occur too slowly in a standard MD run based only on random thermal fluctuations.[59–61] The repertoire of macromolecules relevant to cancer that can be studied by these methods has also expanded enormously in the last 20 years, from just a handful of 'classical' targets (*e.g.* short DNA oligonucleotides, dihydrofolate reductase or thymidylate synthase) to DNA–protein complexes,[62] RNA[63] and whole protein families (*e.g.* growth factor receptors,[64] kinases,[65] and phosphatases[66]), both alone and in complex with ligands.

Researchers arguably have on their laptops and tablets today more computing power than was available on most mainframe computers when MD simulations of macromolecular systems started at the end of the 1970s.[67] Moreover, a realistic alternative to large-scale 'supercomputers' these days is to use massively grid-distributed computing[68] to harness the power of dozens or hundreds of relatively inexpensive personal computers (PCs) toward a common goal such as the folding of a protein[69] or the docking of potential ligands into a protein binding-site.[70] MD simulations in explicit solvent are also massively accelerated when run on single or multiple GPUs.

Spectacular advances in the molecular biology of DNA-binding proteins and the concurrent crystallographic studies of binary and ternary complexes involving drugs, DNA and associated proteins are progressively narrowing the gap between our perception of the structural aspects of drug–DNA complexes and our limited understanding of how these drugs really work in living cells. In the following paragraphs, I will provide a succinct overview of some of these advances in relation to my own research since I was a postdoctoral student in Graham Richards' lab to the present day.

My introduction to molecular modelling and computer simulations originated at the Physical Chemistry Laboratory of Oxford University after completing my doctorate in Madrid in the field of hydrophobic parameters derived from liquid chromatography measurements. Professor Richards was well known in Spain, mostly for his book Quantum Pharmacology, and I had the privilege of listening to one of his talks when both of us coincided in Barcelona on the occasion of a seminar organized by Miguel Martín at Universitat Autònoma. Incidentally, Corwin Hansch also attended that meeting and projected a fascinating video showing atomistic details of the docking of an inhibitor in the active site of an enzyme, possibly alcohol dehydrogenase. I was so struck by the power of these computational methods and the intrinsic beauty of molecular graphics that I decided to try my luck adventuring as a novice in this emerging field. The paper I was given to start with when I was accepted to join Graham's group in April 1987 was one in which the authors reported on the differential binding of some natural and synthetic compounds to DNA tracts rich in adenine and thymine in preference to regions rich in guanines and cytosines.[71] Could theoretical methods shed some light on this issue?

## 5.2 Minor Groove DNA Binders

Netropsin and distamycin (Figure 5.1) are well-characterized DNA minor-groove-binding ligands that have been used extensively as models for the study of drug–DNA interactions in a variety of contexts.[68] Both antibiotics were among the set of small molecules whose discriminatory ability for different DNAs had been evaluated using an ethidium displacement assay that allowed an indirect estimation of drug–DNA association constants.[68] For netropsin, additional spectroscopic and thermodynamic data[72] showed that the free energies for netropsin binding to both poly[d(AT)] · poly[d(AT)] and poly(dA) · poly(dT) were very similar and about 4 kcal $mol^{-1}$ more favourable than the free energy of binding to poly[d(GC)] · poly[d(GC)].

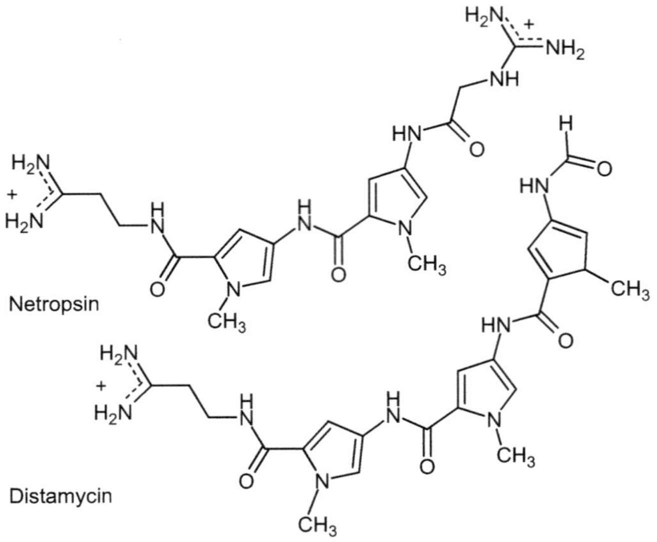


**Figure 5.1** Chemical structures of the natural DNA minor-groove binders, netropsin (top) and distamycin (bottom).

We were aware that an accurate computation of binding free energies is hampered by uncertainties regarding both the particular conformation of the interacting partners in the unbound state and the validity of the energy partitioning schemes, as well as limitations with respect to some energy contributions that are not easily amenable to calculation (*e.g.* hydrophobic and entropic effects). For this reason, we initially focussed our efforts on the binding of these and other compounds (including the dye Hoechst 33258) to dodecanucleotides containing alternating runs of AT and GC. By using a MM force field and performing energy minimization and decomposition, we showed that the differences in interaction energies between the drugs and the two DNA sequences studied could be satisfactorily correlated with the selectivity ratio provided by experiment.[73] Furthermore, when we replaced the pyrrole rings in netropsin and distamycin with imidazole, we could also observe an increased preference for the alternating runs of Gs and Cs due to the new hydrogen bonds between the imidazole nitrogens of these 'lexitropsins' and the amino groups of guanines, although no significant loss of affinity was detected for ATAT stretches. We therefore supported the view that, to have molecules with the potential to bind to predetermined sequences in a highly specific manner, not only 'G,C-reading' but also 'A,T-rejecting' elements had to be incorporated into the ligands. In this respect, a structural factor that we considered relevant to the specificity of the interaction, in agreement with other authors, was the different widths of the minor groove permissible or attainable in the two types of complexes. The same year this work was published, the unanticipated discovery of the first 2 : 1 polyamide:DNA complex paved the way to design strategies that used imidazole in place of pyrrole rings to 'read' the exocyclic amino group of guanine very effectively.[74] Later results suggested that the language of design motifs for polyamide sequence recognition should be extended to include the use of 'words' for recognizing two adjacent base pairs, rather than 'letters' for binding to single base pairs.[75] All in all, this field of molecular recognition of DNA from the perspective of the minor groove is a beautiful example of maturation from serendipity to successful design at the interface of chemistry, biology and (possibly) human medicine.[76]

The major problem with our estimates of binding energies for netropsin and the other drugs was that they did not really capture the essence of the binding process, that is, the difference in energy between drug molecules in solution, interacting with water and ions, and drug molecules in the drug–DNA complex. In fact, computer simulations (now and then) are not capable of calculating this difference directly, especially when large molecules are at play. The difficulty lies in the fact that the binding process should be simulated slowly enough to achieve thermodynamic equilibrium and this involves the generation of a huge number of representative configurations of the system along the reaction coordinate, which is impractical in most situations. However, the non-physical process of perturbing one of the reactants into another (which has been dubbed 'computational alchemy'[77]), both in the free state and in the bound state, is more easily amenable to computation[78,79] and provides an accurate relative free energy change ($\Delta G$) that can be related to the physically relevant one through a thermodynamic cycle (Figure 5.2).

Stimulated by impressive results produced in-house by Graham and his senior postdoctoral student Chris Reynolds in the calculation of redox potentials,[80] and inspired by the illuminating paper on the thermodynamics

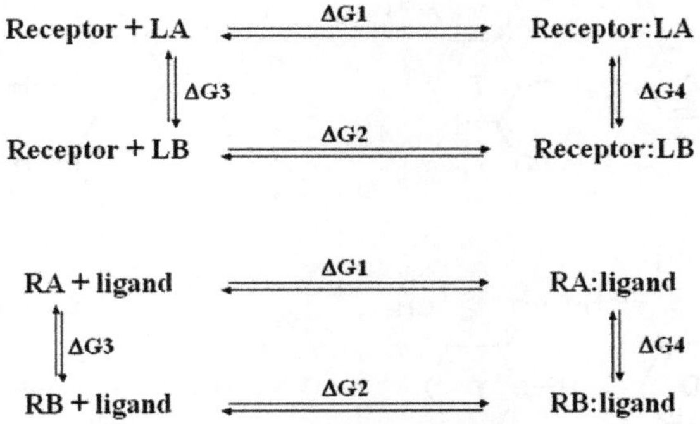


**Figure 5.2** Examples of typical thermodynamic cycles used in free energy perturbation simulations. Top: a ligand A is converted into a ligand B, both binding to a common receptor. Bottom: the same ligand binds with different affinities to receptors A and B. The free energy differences calculated experimentally ($\Delta\Delta G_{\text{binding}} = \Delta G2 - \Delta G1$) must be the same as those calculated from the nonphysical perturbations of A into B ($\Delta\Delta G_{\text{binding}} = \Delta G4 - \Delta G3$).

of netropsin binding to several polynucleotides of defined sequence {which also contained data for poly[d(IC)]·poly[d(IC)]},[69] I realized that we could attempt to apply the same methodology to the problem of DNA–ligand binding. Why not 'mutate' inosines (Figure 5.3) to guanosines (and the reverse) in both the free DNA and its complex with netropsin to assess the effect on the calculated binding energies of growing/removing the exocyclic amino group of guanine? Despite the enormity of the task ahead of us and the lack of precedent for a work of this sort, I was encouraged to go ahead. An advantage over simulations of ligand binding to proteins, which generally need to address substantial molecular reorganization that can span long time scales, was that the conformational changes involved in the alchemical changes studied here were relatively minor, as they did not significantly affect the double-helical structure of the DNA. We managed to borrow computer time from every single machine we could lay our hands on, and the work was completed in little more than one year.

The 'mutation' of the inosine C2 hydrogen into an amino group in the ICIC sequence was made possible with the aid of two 'dummy atoms' (D), which were characterized as having both the point charge and the nonbonded energy parameter set to zero. During the simulations of free and netropsin-bound DNA the parameters for each $HD_2$ group were linearly increased in small increments so that at the end of the perturbation they attained the values corresponding to the standard $NH_2$ group. By doing this, we realized that the DNA minor groove became progressively less deep, more polar and less narrow. In its complex with netropsin, these changes were accompanied by the extrusion of part of the drug molecule into the solvent and a decrease in binding free energy that was in very good agreement with

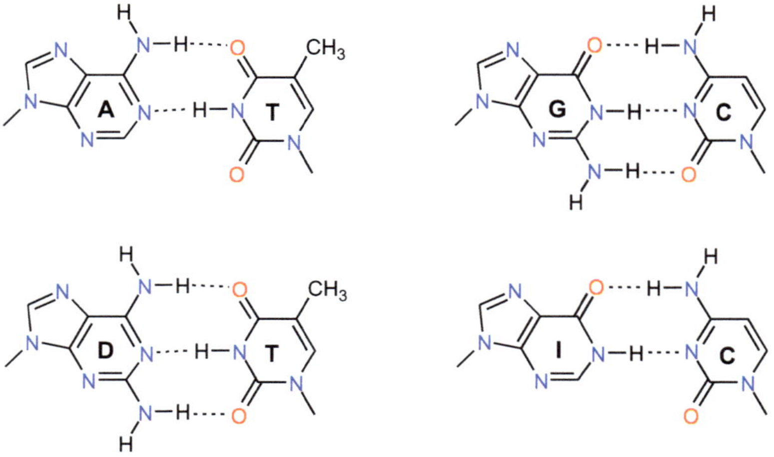


**Figure 5.3** Chemical structures of natural and modified DNA base pairs in the Watson–Crick arrangement. Hydrogen bonds are depicted as broken lines. D stands for 2,6-diaminopurine whereas I stands for the hypoxanthine-containing nucleoside, inosine.

the experimental value.[81] The calculations, therefore, were able to reproduce the observed preference of netropsin for binding to ICIC over GCGC in a quantitative manner while at the same time providing a good rationale for the specificity of the association that pointed, once again, to minor groove width as an important determinant of complex stability.

## 5.3 Natural Bifunctional Intercalators and Hoogsteen Base Pairing

Having first studied ligands with a tendency to bind to AT-rich DNA better than to GC-containing sequences, on my return to Alcalá I became interested in intercalators, which mostly get sandwiched at CpG steps.[82] The process of DNA intercalation allows the insertion of a planar ligand between successive base pairs by increasing their vertical separation (rise) and changing their relative degree of rotation (twist angle) through the introduction of crankshaft motions in the sugar–phosphate backbone.[4] Interest in bifunctional intercalators stems not only from the possibility of enhancing their binding affinities over those of the corresponding monomers but also from the greater opportunities for imposing selective binding to defined sequences that are afforded by the bracketing of two (or possibly more?) base pairs between the intercalator sites. Thus, the size of the site occupied by the ligand increases, *i.e.* for a binding site covering four base pairs, the number of distinguishable sequences is 136 *versus* only 10 unique dinucleotide steps at which monointercalation can take place. The design problem here is to select a suitable linker with the ability to recognize functional groups in either the minor or the major groove. As is usually the case, Nature anticipated by providing us with some extremely elegant and very interesting examples of bifunctional intercalators in the family of quinoxaline antibiotics represented by echinomycin (a.k.a. quinomycin A) and its biosynthetic precursor, triostin A. These agents are primarily produced by *Streptomyces echinatus* and *S. triostinicus*, respectively, and for them a binding site size of four base pairs was early demonstrated and then repeatedly confirmed.[83] These natural bis-intercalators were shown to have a definite preference for binding to a 5′-CpG-3′ core flanked by an A–T base-pair on either side, as opposed to the natural crescent-shaped nonintercalating ligands mentioned above (*e.g.* netropsin and distamycin), which show a strong preference for binding to the minor groove of A–T-rich DNA regions. Rather strikingly, a sequence-dependent rearrangement in the base pairs adjacent to the CpG core was manifest in cases where the base located on the 5′ side of the CpG binding step was a purine: this base rotates 180° about the glycosidic bond to adopt a *syn* orientation relative to the sugar but remains hydrogen-bonded to the opposing thymine making use of a so-called Hoogsteen scheme (Figure 5.4), as first reported by a German investigator bearing this name when he solved the X-ray crystal structure of the 1 : 1 complex of 1-methylthymine with 9-methyladenine.[84] Intriguingly, this base-pairing

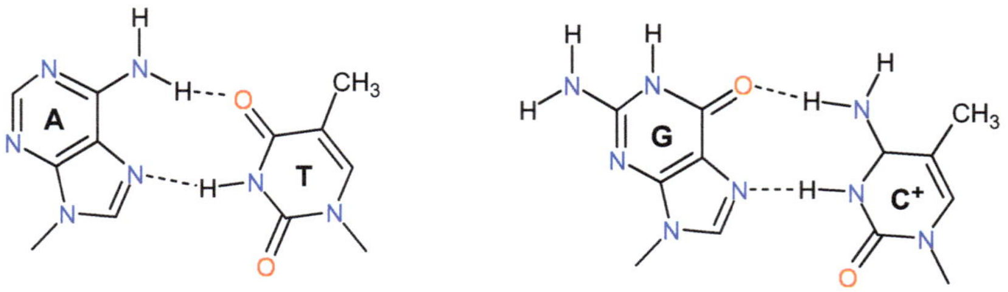


**Figure 5.4** Adenine–thymine (left) and guanine–cytosine base pairs (right) in the Hoogsteen configuration.

arrangement was detected in d(ACGT)$_2$, d(GCGC)$_2$ (at low pH), d(CGTACG)$_2$, d(ACGTACGT)$_2$, d(ACGTATACGT)$_2$, and d(GCGTACGC)$_2$ (again at acidic pH) but not in d(AAACGTTT)$_2$, d(TCGA)$_2$, d(CCGG)$_2$ and d(TCGATCGA)$_2$.

Curious about this incompletely understood phenomenon, I proposed that my first and highly motivated Ph.D. student, José Gallego, should undertake a theoretical investigation of the factors that might be determining these conformational preferences in the base pairs flanking both sides of the CpG binding step. Because Peter Kollman and co-workers had previously focused on the relative stability of the d(CGTACG)$_2$ : (triostin A)$_2$ complex with the central AT base pairs in either Hoogsteen (as found in a crystal structure reported in *Science* magazine in 1984[85]) or Watson–Crick (WC) pairing using MM,[86] we chose to study the complexes between echinomycin and the DNA tetramers d(ACGT)$_2$ and d(TCGA)$_2$, for which conformational information was available from nuclear magnetic resonance (NMR) spectroscopy.[87] Our research tried to address two main issues: (i) why the terminal bases in the d(ACGT)$_2$ complex (hereafter termed $A_H$) appeared to be Hoogsteen paired while those in the d(TCGA)$_2$ one ($T_W$) did not, and (ii) why the former complex was more stable than the latter. To this end we analysed the behaviour of the d(ACGT)$_2$–echinomycin and d(TCGA)$_2$–echinomycin complexes in which both terminal AT base pairs were in either Hoogsteen ($A_H$ and $T_H$) or WC conformation ($A_W$ and $T_W$) by means of four independent MD simulations in aqueous solution each lasting 40 ps (a considerable time in those days for our modest computational resources!).[88]

None of the complexes underwent any large conformational changes with respect to the corresponding initial structure during the simulations in water and the DNA bases remained paired, except for some fraying effects that were more noticeable in the $A_W$ complex. The conformations for the experimentally found complexes, $A_H$ and $T_W$, were in qualitative agreement with the Nuclear Overhauser Effects (NOE) detected in the NMR spectra. We considered and discussed the three major types of interaction that echinomycin establishes with the DNA molecules separately: (i) van der Waals and electrostatic interactions between the depsipeptide part of the drug and the minor groove of the DNA molecules, (ii) hydrogen bonds between the alanine residues of echinomycin and the guanine bases of DNA and (iii) stacking interactions between the quinoxaline chromophores and the adjacent base pairs. In agreement with NMR and X-ray results, the $A_H$ model was found to be stabilized by two strong hydrogen bonds between the amino groups of alanines and N3 of guanines and two more between the carbonyl groups of alanines and the 2-amino groups of guanines, one of which was only slightly weaker due to the asymmetrical binding of echinomycin. In the other three models, one of the hydrogen bonds between the alanine residue and the drug was lost. Although these results argued in favour of the greatest stabilization taking place in the $A_H$ complex they provided no clues about the origin of the WC→Hoogsteen rearrangement.

We then focused our attention on the quinoxaline residues that, per our calculations, were contributing almost 50% of the total interaction energy between echinomycin and the DNA tetramers. We performed a detailed analysis of these interactions in the four models studied because it was generally accepted that the flanking sequence specificity of this antibiotic rested largely on the aromatic stacking interactions of its quinoxaline chromophores[80,89] and also because the dependence of DNA stacking interactions on the *composition* and *orientation* of the nucleic acid bases had been shown to be largely dominated by the electrostatic term.[90] We considered the entire quinoxaline-2-carboxamide coplanar and conjugated system as the drug's molecular entity participating in the stacking interactions rather than solely the quinoxaline rings. We then plotted the dipole moment of this intercalating moiety as the simplest way to represent the polarity of the charge distribution and did the same for each of the adjacent base pairs making up the intercalation site. Because the sum of charges must be zero for the dipole moment to be independent of the origin of coordinates, C1′ atoms of DNA and $C_\alpha$ atoms of serines were included in the calculations as buffers so that each of these systems achieved electrical neutrality.

We made three interesting observations relevant to the problem in hand: (i) in the Hoogsteen arrangement, the dipole moment of an AT base pair increases to more than 5 Debye from the value of about 2 Debye found in the WC conformation, (ii) the quinoxaline-2-carboxamide chromophores of echinomycin have a high dipole moment of 4–5 Debye, with the negative pole located on the pyrazine-2-carboxamide region and the positive one placed on the benzene ring region, and (iii) in the experimentally found $A_H$ complex, the high dipole moment of each AT base pair in a Hoogsteen conformation and that of the stacked echinomycin chromophore have opposite directions, which gives rise to a favourable dipolar interaction; in contrast, the dipole moment of the AT base pair reverses its orientation in the non-experimental Hoogsteen complex $T_H$, so that both dipole vectors point in the same direction and the ensuing dipolar interaction is unfavourable.

Taken together, our interpretation of these results[85] was that a Hoogsteen base-pair scheme should be preferred for binding of echinomycin to the $d(ACGT)_2$ tetramer, whereas the classical WC hydrogen bonding arrangement would be more stable for the TCGA sequence. It was encouraging to see that this was precisely what had been detected experimentally.

When GC rather than AT pairs flank the echinomycin CpG binding step, Hoogsteen base-pairing is likewise observed (at acidic pH) only when the purine base (guanine) is on the 5′ side of the CpG sites, *i.e.* it is detected in $d(GCGC)_2$ and $d(GCGTACGC)_2$,[91,92] but not in $d(CCGG)_2$ regardless of the pH.[93] The need for a low pH is justified because the Hoogsteen hydrogen-bonding scheme in GC pairs requires the protonation of the cytosine base (Figure 5.4).

In an attempt to rationalize these observations, we extended our previous work by modelling and studying, by means of MD simulations in aqueous solution, the complexes of echinomycin with $d(GCGC)_2$ and $d(CCGG)_2$ in which both terminal G–C base pairs adopt either a Hoogsteen (named $G_H$ and $C_H$, respectively) or a WC conformation ($G_W$ and $C_W$). In this case, we found that the conformations of the experimentally detected $G_H$ and $C_W$ complexes were not significantly altered during the course of the 70-ps simulation, as assessed by the low root-mean-square deviation (rmsd) from the initial structure, whereas comparatively larger rmsd values were observed for $C_H$ and $G_W$.[94] Additional calculations also predicted a large increase in the interaction energy between echinomycin and the $d(GCGC)_2$ duplex upon protonation of the terminal cytosines and subsequent Hoogsteen pair formation. In agreement with previous work,[83,85] we found that about 50 per cent of the total interaction energy in all the complexes appeared to be contributed by the van der Waals interactions involving the quinoxaline ring systems alone. This term, however, decreased when the terminal guanine bases in $d(CCGG)_2$ adopted a Hoogsteen conformation. In contrast, in the $G_H$ complex a marked gain in electrostatic stacking interaction energy was apparent between the chromophores of echinomycin and the terminal Hoogsteen-paired bases. We construed that the combination of these two factors could account for the experimental observation that at low pH Hoogsteen base pair formation is only observed for the echinomycin–$d(GCGC)_2$ complex.[84]

All in all, our findings confirmed early theoretical evidence obtained for stacked DNA bases showing that the electrostatic contribution plays a decisive role in the stabilization of the stacking interaction[95] and that this term is very sensitive to the relative orientation of the planar systems involved.[96] In our view, the conformational changes brought about by the binding of echinomycin and triostin A to short oligonucleotides provided an excellent test case for probing the nature and dependence of stacking interactions in DNA–drug complexes using theoretical methods. But could this methodology be extended to the issue of binding selectivity?

## 5.4 Bis-intercalation of Echinomycin and Related Bifunctional Agents in Relation to Binding Sequence Preferences

The origin of the selectivity of triostin A and echinomycin for binding to CpG steps was widely accepted to arise mostly from the hydrogen bonds between the depsipeptide part of these antibiotics and the exocyclic amino group of guanine in the minor groove.[80,82] Traditional work in structure–affinity relationships (SAR) for the quinoxaline family of antibiotics dealt with the effects that introduction of new substituents or removal of existing ones had on the binding properties of a given drug, as assessed mainly by DNA footprinting experiments. The best known examples are probably those provided by des-*N*-tetramethyl triostin A (TANDEM) and [*N*-MeCys$^3$, *N*-MeCys$^7$] TANDEM (CysMeTANDEM), triostin A analogues lacking either all or half of the *N*-methyl groups of cysteines and valines, respectively, that bind better to a central TpA core. This change could also be rationalized as being due to the lack of suitably placed hydrogen bonding functionalities in the modified depsipeptide that could interact with the 2-amino group of guanine. An elegant complementary SAR approach later consisted of changing the nature of the bases making up the DNA target site (Figure 5.3) rather than that of the ligand. This was accomplished in Waring's lab in a series of unequivocal experiments when inosine(I) took the place of guanosine in the DNA molecule (which is tantamount to removing the exocyclic amino group selectively from the minor groove) and when 2,6-diaminopurine (a.k.a. 2-aminoadenine) was used in place of adenine to pair with thymine (which gives rise to a DNA containing an amino group at every step). As expected, echinomycin did not bind to CpI steps, lending further credence to the crucial role played by this group in the interaction of this ligand with the DNA minor groove.[97] Introduction of the extra amino group in the minor groove of DT regions, in contrast, led to a redistribution of binding sites relative to normal DNA, and echinomycin was shown to bind to any pyrimidine–purine combination other than the usual CpG step.[98] Strikingly enough, not only was the selectivity drastically changed but also the affinity of echinomycin for these new binding sites in the modified DNA turned out to be at least one order of magnitude greater than that for normal DNA.

Because our previous studies had highlighted an unfavourable electrostatic interaction between the quinoxaline-2-carboxamide system of echinomycin and the base pairs making up the central CpG step,[85,91] we reasoned that not only hydrogen bonding but also stacking interactions might be playing a role in determining the binding preferences of echinomycin. To test this hypothesis, we generated 13 molecular models of the 1:1 complexes of echinomycin with standard and modified DNA hexamers and evaluated the different contributions to their relative stability.[99] The whole set of sequences contained every combination of binding sites experimentally probed by echinomycin. The aim was to get information both on the stacking interactions between the quinoxaline-2-carboxamide system and any DNA base pair and on all the possible hydrogen bonding arrangements between the depsipeptide and the DNA atoms in the minor groove.

The energy analysis revealed that the interactions involving the alanine residues and the quinoxaline-2-carboxamide chromophores of echinomycin could by themselves account for the observed binding selectivities. Furthermore, by plotting these energy components *versus* the total interaction energy a good linear correlation was detected and the different hexamers appeared to be clustered in two distinct subsets. One subset encompassed a family of good binding sites that presented both a central dinucleotide step endowed with full hydrogen bonding capabilities in the minor groove and an arrangement of base pairs that gave rise to an overall favourable stacking interaction with the antibiotic's chromophores. The sequences in the other subset shared poorer hydrogen bonding possibilities and overall weaker interactions with the quinoxaline-2-carboxamide moieties. The most favourable binding sites were indeed those in which the adenines in the DNA had been replaced with 2,6-diaminopurine (D), and most notably the sequence DTDT, in good agreement with the experimental observations.[95]

It was particularly illustrative to view the footprinting results in the light of calculations that decomposed the binding enthalpy due to stacking interactions with both the sandwiched base pairs and those flanking the bis-intercalation site into van der Waals and electrostatic components. By first examining those complexes with a common central CpG step, it was possible to see that the minor differences between them were, as expected, restricted to interactions with the flanking bases. Nonetheless, the electrostatic component appeared to be more discriminating than the corresponding van der Waals term, as most clearly illustrated by CCGG (apparently, a weaker binding site for echinomycin than ACGT) for which this electrostatic component was found to be slightly repulsive. The fact that the electrostatic interaction between the drug's chromophores and the central CpG step was calculated to be unfavourable strongly suggested to us that this negative effect had to be outweighed by the very favourable electrostatic and hydrogen bonding interactions established with the minor groove. Furthermore, it was indicative that the different charge distribution of the base pairs in a TpD step (which also presents two 2-amino groups in the minor groove) could give rise to an attractive electrostatic stacking interaction with the quinoxaline-2-carboxamide system. This turned out to be the case, and the improved calculated binding energy was in nice accord with the larger experimental association constant. The results rationalizing the enhanced affinity of echinomycin for DTDT over ACGT were reinforced by the interaction energies obtained for 'mixed' sequences for which values half way between those in each of these two complexes were found. The fact that in the modified DNA employed in the experiments the originally protected CpG sites were adjacent to these D-containing high-affinity sites could explain why binding to the canonical CpG steps was precluded.

Directed biosynthesis has produced a number of analogues of echinomycin and triostin containing substituted chromophores. Prominent among these is the bis-quinoline derivative of echinomycin designated 2QN,[100] for which an interesting correlation between spectroscopic and thermodynamic properties had been found.[101] This fact made researchers think that the intercalated ring systems could largely control the sequence selectivity indeed. The footprinting results, however, were not significantly different from those obtained with echinomycin. Thus, the overriding determinant of preferred binding sites for both drugs was the 2-amino group exposed in the minor groove of the DNA double helix and both drugs bound to sequences surrounding TpD steps with much higher affinity than to standard CpG steps. In line with previous calculations, we detected greater stabilization in a 2QN–d(GDTDTC)$_2$ modelled complex containing a central TpD step than in the equivalent 2QN–d(GACGTC)$_2$ complex containing a CpG step.[97] We calculated and displayed the molecular electrostatic potential (MEP) to aid in the visualization of the origin of this difference. The MEPs of both 2QN and echinomycin revealed a distinctive pattern that was considered relevant to DNA binding, namely, a positive region surrounding the intercalating rings and a negative region (which emanates from the sulfur atoms and most of the depsipeptide carbonyl groups) on the opposite side of the molecule (Figure 5.5). The resulting electrostatic asymmetry was deemed to assist the

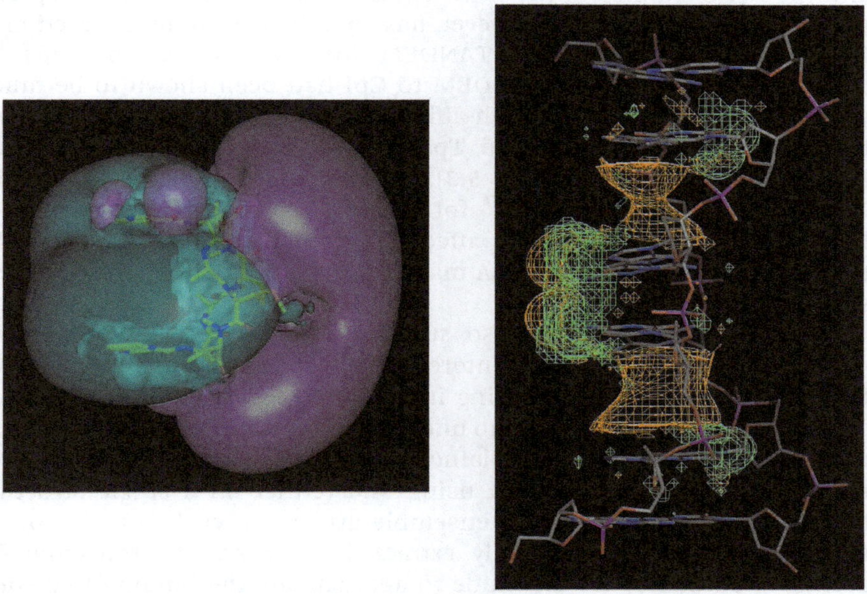

**Figure 5.5** Left: Molecular electrostatic potential (MEP) calculated for echinomycin and displayed as solid contours coloured in pink and cyan for negative and positive regions, respectively. Right: MEP difference map for d(GACGTC)$_2$ and d(GDTDTC)$_2$ in an aqueous medium. The MEP calculated for the latter sequence was subtracted from that calculated for the former to highlight those regions where the electrostatic potential differs between the two DNA molecules. The mesh contours in orange and green delineate those regions where this difference is $\geq 0.5$ or $\leq -0.5$ units, respectively (see text for details).

productive approximation and correct orientation of this type of ligand with respect to the DNA molecule prior to intercalation, possibly compensating for the lack of a net positive charge. In addition, comparison of the MEPs calculated for the two DNA hexanucleotides in the complexed state using a subtractive method revealed substantial differences not only in the major groove, as expected from the reversal in the positions of the O and $NH_2$ groups located in it (Figure 5.3), but also in the spaces between the base pairs that furnish the intercalation cavities and are occupied by the chromophores of the drug upon binding (Figure 5.5). We suggested that these differences could modulate the DNA-binding specificity of echinomycin, 2QN and related analogues acting in concert with the well-established hydrogen bonding interactions involving the exocyclic amino groups in the minor groove.

The specificity of TANDEM and CysMeTANDEM for TpA steps in place of the CpG steps recognized by triostin A, echinomycin and 2QN in standard DNA was initially rationalized as originating from the loss of hydrogen bonding possibilities with the 2-amino group of guanines due to formation of two intramolecular hydrogen bonds between the CO of alanines and the NH of valines in the demethylated analogues. The G→I replacement in the DNA molecule, however, has no influence on the interaction of TANDEM with TpA steps (*i.e.* TANDEM does not bind to CpI steps),[102] and the binding of CysMeTANDEM to CpI had been shown to be much weaker than binding to the high-affinity TpA sites[103] despite the fact that the minor grooves of CpI and TpA steps provide the same hydrogen bonding arrangements (Figure 5.3). Furthermore, the affinity of CysMeTANDEM[80,104] and TANDEM[80] for binding TpA-containing sequences had been shown to be clearly affected by the flanking bases, with the tetranucleotides ATAT and TTAA making up the best and poorest binding sites, respectively.

Since these observations also supported the view that the binding specificity of this type of bis-intercalators depends on some additional factors beyond hydrogen bonding interactions between the depsipeptide and the minor groove, we tried to understand their origin using theoretical methods. The problem, as explained in Section 5.2, is that ligand–DNA interaction energies calculated using MM (either on a single structure that is taken to represent the ensemble average of each complex or on multiple snapshots periodically extracted along an MD trajectory for averaging purposes) are too crude to approximate the binding free energies that truly relate to the association constants determined experimentally. Although this relatively simple approach had shown its merits in detecting trends[96] and can be complemented with continuum methods that consider the desolvation effects that oppose the favourable electrostatic interactions,[105] it neglects entropic contributions as well as the changes in internal energy undergone by both the drug and the DNA molecule upon binding. For the ligands, these changes are generally assumed to be small and of similar magnitude for all the complexes considered, but the situation can be rather different for the DNA molecules due to the sequence-dependent microheterogeneity mentioned in the introduction. For example, it is conceivable that intercalation at certain sequences might be favoured simply because of the inherent tendency of some dinucleotide steps to underwind and roll.[120]

To further test the hypothesis that stacking interactions can play a decisive role in modulating the preferential binding of these bis-intercalating ligands, the computational tools of choice are the free energy perturbation and thermodynamic integration methods.[75,76] However, these studies were not undertaken in my lab until another highly talented Ph.D. student, Esther Marco, joined my group and was later assisted by a new enthusiastic recruit, Ana Negri. We then managed, with the inestimable help of my colleague and friend F. Javi Luque, to complete several thermodynamic cycles in our attempt to understand the origins of both the remarkably increased affinity of 2QN for DTDT relative to GCGC sites and the notable loss of binding affinity between CysMeTANDEM and ICIC compared with ATAT. These studies encompassed the perturbation of whole base pairs in normal DNA and inosine- or 2,6-diaminopurine-containing DNA. Besides, the calculations were performed in such a way that the contributions of the base pairs on either side of the intercalated chromophores could be separately assessed.[106] The calculated differences in binding free energy for the two DNA bis-intercalating agents, 2QN and CysMeTANDEM, were found to be in very good qualitative agreement with the experimental data. Thus, binding of 2QN to $d(GCGGCGCCGC)_2$ was disfavoured over binding to $d(GCGDTDTCGC)_2$ by 1.8 kcal mol$^{-1}$, in accordance with the ~100-fold increase in affinity for 2QN binding to a DTDT site compared with a GCGC site in DNA[97] whereas binding of CysMeTANDEM to $d(CTCICICCAG)_2$ appeared to be even more markedly disfavoured (by 4.4 kcal mol$^{-1}$) over binding to $d(CTCATATCAG)_2$, in consonance with the fact that this ligand binds with very high affinity to TpA steps but does not significantly bind to CpI steps.[100] Furthermore, our calculations clearly pointed to the electrostatic contribution to the stacking interactions involving the distal bases as the critical determinant of binding selectivity for these two bis-intercalating ligands. This finding reinforces the view that the recognition site for these bis-intercalators definitely extends beyond the central dinucleotide step to cover four base pairs extended to sulfur-rich thicoraline, whose ability to stabilize the double helical structure of several custom-designed, fluorophore-labelled oligodeoxynucleotides ('molecular beacons') was studied using standard real-time polymerase chain reaction (PCR) instrumentation.[107]

In conclusion, although the hydrogen bonds between the depsipeptide of the bis-intercalating drug and the functional groups present in the DNA minor groove need to be correct (because there is a penalty when they are missing or wrong), the selectivity patterns can be adequately explained only when the stacking interactions emanating from the base pairs that make up the intercalating site on both sides of the chromophore are also taken into account.

## 5.5 Binding Preferences of Synthetic Pyridocarbazole Bis-intercalators

Ditercalinium and Flexi-Di are the most studied representatives of a class of synthetic bis-intercalators that, in contrast to the natural bis-intercalators just described, bind to DNA from the major groove. Both ligands are based on the dimerization of a quaternized 10-methoxy-7*H*-pyrido[4,3-*c*]carbazolium chromophore[108] by use of either a rigid or a flexible linker (Figure 5.6). It had been shown by NMR spectroscopy[109] and X-ray crystallography[110] that each of the heteroaromatic rings of ditercalinium was able to intercalate into each of the contiguous CpG steps of a $d(CGCG)_2$ oligonucleotide, whereas binding of the drug to $d(GCGC)_2$ showed one ring intercalated at the CpG step and the other stacked on top of one of the external base pairs.[111] My attention was drawn to the fact that none of these linkers alone would appear able to discriminate between GpC and CpG steps, since N7 and O6 atoms of guanines in both of them lie diagonally across the major groove in equivalent positions thus shaping two regions of very negative electrostatic potential. Besides, the demonstration in the minor groove of TpA steps of linking chains similar to those of ditercalinium and Flexi-Di[112] cast doubts on the reported fundamental role of the spacers in dictating the preference of these drugs for the major groove. In light of our previous findings, we suspected that stacking interactions involving the positively charged chromophores could have an important bearing on these experimental observations and could also possibly account for additional crystallographic evidence suggesting significant bending of the DNA helical axis toward the minor groove. In fact, the crystallographers initially suggested that this unusual deformation of the DNA complexed with ditercalinium ($\sim$15° kink)

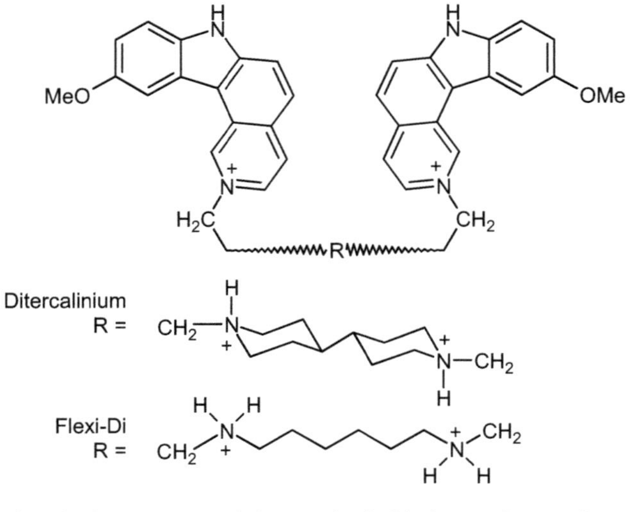


**Figure 5.6** Chemical structures of the synthetic bis-intercalators ditercalinium and Flexi-Di.

could arise from the limited flexibility of the drug's linker but they were surprised to find that the degree of bending was even higher in the crystal structure of the same DNA sequence complexed with Flexi-Di.[113]

To investigate the origins of both the sequence binding preferences and the conformational changes brought about on the DNA molecule upon complexation with these drugs I gave this project to Beatriz de Pascual-Teresa, who had just joined my group after a successful postdoctoral stay with Ken Houk at the University of California. Preliminary calculations on the X-ray crystallographic structures showed that the electrostatic energy component of the stacking interactions between the pyridocarbazolium rings and the DNA bases was favourable for the inner base pairs clamped by them but unfavourable for the base pairs making up the outer boundaries of the bis-intercalation site.[114] Moreover, the electrostatic interaction was found to be attractive for the internal guanines but repulsive for the internal cytosines, which nicely accounted for the experimentally observed poor van der Waals contacts between the latter and the drug chromophores.[106,109] We also established that, as a result of drug binding, the underwound central base pair step was stabilized relative to a GpC step in a standard DNA helix.

We then constructed computer models of the complexes of ditercalinium and Flexi-Di with the alternating hexanucleotide d(GCGCGC)$_2$, in which each drug embraced the central GpC binding site. The complexes were fully solvated and neutralized by addition of sodium counterions, and MD simulations were run for 0.5 ns. To assess the effect of the positively charged spacers, two additional simulations were performed, in which the charges of the linkers were 'switched off', *i.e.* set to zero.

Our results reinforced the idea that the hydrogen-bonding potential of the linkers was effectively reduced in the aqueous medium due to competing interactions with water, their location in the major groove being essentially the result of the marked electrostatically directed orientational preferences displayed by the quaternized pyridocarbazolium chromophores with regard to the highly polarized G–C base pairs.[87]

Taken together, the theoretical results from our work on bis-intercalators provided quantitative support to qualitative descriptions formulated earlier based on the display of the electrostatic characteristics of the stacked systems.[115] Thus, the graphical representation of the quantum-mechanically calculated MEPs on a plane close to the recognition surface can be of help to understand electrostatic complementarity and repulsive interactions. For a G–C pair, the most negative MEP region is found in the surroundings of the N7 and O6 atoms of guanine whereas the most positive MEP region is located around the cytosine ring (Figure 5.7).

## 5.6 Sequence Selectivity of Actinomycin D

The clinical importance of actinomycin D (Figure 5.8) as an antitumor agent and its marked and unusual preference for binding to 5′-GpC-3′ (GpC)

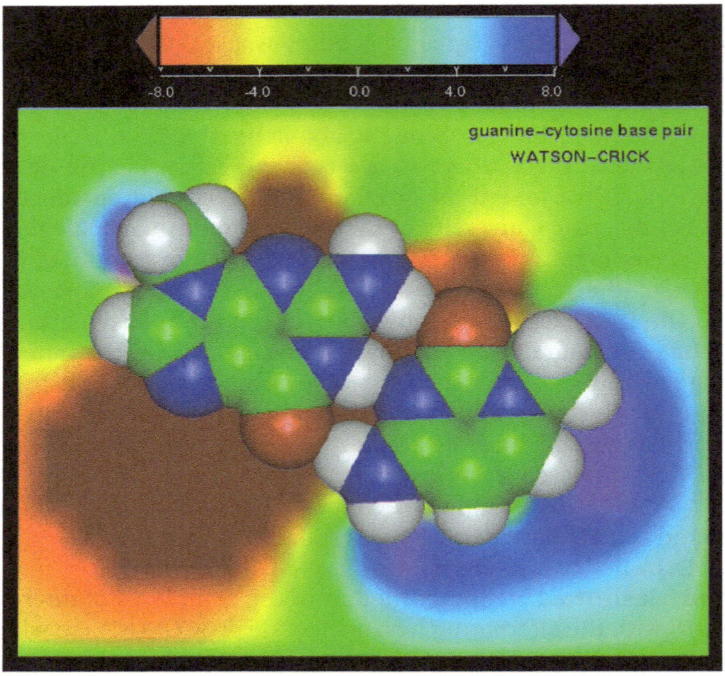


**Figure 5.7** Molecular electrostatic potential of a G–C base-pair as calculated on a plane 1.7 Å below the atom centres, which approximates the recognition surface, and colour-coded such that regions of 8 kcal mol$^{-1}$ appear in blue and regions of −8 kcal mol$^{-1}$ are coloured in red, with intermediate values ramping smoothly. More positive and more negative regions are coloured in violet and brown, respectively. Atoms are represented as CPK spheres: C, green; O, red; N, blue; H, white.

steps and other secondary sites such as GpG (CpC) and GpT (ApC) made this drug an attractive model for our computational studies of ligand–DNA interactions. Binding of actinomycin D to DNA involves intercalation of the 2-aminophenoxazin-3-one planar chromophore between two base pairs and fitting of the two cyclic pentadepsipeptides into the minor groove, each extending for two base pairs on either side of the intercalation site.

The resolution of several crystal structures[116–119] strengthened the belief that the guanine–threonine hydrogen bonds were the factors responsible for the specificity of actinomycin binding to GpC steps, a hypothesis that had been supported by theoretical studies.[120] Nevertheless, this specificity was retained in both actinomine, an actinomycin analogue in which the cyclic pentapeptides are replaced by *N*,*N*-diethylethylen-diamine side chains, and 2-aminophenoxazin-3-one, neither of which can form hydrogen bonds with the 2-amino groups of guanine. These observations lend credence to early reports that pointed to electronic interactions between the

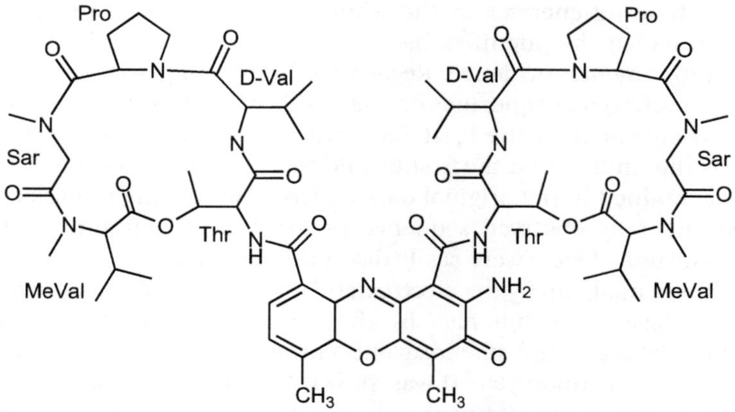


**Figure 5.8** Chemical structure and amino acid composition of actinomycin D.

phenoxazone chromophore and the relatively polarized G–C base pair as the source of specificity for guanine over the other common bases.[121] Furthermore, when 2,6-diaminopurine (D) was used in place of adenine in a DNA fragment, footprinting experiments showed that the sequence specificity of actinomycin was drastically altered and the drug bound to any 5′-purine–pyrimidine-3′ step different from the standard GpC site.[95] Because the minor grooves of GpC and DpT steps present similar hydrogen bonding capabilities, as explained above (Figure 5.3), this interaction alone could not adequately explain the observed change in binding preferences.

To gain a more detailed understanding of the interactions that govern the sequence-specific binding of actinomycin to DNA, we built molecular models of 14 double-helical DNA hexamers that contained a variety of binding sites in the central region, namely, the canonical GpC step, secondary sites such as GpG and GpT, nonpreferred sequences such as CpG, ApT, ApA and TpA and the optimal site DpT. Also, GpC and DpT steps were embedded in different contexts so as to assess the influence of flanking base pairs.[122] The whole set of complexes thus involved all possible hydrogen bonding arrangements between the depsipeptides and the DNA minor groove, as well as all possible combinations of stacking interactions. After energy refinement, intermolecular interaction energies were calculated and partitioned into different van der Waals and electrostatic components. The resulting data matrix was subjected to principal component (PC) analysis as a means to highlight the SAR for this important antitumor antibiotic.

The total interaction energies were found to reflect well the relative affinities of actinomycin D for the different sites: the canonical GpC step was energetically favoured over any other central step in the family of natural sequences, followed by the secondary sites GpG and GpT, although the most favourable binding energies in the whole set were those belonging to sequences in which the adenines had been replaced with D, in accordance with the experimental evidence. Regarding the flanking sequences, pyrimidines were preferred over purines on the 5′ side of the GpC intercalation site, and GGCC appeared as the least favoured sequence. Perhaps more interesting was the finding that the first two PC readily explained over 90% of the variance contained in the original data matrix, reproducing remarkably well the experimentally observed sequence preferences of actinomycin D. The different clusters of sites were easily discriminated along the axis of the first PC, which was made up by the electrostatic interactions of the peptides with the central base pairs but also by the van der Waals and electrostatic interactions between the phenoxazone chromophore and the same base pairs. Therefore, actinomycin D was shown to discriminate among the different sequences by an interplay of hydrogen bonding and stacking interactions. The importance of the stacking term increases for distinguishing amongst G–C-containing sites, such as GpC, GpG and CpG, and is crucial for favouring DpT over GpC central steps. These results indicate that the presence of the 2-amino groups not only alters the hydrogen bonding potential or the geometry of the minor groove, as discussed above, but also imparts distinct stacking properties to the base pairs, both on steric and electrostatic grounds.[111] The simple and intuitive model of like charges repelling one another and unlike charges attracting one another is also valid for molecules and molecular fragments, and the energy minimum found in each of the intercalated ligand–DNA complexes discussed so far can be thought of as the balance point of minimized repulsive and maximized attractive stacking interactions between electric quadrupole moments.[123] Invited by Professor Pier de Benedetti (Università degli Studi di Modena, Italy), in 1998 I contributed a review on stacking interactions and intercalative DNA binding that was published in *Methods*, a companion to *Methods in Enzymology*.[124]

## 5.7 Binding of the Potent Antitumor Agent Trabectedin to DNA

The amino group of guanine was again the centre of my attention when, at the turn of the century, I became involved in the study of the mechanisms of action of several interesting natural products of marine origin under the auspices of the pharmaceutical and biotechnological company PharmaMar, based in Madrid (Spain) and Cambridge (USA). The compound most advanced into clinical use was trabectedin (Figure 5.9), a potent antitumor ecteinascidin originally isolated from the sea squirt *Ecteinascidia turbinata*[125,126] that first enjoyed orphan drug status for the treatment of soft tissue sarcomas in both Europe and the USA and is currently approved for human use and marketed with the name of Yondelis®. Ecteinascidins, which are structurally related to microbially derived safracins and saframycins, undergo a nucleophilic attack by the exocyclic 2-amino group of

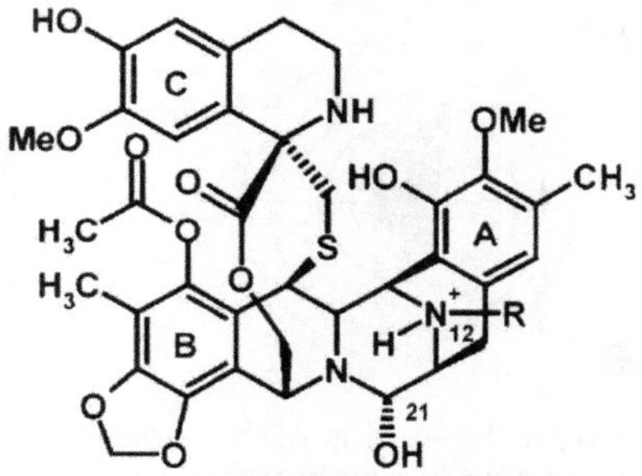


**Figure 5.9** Chemical structures of trabectedin (Yondelis®) and its main metabolite in humans, ET 729. The three fused tetrahydroisoquinoline rings in the polycyclic skeleton are labelled as A, B, and C. Note the positions of the reactive 21-carbinolamine (hemiaminal) group and the protonated N12 that is used to establish a hydrogen bond with an acceptor atom (O2 or N3) located 5′ to the modified guanine in the opposite strand.

guanine in the minor groove of double-stranded DNA when this base is found in a suitable triplet context.[127] Sequence selectivity was shown to operate predominantly through a set of well-defined hydrogen-bonding rules such that the preferred target triplets are 5′-R$\underline{G}$C and 5′-Y$\underline{G}$G, where R and Y stand for purine and pyrimidine, respectively, and the underlined base is the guanine to which the drug is covalently bonded.

Unrestrained MD simulations in aqueous solution were first used by my student Raquel García-Nieto to explore the stability and behaviour of the pre-covalent complexes between trabectedin and two DNA nonamers containing either 5′-AGC or 5′-CGA in the central region.[30] The AGC-containing complex was very stable for the whole length of the 1.4-ns simulation and led to a suitable geometry for adduct formation. In contrast, the complex containing the disfavoured CGA triplet (which contains a single mismatch relative to CGG) did not result in a stable association as the drug slid one step 'up-stream' along the minor groove. These results lent support to the proposed role of a well-defined hydrogen-bonding network in the stabilization of these complexes.[123]

We then simulated the covalent complexes between trabectedin and DNA nonamers containing the 5′-AGC and 5′-CGG target sites. In both cases the bonded drug displayed the predicted binding mode, gave rise to a widening of the minor groove, reduced the twist angle at ApG and CpG steps, and introduced a significant positive roll (relative rotation of the base pairs about the long axis of the base step) at the base pair step involved in covalent bond formation. This increase in roll caused smooth bending of the helix toward the major groove, in very nice qualitative and quantitative agreement with independent results from gel electrophoresis experiments.[128] We found the so-called 'bending dials',[129] which are polar plots that display both the

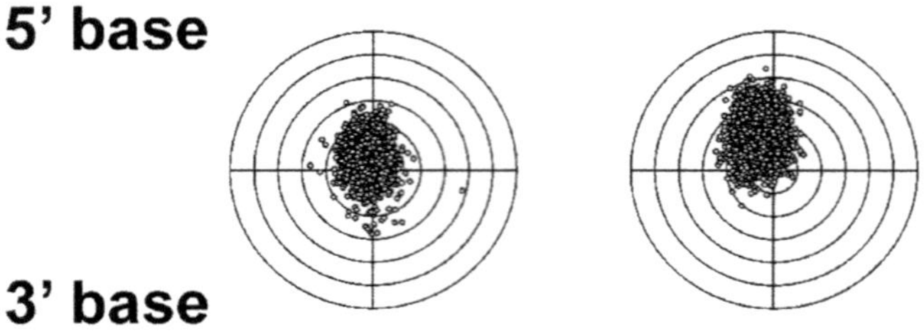


**Figure 5.10** Example of typical 'bending dials' showing the introduction of positive roll and negative tilt in a trabectedin-bonded DNA step (right) relative to the same step in a free DNA duplex (left). The radial and angular coordinates are, respectively, $\theta$ and $\varphi$ (running clockwise from the top), as defined in the text, and each concentric ring indicates a 10° deflection of the helical axis. Individual points are for structures separated by 1 ps throughout the course of a 1.5-ns MD trajectory so that each ensemble of 1500 points constitutes a probability density.

magnitude of the axis deflection angle ($\theta$) and its orientation relative to the major groove ($\varphi$), the most convenient pictorial representation (Figure 5.10) of the local helicoidal parameters, roll and tilt, as calculated by the program CURVES developed in Lavery's group[130] (and today extended as CURVES+ and implemented in a web server[131]). Dots on the upper half of the dial clearly reflect bending compressing the major groove whereas those on the left-hand side of the vertical line indicate negative tilt that compresses the sugar–phosphate backbone of the complementary strand.

In addition, by comparing the dynamic behaviour of these duplexes in the absence and in the presence of bonded drug, we assessed the distinct intrinsic bendabilities of AGC and CGG tracts, which were found to be in consonance with their intrinsic twisting and bending propensities as deduced from crystal structures.[10] Thus, the reported significant bias of CpG steps toward bending into the major groove was already clearly apparent in the simulation of the free DNA whereas the described undertwisting of ApG steps was much stronger in the presence of the drug, which also induced bending by increasing a previously inexistent roll (Figure 5.10).

Quite unexpectedly, we later discovered that the peculiar distortion that trabectedin inflicts on the DNA molecule is strikingly similar to that observed in the complexes of DNA with the zinc fingers of typical transcription factors such as EGR-1 (a.k.a. Zif268).[132] In our view, this raised the interesting possibility that trabectedin might be preferentially targeting the minor groove of those DNA regions in which a protein that contains one or more zinc fingers in its DNA-binding domain is already bound in the major groove. The rationale was that the protein-induced DNA structural rearrangement might be facilitating the docking of trabectedin over binding to a naked DNA site. This hypothesis led us to model the DNA-binding domain of the ubiquitous transcription factor Sp1, which consists of a tandem array of three $Cys_2$–$His_2$ zinc finger modules,[59] as well as to the study of a covalent complex between three trabectedin molecules and an oligonucleotide containing three adjacent target sites during 2 ns of MD simulation.[133] The representative sequence chosen was TGGCGGCGG, which is found in the 5′-upstream promoter region of the human p73 gene, which encodes a protein with structural and functional homologies with the p53 tumour suppressor protein.

The results indicated that tandem binding of several trabectedin molecules to suitable adjacent DNA sites is sterically and energetically feasible, which made us think about the (as yet untested) possibility of cooperative binding. This complex displayed a very stable behaviour, in agreement with the fact that both TGG and CGG triplets represent equally optimal binding sites for the drug. The expected intermolecular hydrogen-bonding scheme between each trabectedin molecule and the Pyr–G–G sites was maintained for the whole length of the simulation. The DNA minor groove was notably wider than in the parallel simulation of the free oligonucleotide, and the ensuing compression of the major groove was similar to that found in previously studied complexes: increases in positive roll were observed at successive steps and were interrupted only at GpC steps, while negative tilt values were calculated for the CpG and TpG steps, indicative of bending into the sugar–phosphate backbone of their complementary strand. Both roll and tilt components contributed to bending of the double helix but, given the adjacency of binding sites around one turn of the helix, the alternation of positive roll at Pyr/Pur and Pur/Pur steps with zero roll at GpC steps on the 3′ side led to a continuous gentle writhe that did not change the overall direction of the helix. This means that the local curvature induced by just one molecule, as reported previously,[30,123] is cancelled by binding of the others, resulting in a DNA conformation that is best described as intermediate between A- and B-form DNA, such as that found in DNA–RNA hybrids and in the complexes of DNA with zinc fingers. To assess the possible dependency of the observed conformation of the free oligonucleotide on the starting geometry, this was built in both A and B forms,[134] and each was simulated under identical conditions. Interestingly, the A-form evolved toward a conformation remarkably similar to that obtained when the simulation started from B-DNA, which was intermediate between A- and B-type DNA as well.[135]

Triplets recognized by the zinc fingers of EGR1 and the related Sp1-like transcription factors include GCG, GAG, GGG and TGG, whereas favoured sites for trabectedin are CGG, TGG, AGC and GGC.[123] Although different combinations of these triplets afford binding sites to both proteins and drugs, the majority of optimal trabectedin sites would be located at the junctions between the triplets that are recognized by successive zinc fingers rather than at the triplets themselves (TGG would be the most obvious exception). Remarkably, direct superposition of the trabectedin-bonded TGG, CGG and CGG triplets onto the corresponding DNA consensus triplets recognized by each EGR1 zinc finger (*i.e.*, GCG, GGG and GCG) showed a good match (rmsd = 1.1 Å using the C1′ atoms) for the three triplets. Moreover, equally good structural similarity (rmsd = 0.9 Å) could be obtained when one complex was shifted one base pair upstream or downstream for the superposition, the best match ultimately depending on the particular sequences that can be simultaneously recognized by the drug and the different zinc fingers of a particular transcription factor or DNA-binding protein. In any case, our findings lent further support to the putative existence of a protein–DNA–trabectedin ternary complex[123] and opened new avenues for research.

In the last case presented, binding of trabectedin took place in a head-to-tail fashion on only one of the DNA strands. However, when we realized that the activity parameters of trabectedin on the panel of 60 human tumour cell lines of the National Cancer Institute's (NCI) Anticancer Drug Screen were highly correlated with those of chromomycin $A_3$ using the COMPARE algorithm,[136] and knowing that this latter anticancer antibiotic binds as a dimer to adjacent TGG sequences on different strands, we decided to model and simulate the structure of the complex formed between the self-complementary dodecanucleotide $d(GTATGGCCATAC)_2$ and two trabectedin molecules, each covalently bonded to a different strand in a tail-to-tail fashion.[137]

For comparison and completeness, the same sequence was also simulated under identical conditions in the absence of any drug. This allowed us to assess the conformational preferences of (i) the TGG triplet, which is ubiquitously over-represented in human viral and eukaryotic sequences, (ii) its complementary sequence, *i.e.* CCA, which is an important component of the so-called 'CCAAT box' that is present in a large number of gene promoters[138] and (iii) the juxtaposition of the two (*i.e.*, TGGCCA), which has been dubbed by some X-ray crystallographers as a 'natural bending element'.[139]

In line with previous observations,[128] bonding of trabectedin resulted in increased DNA stabilization as assessed by smaller rmsd fluctuations in the simulation of the complex relative to the simulation of the free oligonucleotide. In the 2 : 1 complex, each subunit A of trabectedin protrudes perpendicularly off the helix in front of the guanine to which it is bonded, the B subunits stack over the sugar rings of the complementary cytosines in a manner reminiscent of that of typical non-intercalative minor groove binders, and the C subunits expose one flat side to the solvent, whereas the other side makes extensive contacts with the sugar–phosphate backbone of the two nucleosides downstream of each covalently bonded guanine. Also, in this type of tail-to-tail arrangement, both B subunits establish favourable van der Waals contacts at the junction of both DNA triplets.

The greatest unwinding resulting from trabectedin bonding takes place at the TpG (=CpA) steps, which also show the largest values of roll angle, in good accord with the known inverse relationship between this parameter and helical twist.[10] In contrast, the GpC step displays both negative roll and the largest helical twist within the TGGCCA sequence. Overall, the conformational features of the complex were reminiscent of those previously found in the 3 : 1 head-to-tail complex described above and correspond, once more, to an intermediate form between A- and B-DNA.

Remarkably, this 2 : 1 trabectedin complex and the X-ray crystal structure of the complex between a $Mg^{2+}$-chelated chromomycin $A_3$ dimer and a double-stranded $d(TTGGCCAA)_2$ octanucleotide[140] revealed several features in common such as (i) an unwound DNA with a considerably widened minor groove in the virtually superimposable target TGGCCA region (rmsd of only 2.1 Å over phosphate backbone atoms); (ii) a similar pattern of roll values consisting of increased positive roll at TpG steps and small positive roll at GpG steps, together with negative roll at the central GpC step, which was more marked in the trabectedin complex; and (iii) negative slide at both GpG steps, in good agreement with the finding of glycosyl torsional angles, $\chi$, closer to those of A-DNA.

Our simulation of the free dodecamer showed two distinct bends produced by rolling at each TpG step but no bending at the junction between TGG and CCA, in good agreement with results from a two-dimensional NMR spectroscopy study reporting that GGC in the related self-complementary decamer $d(CATGGCCATG)_2$ forms a tight stack with parallel bases, and that high positive roll is present at both TpG steps.[141] This result, however, was at odds with earlier X-ray crystallography observations showing that stacked B-DNA double helices of general sequence C–C–A–x–x–x–x–T–G–G exhibit the same 23° bend across the –T–G–G C–C–A– nonbonded junction in the crystal lattice[134] that is encountered in the middle of the C–A–T–G–G–C–C–A–T–G decamer.[142] Since high roll at GGC is not seen in GpC steps from other sequences, the most likely explanation is that this structural effect, together with the unusually high positive slide adopted by the CpA(=TpG) steps and the nonstandard $B_{II}$ backbone conformation in these crystal forms, are due to strong intermolecular interactions in the crystal lattice.

Finally, both trabectedin and chromomycin $A_3$ have been shown to exert at least part of their cytotoxicity by interfering with cell replication and transcription. In light of the structural information obtained, we proposed that these effects can be largely due to the stalling of replication and transcription forks through stabilization of the helical structure of duplex DNA and prevention of strand separation in a way reminiscent of that of classical DNA interstrand cross-linkers (ICL). In this respect, several proposals of the likely existence of one or several cytotoxic trabectedin–DNA–protein ternary complexes involving members of the DNA repair machinery prompted us to model some of these macromolecular ensembles.[143,144]

## 5.8 Other Examples of DNA Minor-groove-bonding Tetrahydroisoquinoline Antibiotics

When we studied the sequence binding preferences of the N12-demethylated analogue of trabectedin, ET729 (Figure 5.9), we found that this compound has an extended repertoire of binding sites compared with trabectedin, including the CGA triplet.[145] This was of interest, not only because ET729 is one of the main metabolites of trabectedin in humans (through the action of

CYP3A4) and can therefore contribute to some aspects of the clinical activity of the parent molecule, but also because its study could help clarify mechanistic aspects of the bonding reaction. Indeed, we found that CGG triplets provide good binding sites for trabectedin because anchoring of the methylenedioxy group by the amino group present in the minor groove on the 3′ side of the triplet determines a subtle change in overall orientation that is enough to counterbalance the steric clash between the methyl group on N12 and the amino group from the base pair on the 5′ side. When this hydrogen-bonding interaction is lacking because an A–T base pair is found on the 3′ side (*e.g.*, CGA), trabectedin discriminates against a G–C base pair on the 5′ side of the covalently modified guanine. This interpretation also provides a rationale for the finding that CGT, GGT and GGA are non-favoured sites for trabectedin. In contrast, the lack of the N12 methyl group in ET729 allows this compound to achieve the optimal juxtaposition required to promote its activation by guanine at a CGA site. The fact that these understated differences in binding orientation and geometry were sufficient to rationalize the experimentally found behaviours revealed, in our view, the power of state-of-the-art MD simulations to probe the details of ligand–DNA interactions in the absence of experimental restraints.

PM00104 (Zalypsis®) is a chemically engineered analogue of trabectedin that contains two fused tetrahydroisoquinoline rings and a more linearly extended substituent in place of the C ring. In common with Yondelis®, it also forms covalent bonds with the amino group of selected guanines in double-stranded DNA, both in the absence and in the presence of methylated cytosines, by virtue of a reactive hemiaminal moiety. The results of fluorescence-based thermal denaturation experiments demonstrated that the most favourable DNA triplets for covalent adduct formation are AGG, TGG, CGG, GGC and AGC, and also that the average increase in DNA melting temperature that its adduct brings about on the double helix is about half that achieved by trabectedin and similar to that attained with echinomycin or thiocoraline. The binding site preferences and the distinct stabilization of the DNA double strand could be rationalized by means of molecular modelling of the precovalent complexes between Zalypsis® and a series of oligonucleotides of general sequence 5′-d(ATAATAXYZATAATA)/5′-d(TATTATZ′Y′X′TATTAT), where XYZ/Z′Y′X′ stands for AGA/TCT, AGG/CCT and AGC/GCT.[146]

In lurbinectedin (formerly known as PM01183), the replacement of the tetrahydroisoquinoline C subunit of trabectedin with a tetrahydro β-carboline confers this analogue some pharmacokinetic and intrinsic activity benefits.[147] Modelling and simulation by means of MD of the pre-covalent and covalent complexes between PM01183 and an oligonucleotide of sequence 5′-d(CAATACGGATAAG)/5′-d(CTTATCCGTATTG) revealed a widened minor groove and a major groove compressed in the central region due to significant increases in roll at C6/G7 and G7/G8 steps. Occupancy of the minor groove and solvent exposure of some functional groups were thought likely to have an effect on the ability of this slightly bent binary complex to associate or interfere with one or more DNA-binding proteins involved in transcription, replication and/or repair within the cell, but not necessarily the same ones that are involved in the recognition of trabectedin–DNA adducts.

## 5.9 Melting DNA on the Computer

I have briefly discussed already that the difference in melting temperature of a double-stranded DNA molecule in the absence and presence of noncovalently bound (or covalently bonded) ligands can provide experimental information about the thermal stabilization brought about by the ligand. A pending question in my mind was whether this could also be achieved *in silico* (*i.e.* on the computer). A literature search revealed that this had been accomplished and published by B. Montgomery ('Monte') Pettitt's group, at the University of Houston, Texas, who had performed two 100-ns MD simulations at 400 K on a homopolymeric DNA duplex of sequence $d(A_{12}) \cdot d(T_{12})$ in explicit salt water.[148] The choice of sequence obeyed considerations of both simplicity and practicality since the melting temperature of this dodecamer in their simulated solution environment was estimated to be as low as 318 K, which hinted at the possibility of success in a computationally feasible time frame. The two simulations differed only in how random velocities were assigned but (i) fraying (lack of WC hydrogen bonds at the ends of the duplex), (ii) peeling (motions where untwisting is augmented by noncanonical hydrogen bonds with the backbone of a strand), (iii) searching (incremental swapping of WC hydrogen bonds resulting in the change in relative sequence registry of one strand with another) and (iv) reannealing events were apparent in both cases.

Encouraged by this precedent but slightly worried about whether our computational resources (which included some limited access to the Barcelona Supercomputer 'Mare Nostrum') would be up to the task, I handed over this challenge to my new Ph.D. student Juan A. Bueren-Calabuig, who gladly took in his stride the simulation of the dynamic behaviour (and hopefully denaturation) of a duplex of sequence 5′-d(TAATAACGGATTATT) · 5′-d(AATAATCCGTTATTA) in 0.1 M NaCl aqueous solution at 400 K. Fortunately enough, in slightly less than 200 ns of simulation we were able to characterize in atomic detail the complete thermal denaturation profile of this mixed-sequence A–T-rich DNA 15-mer containing a central CGG triplet suitable for drug binding. This meant that we were rightly positioned to try and simulate the monoadducts formed between the underlined guanine in this oligonucleotide and different covalent binders, namely Yondelis®, Zalypsis® and PM01183, as well as the bisadduct formed at the central CpG step with the prototypical ICL mitomycin C (MMC).

A striking asymmetry of melting was observed on both sides of the central CGG triplet and the strand separation process was unequivocally affected by the presence of the drug adducts. Thus, whereas complete helix unzipping and loss of WC base pairing was apparent for the drug-free 15-mer after

190 ns of simulation, reannealing events were much more conspicuous and long-lasting in the monoadduct-containing duplexes, whose central regions still remained hydrogen-bonded to significant extents at the final time of 200 ns.[149] In our view, these significant differences attested to the demonstrated ability of these drugs to stabilize dsDNA, stall replication and transcription forks and recruit DNA repair proteins. This stabilization, quantified here in terms of undisrupted base pairs, also supported our hypothesis that a monoadduct of this sort can functionally mimic a DNA interstrand crosslink despite the fact that these drugs are covalently bonded to only one strand.

Contrary to our expectations, however, the oligonucleotide containing the MMC bisadduct followed a very similar course to that of the drug-free DNA, the main difference being that the two DNA strands could not physically separate completely due to the existence of the interstrand crosslink between the two guanines making up the central CpG step. Thus, only these two G–C base pairs remained WC hydrogen-bonded at the end of the simulation. Of further interest was the characterization in this complex of a number of unique hydrophobically collapsed intermediates in which the MMC ring system appeared pseudointercalated in transient DNA secondary structures. These observations are suggestive of distinct challenges posed to the DNA repair machinery that might account for some of the differences observed experimentally in mammalian cells between the actions of MMC and the DNA-binding antitumour agents tetrahydroisoquinolines.[150]

## 5.10 Mitomycin Bis-adduct Formation as a Test Case for QM/MM Methods

As hybrid QM/MM methods continued to mature and were implemented in the AMBER suite of programmes traditionally employed in my lab, I was irrevocably tempted to try and simulate the formation of a covalent bond between a ligand and a DNA molecule. What better choice than MMC, a microbially produced antitumour prodrug experimentally studied over the years—most notably by Maria Tomasz and co-workers—and regularly used as a traditional ICL? Furthermore, sensitivity to MMC was used to classify Fanconi anemia (FA) patients into different complementation groups and has been instrumental to delineate the so-called FA/BRCA pathway of the DNA damage response.[151]

MMC's unique ability to crosslink the complementary strands of the double helix specifically at 5′-CG-3′ (CpG) steps in hypoxic cells is especially remarkable given its constrained architecture and rests on the following sequence of events: (i) intracellular reduction of its quinone ring and opening of its alkylating aziridine ring involving C1 and C2 atoms, (ii) formation of a covalent bond with the 2-amino group of selected guanines in the minor groove of dsDNA, and (iii) evolution of the monofunctional adduct, but exclusively at CpG steps, towards a more deleterious bifunctional lesion known as an interstrand crosslink (Figure 5.11). Examination of the

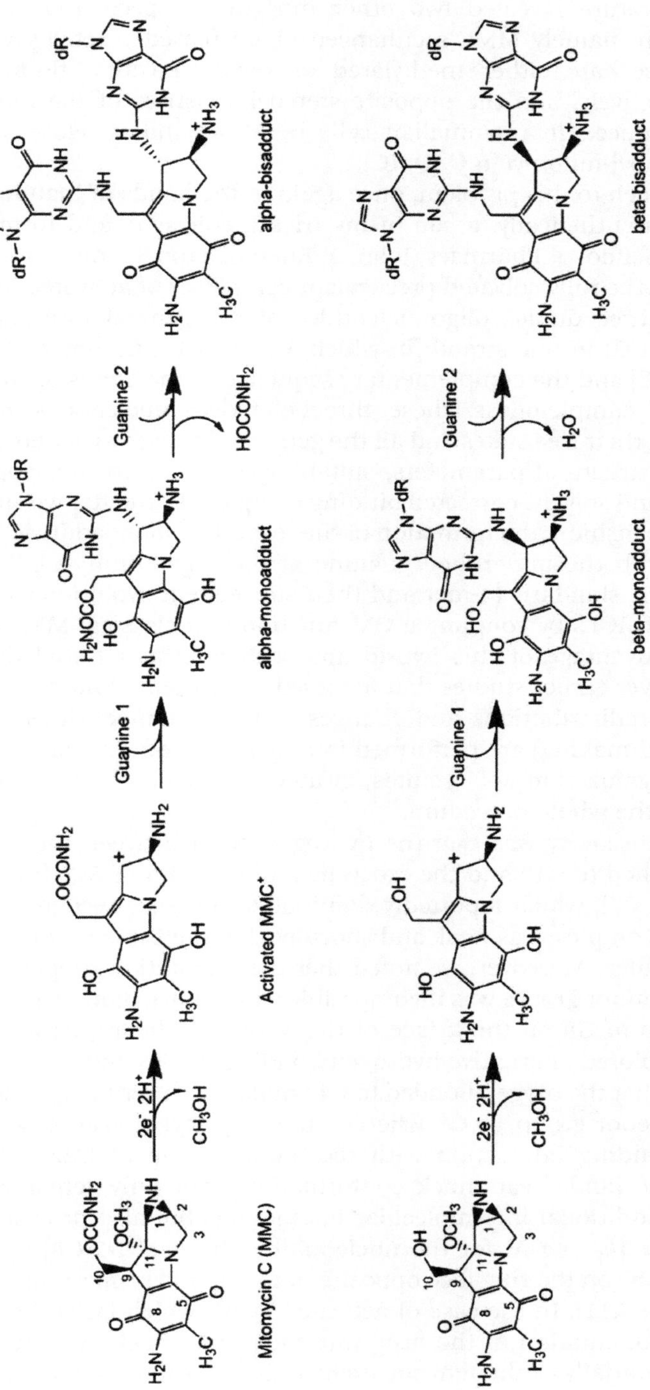


**Figure 5.11** Schematics showing the reaction pathways that lead to formation of the α-MMC–DNA (top) and β-DMC–DNA monoadducts (bottom) and their corresponding bisadducts (ICLs) of opposite stereochemistry. Note that the DNA-alkylating functionalities of MMC and DMC are unmasked in living cells only upon bioreductive activation.

relevant literature revealed two other intriguing aspects that also drew our attention, namely MMC's enhanced or decreased reactivity when the cytosine bases are either methylated or replaced with 5-fluorocytosine (5FC), respectively, and the opposite stereochemistries of the major DNA adducts produced in mammalian cells by MMC and its close analogue, 10-decarbamoyl-mitomycin C (DMC).

Our approach to this problem, once again in the hands of Juan A. Bueren-Calabuig (parenthetically, a son of my friend, colleague and former classmate at the School of Pharmacy, Juan. A. Bueren, a top Spanish expert in FA) was to study the fully solvated precovalent complexes of activated MMC and DMC with three duplex oligonucleotides of the general formula d(CGA-TAA***X***GCTAACG) in one strand [in which ***X*** stands for C, 5-methyl cytosine (5MC) or 5FC] and the complementary sequence in the opposite strand. For the sake of completeness, these three oligodeoxynucleotides were also simulated in their free states and all the generated trajectories were analysed in terms of structural parameters, suitable geometries for nucleophilic attack, MEPs and solvent-corrected binding energies. Thereafter we simulated, in a stepwise fashion, the formation of the respective monoadducts for MMC and DMC with the underlined guanine at the high-affinity CGC site embedded in the standard 14-mer and their subsequent evolution to the final (and distinct) ICLs by coupling a QM function to a classical MM potential. The main advantage of this hybrid approach for the solvated drug–DNA complexes over earlier studies that focussed on simplified model systems is that charge redistributions and changes in atom connectivities (*i.e.* bond breaking and making) are performed in a dynamic context that realistically allows reorganization of solutes, water molecules and counterions throughout the whole procedure.[152]

Our first discovery was that the hydrogen bond between the carbamate oxygen attached to C10 and the exocyclic amino group of G22 (the guanine pairing with C7), which reportedly dominated sequence recognition in the precovalent complex, was weak and short-lived in the presence of competing water molecules. Moreover, we noted that the orientation proposed for the drug in the minor groove was incompatible with a nucleophilic attack by the amino group of G8 on the $\alpha$ face of the MMC tetrahydropyrrole ring. We therefore explored alternative hydrogen bonding modes and found that it is more likely that the oxygen bonded to C10 makes an internal hydrogen bond with the phenol group at C8 whereas the carbonyl oxygen engages in a hydrogen-bonding interaction with the amino group of G22. This novel binding pose (and 'near-attack conformation') not only remained stable because of additional intermolecular interactions but also provided a suitable distance (*i.e.* $\leq$4 Å) for the nucleophilic attack of N2(G8) onto C1 of activated MMC on the ring face opposite to that of the primary amino group at C2 (Figure 5.11). In the case of activated DMC, which lacks the carbonyl oxygen, the orientation of the drug was such that attack by N2(G8) would occur preferentially, although not exclusively, on the $\beta$ face of the drug's tetrahydropyrrole ring. Taken together, our results strongly indicated that the very distinct relative occurrence of α and β DNA monoadducts caused by MMC and DMC is largely dictated by the nature of the hydrogen bond involving either the carbamate or hydroxyl oxygen of both drugs in their active form.

Once the putative binding mode for both drugs in the minor groove of the central C7–G22/G8–C21 region was identified, we carried out similar MD simulations with the oligodeoxynucleotides in which C7 and C21 had been replaced with either 5MC or 5FC. Calculation of the binding energies throughout the trajectories using the MM–GBSA method, as implemented in AMBER,[153] and energy decomposition analyses showed that a methylated C21, *i.e.* the cytosine complementary to the guanine undergoing alkylation, significantly improved its electrostatic interaction with the positively charged drug over a standard C or 5FC, in good accord with the overtly distinct MEP maps. Inspection of the stacking geometries between neighbouring G̲8–C21 and C9–G20 base pairs also revealed a significant displacement towards negative tilt values relative to free DNA in the precovalent complexes, a characteristic that is naturally present, in the absence of bound drug, when C9 and C21 are methylated and argues in favour of an intrinsic structural preorganization for MMC binding at 5MCpG sites. Altogether, these modelling results are in very good agreement with the experimental evidence that the 5MC-induced rate enhancement of G monoalkylation is due to the base-paired 5MC rather than the 5′-flanking one.[154]

The QM reaction coordinate leading to covalent bond formation consisted of progressively shortening the distance between N2 of G̲8 and the drugs' C1 using different snapshots from the MD trajectories of the precovalent complexes. The consistent picture obtained was that the opposite α or β stereochemistry at C1 of the resulting adduct (Figure 5.11) is entirely dependent on the conformation-driven positioning of the activated drug in the DNA minor groove, as discussed above.

The structures of the resulting MMC–DNA and DMC–DNA complexes containing, respectively, the α- and β-monoadducts were relaxed upon relocation of the extra H22 on N2(G̲8) to the amino group of the bonded drug and equilibrated for 10 ns of unrestrained MD simulations. The major characteristics of the MMC–DNA complex were (i) an absence of significant deviations in helical parameters from those of free DNA and (ii) two possible off-centre locations of the covalently bonded drug inside a slightly widened minor groove depending on the orientation of the carbamate side chain. In contrast, the DNA minor groove of the central region lodging the DMC β-adduct was slightly compressed due to the simultaneous interaction of the drug with the sugar–phosphate backbone of both strands and the sustained hydrogen-bonding interaction between O2(T23) and the C10 hydroxyl group of DMC.

Once the structures of the duplexes containing the α- and β-monoadducts were equilibrated, we proceeded to (i) generate the second alkylating species by removing the drug's carbamate (in the case of MMC) or hydroxyl (in the case of DMC), (ii) equilibrate for 10 ns using standard MD and (iii) switch to the QM/MM method to shorten the N2–C10 distance and create the second covalent bond. The resulting bisadducts were likewise equilibrated for 20 ns and the respective trajectories showed that the DNA can accommodate this dual covalent modification (on G8 in one strand and G22 in the opposite strand) undergoing minimal distortion, as originally reported,[155] only in the case of the C1α-ICL. In the C1β-ICL, in contrast, we found that the rise, shift and roll parameters at the C7pG8 step significantly deviated from the values in free DNA. Interestingly, the negative roll indicates bending of the helix toward the minor groove, in contrast to what is observed in DNA duplexes containing adducts with the monofunctional trabectedin or PM01183 (see Section 5.7. above), which appear to behave functionally as ICL mimics.[141]

We concluded this work by stating that the reported conformational differences in these adducts, which appear in different proportions in the DNA of cells that have been exposed to either MMC or DMC, are likely to pose dissimilar challenges to the DNA repair machineries. This may account for the experimental findings that FA cells are hypersensitive to MMC but have normal sensitivity to DMC whereas in other cell types DMC is more cytotoxic than MMC.[156]

## 5.11 Lamellarins as Topoisomerase I Poisons

A long-standing goal of molecular modelling studies for many years was to understand the way camptothecin (CPT) and other related compounds, such as the clinically used antineoplastic agent topotecan (Figure 5.12), specifically block the religation step that follows single-strand nicking of DNA by eukaryotic topoisomerase I (Top1), an enzyme that mediates relaxation of supercoiled DNA in cells. However, the precise atomic details of the interaction with CPT-like Top1 inhibitors were fully understood only when topotecan was cocrystallized with the 'covalent binary complex',[157] despite the previous existence of a variety of proposed models[158] and the availability of X-ray crystal structures of Top1 interacting both covalently and non-covalently with DNA.[159] These notable advances in structural biology fuelled my interest in this anticancer target and paved the way to theoretical studies dealing with a large protein–DNA–ligand ensemble that could be treated in the presence of explicit water molecules thanks to concomitant improvements in the capacity and power of computers. Thus, when Christian Bailly proposed a collaboration to look at some novel agents that were also capable of converting Top1 into a cellular poison, I was very keen on accepting. The molecules belonged to the class of 6*H*-[1]benzopyrano-[4,3;4,5]pyrrolo[2,1-*a*]isoquinoline alkaloids known as lamellarins, which had been originally isolated from a prosobranch mollusc (*Lamellaria* sp.), but were later identified in several other marine organisms. One of the most cytotoxic compounds in this series is lamellarin D (LMD) (Figure 5.12), some derivatives of which were selected by PharmaMar for preclinical development.[160]

We used the crystal structure of human Top1 covalently linked to a double-stranded DNA molecule with bound topotecan, at 2.10 Å resolution

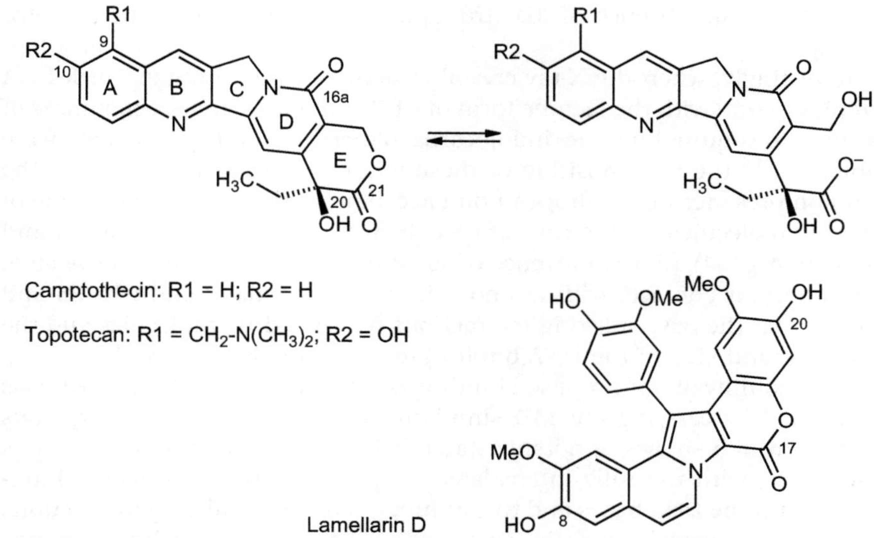


**Figure 5.12** Chemical structures of camptothecin, topotecan and lamellarin D.

(PDB code: 1k4t),[138] to model the drug-free covalent complexes between Top1 and two different DNA 15-mers containing either a CpG or a TpG intercalation site. These were the sites of cleavage induced by LMD and/or CPT in the Top1-mediated DNA cleavage assays.[161] Our work aimed to validate a previously proposed mode of LMD binding to the binary complex[162] using automated docking techniques and MD simulations, and to delineate the structure–activity relationships of lamellarins. For comparison purposes, the same MD protocol was applied to a similar complex containing CPT, for which no such information was available at the time we started our studies.

LMD and CPT were each unbiasedly docked into the intercalation site within the DNA–Top1 complex using the Lamarckian genetic algorithm implemented in AutoDock 3.0.[163] In the orientation best scored by the program, the ring skeleton of CPT was practically superimposable (rmsd of 0.3 Å) on that of topotecan, as found in the X-ray crystal structures, and the 20-OH was found to be engaged in a hydrogen-bonding interaction with the side chain carboxylate of Asp533. The orientation chosen for LMD as representative of the mode of binding of this drug to the DNA–Top1 complex presented (i) the exocyclic phenyl ring in the major groove, (ii) the C8 and C20 hydroxyl groups at hydrogen-bonding distances from the side chains of Asn722 and Glu356, respectively, and (iii) the keto group (O17) facing the guanidinium group of Arg364. Thus, C10 and C16a of CPT appear to be positionally equivalent to C20 and C8 in LMD despite the fact that the concave and convex sides of both drugs are found in different grooves of the DNA molecule. This result strongly suggests that docking and modelling approaches based simply on molecular similarity can be very misleading, especially in the absence of 3D structural information about the receptor binding site.

Interestingly, when the X-ray crystal structures of the human Top1–DNA complex bound with the lactone form of CPT and representative members of the indenoisoquinoline and indolocarbazole classes of Top1 poisons were published,[164] the superposition of these ternary complexes revealed (i) the common presence of a hydrogen-bond acceptor on the minor groove side of the drug molecules (in the case of CPT, between the pyridine N1 atom and the Nε of Arg364), (ii) the presence of substituents on the major groove side, (iii) important contacts with amino acid residues Asn352 and Glu356 and also (iv) that the base-stacking interactions between drug molecules and the cleaved strand side of the DNA duplex are not spatially conserved.

The feasibility of our proposed binding orientations for CPT and LMD was ascertained by carrying out MD simulations of the complexes in aqueous solution, which showed a notably stable behaviour during the whole 1.5-ns trajectories, with the fully intercalated drugs giving rise to stacking interactions with the DNA bases and to a number of other stabilizing interactions with both the protein and the surrounding solvent. The hydrogen bonds involving the hydroxyls at C8 and C20 of LMD were often broken and re-formed as they exchanged with similar interactions with neighbouring water molecules.

The binding energy analysis of the Top1–DNA–CPT complex and the two Top1–DNA–LMD complexes (with the drug intercalated at TpG or CpG) revealed a predominance of van der Waals interactions involving the nucleic bases making up the intercalation site and electrostatic interactions with some crucial protein residues (Arg488, Lys532, Asp533, Arg364, Glu356 and Asn722). In the three cases studied, the most favourable and quantitatively more important interaction was with $G_{+1}$, although in the CG–LMD complex the magnitude of this van der Waals term was similar to that between LMD and the guanine complementary to $C_{-1}$. Remarkably, the interaction with $T_{-1}$ was more favourable for CPT (by about 3 kcal mol$^{-1}$) than it was for LMD. In addition, the lack of detectable Top1-induced C$^{\downarrow}$G sites in the presence of CPT could be explained by the steric clash that would arise between the exocyclic 4-amino group of $C_{-1}$ and the sp$^3$ carbon in ring C. In contrast, both TpG and CpG sites provided equally good binding sites for LMD, in good accord with the experimental findings.

Since our simulations consistently showed the 20-OH of LMD engaged in a direct or water-mediated hydrogen bond with the carboxylate of Glu356, we decided to test whether the loss of this interaction might account for the deleterious effect that removing the 20-OH group in LMD has both on Top1 inhibition and cytotoxicity.[142] To this end, we made use of a thermodynamic cycle (Figure 5.2) that allowed us to calculate the difference in binding free energies for these two molecules by utilizing statistical mechanical information generated from a set of MD simulations during which this hydroxyl was converted to a hydrogen when the drug was free in solution and also when it was part of each of the two ternary complexes studied. The calculated free energy differences were 2.6 and 2.9 kcal $mol^{-1}$ for the CG and TG Top1–DNA complexes, respectively, favouring LMD over the analogue without the 20-OH, which nicely accounted for the decreased potency detected in the experimental assays. This finding lent credence to the proposed binding mode, which was additionally supported by the observation that incorporation of a hydroxyl at the C10 position of CPT (positionally equivalent to the 20-OH of LMD), as in topotecan (Figure 5.12) and other simpler analogues, results in greatly improved Top1 inhibitory activity.

## 5.12 Concluding Remarks

I believe that molecular modelling and computer simulation techniques have come of age and have proved their value in an enormity of research fields, including the study of DNA–ligand and DNA–protein interactions. In this chapter, I have tried to provide a summary of my research activity in this area, which has naturally evolved in consonance with the increased availability of structural data on biological macromolecules, advances in force fields (including improved treatment of long-range electrostatic interactions), and amazing developments in computer technologies, most significantly processing speeds, multi-processor programming and data storage capacity. I consider myself fortunate to have been offered, as a molecular modelling abecedarian, to study the DNA double helix, which despite its overt simplicity continues to yield surprises and awaits full understanding of its encoded messages in all living organisms. It was also a happy coincidence that the field I got involved in after my doctorate in liquid chromatography has continuously benefited since then from a vast array of web servers, standalone programs, interdisciplinary knowledge and expertise contributed by many talented scientists from all over the world.

For many years, I also had the privilege to assist my colleagues at the Organic and Pharmaceutical Chemistry Department in their attempts to synthesize new DNA-interacting compounds with potential anti-cancer activity.[165–168] This collaboration has had a major impact on my career because it has highlighted the tremendous challenges involved in the developments of new drugs. In parallel, I have also marvelled at the great chemical diversity and tremendous potential of natural products, particularly from the sea, which can yield very promising drugs with unique mechanisms of action. While it is true that we have learnt many lessons from past studies, it is not less certain that the discoveries of today are posing even more challenging questions for tomorrow. I can only trust that we will continue forming new generations of enthusiastic and gifted students capable of answering them and passing the baton on to the next relay of scientists.

## Acknowledgements

For the personal work described in this article I was assisted by a gifted and dedicated group of co-workers and collaborators who have enriched this entire research effort. I thank all of them for their enthusiasm and helpful discussions over the years. I am also grateful to the Spanish Comisión Interministerial de Ciencia y Tecnología, the National Foundation for Cancer Research, Comunidad Autónoma de Madrid, and PharmaMar Laboratories for continued financial support.

## References

1. L. H. Hurley and F. L. Boyd, DNA as a target for drug action, *Trends Pharmacol. Sci.*, 1988, **9**, 402–407.
2. L. H. Hurley, DNA and associated targets for drug design, *J. Med. Chem.*, 1989, **32**, 2027–2033.
3. L. H. Hurley, DNA and its associated processes as targets for cancer therapy, *Nat. Rev. Cancer*, 2002, **2**, 188–200.
4. W. Saenger, *Principles of Nucleic Acid Structure*, Springer-Verlag, New York, 1984.
5. L. H. Hurley, Secondary DNA structures as molecular targets for cancer therapeutics, *Biochem. Soc. Trans.*, 2001, **29**, 692–696.
6. N. M. Luscombe, R. A. Laskowski and J. M. Thornton, Amino acid-base interactions: a three-dimensional analysis of protein-DNA interactions at an atomic level, *Nucleic Acids Res.*, 2001, **29**, 2860–2874.
7. M. Suzuki, S. E. Brenner, M. Gerstein and N. Yagi, DNA recognition code of transcription factors, *Protein Eng.*, 1995, **8**, 319–328.
8. D. S. Goodsell, Sequence recognition of DNA by lexitropsins, *Curr. Med. Chem.*, 2001, **8**, 509–516.
9. X. L. Yang and A. H. Wang, Structural studies of atom-specific anticancer drugs acting on DNA, *Pharmacol. Ther.*, 1999, **83**, 181–215.
10. A. A. Gorin, V. B. Zhurkin and W. K. Olson, B-DNA twisting correlates with base-pair morphology, *J. Mol. Biol.*, 1995, **247**, 34–48.
11. M. A. Young, G. Ravishanker, D. L. Beveridge and H. M. Berman, Analysis of local helix bending in crystal structures of DNA oligonucleotides and DNA-protein complexes, *Biophys. J.*, 1995, **68**, 2454–2468.
12. J. D. Watson and F. H. C. Crick, Molecular structure of nucleic acids, *Nature*, 1953, **171**, 737–738.
13. H. M. Berman, J. Westbrook, Z. Feng, G. Gilliland, T. N. Bhat, H. Weissig, I. N. Shindyalov and P. E. Bourne, The Protein Data Bank, *Nucleic Acids Res.*, 2000, **28**, 235–242.
14. H. M. Berman, W. K. Olson, D. L. Beveridge, J. Westbrook, A. Gelbin, T. Demeny, S.-H. Hsieh, A. R. Srinivasan and B. Schneider, The Nucleic Acid Database: a comprehensive relational database of three-dimensional structures of nucleic acids, *Biophys. J.*, 1992, **63**, 751–759.
15. R. Sayle and E. J. Milner-White, RASMOL: biomolecular graphics for all, *Trends Biochem. Sci.*, 1995, **20**, 374–376.
16. http://spdbv.vital-it.ch/.
17. http://www.ks.uiuc.edu/Research/vmd/.

18. W. L. DeLano, *The PyMOL Molecular Graphics System*, DeLano Scientific LLC, San Carlos, CA, 2004. http://pymol.sourceforge.net/.
19. R. M. Hanson, *Jmol* – a paradigm shift in crystallographic visualization, *J. Appl. Crystallogr.*, 2010, **43**, 1250–1260.
20. https://www.umass.edu/microbio/chime/getchime.htm.
21. http://jmol.sourceforge.net/index.en.html.
22. D. C. Phillips, In *Biomolecular Stereodynamics, Vol. II*; ed. R. H. Sarma, Adenine Press, Guilderland, NY, 1981, p. 497.
23. F. Gago, Modelling and simulation: a computational perspective in anticancer drug discovery, *Curr. Med. Chem.: Anti-Cancer Agents*, 2004, **4**, 401–403.
24. J. Ma, Usefulness and limitations of normal mode analysis in modeling dynamics of biomolecular complexes, *Structure*, 2005, **13**, 373–380.
25. M. Karplus, Molecular dynamics simulations of biomolecules, *Acc. Chem. Res.*, 2002, **35**, 321–323.
26. R. P. Feynman, R. B. Leighton and M. Sands *The Feynman Lectures in Physics*, Addison-Wesley, Reading, MA, 1963, vol. I, p. 36.
27. T. E. Cheatham, P. Cieplak and P. A. Kollman, A modified version of the Cornell et al. force field with improved sugar pucker phases and helical repeat, *J. Biomol. Struct. Dyn.*, 1999, **16**, 845–862.
28. T. E. Cheatham, J. L. Miller, T. Fox, T. A. Darden and P. A. Kollman, Molecular dynamics simulation on solvated biomolecular systems: the particle mesh Ewald method leads to stable trajectories of DNA, RNA and proteins, *J. Am. Chem. Soc.*, 1995, **117**, 4193–4194.
29. R. Kazlauskas, Modeling – A tool for experimentalists, *Science*, 2001, **293**, 2277–2279.
30. D. Strahs and T. Schlick, A-tract bending: Insights into experimental structures by computational models, *J. Mol. Biol.*, 2000, **301**, 643–663.
31. R. García-Nieto, I. Manzanares, C. Cuevas and F. Gago, Bending of DNA upon binding of ecteinascidin 743 and phthalascidin 650 studied by unrestrained molecular dynamics simulations, *J. Am. Chem. Soc.*, 2000, **122**, 7172–7182.
32. D. Barsky, N. Foloppe, S. Ahmadia, D. M. Wilson and A. D. MacKerell, New insights into the structure of abasic DNA from molecular dynamics simulations, *Nucleic Acids Res.*, 2000, **28**, 2613–2626.
33. E. Giudice, P. Varnai and R. Lavery, Base pair opening within B-DNA: free energy pathways for GC and AT pairs from umbrella sampling simulations, *Nucleic Acids Res.*, 2003, **31**, 1434–1443.
34. N. Spackova, E. Cubero, J. Sponer and M. Orozco, Theoretical study of the guanine → 6-thioguanine substitution in duplexes, triplexes, and tetraplexes, *J. Am. Chem. Soc.*, 2004, **126**, 14642–14650.
35. P. Varnai and K. Zakrzewska, DNA and its counterions: a molecular dynamics study, *Nucleic Acids Res.*, 2004, **32**, 4269–4280.
36. D. L. Beveridge, G. Barreiro, K. S. Byun, D. A. Case, T. E., Cheatham, S. B. Dixit, E. Giudice, F. Lankas, R. Lavery, J. H. Maddocks, R. Osman, E. Seibert, H. Sklenar, G. Stoll, K. M. Thayer, P. Varnai and M. A. Young,

Molecular dynamics simulations of the 136 unique tetranucleotide sequences of DNA oligonucleotides. I. Research design and results on d(CpG) steps, *Biophys. J.*, 2004, **87**, 3799–3813.
37. S. Y. Ponomarev, K. M. Thayer and D. L. Beveridge, Ion motions in molecular dynamics simulations on DNA, *Proc. Natl. Acad. Sci. U. S. A.*, 2004, **101**, 14771–14775.
38. N. K. Banavali and B. Roux, Free energy landscape of A-DNA to B-DNA conversion in aqueous solution, *J. Am. Chem. Soc.*, 2005, **127**, 6866–6876.
39. M. Pasi, J. H. Maddocks, D. Beveridge, T. C. Bishop, D. A. Case, T., Cheatham, P. D. Dans, B. Jayaram, F. Lankas, C. Laughton, J. Mitchell, R. Osman, M. Orozco, A. Pérez, D. Petkevičiūtė, N. Spackova, J. Sponer, K. Zakrzewska and R. Lavery, μABC: a systematic microsecond molecular dynamics study of tetranucleotide sequence effects in B-DNA, *Nucleic Acids Res.*, 2014, **42**, 12272–12283.
40. T. Schlick, D. A. Beard, J. Huang, D. A. Strahs and X. Qian, Computational challenges in simulating large DNA over long times, *Comput. Sci. Eng.*, 2000, 38–51.
41. T. E., Cheatham, Simulation and modeling of nucleic acid structure, dynamics and interactions, *Curr. Opin. Struct. Biol.*, 2004, **14**, 360–367.
42. R. Lavery, J. H. Maddocks, M. Pasi and K. Zakrzewska, Analyzing ion distributions around DNA, *Nucleic Acids Res.*, 2014, **42**, 8138–8149.
43. M. Karplus, Molecular dynamics simulations of biomolecules, *Acc. Chem. Res.*, 2002, **35**, 321–322.
44. T. E. Cheatham, P. Cieplak and P. A. Kollman, A modified version of the Cornell *et al.* Force field with improved sugar pucker phases and helical repeat, *J. Biomol. Struct. Dyn.*, 1999, **16**, 845–862.
45. A. D. MacKerell and N. K. Banavali, All-atom empirical force field for nucleic acids: II Application to molecular simulations of DNA and RNA in solution, *J. Comput. Chem.*, 2000, **21**, 105–120.
46. M. Feig and B. M. Pettitt, Sodium and chlorine ions as part of the DNA solvation shell, *Biophys. J.*, 1999, **77**, 1769–1781.
47. T. E. Cheatham and P. A. Kollman, Molecular dynamics simulation of nucleic acids, *Annu. Rev. Phys. Chem.*, 2000, **51**, 435–471.
48. C. Sagui and T. A. Darden, Molecular dynamics simulations of biomolecules: long-range electrostatic effects, *Annu. Rev. Biophys. Biomol. Struct.*, 1999, **28**, 155–179.
49. E. Giudice and R. Lavery, Simulations of nucleic acids and their complexes, *Acc. Chem. Res.*, 2002, **35**, 350–357.
50. T. E. Cheatham, Simulation and modeling of nucleic acid structure, dynamics and interactions, *Curr. Opin. Struct. Biol.*, 2004, **14**, 360–367.
51. Y. Duan and P. A. Kollman, Pathways to a protein folding intermediate observed in a1-microsecond simulation in aqueous solution, *Science*, 1998, **282**, 740–744.
52. R. Elber, A. Ghosh and A. Cardenas, Long time dynamics of complex systems, *Acc. Chem. Res.*, 2002, **35**, 396–403.

53. D. Bashford and D. Case, Generalized Born models of macromolecular solvation effects, *Annu. Rev. Phys. Chem.*, 2000, **51**, 129–152.
54. W. L. Jorgensen, J. Chandrasekhar and J. D. Madura, Comparison of simple potential functions for simulating liquid water, *J. Chem. Phys.*, 1983, **79**, 926–935.
55. V. Tozzini, Coarse-grained models for proteins, *Curr. Opin. Struct. Biol.*, 2005, **15**, 144–150.
56. M. A. Young, S. Gonfloni, G. Superti-Furga, B. Roux and J. Kurigan, Dynamic coupling between SH2 and SH3 domains of c-Src and Hck underlies their inactivation by C-terminal tyrosine phosphorylation, *Cell*, 2001, **105**, 115–126.
57. D. G. Truhlar, J. Gao, C. Alhambra, M. García-Viloca, J. Corchado, M. L. Sánchez and J. Villa, The incorporation of quantum effects in enzyme kinetics modeling, *Acc. Chem. Res.*, 2002, **35**, 341–349.
58. J. Mendieta, S. Martín-Santamaría, E.-M. Priego, J. Balzarini, M.-J. Camarasa, M.-J. Pérez-Pérez and F. Gago, Role of His-85 in the catalytic mechanism of thymidine phosphorylase as assessed by targeted molecular dynamics simulations and quantum mechanical calculations, *Biochemistry*, 2004, **43**, 405–414.
59. B. Isralewitz, M. Gao and K. Schulten, Steered molecular dynamics and mechanical functions of proteins, *Curr. Opin. Struct. Biol.*, 2001, **11**, 224–230.
60. J. Ma and M. Karplus, Molecular switch in signal transduction: reaction paths of the conformational changes in ras p21, *Proc. Natl. Acad. Sci. U. S. A.*, 1997, **94**, 11905–11910.
61. F. Rodríguez-Barrios, J. Balzarini and F. Gago, The molecular basis of resilience to the effect of the Lys103Asn mutation in non-nucleoside HIV-1 reverse transcriptase inhibitors studied by targeted molecular dynamics simulations, *J. Am. Chem. Soc.*, 2005, **127**, 7570–7578.
62. E. Marco, R. García-Nieto and F. Gago, Assessment by molecular dynamics simulations of the structural determinants of DNA binding specificity for transcription factor Sp1, *J. Mol. Biol.*, 2003, **328**, 9–32.
63. G. J. R. Zaman, P. J. Michiels and C. A. van Boeckel, Targeting RNA: new opportunities to address drugless targets, *Drug Discovery Today*, 2003, **8**, 297–306.
64. T. Hunter, Signaling–2000 and beyond, *Cell*, 2000, **100**, 113–127.
65. S. R. Hubbard, Protein tyrosine kinases: autoregulation and small-molecule inhibition, *Curr. Opin. Struct. Biol.*, 2002, **12**, 735–741.
66. S. Klumpp and J. Krieglstein, Serine/threonine protein phosphatases in apoptosis, *Curr. Opin. Pharmacol.*, 2002, **2**, 458–462.
67. J. A. McCammon, B. R. Gelin and M. Karplus, Dynamics of folded proteins, *Nature*, 1977, **267**, 585–590.
68. A. Chien, I. Foster and D. Goddette, Grid technologies empowering drug discovery, *Drug Discovery Today*, 2002, **15**, S176–S180.
69. Folding@home project: http://www.stanford.edu/group/pandegroup/folding/.

70. E. K. Davies and W. G. Richards, The potential of Internet computing for drug discovery, *Drug Discovery Today*, 2002, **7**, S99–S103.
71. B. C. Baguley, Nonintercalative DNA-binding antitumour compounds, *Mol. Cell. Biochem.*, 1982, **43**, 167–181.
72. L. A. Marky and K. J. Breslauer, Origins of netropsin binding affinity and specificity: correlations of thermodynamic and structural data, *Proc. Natl. Acad. Sci. U. S. A.*, 1987, **84**, 4359–4363.
73. F. Gago, C. A. Reynolds and W. G. Richards, The binding of non-intercalative drugs to alternating DNA sequences, *Mol. Pharmacol.*, 1989, **35**, 232–241.
74. P. B. Dervan and B. S. Edelson, Recognition of the DNA minor groove by pyrrole-imidazole polyamides, *Curr. Opin. Struct. Biol.*, 2003, **13**, 284–299.
75. K. L. Buchmueller, A. M. Staples, C. M. Howard, S. M. Horick, P. B. Uthe, N. M. Le, K. K. Cox, B. Nguyen, K. A. Pacheco, W. D. Wilson and M. Lee, Extending the language of DNA molecular recognition by polyamides: unexpected influence of imidazole and pyrrole arrangement on binding affinity and specificity, *J. Am. Chem. Soc.*, 2005, **127**, 742–750.
76. P. B. Dervan, Molecular recognition of DNA by small molecules, *Bioorg. Med. Chem.*, 2001, **9**, 2215–2235.
77. T. P. Straatsma and J. A. McCammon, Computational alchemy, *Annu. Rev. Phys. Chem.*, 1992, **43**, 407–435.
78. P. Kollman, Free energy calculations: applications to chemical and biochemical phenomena, *Chem. Rev.*, 1993, **93**, 2395–2417.
79. T. Rodinger and R. Pomès, Enhancing the accuracy, the efficiency and the scope of free energy simulations, *Curr. Opin. Struct. Biol.*, 2005, **15**, 164–170.
80. C. A. Reynolds, P. M. King and W. G. Richards, Computed redox potentials and the design of bioreductive agents, *Nature*, 1988, **334**, 80–82.
81. F. Gago and W. G. Richards, Netropsin binding to poly[d(IC)]·poly[d(IC)] and poly[d(GC)]·poly[d(GC)]: a computer simulation, *Mol. Pharmacol.*, 1990, **37**, 341–346.
82. *Molecular Basis of Specificitiy in Nucleic Acid-Drug Interactions*, ed. B. Pullman and J. Jortner, Kluwer Academic Publishers, Dordrecht, 1990.
83. M. J. Waring The molecular basis of specific recognition between echinomycin and DNA, in *Molecular Basis of Specificitiy in Nucleic Acid-Drug Interactions*, ed. B. Pullman and J. Jortner, Kluwer Academic Publishers, Dordrecht, 1990, pp. 225–245.
84. K. Hoogsteen, The structure of crystals containing a hydrogen-bonded complex of 1-methylthymine and 9-methyladenine, *Acta Crystallogr.*, 1959, **2**, 822–823.
85. A. H.-J. Wang, G. Ughetto, G. J. Quigley, T. Hakoshima, G. A. van der Marel, J. H. van Boom and A. Rich, The molecular structure of a DNA-triostin A complex, *Science*, 1984, **225**, 1115–1121.
86. U. C. Singh, N. Pattabiraman, R. Langridge and P. A. Kollman, Molecular mechanical studies of $d(CGTACG)_2$: complex of triostin A

with the middle AT base pairs in either a Hoogsteen or Watson-Crick pairing, *Proc. Natl. Acad. Sci. U. S. A.*, 1986, **83**, 6402–6406.
87. X. Gao and D. J. Patel, NMR Studies of echinomycin bisintercalation complexes with d(A1-C2-G3-T4) and d(T1-C2-G3-A4) duplexes in aqueous solution: sequence-dependent formation of Hoogsteen A1 · T4 and Watson-Crick T1 · A4 base pairs flanking the bisintercalation site, *Biochemistry*, 1988, **27**, 1744–1751.
88. J. Gallego, A. R. Ortiz and F. Gago, A molecular dynamics study of the bis-intercalation complexes of echinomycin with $d(ACGT)_2$ and $d(TCGA)_2$ : rationale for sequence-specific Hoogsteen base-pairing, *J. Med. Chem.*, 1993, **36**, 1548–1561.
89. T. V. Alfredson and A. H. Maki, Phosphorescence and optically detected magnetic resonance studies of echinomycin-DNA complexes, *Biochemistry*, 1990, **29**, 9052–9064.
90. R. Rein Studies of biomolecular interactions: principles of nucleic acid structure and function from the point of view of constituent interactions, in *Perspectives in Quantum Chemistry and Biochemistry, vol. II*, ed. B. Pullman, John Wiley & Sons, New York, 1978, pp. 307–362.
91. A. H.-J. Wang, G. Ughetto, G. J. Quigley and A. Rich, Interactions of quinoxaline antibiotic and DNA: the molecular structure of a triostin A – d(GCGTACGC) complex, *J. Biomol. Struct. Dyn.*, 1986, **4**, 319–342.
92. J. A. Cuesta-Seijo and G. M. Sheldrick, Structures of complexes between echinomycin and duplex DNA, *Acta Crystallogr., Sect. D: Biol. Crystallogr.*, 2005, **61**, 442–448.
93. X. Gao and D. J. Patel, Antitumour drug-DNA Interactions: NMR studies of echinomycin and chromomycin complexes, *Q. Rev. Biophys.*, 1989, **22**, 93–138.
94. J. Gallego, F. J. Luque, M. Orozco and F. Gago, Binding of echinomycin to $d(GCGC)_2$ and $d(CCGG)_2$: distinct stacking interactions dictate the sequence-dependent formation of Hoogsteen base pairs, *J. Biomol. Struct. Dyn.*, 1994, **12**, 111–129.
95. M. Aida and C. Nagata, An *ab initio* molecular orbital study on the stacking interaction between nucleic acid bases: dependence on the sequence and relation to the conformation, *Int. J. Quantum Chem.*, 1986, **29**, 1253–1261.
96. R. L. Ornstein, R. Rein, D. L. Breen and R. D. MacElroy, An optimized potential function for the calculation of nucleic acid interaction energies. I. Base stacking, *Biopolymers*, 1978, **17**, 2341–2360.
97. C. Marchand, C. Bailly, M. McLean, S. E. Moroney and M. J. Waring, The 2-amino group of guanine is absolutely required for specific binding of the anti-cancer antibiotic echinomycin to DNA, *Nucleic Acids Res.*, 1992, **20**, 5601–5606.
98. C. Bailly, C. Marchand and M. J. Waring, New binding sites for antitumor antibiotics created by relocating the purine 2-amino group in DNA, *J. Am. Chem. Soc.*, 1993, **115**, 3784–3785.

99. J. Gallego, F. J. Luque, M. Orozco, C. Burgos, J. Alvarez-Builla, M. M. Rodrigo and F. Gago, DNA sequence-specific reading by echinomycin: role of hydrogen bonding and stacking interactions, *J. Med. Chem.*, 1994, **37**, 1602–1609.
100. C. Bailly, S. Echepare, F. Gago and M. J. Waring, Recognition elements that determine affinity and sequence-specific binding to DNA of 2QN, a biosynthetic bis-quinoline analogue of echinomycin, *Anti-Cancer Drug Des.*, 1999, **14**, 291–303.
101. T. V. Alfredson, A. H. Maki and M. J. Waring, Optically detected triplet-state magnetic resonance studies of the DNA complexes of the bisquinoline analogue of echinomycin, *Biochemistry*, 1991, **30**, 9665–9675.
102. C. Bailly and M. J. Waring, DNA recognition by quinoxaline antibiotics: use of base-modified DNA molecules to investigate determinants of sequence-specific binding of triostin A and TANDEM, *Biochem. J.*, 1998, **330**, 81–87.
103. K. J. Addess and J. Feigon, Sequence specificity of quinoxaline antibiotics. 2. NMR studies of the binding of [*N*-MeCys$^3$,*N*-MeCys$^7$]-TANDEM and triostin A to DNA containing a CpI step, *Biochemistry*, 1994, **33**, 12397–12404.
104. M. Lavesa and K. R. Fox, Preferred binding sites for [*N*-MeCys$^3$, *N*-MeCys$^7$]TANDEM determined using a universal footprinting substrate, *Anal. Biochem.*, 2001, **293**, 246–250.
105. B. Honig and A. Nicholls, Classical electrostatics in biology and chemistry, *Science*, 1995, **268**, 1144–1149.
106. E. Marco, A. Negri, F. J. Luque and F. Gago, Role of stacking interactions in the binding sequence preferences of DNA bis-intercalators: insight from thermodynamic integration free energy simulations, *Nucleic Acids Res.*, 2005, **33**, 6214–6224.
107. A. Negri, E. Marco, V. García-Hernández, A. Domingo, A. L. Llamas-Saiz, S. Porto-Sandá, R. Riguera, W. Laine, M. H. David-Cordonnier, C. Bailly, L. F. García-Fernández, J. J. Vaquero and F. Gago, Antitumor activity, X-ray crystal structure, and DNA binding properties of thiocoraline A, a natural bis-intercalating thiodepsipeptide, *J. Med. Chem.*, 2007, **50**, 3322–3333.
108. D. Pelaprat, A. Delbarre, I. Le Guen and B. P. Roques, DNA intercalating compounds as potential antitumor agents. 2. Preparation and properties of 7*H*-pyridocarbazole dimers, *J. Med. Chem.*, 1980, **23**, 1336–1343.
109. A. Delbarre, M. Delepierre, C. Garbay, J. Igolen, J.-B. Le Pecq and B. P. Roques, Geometry of the antitumor drug ditercalinium bisintercalated into d(CpGpCpG)$_2$ by $^1$H NMR, *Proc. Natl. Acad. Sci. U. S. A.*, 1987, **84**, 2155–2159.
110. Q. Gao, L. D. Williams, M. Egli, D. Rabinovich, S. L. Chen, G. J. Quigley and A. Rich, Drug-induced DNA repair: X-ray structure of a DNA-ditercalinium complex, *Proc. Natl. Acad. Sci. U. S. A.*, 1991, **88**, 2422–2426.

111. M. Delepierre, C. Milhe, A. Namane, T. H. Dinh and B. P. Roques, $^{1}$H- and $^{31}$P-NMR studies of ditercalinium binding to a d(GCGC)$_2$ and d(CCTATAGG)$_2$ minihelices: a sequence specificity study, *Biopolymers*, 1991, **31**, 331–353.
112. H. P. Spielmann, D. E. Wemmer and J. P. Jacobsen, Solution structure of a DNA complex with the fluorescent bis-intercalator TOTO determined by NMR spectroscopy, *Biochemistry*, 1995, **34**, 8542–8553.
113. M. E. Peek, L. A. Lipscomb, J. A. Bertrand, Q. Gao, B. P. Roques, C. Garbay-Jaureguiberry and L. D. Williams, DNA distortion in bis-intercalated complexes, *Biochemistry*, 1994, **33**, 3794–3800.
114. B. de Pascual-Teresa, J. Gallego, A. R. Ortiz and F. Gago, Molecular dynamics simulations of the bis-intercalated complexes of ditercalinium and flexi-di with the hexanucleotide d(GCGCGC)$_2$: theoretical analysis of the interaction and rationale for the sequence binding specificity, *J. Med. Chem.*, 1996, **39**, 4810–4824.
115. J. Gallego, B. de Pascual-Teresa, A. R. Ortiz, M. T. Pisabarro and F. Gago Molecular electrostatic potentials of DNA base pairs and drug chromophores in relation to DNA conformation and bis-intercalation by quinoxaline antibiotics and ditercalinium, in *QSAR and Molecular Modelling: Concepts, Computational Tools and Biological Applications*, ed. F. Sanz, J. Giraldo, F. Manaut and J. R. Prous, Barcelona, 1995, pp. 274–281.
116. H. M. Sobell, S. C. Jain, T. D. Sakore and C. E. Nordman, Stereochemistry of actinomycin–DNA binding, *Nature, New Biol.*, 1971, **231**, 200–205.
117. F. Takusagawa, M. Dabrow, S. Neidle and H. M. Berman, The structure of a pseudo intercalated complex between actinomycin and the DNA binding sequence d(GpC), *Nature*, 1982, **296**, 466–469.
118. S. Kamitori and F. Takusagawa, Crystal structure of the 2:1 complex between d(GAAGCTTC) and the anticancer drug actinomycin D, *J. Mol. Biol.*, 1992, **225**, 445–456.
119. S. Kamitori and F. Takusagawa, Multiple binding modes of anticancer drug actinomycin D: X-ray, molecular modeling and spectroscopic studies of d(GAAGCTTC)$_2$–actinomycin D complexes and its host DNA, *J. Am. Chem. Soc.*, 1994, **116**, 4154–4165.
120. T. P. Lybrand, S. C. Brown, S. Creighton, R. H. Shafer and P. A. Kollman, Computer modeling of actinomycin D interactions with double-helical DNA, *J. Mol. Biol.*, 1986, **191**, 495–507.
121. W. Müller and D. M. Crothers, Studies of the binding of actinomycin and related compounds to DNA, *J. Mol. Biol.*, 1968, **35**, 251–290.
122. J. Gallego, A. R. Ortiz, B. de Pascual-Teresa and F. Gago, Structure-affinity relationships for the binding of actinomycin D to DNA, *J. Comput.-Aided Mol. Des.*, 1997, **11**, 114–128.
123. J. H. Williams, The molecular electric quadrupole moment and solid-state architecture, *Acc. Chem. Res.*, 1993, **26**, 593–598.

124. F. Gago, Stacking interactions and intercalative DNA binding, *Methods*, 1998, **14**, 277–292.
125. K. L. Rinehart, Antitumor compounds from tunicates, *Med. Res. Rev.*, 2000, **20**, 1–27.
126. I. Manzanares, C. Cuevas, R. García-Nieto, E. Marco and F. Gago, Advances in the chemistry and pharmacology of ecteinascidins, a promising new class of anticancer agents, *Curr. Med. Chem.: Anti-Cancer Agents*, 2001, **1**, 257–276.
127. F. Gago and L. H. Hurley, Devising a structural basis for the potent cytotoxic effects of ecteinascidin 743, in *Small Molecule DNA and RNA Binders: From Synthesis to Nucleic Acid Complexes*, ed. M. Demeunynck, C. Bailly and W. D. Wilson, Wiley-VCH, Weinheim, Germany, 2002, pp. 643–675.
128. M. Zewail-Foote and L. H. Hurley, Ecteinascidin 743: a minor groove alkylator that bends DNA toward the major groove, *J. Med. Chem.*, 1999, **42**, 2493–2497.
129. G. Ravishanker, S. Swaminathan, D. L. Beveridge, R. Lavery and H. Sklenar, Conformational and helicoidal analysis of 30 ps of molecular dynamics on the d(CGCGAATTCGCG) double helix: "curves", dials and windows, *J. Biomol. Struct. Dyn.*, 1989, **6**, 669–699.
130. R. Lavery and H. Sklenar, The definition of generalized helicoidal parameters and of axis curvature for irregular nucleic acids, *J. Biomol. Struct. Dyn.*, 1988, **6**, 63–91.
131. C. Blanchet, M. Pasi, K. Zakrzewska and R. Lavery, CURVES+ web server for analyzing and visualizing the helical, backbone and groove parameters of nucleic acid structures, *Nucleic Acids Res.*, 2011, **39**(Web Server issue), W68–W73.
132. R. García-Nieto, I. Manzanares, C. Cuevas and F. Gago, Increased DNA binding specificity for antitumor ET743 through protein-DNA interactions?, *J. Med. Chem.*, 2000, **43**, 4367–4369.
133. E. Marco, R. García-Nieto, J. Mendieta, I. Manzanares, C. Cuevas and F. Gago, A 3 · (ET743)-DNA complex that both resembles an RNA-DNA hybrid and mimics zinc finger-induced DNA structural distortions, *J. Med. Chem.*, 2002, **45**, 871–880.
134. S. Arnott and D. W. Hukins, Optimised parameters for A-DNA and B-DNA, *Biochem. Biophys. Res. Commun.*, 1972, **47**, 1504–1509.
135. H. L. Ng, M. L. Kopka and R. E. Dickerson, The structure of a stable intermediate in the A↔B DNA helix transition, *Proc. Natl. Acad. Sci. U. S. A.*, 2000, **97**, 2035–2039.
136. K. D. Paull, R. H. Shoemaker, L. Hodes, A. Monks, D. A. Scudiero, L. Rubinstein, J. Plowman and M. R. Boyd, Display and analysis of patterns of differential activity of drugs against human tumor cell lines: development of mean graph and COMPARE algorithm, *J. Natl. Cancer Inst.*, 1989, **81**, 1088–1092.
137. E. Marco and F. Gago, DNA structural similarity in the 2:1 complexes of the antitumor drugs Yondelis™ (trabectedin) and chromomycin $A_3$ with

an oligonucleotide sequence containing two adjacent TGG binding sites on opposing strands, *Mol. Pharmacol.*, 2005, **68**, 1–9.

138. R. Mantovani, A survey of 178 NF-Y binding CCAAT boxes, *Nucleic Acids Res.*, 1998, **26**, 1135–1143.
139. K. Grzeskowiak, D. S. Goodsell, M. Kaczor-Grzeskowiak, D. Cascio and R. E. Dickerson, Crystallographic analysis of C-C-A-A-G-C-T-T-G-G and its implications for bending in B-DNA, *Biochemistry*, 1993, **32**, 8923–8931.
140. M.-H. Hou, H. Robinson, Y.-G. Gao and A. H.-J. Wang, Crystal structure of the [$Mg^{2+}$-(chromomycin $A_3)_2$]-d$(TTGGCCAA)_2$ complex reveals GGCC binding specificity of the drug dimer chelated by a metal ion, *Nucleic Acids Res.*, 2004, **32**, 2214–2222.
141. U. Dornberger, J. Flemming and H. Fritzsche, Structure determination and analysis of helix parameters in the DNA decamer d(CATGGC-CATG)$_2$ comparison of results from NMR and crystallography, *J. Mol. Biol.*, 1998, **284**, 1453–1463.
142. D. S. Goodsell, M. L. Kopka, D. Cascio and R. E. Dickerson, Crystal structure of CATGGCCATG and its implications for A-tract bending models, *Proc. Natl. Acad. Sci. U. S. A.*, 1993, **90**, 2930–2934.
143. A. B. Herrero, C. Martín-Castellanos, E. Marco, F. Gago and S. Moreno, Cross-talk between nucleotide excision and homologous recombination DNA repair pathways in the mechanism of action of antitumor trabectedin, *Cancer Res.*, 2006, **66**, 8155–8162.
144. S. Feuerhahn, C. Giraudon, M. Martínez-Díez, J. A. Bueren-Calabuig, C. M. Galmarini, F. Gago and J. M. Egly, XPF-dependent DNA breaks and RNA polymerase II arrest induced by antitumor DNA interstrand crosslinking-mimetic tetrahydroisoquinoline alkaloids, *Chem. Biol.*, 2011, **18**, 988–999.
145. E. Marco, M.-H. David-Cordonnier, C. Bailly, C. Cuevas and F. Gago, Further insight into the DNA recognition mechanism of trabectedin from the differential affinity of its demethylated analog ecteinascidin ET729 for the triplet binding site CGA, *J. Med. Chem.*, 2006, **49**, 6925–6929.
146. J.-F. M. Leal, V. García-Hernández, V. Moneo, D. Domingo, J. A. Bueren-Calabuig, A. Negri, F. Gago, M. J. Guillén-Navarro, P. Avilés, C. Cuevas, L. F. García-Fernández and C. M. Galmarini, Molecular pharmacology and antitumor activity of *Zalypsis*® in several human cancer cell lines, *Biochem. Pharmacol.*, 2009, **78**, 162–170.
147. J. F. M. Leal, M. Martínez-Díez, V. García-Hernández, V. Moneo, A. Domingo, J. A. Bueren-Calabuig, A. Negri, F. Gago, M. J. Guillén-Navarro, P. Avilés, C. Cuevas, L. F. García-Fernández and C. M. Galmarini, PM01183, a new DNA minor-groove covalent binder with potent in vitro and in vivo antitumor activity, *Br. J. Pharmacol.*, 2010, **161**, 1099–1110.
148. K. Y. Wong and B. M. Pettitt, The pathway of oligomeric DNA melting investigated by molecular dynamics simulations, *Biophys. J.*, 2008, **95**, 5618–5626.

149. J. A. Bueren-Calabuig, C. Giraudon, C. M. Galmarini, J. M. Egly and F. Gago, Temperature-induced melting of double-stranded DNA in the absence and presence of covalently bonded antitumour drugs: insight from molecular dynamics simulations, *Nucleic Acids Res.*, 2011, **39**, 8248–8257.
150. S. Martínez, L. Pérez, C. Galmarini, M. Aracil, J. C. Tercero, F. Gago, B. Albella and J. A. Bueren, Inhibitory effects of marine-derived DNA-binding antitumour tetrahydroisoquinolines on the Fanconi anemia pathway, *Br. J. Pharmacol.*, 2013, **170**, 871–882.
151. K. D. Mirchandani and A. D. D'Andrea, The Fanconi anemia/BRCA pathway: A coordinator of cross-link repair, *Exp. Cell Res.*, 2006, **312**, 2647–2653.
152. J. A. Bueren-Calabuig, A. Negri, A. Morreale and F. Gago, Rationale for the opposite stereochemistry of the major monoadducts and interstrand crosslinks formed by mitomycin C and its decarbamoylated analogue at CpG steps in DNA and the effect of cytosine modification on reactivity, *Org. Biomol. Chem.*, 2012, **10**, 1543–1552.
153. P. A. Kollman, I. Massova, C. Reyes, B. Kuhn, S. Huo, L. Chong, M. Lee, T. Lee, Y. Duan, W. Wang, O. Donini, P. Cieplak, J. Srinivasan, D. A. Case and T. E. Cheatham, Calculating structures and free energies of complex molecules: combining molecular mechanics and continuum models, *Acc. Chem. Res.*, 2000, **33**, 889–897.
154. W. S. Johnson, Q. Y. He and M. Tomasz, Selective recognition of the m5CpG dinucleotide sequence in DNA by mitomycin C for alkylation and cross-linking, *Bioorg. Med. Chem.*, 1995, **3**, 851–860.
155. M. Tomasz, R. Lipman, D. Chowdary, J. Pawlak, G. L. Verdine and K. Nakanishi, Isolation and structure of a covalent cross-link adduct between mitomycin C and DNA, *Science*, 1987, **235**, 1204–1208.
156. M. M. Paz, S. Ladwa, E. Champeil, Y. Liu, S. Rockwell, E. K. Boamah, J. Bargonetti, J. Callahan, J. Roach and M. Tomasz, Mapping DNA adducts of mitomycin C and decarbamoyl mitomycin C in cell lines using liquid chromatography/electrospray tandem mass spectrometry, *Chem. Res. Toxicol.*, 2008, **21**, 2370–2378.
157. B. L. Staker, K. Hjerrild, M. D. Feese, C. A. Behnke, A. B. Burgin and L. Stewart, The mechanism of topoisomerase I poisoning by a camptothecin analog, *Proc. Natl. Acad. Sci. U. S. A.*, 2002, **99**, 15387–15392.
158. C. J. Thomas, N. J. Rahier and S. M. Hecht, Camptothecin: current perspectives, *Bioorg. Med. Chem.*, 2004, **12**, 1585–1604.
159. M. R. Redinbo, L. Stewart, P. Kuhn, J. J. Champoux and W. G. J. Hol, Crystal structures of human topoisomerase I in covalent and noncovalent complexes with DNA, *Science*, 1998, **279**, 1504–1513.
160. C. Tardy, M. Facompré, M. Laine, B. Baldeyrou, D. García-Grávalos, A. Francesch, C. Mateo, A. Pastor, J. A. Jiménez, I. Manzanares, C. Cuevas and C. Bailly, Topoisomerase I-mediated DNA cleavage as a guide to the development of antitumor agents derived from the marine

alkaloid lamellarin D: Triester derivatives incorporating amino acid residues, *Bioorg. Med. Chem.*, 2004, **12**, 1697–1712.

161. E. Marco, W. Laine, C. Tardy, A. Lansiaux, M. Iwao, F. Ishibashi, C. Bailly and F. Gago, Molecular determinants of topoisomerase I poisoning by lamellarins: comparison with camptothecin and structure-activity relationships, *J. Med. Chem.*, 2005, **48**, 3796–3807.
162. M. Facompré, C. Tardy, C. Bal-Mayeu, P. Colson, C. Pérez, I. Manzanares, C. Cuevas, C. Bailly and D. Lamellarin, A novel potent inhibitor of topoisomerase I, *Cancer Res.*, 2003, **63**, 7392–7399.
163. G. M. Morris, D. S. Goodsell, R. Huey, W. E. Hart, S. Halliday, R. Belew and A. J. Olson *AutoDock: Automated Docking of Flexible Ligands to Receptors, Version 3.0*, The Scripps Research Institute, La Jolla, CA, 1999.
164. B. L. Staker, M. D. Feese, M. Cushman, Y. Pommier, D. Zembower, L. Stewart and A. B. Burgin, Structures of three classes of anticancer agents bound to the human topoisomerase I-DNA covalent complex, *J. Med. Chem.*, 2005, **48**, 2336–2345.
165. A. Molina, J. J. Vaquero, J. L. García-Navío, J. Alvarez-Builla, B. de Pascual-Teresa, F. Gago, M.-M. Rodrigo and M. Ballesteros, Synthesis and DNA binding properties of γ-carbolinium derivatives and benzologues, *J. Org. Chem.*, 1996, **61**, 5587–5599.
166. J. Pastor, J. Siro, J. L. García-Navío, J. J. Vaquero, J. Alvarez-Builla, F. Gago, B. de Pascual-Teresa, M. Pastor and M. M. Rodrigo, Azino fused benzimidazolium salts as DNA intercalating agents, *J. Org. Chem.*, 1997, **62**, 5476–5483.
167. A. Molina, J. J. Vaquero, J. L. García-Navío, J. Alvarez-Builla, B. de Pascual-Teresa, F. Gago and M. M. Rodrigo, Novel DNA intercalators based on the pyridazino[1′,6′:1,2]pyrido[4,3-*b*]indol-5-inium system, *J. Org. Chem.*, 1999, **64**, 3907–3915.
168. V. Martínez, C. Burgos, J. Alvarez-Builla, G. Fernández, A. Domingo, R. García-Nieto, F. Gago, I. Manzanares, C. Cuevas and J. J. Vaquero, Benzo[f]azino[2,1-a]phthalazinium cations: novel DNA intercalating chromophores with antiproliferative activity, *J. Med. Chem.*, 2004, **47**, 1136–1148.

CHAPTER 6

# Binding of Small Molecules to Trinucleotide DNA Repeats Associated with Neurodegenerative Diseases

MING-HON HOU,*[a,b] ROSHAN SATANGE[a,b] AND CHUNG-KE CHANG[c]

[a] Institute of Genomics and Bioinformatics; National Chung Hsing University, Taichung, Taiwan; [b] PhD Program in Medical Biotechnology, National Chung Hsing University, Taichung, Taiwan; [c] Institute of Biomedical Sciences, Academia Sinica, Nangang, Taipei, Taiwan
*Email: mhho@dragon.nchu.edu.tw

## 6.1 Introduction

The eukaryotic genome contains a large number of repetitive DNA sequences dispersed throughout the genetic material which play major roles in the maintenance of normal function and pathology.[1] The DNA repeats undergo frequent length changes due to insertions in the repeat units (generally termed repeat expansion) that may disrupt cellular replication, repair and recombination processes and ultimately lead to altered gene expression. The association of a number of human disorders with expansion of repetitive DNA has been well established.[1,2] The correlation between unusual repetitive sequences and the aetiology of human disease, and the mechanism by which these repeats are expanded, have been extensively studied in the past

Chemical Biology No. 7
DNA-targeting Molecules as Therapeutic Agents
Edited by Michael J. Waring

Published by the Royal Society of Chemistry, www.rsc.org

two decades. Over 30 different hereditary disorders have been associated with an increase in the number of copies of simple repeats in human genomic DNA. Neurological diseases, in particular, comprise a large portion of these disorders. Among these, expansion of trinucleotide repeats (TNRs) is arguably the most important, accounting for at least 16 diseases including some of the most debilitating ones such as Huntington's and fragile X syndrome. Unusual expansion of TNRs located in the coding region (exons) as well as in non-coding regions (introns or UTR's) can cause these diseases.[3] For instance, the expansion of CGG TNRs in the coding sequence of the FMR1 (fragile mental retardation 1) gene is known to produce an aberrant protein that plays a critical role in the pathogenesis of this disease.[4,5] On the other hand, expansion of a CTG TNR in the 3′-untranslated region of DMPK, the gene encoding myotonic dystrophy protein, causes the dominantly inherited neuromuscular disorder myotonic dystrophy. Table 6.1 summarizes different types of trinucleotide repeat expansion disease as well as their association with the human genome.[5–7]

As mentioned above, the TNRs are of special importance when they are implicated in some of the most devastating neurological disorders, such as Huntington's disease, and are the best studied of all DNA repeat expansions. Recent advances in the field of structural and molecular biology have allowed researchers to obtain mechanistic insights into how such expansions might occur.[8] Availability of TNR structures at atomic resolution further boosted the study of the interaction between DNA repeats and small molecules, many of which could have therapeutic or diagnostic value. Small molecules which bind to TNR DNA in a sequence-specific manner have the potential to inhibit the activity of such genes and consequently may be useful for treating a variety of diseases. In fact, next-generation drug development against various neurological diseases is mainly focused on the discovery of small molecules that bind to DNA repeats.[9,10] This chapter outlines the mechanism of TNR expansion along with our current knowledge of how various DNA-binding ligands interact with TNRs—knowledge which may ultimately lead to potential applications in clinical settings.

### 6.1.1 Trinucleotide Repeat DNA

While over 15 degenerative diseases have been associated with the abnormal expansion of trinucleotide repeats, only a few of all the possible triplet expansions, namely CAG, CGG, GCC, CTG and GAA, have been associated with known pathological disorders (Table 6.1). This raises a basic question at the molecular level. What are the structural features that make particular DNA repeats pathogenic? What is the mechanism for their expansion and how does it depend on the triplet itself? Baldi *et al.* tried to probe these questions using computational methods.[6] It has been shown that the triplets in TNRs fall into extreme classes in terms of structural parameters when compared with normal DNA. These extreme properties pertain to the triplet repeat

**Table 6.1** List of human diseases associated with trinucleotide DNA repeats (TNRs) expansion. Reprinted from *DNA Repair*, **32**, Zhao and Usdin, The Repeat Expansion Diseases: The dark side of DNA repair, 96–105, Copyright 2015, with permission from Elsevier.[a]

| Disease name | Repeat type | Normal repeat length | Disordered repeat length | Gene/locus name | Repeat location |
|---|---|---|---|---|---|
| Huntington's disease | CAG | 9–35 | 36–121 | HTT/IT15 | Coding |
| SCA type 1 | | 6–42 | 39–83 | ATXN1/SCA1 | Coding |
| SCA type 2 | | 14–31 | 32–200 | ATXN2/SCA2 | Coding |
| SCA type 3/Machado Joseph disease | | 12–44 | 52–86 | ATXN3/SCA3 | Coding |
| SCA type 6/Episodic ataxia type 2 | | 4–18 | 21–33 | CACNA1A | Coding |
| SCA type 7 | | 4–35 | 37–306 | ATXN7/SCA7 | Coding |
| SCA type 12 | | 7–28 | 55–78 | PPP2R2B | Non-coding (5′ UTR) |
| SCA type 17 | | 25–42 | 45–66 | TBP | Coding |
| DRPLA/Haw River syndrome | | 7–35 | 49–93 | ATN1/DRPLA (B37) | Coding |
| Kennedy's diseases/SBMA | | 9–34 | 38–62 | AR | Coding |
| Fragile X syndrome (FXS) | CGG | <55 | 60–200 | FMR1/FRXA | Non-coding (5′ UTR) |
| Myotonic dystrophy 1 (DM1) | CTG | 5–38 | >50 | DMPK | Non-coding (3′ UTR) |
| SCA type 8 | | 15–37 | >74 | ATXN8/SCA8 | Non-coding (3′ UTR) |
| FRAXE MR | GCC | 6–25 | >200 | AFF/FMR2 (FRAXE) | Non-coding (5′ UTR) |
| Friedreich's ataxia (FRDA) | GAA | 8–33 | >80–90 | FXN | Non-coding |

[a]Abbreviations: SCA, Spinocerebellar ataxia; DRPLA, Dentatorubral–pallidoluysian atrophy; SBMA, Spinal and bulbar muscular atrophy; FRAXE MR, Fragile XE mental retardation syndrome.

class, rather than the triplet alone. A triplet that is not extreme for a given scale may become extreme once its two shifted versions are considered. To exemplify, AGC has relatively low bendability when taken alone, but corresponds to the most bendable class when GCA and CAG classes are taken into account in the following repeating sequence: AGCAGCAGC. Furthermore, computational and experimental evidence showed that $(CTG)_n$.$(CAG)_n$ repeats are highly flexible compared to all other classes of repeats, which may explain the association of more diseases with their abnormal expansion.[11,12] Studies have also shown that the length of repeats is one of the most important determinants of onset of disease.[13] From a structural viewpoint, these repeat sequences can form a variety of secondary structures, such as hairpins, intramolecular triplexes, quadruplexes and slipped-strand structures, that may be important intermediates, resulting in notable genetic instability.[14]

### 6.1.2 Diseases Associated with Expansion of Repetitive DNA

Since the 1990s triplet repeat disorders have become some of the most widely studied inherited diseases. In particular, neurological diseases comprise a large portion of these disorders.[15] These include diseases caused by abnormal expansion of TNRs, *e.g.* $(CGG)_n$/$(CCG)_n$ in fragile X syndrome,[16,17] $(CAG)_n$/$(CTG)_n$ in Huntington's disease,[18,19] $(GAA)_n$/$(TTC)_n$ in Friedreich's ataxia and $(GCN)_n$/$(NGC)_n$ in a variety of diseases.[20,21] The age of incidence for these diseases ranges from early childhood, such as X-linked mental retardation, to the later stages of life, as in cases of Huntington's disease or the inherited ataxias.

Most of these diseases show a broad spectrum of phenotypes along with a unique mechanism of pathogenesis, which can be attributed to the difference in the size of the repeats. Other important factors pertaining to the pathogenesis of diseases may include the sequence of trinucleotide repeats, their position in a particular gene, as well as other environmental factors. Studies have shown that the pathogenesis of trinucleotide repeat disorders is highly complex and might involve more than one molecular mechanism.[22] The next sub-section will discuss possibilities.

### 6.1.3 Molecular Mechanism of TNR Expansion

Several models have been proposed to explain TNR expansion. The first was based on the idea of occasional strand slippage during replication. Several repeats might be 'looped out' in the nascent DNA strand which would convert the loop into expansions after subsequent rounds of replication.[8,23] However, this hypothesis failed to explain why only a handful of all repeats expand, or what determines the threshold length and the large-scale character of the expansions. Furthermore, instead of a simple loop structure, it has been reported that the expandable repeats adopt unusual structural features in either the single-stranded or double-stranded state in the cell,[24] which may affect the outcome of the DNA strand slippage model. For example, single-stranded $(CNG)_n$ repeats are able to form hairpin DNA structures that consist of both Watson–Crick base pairs and mismatched base pairs. The stability of these hairpin DNA structures changes in the order: CGG > CCG ~ CTG > CAG. Thus, the original DNA slippage model of repeat expansion has been amended to take into consideration the physical and chemical properties of the TNRs. In the modified model, the transient intra-strand hairpin structure containing mismatched base pairs can promote DNA slippage in a sequence-dependent fashion and is a causative factor for DNA expansion.

Extreme TNR expansions, such as those observed in many neurological diseases, can be attributed to the combinatorial effects of the transient formation of hairpins along with slipped registers in the complementary DNA strands during replication.[25] Negative supercoiling of DNA introduces strain that may allow triplet repeats to easily adopt conformations different from the canonical duplex.[26] For instance, the duplex structure of $(CGG)_n$ repeats may contain G–G mismatch pairs and the duplex structure associated with the $(CCG)_n$ repeat appears to have C–C mismatch pairs in both intra-helical and extra-helical conformations. However, single-stranded $(CGG)_n$ and $(CCG)_n$ repeats can also fold into tetra-helical structures stabilized by intertwining G quartets and i-motifs.[27,28] These alternative conformations may force DNA polymerase to pause and, along with primer relocation, may ultimately result in the elongation of integral numbers of triplets to produce TNR expansions. This suggests that the interaction between TNR structure and DNA polymerases may also play a central role in the generation of genetic instabilities leading to repeat expansion.

Regardless of the model, the current hypotheses for the mechanism of abnormal TNR expansion are all centred around the generation of DNA replication errors (Figure 6.1). A caveat here is that our understanding of the exact mechanism of TNR expansion is limited to date; however, with the advancement of newer techniques it may be possible to deduce more exact mechanisms behind abnormal (or unusual) TNR expansion. Elucidating these mechanisms will definitely help towards designing novel therapeutic molecules or modifying existing therapies. In subsequent sections, we will elaborate on our current knowledge about existing small molecules that bind to TNRs as well as their mechanisms of interaction.

## 6.2 Interaction of DNA-binding Drugs with Triplet Repeats Connected with Neurological Diseases

Current treatments for neurological diseases associated with TNR expansion are based on the use of drugs which can provide symptomatic relief; however there is no drug available that can be used to provide a fundamental cure for these diseases. For example, antipsychotic drugs, such as

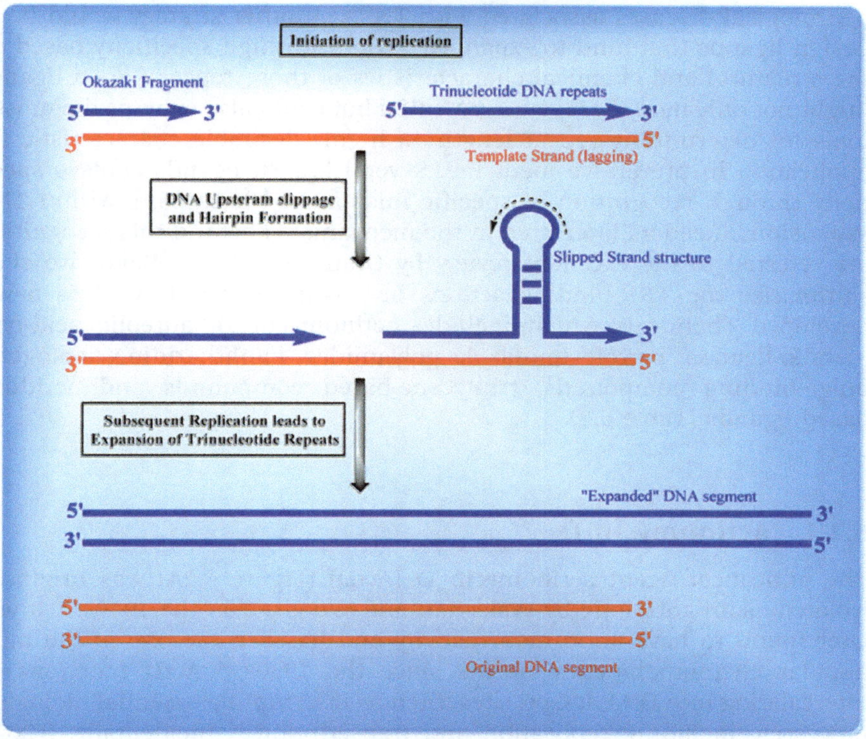


**Figure 6.1** Proposed mechanism of abnormal TNR DNA expansion. During replication, part of the TNRs on the newly synthesized strand (blue) may dissociate from the template ('slip') and form a hairpin structure stabilized by intra-strand base pairing (short blue lines). This results in shortening of the new strand and allows for its continued extension, thus incorporating a larger number of TNRs than are present in the parent strand (red). The final double-stranded product contains a heteroduplex within the TNR region, with expanded TNRs on the new strand and the normal number of TNRs on the parent strand. Since the new strand containing expanded TNRs can act as a template during subsequent rounds of replication, eventually the replicated DNA products will contain expanded TNRs on both strands of the duplex.

olanzapine and valproate, are effective against neurologic symptoms and involuntary movement.[29,30] As described previously, the length of the TNR is the most important determinant of these neurological disorders, so drugs that reduce the repeat length or impair repeat expansion could be used to delay onset and reduce the severity of these diseases. Intriguingly, many important anticancer drugs and antibiotics possess desirable qualities when interacting with DNA that may help to combat TNR-related disorders. It has been established that such drugs can inhibit replication or block transcription by interfering with the template function of DNA, and thus may form the basis for development of novel therapeutics against neurological diseases associated with TNRs. Another strategy would be to design ligands that bind to expanded TNRs with high specificity based on the structural and chemical characteristics of those repeats. Such ligands might not only have therapeutic potential but could also serve as diagnostic tools for determining repeat length—a highly desirable characteristic for application in preventive medicine. Several known ligands achieve specificity through recognition of specific mismatched base pairs within TNR expansions. Readers interested in the mechanisms behind this recognition are referred to the excellent review by Granzhan *et al.*[31] Below we shall summarize the TNR-binding actions of existing drugs as well as newly developed chemicals which include: actinomycin D, aureolic acid-type metallo-ligands, pyrrole–imidazole polyamides, naphtharidine analogues, bulge-binding compounds, triptycene-based compounds and acridine-based ligands (Table 6.2).

### 6.2.1 Actinomycin D

The anticancer agent actinomycin D (ActD) (Figure 6.2A) was originally isolated as an antibiotic by Waksman and Woodruff in the 1940s.[32] It was later found to have antitumour activity and has become one of the most popular chemotherapeutic agents since the 1960s.[33] ActD functions by intercalating into DNA at GpC sites, thereby interrupting essential biological processes, including replication and transcription.[34] Incidentally, $(CXG)_n$ TNRs have extensive clustering of GpC sequences, which consequently provide potent binding targets for ActD. Previously it has been demonstrated that the binding affinity at a GpC site is influenced by its flanking sequences. The crystal structure of the ActD–$(GATGCTTC)_2$ complex showed that a T–T mismatch adjacent to a GpC step provides an excellent binding site for ActD by creating a cavity at the TpG step, which allows the *N*-methyl group of *N*-methylvaline (MeVal) to fit in.[35] Conversely, we showed through thermal denaturation circular dichroism analyses of $(CTG)_n$ TNRs that ActD favours binding to GpC sequences flanking a single T–T mismatch over those flanking a single or multiple Watson–Crick base pairs, and that the binding induces conformational transition of the DNA to an A-type structure. The structural basis of this preference was elucidated from the crystal structure of ActD bound to a $(ATGCTGCAT)_2$ duplex (Figure 6.2B), which contains a CTG triplet sequence.[36] Remarkably, two individual ActD molecules were found to bind to the two GpC sites separated by the T–T mismatch in the complex. The proximity of the two ActDs in the minor groove results in contacts between the two β peptide rings, each from one ActD molecule, and induces a slight bending of the nonameric DNA duplex in which the two halves are tilted towards the major groove. The helix axis of the top half bends $\sim 15^\circ$ away from that of the bottom half. Extensive hydrophobic interactions between the two β peptide rings stabilise the two ActD

**Table 6.2** List of currently available compounds acting on Trinucleotide DNA Repeats (TNRs) Expansion.

| Ligand | Classification | Target repeats | Function | Ref. | PDB_ID |
|---|---|---|---|---|---|
| Actinomycin D | DNA-binding antibiotic | CTG, CGG | Inhibition of DNA replication | 36, 38 | 4HIV, 1MNV |
| $Co^{II}(Chro)_2$ | DNA-binding antibiotic | CCG | Inhibition of DNA replication | [a] | 5XJW, 5YZE |
| $Ni^{II}(Chro)_2$ | DNA-binding antibiotic | CCG | Detection of triplet repeat sequences | 44 | 5XEW |
| Py–Im polyamide–Pyrene | Pyrrole–imidazole polyamide | CAG | Detection of triplet repeat sequences | 47 | – |
| Py–Im polyamide–chlorambucil | Pyrrole–imidazole polyamide | CAG/CTG | Inhibition of DNA alkylation | 50 | – |
| Naphthyridine azaquinolone (NA) | Naphthyridine analog | CAG | Inhibition of DNA replication | 51 | 1X26 |
| Naphthyridine carbamate dimer (NCD) | Naphthyridine analog | CGG | Inhibition of DNA replication | 52, 53, 55 | – |
| Naphthyridine tetramers (Z-NCTS) | Naphthyridine analog | CGG | Inhibition of DNA replication | 55 | – |
| Methylcarbamoylnaphthyridine dimer (MCND) | Naphthyridine analog | GAA | Inhibition of DNA replication | 57 | – |
| FecNCD1 | Naphthyridine analog | CGG | Detection of triplet repeat sequences | 59 | – |
| FecNCD2 | Naphthyridine analog | CGG | Detection of triplet repeat sequences | 59 | – |
| 2-amino-5,6,7-trimethyl-1,8-naphthyridine (ATMND) | Naphthyridine analog | CCG | Detection of triplet repeat sequences | 60 | – |
| Aminobenzo[c][1,8]naphthyridine derivative (Am-BzN) | Naphthyridine derivative | CCG | Detection of triplet repeat sequences | 61 | – |
| CMBL4 | Naphthyridine derivative | TGG | Recognition of the T/GG site | 58 | – |

**Table 6.2** (*Continued*)

| Ligand | Classification | Target repeats | Function | Ref. | PDB_ID |
|---|---|---|---|---|---|
| Double-decker intercalator (DDI) | NCSi-gb mimic | ATT, AAT, CTG | Stimulation of DNA slippage synthesis | 66 | – |
| DDI-1A | NCSi-gb mimic | AAT, CAG | Stimulation of DNA slippage synthesis | 63 | – |
| DDI-1B | NCSi-gb mimic | AAT, CAG | Stimulation of DNA slippage synthesis | 63 | – |
| Trip3, Trip4 | Triptycene-based ligand | CAG/CTG | Nucleic acid junction binding | 71 | – |
| Triaminotriazine-acridine conjugate | Acridine based ligand | CTG, CUG | Inhibition of replication and transcription | 72 | – |
| 1*H*-pyrrolo [3,2-*h*]uinolone-8-amine (PQA) | Pyrrolo quinolone amine | CTG | Binding to T bulge DNA | 73 | – |

[a]Unpublished data.

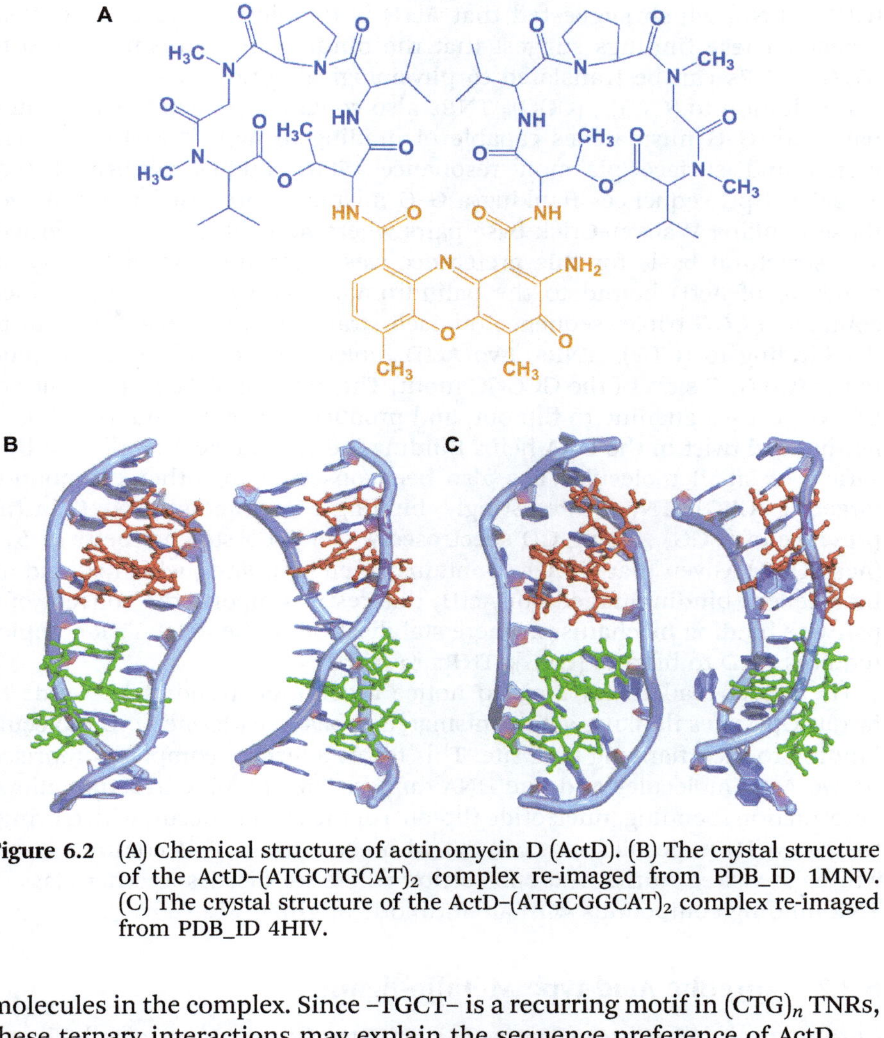


**Figure 6.2** (A) Chemical structure of actinomycin D (ActD). (B) The crystal structure of the ActD–$(ATGCTGCAT)_2$ complex re-imaged from PDB_ID 1MNV. (C) The crystal structure of the ActD–$(ATGCGGCAT)_2$ complex re-imaged from PDB_ID 4HIV.

molecules in the complex. Since –TGCT– is a recurring motif in $(CTG)_n$ TNRs, these ternary interactions may explain the sequence preference of ActD.

Siboni *et al.* tested the effect of ActD on cell and mouse models of the inherited disease myotonic dystrophy type 1 (DM1).[37] DM1 patients have large CTG expansions that are toxic when transcribed and affect the ability of muscles to relax after contraction. They hoped that ActD would be able to reduce or eliminate transcription of CTG repeats in the cells. Siboni *et al.* demonstrated that ActD did decrease CUG transcript levels in a dose-dependent manner in DM1 cell and mouse models at significantly lower concentrations (nanomolar) than those used as a general transcription inhibitor or cancer chemotherapeutic. Their ITC data also supported an allosteric model of ActD binding to CTG repeat DNA, consistent with the previously mentioned crystal structure results. On the other hand, ActD did not produce an enthalpic change when titrated into

$(CUG)_4$ RNA, which suggested that ActD is unable to bind to CUG RNA repeats. These findings suggest that the binding mechanism of ActD to $(CTG)_n$ TNRs can be translated to physiological conditions.

In addition to $(CTG)_n$, $(CGG)_n$ TNRs also contain GpC clusters that alternate with G–G mismatches capable of binding to ActD.[38] Melting experiments and surface plasmon resonance (SPR) studies confirmed that adjacent GpC sequences flanking a G–G mismatch are also preferred over those flanking Watson–Crick base pairs where ActD binding is concerned. The structural basis for this preference was determined from the crystal structure of ActD bound to the palindromic ATGCGGCAT duplex, which contains a CGG triplet sequence on each strand (Figure 6.2C).[38] Similar to the binding to $(CTG)_n$ TNRs, two ActD molecules were observed to bind to the two GpC sites of the GCGGC motif. The binding of ActD molecules to GCGGC causes guanine to flip out, and produces a sharp bend as well as a left-handed twist in the DNA helix. Binding-induced nucleotide flip-out by a variety of small molecules has also been observed for other compounds targeting $(CXG)_n$ TNRs. Interestingly, binding experiments on ActD in the presence of $(CGG)_{16}$ using CD spectroscopy yielded a stoichiometry of 6 : 1 (ActD:DNA). Given that $(CGG)_{16}$ contains seven GpC sites, which should all be excellent binding targets for ActD, this result supports the concept of a pairwise binding mechanism, where stabilisation of the ActD–TNR complex requires ActD to bind to $(CXG)_n$ TNRs in pairs.

The astute reader may start to notice a common trend: ActD tends to favour GpC sites flanking single mismatched bases, with one ActD molecule binding to each flanking GpC site. This forms a ternary complex comprised of two ActD molecules and one DNA duplex. The complex structure allows deformation (bending, nucleotide flip-out) of the DNA to occur, which brings the two ActD molecules into contact with each other. This common trend will be re-visited in the next sub-section where we discuss another class of TNR-binding compounds with an antibiotic origin.

### 6.2.2 Aureolic Acid-type Metallo-ligands

Chromomycin A3 (Chro) (Figure 6.3), an aureolic acid-type metallo-ligand isolated from *Streptomyces griseus*, was originally discovered for its antibiotic and antitumour activities in the 1960s.[39] It was later demonstrated by optical spectroscopy and structural studies that Chro was able to bind to DNA duplexes at GC-rich sequences that are at least three base pairs (bp) long.[40] This DNA-binding activity requires the presence of divalent ions. Chro contains di- and trisaccharide moieties connected to a β-ketophenol chromophore *via* *O*-glycosidic bonds arranged in a 2,6 relationship around the anthracene ring. The crystal structure of a Chro–DNA complex revealed that the metal-coordinated dimer of Chro causes a kink in the DNA on the minor groove side and induces a local conformational change in the duplex.[41] Dimerization of Chro through the metal ion allows the two Chro monomers to clasp onto both strands of the duplex, resulting in enhanced stability.

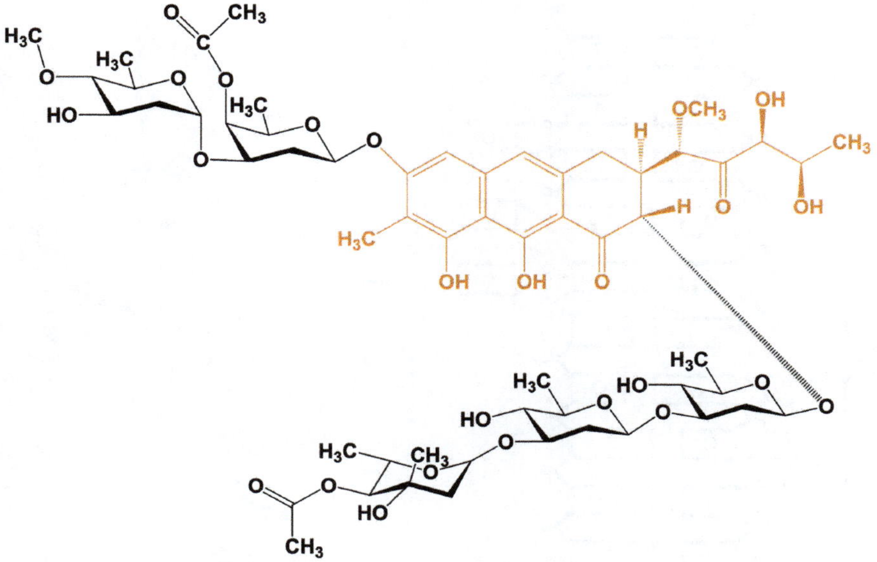

**Figure 6.3** Chemical structure of chromomycin $A_3$.
Adapted with permission from Hou *et al.*, *Biochemistry*, 2008, **47**(20), 5493–5502. Copyright 2008 American Chemical Society.

Unlike ActD, which is an intercalator, Chro belongs to the groove-binding class of DNA-interacting compounds.[42] However, similar to ActD, the binding affinity of Chro towards GpC sites is also affected by the flanking base pair.[43] The binding of the Co(II) complex of dimeric Chro, $Co^{II}(Chro)_2$, to DNA with $(CXG)_n$ TNRs has also been reported.[40] The results show that for short DNA hairpins, GpC sites located in the stem region with flanking G–G mismatches provide excellent binding sites for $Co^{II}(Chro)_2$ when compared with GpC sites with flanking A–A, T–T, or C–C mismatches. $Co^{II}(Chro)_2$ stabilised the cruciform conformation of $(CGG)_{16}$ and two $Co^{II}(Chro)_2$ molecules may bind at two GpC sites separated by at least one GpC step in the hairpin structure of $(CGG)_{16}$. In a gel electrophoretic expansion assay using the synthetic DNA sequence 5′-$(CGG)_{16}(CCG)_6$-3′, $Co^{II}(Chro)_2$ was found to interfere with the expansion process of CGG triplet repeats, suggesting its potential as a therapeutic lead.

Recent studies from our group also showed that $Ni^{II}(Chro)_2$ greatly stabilised the structure of a hairpin duplex with $(CCG)_n$ repeats over other duplex CXG TNRs (where X = A, T or G).[44] The crystal structure of the $Ni^{II}(Chro)_2$–CCG TNR complex revealed a number of DNA deformations, including many flipped-out cytosines (Figure 6.4), that result in the formation of a central GGCC tetranucleotide binding core with a DNA kink. The drastic changes in DNA structure upon $Ni^{II}(Chro)_2$ binding suggest that $Ni^{II}(Chro)_2$ employs a 'forced' induced-fit mechanism to bind to CCG TNRs with high specificity. Because $(CCG)_n$ TNRs represent a major signature of fragile X

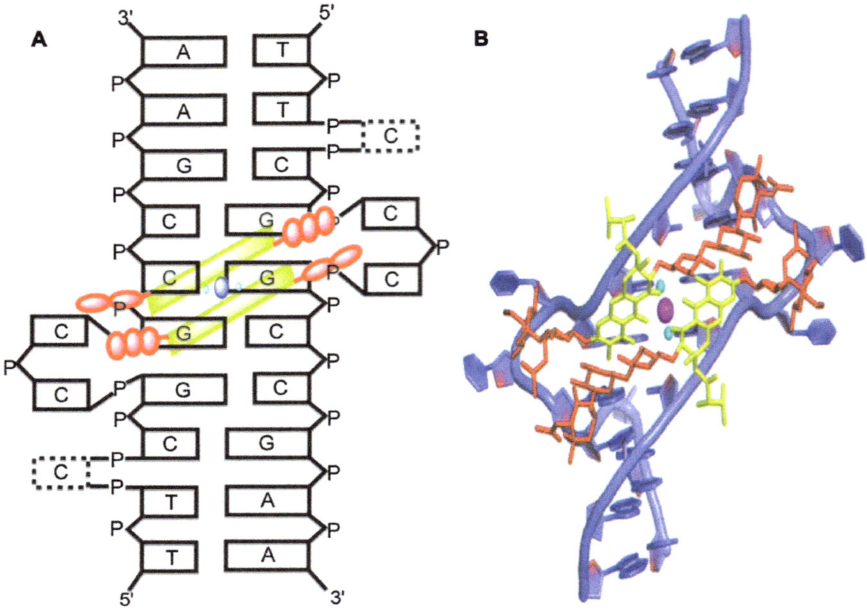


**Figure 6.4** (A) Schematic diagram of $Ni^{II}(Chro)_2$–d(TT(CCG)$_3$AA) complex showing central GGCC tetranucleotide binding core with flipped out cytosine. (B) Crystal structure of $Ni^{II}(Chro)_2$–d(TT(CCG)$_3$AA) complex redrawn from PDB_ID 5XEW.
Figure reproduced with permission from W. H. Tseng *et al.*, Induced-fit recognition of CCG Nucleotide repeats by a nickel chromomycin complex results in large-scale DNA deformation, *Angew. Chem., Int. Ed.*, John Wiley & Sons, Inc. © 2017 WILEY-VCH Verlag GmbH & Co. KGaA, Weinheim.

syndrome, the unique binding properties of $Ni^{II}(Chro)_2$ may be exploited to enable detection of this disease. In fact, $Ni^{II}(Chro)_2$ also has unique fluorescence properties, which may provide a convenient read-out, and preliminary assays based on these properties were able to distinguish between normal and fragile X syndrome subjects in clinical specimens.

### 6.2.3 Pyrene-functionalized Pyrrole–Imidazole Polyamides

Whilst actinomycin D and chromomycin A3 represent natural molecules capable of binding to TNRs, scientists have long embraced the possibility of producing synthetic molecules capable of detecting TNRs containing specific motifs. In this context *N*-methylpyrrole (Py)–*N*-methylimidazole (Im) polyamides have become a favourite starting point because of their ability to recognize specific DNA sequences (Figure 6.5A).[45,46] Although the binding of Py–Im polyamides to DNA has been studied since the mid-1990s, it was Sugiyama and co-workers who designed various Py–Im polyamide analogues to recognize specific TNR sequences at the turn of this century. Of particular

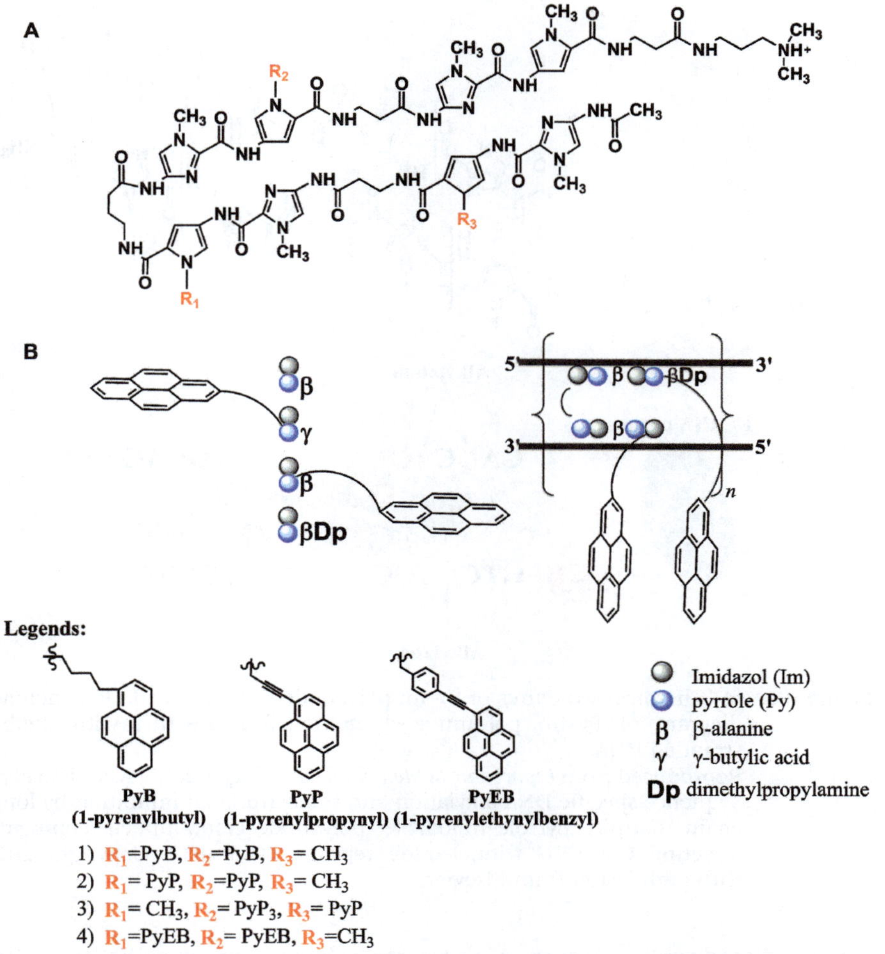


**Figure 6.5** (A) Chemical structures of bis-pyrenyl Py–Im polyamides. $R_1$, $R_2$ and $R_3$ represent different functional groups for each ligand. These functional groups are listed at the bottom of the figure. (B) Schematic diagram of Py–Im polyamides interacting with double-stranded DNA. Reproduced from *Bioorganic & Medicinal Chemistry*, **14**(22), T. Bando, *et al.*, Detection of CAG repeat DNA sequences by pyrene-functionalized pyrrole-imidazole polyamides, 6937–6942. Copyright 2007, with permission from Elsevier.

interest are the Py–Im polyamide–pyrene fluorophores, which specifically recognise CAG repeats (Figure 6.5B).[47]

The pyrene-based fluorophores are particularly attractive and efficient DNA-binding probes because of their large extinction coefficients, excellent quantum yields, and good stability in aqueous solution. DNA-damaging agents, such as alkylating agents, UV irradiation and oxidation, increase the rate of triplet repeat deletion and may prove effective against TNR-related

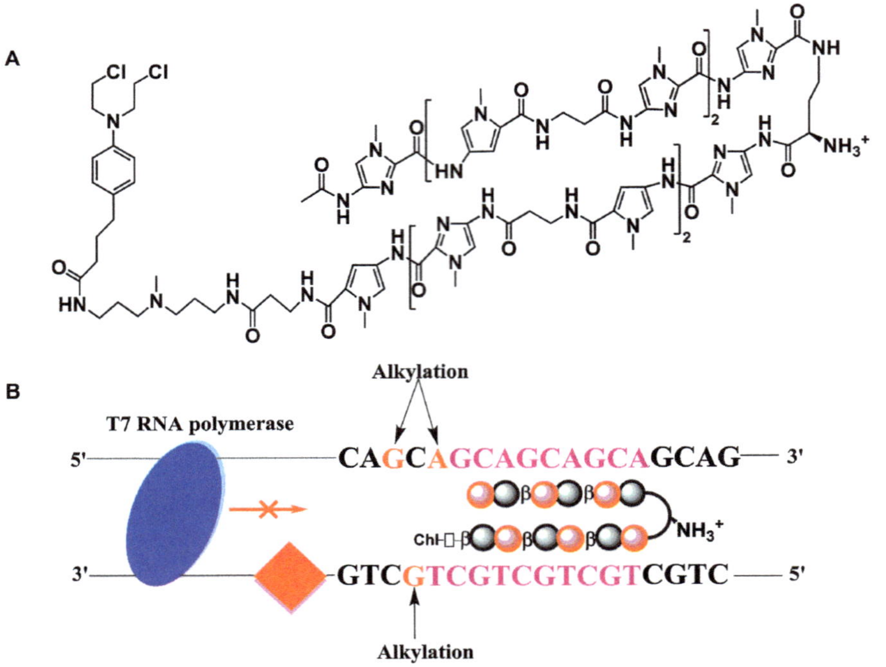


**Figure 6.6** (A) Chemical structures of Py–Im polyamide–chlorambucil. (B) Schematic diagram of Py–Im polyamide–chlorambucil interacting with double-stranded DNA.
Reproduced from *Bioorganic & Medicinal Chemistry*, **22**(17), Asamitsu *et al.*, Sequence-specific DNA alkylation and transcriptional inhibition by long--chain hairpin pyrrole–imidazole polyamide–chlorambucil conjugates targeting CAG/CTG trinucleotide repeats, 4646–4657. Copyright 2014, with permission from Elsevier.

genetic disorders.[48,49] However, these are relatively nonspecific and often have undesired side effects. To increase the sequence specificity of alkylating agents, Py–Im polyamides conjugated to a DNA-alkylating agent, chlorambucil (Chl) (Figure 6.6A), to target CAG/CTG TNRs with different lengths or structures have been designed and synthesized by Sugiyama's group.[50] Sequencing gel analysis demonstrated that a long-chain hairpin Py–Im polyamide conjugated to Chl alkylated the N3 of adenines or guanines of CAG/CTG TNRs. Sequence-specific alkylation of TNRs by the Py–Im polyamide–Chl conjugate also inhibited the elongation of transcripts by RNA polymerase at the alkylating sites in transcription assays *in vitro* (Figure 6.6B).

## 6.2.4 Naphthyridine and Its Analogues

A relatively recent development is the design of naphthyridine analogues that selectively bind to CXG repeats by Nakatani's group.[51–56] These surprisingly versatile analogues incorporate two heterocyclic rings that intercalate into DNA and form specific hydrogen bonds with the nitrogenous bases. One such analogue, a naphthyridine–azaquinolone (NA) conjugate (Figure 6.7A), stabilised $(CAG)_n$ hairpin formation and inhibited polymerase-mediated DNA synthesis.[51] The NMR structure of NA in complex with a CAG/CAG duplex revealed that NA protruded into the A–A mismatch and the flanking G–C base pairs, causing the extrusion of a cytidine nucleotide from the DNA helix (Figure 6.7B).[51] The hydrogen bonding partners between NA and DNA, which include the naphthyridine–guanine and 8-azaquinolone–adenine pairs, were well stacked in the DNA helix and showed structural mimicry of Watson–Crick base pairing (Figure 6.7C). A sensor with immobilized NA has proved useful as a rapid diagnostic tool for examining $(CAG)_n$

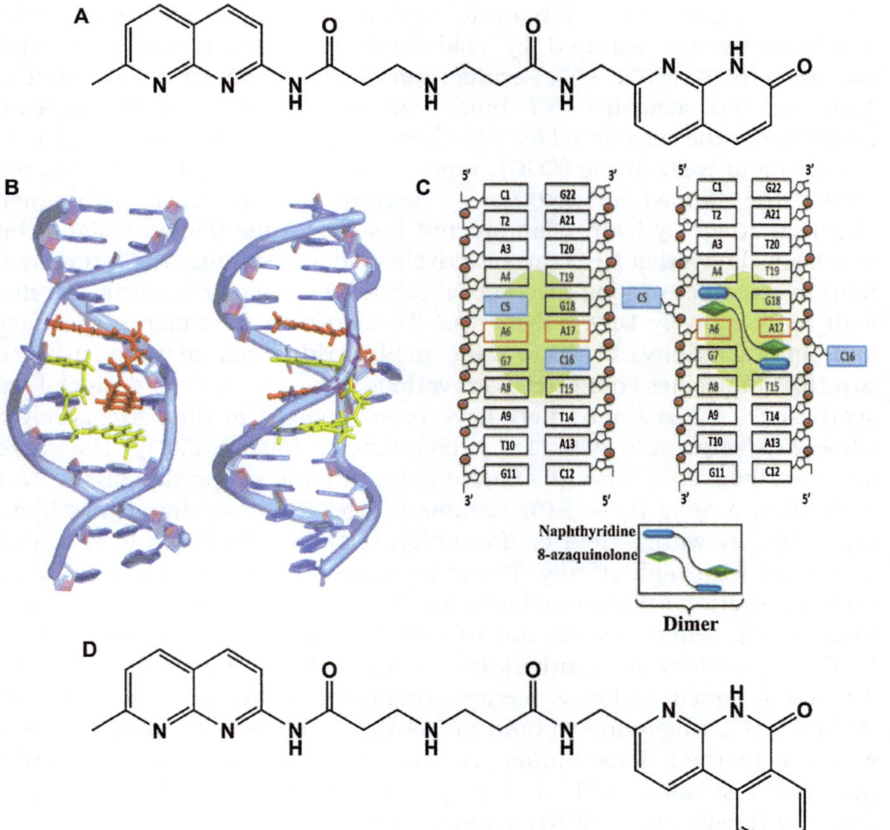


**Figure 6.7** Binding of naphthyridine–azaquinolone (NA) to CAG/CAG complex. (A) Chemical structure of NA. (B) Solution structure of the NA–CAG/CAG complex re-imaged from PDB_ID 1X26. (C) Schematic representation of NA dimer interacting with CAG/CAG complex DNA. (D) Chemical structure of NA analogue NBzA.
Adapted by permission from Macmillan Publishers Ltd: *Nature Chemical Biology* (ref. 51), copyright 2005.

repeat length. On the other hand, one derivative of NA, NBzA (Figure 6.7D), modified by incorporating an additional ring on to the azaquinolone was found to interact with both r(CAG)$_9$ and d(CAG)$_9$.[10] The binding of NBzA to d(CAG)$_9$ and r(CAG)$_9$, similarly to NA binding, proceeded in pairwise and stepwise manners respectively, as assessed by electron spray ionization–time-of-flight (ESI–TOF) mass spectrometry. Although the tricyclic system in NBzA did not show significant effects on the binding and selectivity towards CAG repeats, it provides a large chemical space for further modification of NA analogues to achieve higher affinity and selectivity for DNA.

Nakatani's group also synthesized a series of compounds which bind to G–G mismatches in the DNA duplex. Naphthyridine carbamate dimer (NCD) (Figure 6.8A) is one such compound, composed of 2-amino-1,8-naphthyridine and a carbamate linker.[52] It binds to the CGG/CGG duplex with exclusive 2 : 1 stoichiometry as measured by cold spray ionization–time-of-flight mass spectrometry (CSI–TOF MS). Results from melting temperature ($T_m$) shift experiments indicated that NCD binds selectively to the CGG/CGG duplex as evidenced by the substantial increase in melting temperature ($\Delta T_m = 23.1$ °C).

Binding of NCD to the (CGG)$_n$ repeat disrupts the G–C base pairing and causes the flipped-out cytosine to become susceptible to subsequent chemical attack by hydroxylamine and hot piperidine (Figure 6.8D). It has also been shown that NCD can selectively bind and stabilise the intra-strand hairpin structure of long CGG repeat sequences, which effectively interfere with replication by DNA polymerase. To increase the sequence specificity and binding affinity of NCD to TNRs, naphthyridine tetramers consisting of two NCD molecules connected with various linkers have been designed and synthesized.[55] These tetramers have been reported to bind exclusively to CGG/CGG sequences with 1 : 1 stoichiometry (Figure 6.8D). The linker structure had a significant effect on the binding properties of the NCD molecules. Among these NCD tetramers, Z-NCTS possessing a (Z)-stilbene linker moiety with a rigid conformation (Figure 6.8B) binds to CGG/CGG sequences with high affinity. Thermodynamic parameters obtained by isothermal titration calorimetry indicated that the stronger binding of Z-NCTS to CGG/CGG sequences was due to a smaller binding entropy penalty. The binding specificity of naphthyridine analogues is not limited to CXG TNRs. A newly designed analogue, methylcarbamoylnaphthyridine dimer (MCND) (Figure 6.8C), was found to bind to (GAA)$_{10}$.[57] Subsequent characterisation with UV thermal denaturation, circular dichroism spectroscopy, surface plasmon resonance and cold-spray-ionization–time-of-flight mass spectrometry revealed that MCND stabilised the hairpin structure of the (GAA)$_n$ repeat with a binding stoichiometry of 3 : 1. The G–G mismatch flanked by two A–A mismatches was identified as the most probable binding site for MCND. DNA polymerase stop assays showed that MCND bound to the (GAA)$_n$ repeat interfered with primer extension at the first two GAA sites of the template.[57]

Since the T/GG site is an important structural element of (TGG)$_n$ repeats and may cause microdeletions, CMBL4, a cyclic bis-naphthyridine carbamate

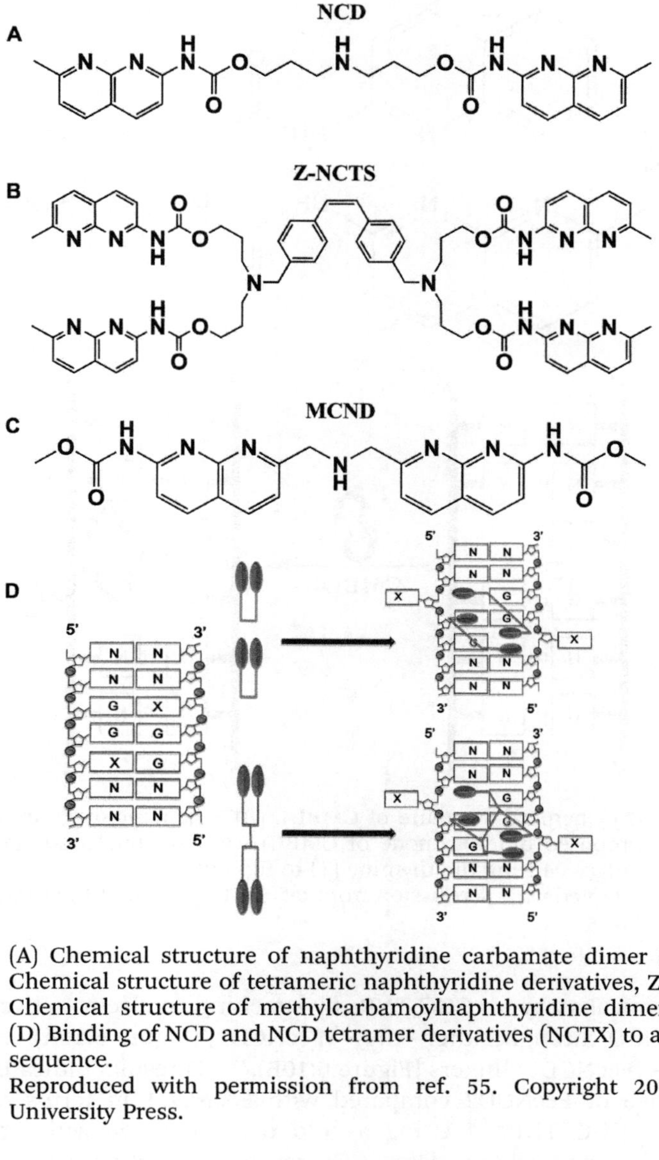


**Figure 6.8** (A) Chemical structure of naphthyridine carbamate dimer (NCD). (B) Chemical structure of tetrameric naphthyridine derivatives, Z-NCTS. (C) Chemical structure of methylcarbamoylnaphthyridine dimer (MCND). (D) Binding of NCD and NCD tetramer derivatives (NCTX) to a XGG/XGG sequence.
Reproduced with permission from ref. 55. Copyright 2011 Oxford University Press.

dimer with limited conformational flexibility, has been designed to bind selectively to the sequence 5′-T-3′/5′-GG-3′ (T/GG) with 1 : 1 stoichiometry (Figure 6.9A).[58] Potassium permanganate oxidation studies on CMBL4–DNA duplex complex containing the T/GG site revealed that binding of CMBL4 accelerated the oxidation of thymine at that site, which suggests that the thymine base originally in the π-stack has been flipped out (Figure 6.9B). Besides their therapeutic potential, naphthyridine analogues may also be used as diagnostic tools for abnormal TNRs. He *et al.* developed a label-free

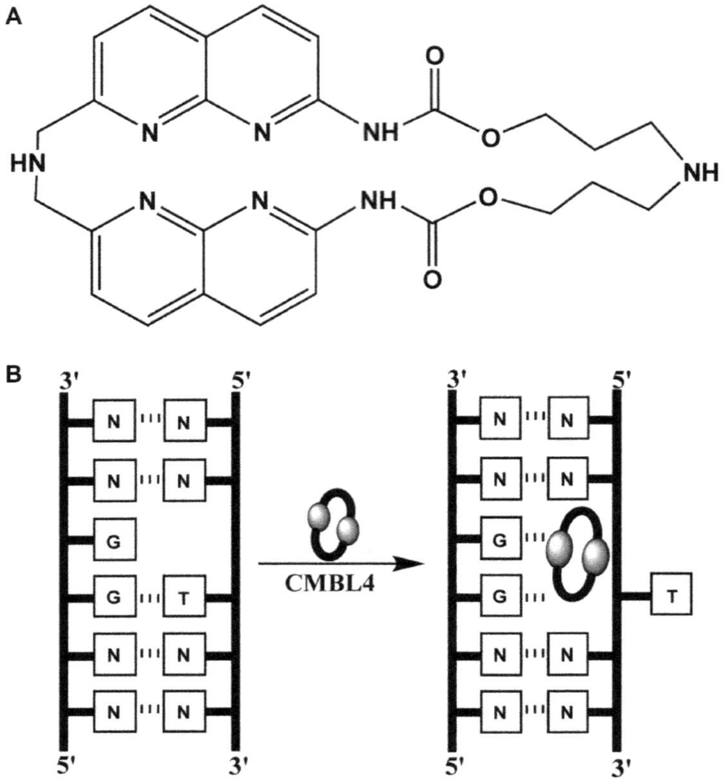


**Figure 6.9** (A) Chemical structure of CMBL4. (B) Schematic diagram showing the proposed binding mode of CMBL4 with two nucleotide guanine (GG) bulges causing the thymine (T) to flip out.
Adapted with permission from ref. 58. Copyright 2016 Oxford University Press.

electrochemical assay for CGG TNRs by attaching an electro-active ferrocenyl group to NCD through either $-CH_2-$ (FecNCD1) (Figure 6.10A) or $-CO-NH-CH_2-CH_2-$ (FecNCD2) linkers (Figure 6.10B).[59] CD results indicated superior performance of FecNCD2 compared with FecNCD1 in terms of selective binding to CGG TNRs.[59] Using a gold disk electrode with immobilized FecNCD2, it was possible to detect CGG repeats in solution using competitive experiments. This new approach offers significant potential for the development of a versatile and low-cost biosensor for early diagnosis of neurodegenerative diseases associated with TNRs. Another popular class of naphthyridine-based compounds geared towards diagnosis development is based on the 2-amino-1,8-naphthyridine chromophore (Am-N). Am-N binds selectively to cytosines at neutral pH when its N1 position is protonated. Am-N dimers bind strongly to the C–C mismatches regardless of flanking sequences with association constants in the order of $10^6$ $M^{-1}$. Sato *et al.* have shown that

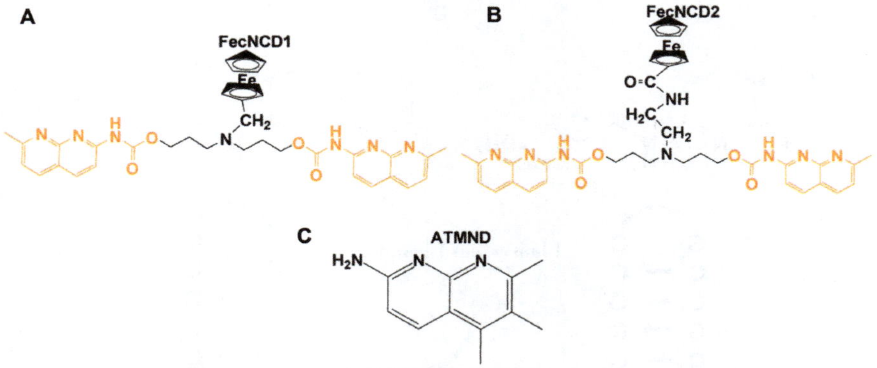


**Figure 6.10** Chemical structures of naphthyridine analogues (A) FecNCD1, (B) FecNCD2, and (C) 2-amino-5,6,7-trimethyl-1, 8-naphthyridine (ATMND). Reproduced from *Biosensors and Bioelectronics*, **49**, He *et al.*, Facile electrochemical biosensor based on a new bifunctional probe for label-free detection of CGG trinucleotide repeat, 282–289. Copyright 2013 with permission from Elsevier.

the fluorescent compound 2-amino-5,6,7-trimethyl-1,8-naphthyridine (ATMND) (Figure 6.10C) selectively binds to C–C mismatches present in the hairpin structures of $(CCG)_n$ TNR, which is accompanied by significant quenching of ATMND fluorescence.[60]

Shibata *et al.* added an aromatic ring to the bicyclic parent molecule Am-N (Figure 6.11A) to produce a tricyclic aminobenzo[c][1,8]naphthyridine derivative (Am-BzN) (Figure 6.11B).[61] They anticipated that the binding affinity of the Am-BzN molecules would be enhanced by favourable stacking with neighbouring bases when bound to the C–C mismatch in the CCG/CCG motif of a DNA duplex without losing their selectivity and photoemission characteristics. The Am-BzN ligand indeed showed fluorescence quenching when bound to CCG TNRs, and the amount of quenching was sensitive to the size of the CCG repeat (Figure 6.11C). The versatility of the naphthyridine analogues is further highlighted in the design of an NCD-based spin label. Although originally developed for quantum computing, nitronyl nitroxide radicals attached to NCDs can bind to a CGG/CGG triplet in a noncovalent fashion and act as reporters of structural and dynamic processes.[62] Since naphthyridine analogues have been designed to target different TNR sequences, similar approaches may be applied to produce various spin labels as experimental tools for future research.

### 6.2.5 Bulge-binding Agents

Bulge structures have been postulated to act as intermediates in DNA strand-slippage synthesis associated with the expansion of nucleotide repeats in various neurological diseases.[63–65] Compounds which bind specifically to such structures make useful probes to obtain insights into the mechanism

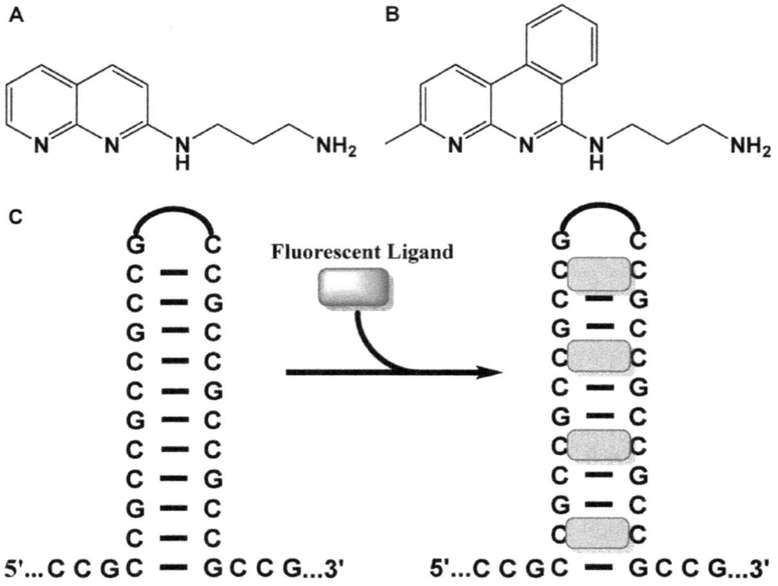


**Figure 6.11** Chemical Structures of: (A) 2-amino-1,8-naphthyridine chromophore [Am-N] and (B) aminobenzo[c][1,8]naphthyridine derivative [Am-BzN]. (C) Diagrammatic representation of fluorescence change of ligand after binding to trinucleotide repeat DNA.
Reproduced from Shibata and Nakatani, Fluorescence Probe for Detecting CCG Trinucleotide Repeat DNA Expansion and Slip-Out, *ChemBioChem*, John Wiley and Sons, © 2014 WILEY-VCH Verlag GmbH & Co. KGaA, Weinheim.

of nucleotide expansion.[66] Goldberg and co-workers have produced a variety of bulge-binding compounds derived from the ene-diyne antitumour antibiotic neocarzinostatin chromophore (NCS-chrom) (Figure 6.12A).[66–68] NCSi-gb (Figure 6.12B), a general-base post-activated NCS-chrom naturally obtained in the absence of thiols, has been shown to target bulge DNA by combining the following structural features: a spirocyclic ring junction with an appropriate angle, two discrete aromatic moieties for $\pi$-stacking interaction, and a pendant amino sugar group that increases binding at the bulge site. A synthetic NCSi-gb mimic, designated DDI ('double-decker intercalator') (Figure 6.12C),[69] binds to nucleic acid bulge structures with high affinity and promotes DNA strand slippage and consequent expansion of nucleotide repeats during replication by DNA polymerase I.[66] Interestingly, whilst NCSi-gb binding involves the major groove of the DNA, DDI preferentially binds to the bulge site *via* the minor groove. DDI can overcome the inhibitory effect of compounds that normally prevent formation of slippage-induced DNA products, such as ethidium bromide, distamycin and doxorubicin, presumably by destabilizing the binding sites for these other agents on duplex DNA.

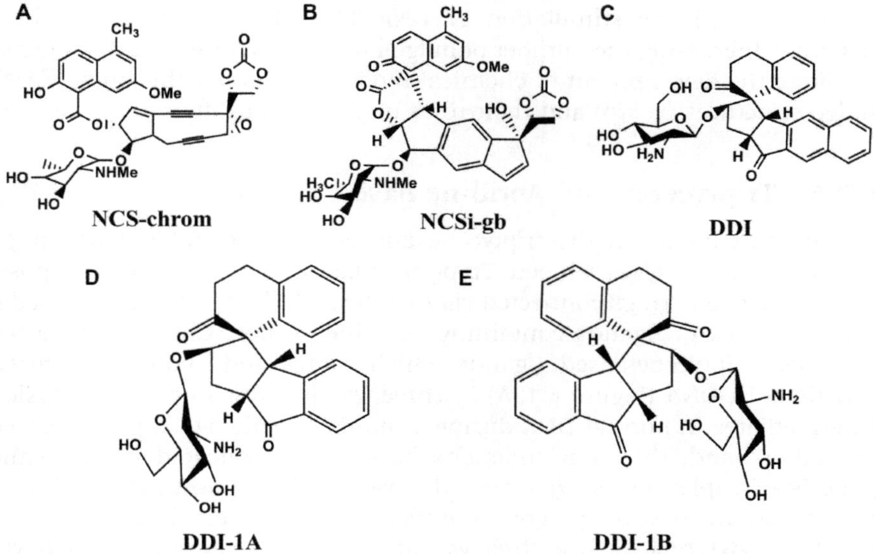


**Figure 6.12** Chemical structures of NCSi-gb mimics. (A) NCS-chrom. Adapted with permission from ref. 67. Copyright 2007 American Chemical Society. (B) NCSi-gb, (C) DDI, (D) DDI-1A and (E) DDI-1B. Reproduced with permissions from Ouyang *et al.*, *In vitro* expansion of DNA triplet repeats with bulge binders and different DNA polymerases, *FEBS Journal*, John Wiley and Sons, © 2014 WILEY-VCH Verlag GmbH & Co. KGaA, Weinheim.

Several simpler analogues of DDI with a spirocyclic backbone, DDI-1A (Figure 6.12D) and DDI-1B (Figure 6.12E), have also been designed and synthesized to explore the specificity of bulge-site binding and the enhancement of DNA strand slippage synthesis *in vitro*.[63] The geometry of these compounds mimics the helix bases through a spirolactone moiety containing two arene units. The spiroalcohol backbone of DDI-1A and DDI-1B adopts a wedge-shaped configuration that is essential to their bulge-binding activity. CD spectra and UV thermal stability studies showed that both DDI-1A and DDI-1B exhibit selective binding to the DNA bulge and induce a significant conformational change in bulge DNA. The stimulating activity of these compounds towards DNA strand slippage suggests that they may bind to or induce the formation of a non-Watson–Crick structure during *in vitro* replication of DNA TNRs.[70] DDI-1A and DDI-1B stimulate slippage in a variety of repeat sequences under *in vitro* conditions, including AAT, ATT, CTG, CAG, GT and CA repeats. The simpler chemical structure of DDI-1A and DDI-1B compared with DDI may provide a good starting point towards the development of slippage inhibitors that target bulge structures with high specificity.

Liu *et al.* synthesized a phenothiazine amino sugar containing a novel wedge-shaped topology and a large dihedral angle (~140°) between two benzene rings which also binds to the DNA bulge with ~42 μM affinity and showed remarkable stimulation of DNA TNR strand slippage synthesis *in vitro*.[70] Increasing the number of bulge-binding compounds available may facilitate the development of chemical model systems for the study of DNA bulge structures *in vitro* and their roles in TNR expansion.

### 6.2.6 Triptycene- and Acridine-based Ligands

Compounds based on the triptycene and acridine scaffolds which target TNRs have also been developed. Tryptycene has a rigid framework composed of three benzene rings connected *via* bicyclic [2.2.2] octane and was used to build a series of ligands containing pyrazine groups. Chenoweth's group designed triptycene-based ligands which target and stabilise three-way junctions in DNA (Figure 6.13A).[71] Three-way junctions occur as transient intermediates in normal DNA during a number of biological processes. On the other hand, three-way junctions have been implicated in the pathogenesis of triplet-repeat-expansion diseases such as Huntington's disease and myotonic dystrophy. For example, dynamic slipped-out regions in $(CAG)_n \cdot (CTG)_n$ repeats form three-way junctions that may be associated with

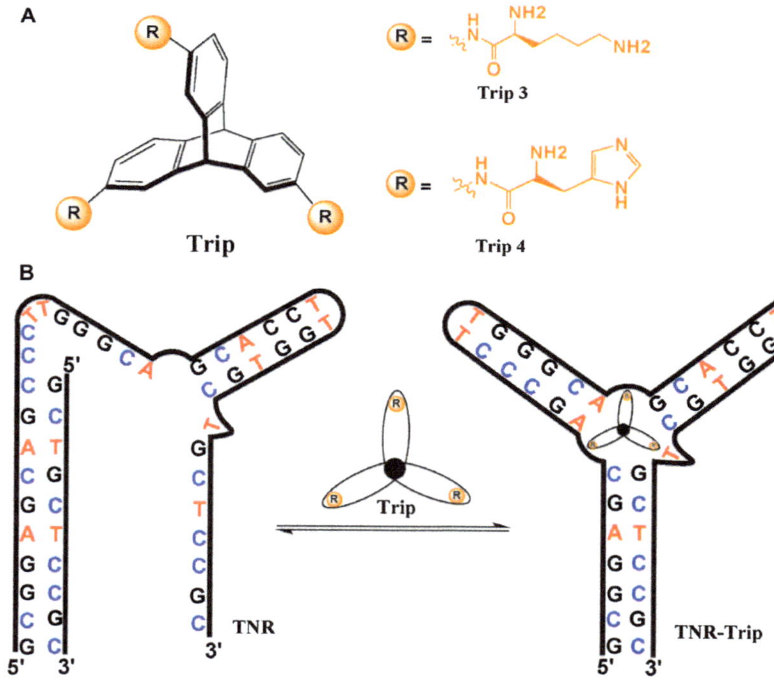


**Figure 6.13** (A) Chemical structure of triptycene derivatives: Trip3 and Trip4. (B) Diagrammatic representation of gel shift assay. Addition of triptycene causes reformation of the junction (TNR–Trip).
Adapted from ref. 71 with permission from The Royal Society of Chemistry.

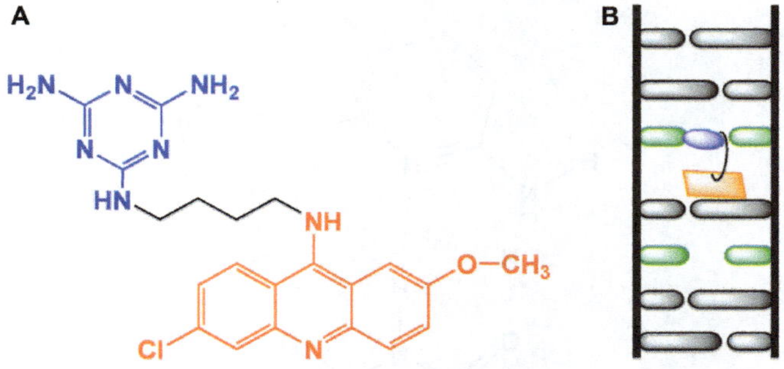


**Figure 6.14** (A) Chemical structure of Triaminotriazine-acridine ligand 1. (B) Schematic illustration showing the possible mode of binding to the mismatched DNA.
Reproduced with permission from Arambula *et al.* (2009), A simple ligand that selectively targets CUG trinucleotide repeats and inhibits MBNL protein binding, *Proc. Natl. Acad. Sci.*, **106**(38), 16068–16073.

TNR expansion. Based on gel mobility shift assays, fluorescence-quenching experiments and CD spectroscopy, it has been proposed that the compounds Trip 3 and Trip 4 enhance the helicity of the slipped-out junctions in $(CAG)_n \cdot (CTG)_n$ repeats. Triptycene-based compounds also provide significant stabilisation of DNA and RNA three-way junctions (Figure 6.13B), and may have potential as a novel class of nucleic acid junction binders with enhanced recognition properties.

Acridine-based compounds represent another way to design TNR-targeting compounds based on a rigid scaffold. Zimmerman's group developed a simple triaminotriazine-acridine conjugate presenting a Janus-wedge hydrogen bonding surface to thymines in T–T mismatches which selectively targets CTG repeats with sub-micromolar affinity.[72] ITC studies of ligand 1 (Figure 6.14A) with various sequences showed a preference toward the T–T mismatch ($K_d$ of $390 \pm 80$ nM) with a 13-, 169- and 85-fold reduction in affinity toward single C–C, A–A and G–G mismatches respectively. In addition to DNA, ligand 1 also recognizes U–U mismatches in duplex RNA containing CUG repeats. The recognition is presumably provided by the two edges of the triazine heterocycle which have the potential to form simultaneously a full set of hydrogen bonds with the poorly paired thymine or uracil bases (Figure 6.14B). Since both $(CTG)_n$ DNA and $(CUG)_n$ RNA repeats are involved in a common form of muscular dystrophy, myotonic dystrophy type 1 (DM1), the ability of ligand 1 to recognise closely related DNA and RNA sequences and potentially inhibit both replication and transcription of the expanded TNR makes it a powerful lead compound for therapeutic development.

## 6.3 Conclusion

In this chapter we have reviewed many classes of small molecules capable of binding to TNRs of varying sequences with high affinity. Some of these

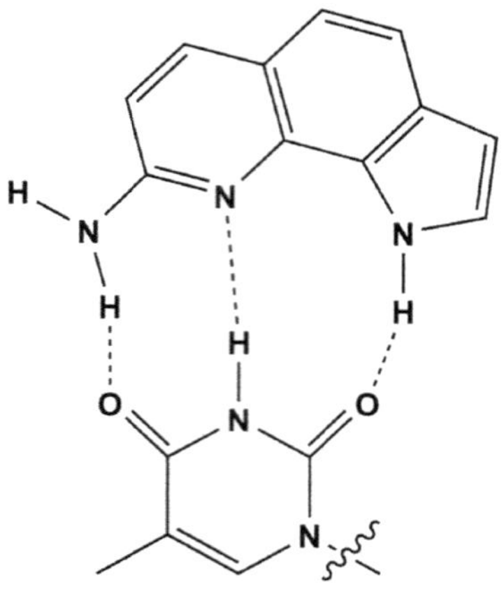


**Figure 6.15** Chemical structure of PQA moiety and possible hydrogen bonding with thymine base.
Reprinted from *Bioorganic & Medicinal Chemistry Letters*, **26**(15), Matsumoto *et al.*, Synthesis of 1*H*-pyrrolo[3,2-h]quinoline-8-amine derivatives that target CTG trinucleotide repeats. Copyright 2016 with permission from Elsevier.

small molecules, such as ActD and Chro, have a long history and were originally developed for other purposes, but were later found to have novel TNR-binding activities. Others were synthesized from different chemical scaffolds with the sole purpose of binding to specific TNR sequences. New molecules are also constantly being developed by many groups. For example, Matsumoto *et al.* recently synthesised a series of 1*H*-pyrrolo [3,2-*h*]-quinolone-8-amine (PQA) derivatives that recognise CTG repeats by forming hydrogen bonds with thymine (Figure 6.15).[73] Since TNR expansions are strongly associated with so many neurodegenerative diseases, there is widespread interest in harnessing the potential of these molecules as diagnostic and therapeutic agents, particularly in aging societies.

A major bottleneck for the medical development of all current TNR-binding compounds is their lack of extreme selectivity. Although these compounds are considered specific against certain TNR sequences, almost all of them still exhibit nonspecific binding to other parts of the DNA. Given the sheer abundance of noncognate *versus* cognate sequences under physiological conditions, specific binding is often overwhelmed by nonspecific interactions. The structures of DNA (or RNA) triplet–small-molecule complexes could provide clues to the solution of this problem. All TNR-binding compounds induce TNRs to form unusual structures at the mismatch positions, and elucidation of this structural code could form the basis for rational design and optimisation of compounds with high selectivity for defined TNR sequences. The complex pathogenesis of neurodegenerative diseases may also vitiate the efficacy of any single-agent/single-target therapeutic strategy. Multi-target drug discovery (MTDD) has already been used

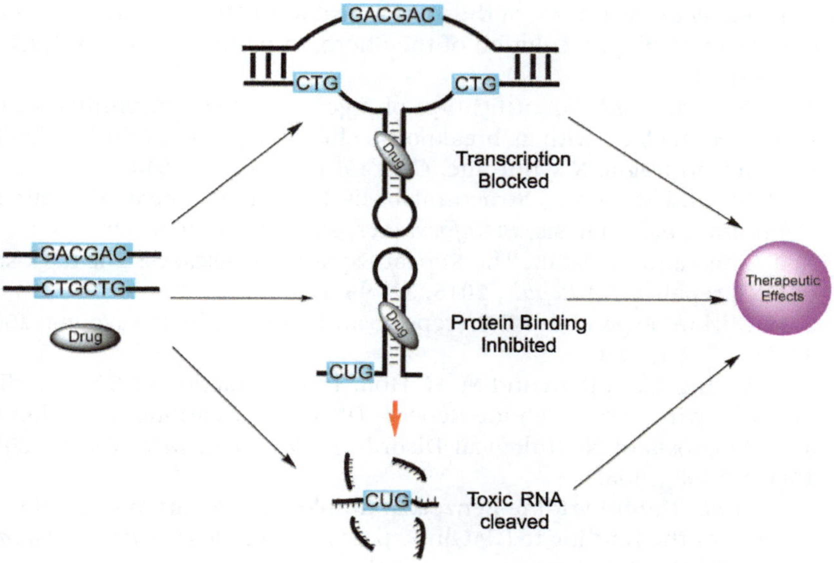


**Figure 6.16** Conceptual diagram of multi-target drug treatment against TNR-related disorders.
Adapted with permission from L. Nguyen *et al.*, *J. Am. Chem. Soc.*, 2015, **137**(44), 14180. Copyright 2015 American Chemical Society.

to design a drug that modulates multiple targets simultaneously in TNR diseases (Figure 6.16). For example, Zimmerman's group developed several small ligands that target DM1 *in vitro* in three distinct ways: first, by inhibiting transcription through DNA binding; second, by inhibiting aberrant protein binding to the toxic RNA through RNA binding; and third, by acting as ribonuclease mimics to degrade the toxic RNA.[74]

Overall, expanding the repertoire of TNR-binding compounds and understanding their physicochemical characteristics will provide researchers with new tools to probe the molecular mechanisms of TNR-associated neurological disorders. We are hopeful that future derivatives may one day become useful diagnostic tools or part of an effective therapeutic regimen against these diseases.

## References

1. B. Charlesworth, P. Sniegowski and W. Stephan, The evolutionary dynamics of repetitive DNA in eukaryotes, *Nature*, 1994, **371**(6494), 215–220.
2. S. M. Mirkin, DNA structures, repeat expansions and human hereditary disorders, *Curr. Opin. Struct. Biol.*, 2006, **16**(3), 351–358.
3. R. P. Bowater and R. D. Wells, The intrinsically unstable life of DNA triplet repeats associated with human hereditary disorders, *Prog. Nucleic Acid Res. Mol. Biol.*, 2001, **66**, 159–202.

4. Y. H. Fu, *et al.*, Variation of the CGG repeat at the fragile X site results in genetic instability: resolution of the Sherman paradox, *Cell*, 1991, **67**(6), 1047–1058.
5. A. J. Verkerk, *et al.*, Identification of a gene (FMR-1) containing a CGG repeat coincident with a breakpoint cluster region exhibiting length variation in fragile X syndrome, *Cell*, 1991, **65**(5), 905–914.
6. P. Baldi and B. Søren, Structural basis for triplet repeat disorders: A Computational Analysis, *Bioinformatics*, 1999, **11**(6), 918–929.
7. X. N. Zhao and K. Usdin, The Repeat Expansion Diseases: The dark side of DNA repair, *DNA Repair*, 2015, **32**, 96–105.
8. S. M. Mirkin, Expandable DNA repeats and human disease, *Nature*, 2007, **447**(7147), 932–940.
9. C. K. Chang, C. R. Jhan and M. H. Hou, The Interaction of DNA-Binding Ligands with Trinucleotide-Repeat DNA: Implications for Therapy and Diagnosis of Neurological Disorders, *Curr. Top. Med. Chem.*, 2015, **15**(14), 1398–1408.
10. J. Li, *et al.*, Naphthyridine-Benzoazaquinolone: Evaluation of a Tricyclic System for the Binding to (CAG)n Repeat DNA and RNA, *Chem. – Asian J.*, 2016, **11**(13), 1971–1981.
11. A. Bacolla, *et al.*, Flexible DNA: genetically unstable CTG.CAG and CGG.CCG from human hereditary neuromuscular disease genes, *J. Biol. Chem.*, 1997, **272**(27), 16783–16792.
12. P. D. Chastain and R. R. Sinden, CTG repeats associated with human genetic disease are inherently flexible, *J. Mol. Biol.*, 1998, **275**(3), 405–411.
13. N. J. Veitch, *et al.*, Inherited CAG.CTG allele length is a major modifier of somatic mutation length variability in Huntington disease, *DNA Repair*, 2007, **6**(6), 789–796.
14. C. E. Pearson and R. R. Sinden, Trinucleotide repeat DNA structures: dynamic mutations from dynamic DNA, *Curr. Opin. Struct. Biol.*, 1998, **8**(3), 321–330.
15. H. T. Orr and H. Y. Zoghbi, Trinucleotide repeat disorders, *Annu. Rev. Neurosci.*, 2007, **30**, 575–621.
16. S. Yu, *et al.*, Fragile X genotype characterized by an unstable region of DNA, *Science*, 1991, **252**(5009), 1179–1181.
17. R. I. Richards, *et al.*, Fragile X syndrome unstable element, p(CCG)n, and other simple tandem repeat sequences are binding sites for specific nuclear proteins, *Hum. Mol. Genet.*, 1993, **2**(9), 1429–1435.
18. S. E. Andrew, *et al.*, The relationship between trinucleotide (CAG) repeat length and clinical features of Huntington's disease, *Nat. Genet.*, 1993, **4**(4), 398–403.
19. J. Petruska, M. J. Hartenstine and M. F. Goodman, Analysis of strand slippage in DNA polymerase expansions of CAG/CTG triplet repeats associated with neurodegenerative disease, *J. Biol. Chem.*, 1998, **273**(9), 5204–5210.
20. S. T. Warren, Polyalanine expansion in synpolydactyly might result from unequal crossing-over of HOXD13, *Science*, 1997, **275**(5298), 408–4409.

21. B. L. Heidenfelder, A. M. Makhov and M. D. Topal, Hairpin formation in Friedreich's ataxia triplet repeat expansion, *J. Biol. Chem.*, 2003, **278**(4), 2425–2431.
22. C. J. Cummings and H. Y. Zoghbi, Fourteen and counting: unrevealing trinucleotide repeat diseases, *Hum. Mol. Genet.*, 2000, **9**(6), 909–916.
23. G. B. Panigrahi, *et al.*, Slipped (CTG)*(CAG) repeats can be correctly repaired, escape repair or undergo error-prone repair, *Nat. Struct. Mol. Biol.*, 2005, **12**(8), 654–662.
24. J. R. Gatchel and H. Y. Zoghbi, Diseases of unstable repeat expansion: mechanisms and common principles, *Nat. Rev. Genet.*, 2005, **6**(10), 743–755.
25. A. Pluciennik, *et al.*, Tandem duplication. A novel type of triplet repeat instability, *J. Biol. Chem.*, 2000, **275**(37), 28386–28397.
26. J. P. Jakupciak and R. D. Wells, Genetic instabilities in (CTG.CAG) repeats occur by recombination, *J. Biol. Chem.*, 1999, **274**(33), 23468–23479.
27. Y. W. Chen, *et al.*, Structural basis for the identification of an i-motif tetraplex core with a parallel-duplex junction as a structural motif in CCG triplet repeats, *Angew. Chem., Int. Ed. Engl.*, 2014, **53**(40), 10682–10686.
28. B. Moore, *et al.*, Decoding of tandem quadruplets by adjacent tRNAs with eight-base anticodon loops, *Nucleic Acids Res.*, 2000, **28**(18), 3615–3624.
29. H. C. Dipple, The use of olanzapine for movement disorder in Huntington's disease: a first case report, *J. Neurol. Neurosurg. Psychiatry*, 1999, **67**(1), 123–124.
30. V. E. Grove, J. Quintanilla and G. T. DeVaney, Improvement of Huntington's disease with olanzapine and valproate, *N. Engl. J. Med.*, 2000, **343**(13), 973–974.
31. A. Granzhan, N. Kotera and M. P. Teulade-Fichou, Finding needles in a basestack: recognition of mismatched base pairs in DNA by small molecules, *Chem. Soc. Rev.*, 2014, **43**(10), 3630–3665.
32. S. A. Waksman and H. B. Woodruff, Bacteriostatic and bacteriocidal substances produced by soil actinomycetes, *Proc. Soc. Exp. Biol.*, 1940, **45**, 609–614.
33. U. Hollstein, Actinomycin. Chemistry and mechanism of action, *Chem. Rev.*, 1974, **74**(6), 625–652.
34. S. Y. Wang, *et al.*, Spermine attenuates the action of the DNA intercalator, actinomycin D, on DNA binding and the inhibition of transcription and DNA replication, *PLoS One*, 2012, 7(11), e47101.
35. C. Lian, H. Robinson and A. Wang, Structure of Actinomycin D bound with (GAAGCTTC)2 and (GATGCTTC)2 and Its Binding to the (CAG)n:(CTG)n Triplet Sequence As Determined by NMR Analysis, *J. Am. Chem. Soc.*, 1996, **118**, 8791.
36. M. H. Hou, *et al.*, Crystal structure of actinomycin D bound to the CTG triplet repeat sequences linked to neurological diseases, *Nucleic Acids Res.*, 2002, **30**(22), 4910–4917.

37. R. B. Siboni, *et al.*, Actinomycin D Specifically Reduces Expanded CUG Repeat RNA in Myotonic Dystrophy Models, *Cell Rep.*, 2015, **13**(11), 2386–2394.
38. Y. S. Lo, *et al.*, The structural basis of actinomycin D-binding induces nucleotide flipping out, a sharp bend and a left-handed twist in CGG triplet repeats, *Nucleic Acids Res.*, 2013, **41**(7), 4284–4294.
39. M. Miyamoto, *et al.*, The structure of Chromomycin A3, *Tetrahedron Lett.*, 1964, **5**(34), 2367–2370.
40. Y. W. Chen and M. H. Hou, The binding of the Co(II) complex of dimeric chromomycin A3 to GC sites with flanking G:G mismatches, *J. Inorg. Biochem.*, 2013, **121**, 28–36.
41. M. H. Hou, *et al.*, Crystal structure of the [Mg2 + -(chromomycin A3)2]-d(TTGGCCAA)2 complex reveals GGCC binding specificity of the drug dimer chelated by a metal ion, *Nucleic Acids Res.*, 2004, **32**(7), 2214–2222.
42. M. Waring, Variation of the supercoils in closed circular DNA by binding of antibiotics and drugs: evidence for molecular models involving intercalation, *J. Mol. Biol.*, 1970, **54**(2), 247–279.
43. M. H. Hou, *et al.*, Studies of sequence-specific DNA binding, DNA cleavage, and topoisomerase I inhibition by the dimeric chromomycin A3 complexed with Fe(II), *Biochemistry*, 2008, **47**(20), 5493–5502.
44. W. H. Tseng, *et al.*, Induced-fit recognition of CCG trinucleotide repeats by a nickel chromomycin complex results in large-scale DNA deformation, *Angew. Chem., Int. Ed.*, 2017, **56**(30), 8761–8765.
45. S. White, *et al.*, Recognition of the four Watson-Crick base pairs in the DNA minor groove by synthetic ligands, *Nature*, 1998, **391**(6666), 468–471.
46. P. B. Dervan and B. S. Edelson, Recognition of the DNA minor groove by pyrrole-imidazole polyamides, *Curr. Opin. Struct. Biol.*, 2003, **13**(3), 284–299.
47. T. Bando, *et al.*, Detection of CAG repeat DNA sequences by pyrene-functionalized pyrrole-imidazole polyamides, *Bioorg. Med. Chem.*, 2007, **15**(22), 6937–6942.
48. V. I. Hashem and R. R. Sinden, Chemotherapeutically induced deletion of expanded triplet repeats, *Mutat. Res.*, 2002, **508**(1–2), 107–119.
49. M. Gomes-Pereira and D. G. Monckton, Chemically induced increases and decreases in the rate of expansion of a CAG*CTG triplet repeat, *Nucleic Acids Res.*, 2004, **32**(9), 2865–2872.
50. S. Asamitsu, *et al.*, Sequence-specific DNA alkylation and transcriptional inhibition by long-chain hairpin pyrrole-imidazole polyamide-chlorambucil conjugates targeting CAG/CTG trinucleotide repeats, *Bioorg. Med. Chem.*, 2014, **22**(17), 4646–4657.
51. K. Nakatani, *et al.*, Small-molecule ligand induces nucleotide flipping in (CAG)n trinucleotide repeats, *Nat. Chem. Biol.*, 2005, **1**(1), 39–43.
52. T. Peng and K. Nakatani, Binding of naphthyridine carbamate dimer to the (CGG)n repeat results in the disruption of the G-C base pairing, *Angew. Chem., Int. Ed. Engl.*, 2005, **44**(44), 7280–7283.

53. K. Nakatani, *et al.*, Synthesis of dimeric 2-amino-1,8-naphthyridine and related DNA-binding molecules, *Current Protocols in Nucleic Acid Chemistry*, 2008, Unit 8.6, pp. 1–21.
54. M. Hagihara, H. He and K. Nakatani, Small molecule modulates hairpin structures in CAG trinucleotide repeats, *ChemBioChem*, 2011, **12**(11), 1686–1689.
55. C. Dohno, *et al.*, Naphthyridine tetramer with a pre-organized structure for 1:1 binding to a CGG/CGG sequence, *Nucleic Acids Res.*, 2012, **40**(6), 2771–2781.
56. M. Hagihara, *et al.*, A small molecule regulates hairpin structures in d(CGG) trinucleotide repeats, *Bioorg. Med. Chem. Lett.*, 2012, **22**(5), 2000–2003.
57. H. He, M. Hagihara and K. Nakatani, A small molecule affecting the replication of trinucleotide repeat d(GAA)n, *Chemistry*, 2009, **15**(40), 10641–10648.
58. S. Mukherjee, *et al.*, Cyclic mismatch binding ligand CMBL4 binds to the 5′-T-3′/5′-GG-3′ site by inducing the flipping out of thymine base, *Nucleic Acids Res.*, 2016, **44**(15), 7090–7099.
59. H. He, *et al.*, Facile electrochemical biosensor based on a new bifunctional probe for label-free detection of CGG trinucleotide repeat, *Biosens. Bioelectron.*, 2013, **49**, 282–289.
60. Y. Sato, *et al.*, Fluorescent trimethyl-substituted naphthyridine as a ligand for C-C mismatch detection in CCG trinucleotide repeats, *Chem. Commun.*, 2011, **47**(20), 5885–5887.
61. T. Shibata and K. Nakatani, Fluorescence Probe for Detecting CCG Trinucleotide Repeat DNA Expansion and Slip-Out, *Chembiochem*, 2016, **17**(18), 1685–1688.
62. K. Maekawa, *et al.*, Programmed assembly of organic radicals on DNA, *Chem. Commun.*, 2010, **46**(8), 1247–1249.
63. D. Ouyang, *et al.*, In vitro expansion of DNA triplet repeats with bulge binders and different DNA polymerases, *FEBS J.*, 2008, **275**(18), 4510–4521.
64. J. Volker, *et al.*, DNA repair and DNA triplet repeat expansion: the impact of abasic lesions on triplet repeat DNA energetics, *J. Am. Chem. Soc.*, 2009, **131**(26), 9354–9360.
65. S. L. Lam, *et al.*, The origin of genetic instability in CCTG repeats, *Nucleic Acids Res.*, 2011, **39**(14), 6260–6268.
66. L. S. Kappen, *et al.*, Stimulation of DNA strand slippage synthesis by a bulge binding synthetic agent, *Biochemistry*, 2003, **42**(7), 2166–2173.
67. L. S. Kappen, *et al.*, Probing DNA bulges with designed helical spirocyclic molecules, *Biochemistry*, 2007, **46**(2), 561–567.
68. N. Zhang, *et al.*, Solution structure of a designed spirocyclic helical ligand binding at a two-base bulge site in DNA, *Biochemistry*, 2007, **46**(16), 4793–4803.
69. Z. Xi, D. Ouyang and H. T. Mu, Stimulation on DNA triplet repeat strand slippage synthesis by the designed spirocycles, *Bioorg. Med. Chem. Lett.*, 2006, **16**(5), 1180–1184.

70. L. Liu, *et al.*, A bulge binding agent with novel wedge-shape topology for stimulation of DNA triplet repeat strand slippage synthesis, *Bioorg. Med. Chem. Lett.*, 2008, **18**(23), 6184–6188.
71. S. A. Barros and D. M. Chenoweth, Triptycene-based small molecules modulate (CAG).(CTG) repeat junctions, *Chem. Sci.*, 2015, **6**(8), 4752–4755.
72. J. F. Arambula, *et al.*, A simple ligand that selectively targets CUG trinucleotide repeats and inhibits MBNL protein binding, *Proc. Natl. Acad. Sci. U. S. A.*, 2009, **106**(38), 16068–16073.
73. J. Matsumoto, *et al.*, Synthesis of 1H-pyrrolo[3,2-h]quinoline-8-amine derivatives that target CTG trinucleotide repeats, *Bioorg. Med. Chem. Lett.*, 2016, **26**(15), 3761–3764.
74. L. Nguyen, *et al.*, Rationally designed small molecules that target both the DNA and RNA causing myotonic dystrophy type 1, *J. Am. Chem. Soc.*, 2015, **137**(44), 14180–14189.

CHAPTER 7

# *Parsing the Enthalpy–Entropy Compensation Phenomenon of General DNA–Ligand Interactions by a 'Gradient Determinant' Approach*

ROBIN C. K. YANG,[a] KAREN L. KAO,[a] WEN-CHEN YAO,[a] YU-LING CHEN,[a] CHIA-CHUN HUNG,[a] YU-TING LI,[a] JIA-RONG WU,[a] WEI-KANG HUNG,[a] MICHAEL J. WARING[b] AND LEUNG SHEH*[a]

[a] Department of Chemistry, Tunghai Christian University, Taichung, Taiwan 407, Republic of China; [b] Department of Pharmacology, University of Cambridge, Tennis Court Road, Cambridge CB2 1PD, England, UK
*Email: sheh28@hotmail.com.tw

## 7.1 Introduction

Understanding the complex chemical nature of DNA–ligand interactions in terms of binding affinities, base sequence-specificity, allosteric and thermodynamic features is essential to the advancement of structural biochemistry. Over three decades, considerable efforts from several research teams have been devoted to studies of DNA–small molecule interaction including a large variety of small ligands, namely, antitumour antibiotics,[1–5]

Chemical Biology No. 7
DNA-targeting Molecules as Therapeutic Agents
Edited by Michael J. Waring

Published by the Royal Society of Chemistry, www.rsc.org

lexitropsins,[5–7] synthetic polyamides,[8–12] heterocycles[13] and synthetic peptides.[14–19] Results from recent studies suggest that understanding the molecular basis governing the recognition of specific sequences in target DNA by small synthetic molecules will furnish a useful tool for future gene therapy.[5]

For over a decade we have been investigating the sequence-selective recognition of DNA by small synthetic peptides incorporating the XP(Hyp)RK motif (Hyp = 4-hydroxyproline).[14–19] Peptides incorporating both the XP(Hyp)RK motif and oligomeric units of 4-amino-1-methylpyrrole-2-carboxylic acid residues (Py) possess excellent DNA binding capability, with preference toward sequences containing consecutive arrays of A or T nucleotides. The original design of XPRK-containing peptides was inspired by the finding that an SPXX motif[20] is often found in repeating sequences in histones, steroid hormone receptors, various segmentation gene products as well as some oncogene products. The SPXX motif assumes a β-turn stabilised by two hydrogen bonds, and the side chains of the two basic residues engage in salt bridges with the DNA phosphate groups.

Since allosteric behaviour in macromolecule–small ligand interactions is an important facet of molecular recognition we began focusing our studies on allosteric processes associated with DNA–peptide binding employing the technology of DNA footprinting.[14–19] In 2006 we proposed a network-based DNA–peptide allosteric interaction model interconnecting multiple sites in fragments of the latent membrane protein (LMP-1) gene from a pathogenic Epstein–Barr virus variant derived from nasopharyngeal carcinoma.[16] After numerous footprinting studies we further proposed three different types of network-based allosteric communication between synthetic peptides binding to DNA: circuit type, incomplete circuit type and non-circuit type characterized by interstrand bidentate interactions.[17,18]

The enthalpy–entropy compensation (EEC) phenomenon has been extensively reported and discussed in various fields of chemistry, but still remains controversial and unresolved.[22] Because so much controversy surrounds it, the concept has attracted a good deal of interest. Positive correlation between $\Delta H$ and $\Delta S$ is referred to as compensation whereas negative correlation can be regarded as a form of interference. It has been proposed that apparent EEC phenomena are reflections of energetic differences in the system at the molecular level. When the values of $\Delta H$ and $\Delta S$ in a system are small, artefacts in the determination of EEC may easily occur. In two recent reports we attempted to resolve the EEC paradox as it applies to DNA–peptide interactions.[18,19] Distinctive enthalpy–entropy compensation relationships were demonstrated for the interaction of all twelve of our designed peptides with DNA, affording a straight line of slope close to unity in a plot of $\Delta H$ *versus* $T\Delta S$, with a y-axis intercept (average $\Delta G$) corresponding to $-8.5\ \mathrm{kcal\ mol^{-1}}$, while observed $\Delta G$ ranged from $-8.2$ to $-9.1\ \mathrm{kcal\ mol^{-1}}$ for the individual peptides.[18] The EEC seen with peptide RY-12 binding to the model duplex was shown to persist throughout a range of incubation temperatures. The net compensation of energy between the favourable negative $\Delta H$ and unfavourable negative $\Delta S$ components thus appears to constrain the value of net binding free energy $\Delta G$ within a remarkably constant range, as is clearly visible in a three-dimensional energetic plot.[18] We concluded that the preservation of a rather narrowly defined $\Delta G$ value is central to the EEC in DNA–peptide interactions, illuminating the universal EEC paradox commonly found in diverse biochemical processes.[18] In a second study, the energetic basis of DNA binding by four chlorambucil–peptide conjugates and four parent peptides was investigated by isothermal titration calorimetry, revealing that the binding of both the chlorambucil–peptide conjugates and the parent peptides is overwhelmingly enthalpy-driven.[19] This study also showed the maintenance of a conserved negative binding free energy in DNA–conjugate interactions, seemingly a crucial feature of the universal EEC phenomenon.[19]

To further examine our hypothesis and interpretation regarding the EEC paradox in DNA–peptide interactions, we carried out a study on DNA binding and its energetic aspects with seven newly designed peptides, together with the 12 designed peptides previously reported,[18] affording a total of 19 peptides for isothermal calorimetric experiments. The new peptides arose from our finding that incorporation of a new KP(Hyp)XR motif with four Py residues affords designed peptides with very good DNA-binding affinity as shown by agarose gel electrophoresis. Four peptide amides in this new series were synthesized, differing in the X amino acid residue: Lys-Hyp-His-Arg-$(Py)_4$-Arg-His-Hyp-Lys-$NH_2$ (KHypH-12); Lys-Hyp-Ser-Arg-$(Py)_4$-Arg-Ser-Hyp-Lys-$NH_2$ (KHypS-12); Lys-Pro-Trp-Arg-$(Py)_4$-Lys-Arg-$NH_2$ (PyWK-10); and Lys-Pro-Ser-Arg-$(Py)_4$-Lys-Arg-$NH_2$ (PySK-10). The dodecapeptides KHypH-12 and KHypS-12 were designed with the *C*-terminal tetrapeptide portion as -RXP(Hyp)-Lys-. Three peptides of the sequence Arg-Lys-$(Py)_4$-Lys-Arg-Hyp-X-$NH_2$ were synthesized in which the *C*-terminal tetrapeptide sequence was reversed compared with that of the motif X-Hyp-Arg-Lys-, producing decapeptides Arg-Lys-$(Py)_4$-Lys-Arg-Hyp-His-$NH_2$ (RHyH-10); Arg-Lys-$(Py)_4$-Lys-Arg-Hyp-Met-$NH_2$ (RHyM-10) and Arg-Lys-$(Py)_4$-Lys-Arg-Hyp-Glu-$NH_2$ (RHyE-10).

The choice of Trp, His, Glu and Ser residues in the X position of these peptides was empirical since previous designed peptides containing these residues were known to possess good DNA sequence recognition capability. We grew interested in investigating the sequence selectivity of these peptides containing amino sequences that deviate from that of the rather well investigated XP(Hyp)RK peptides. Thus, we used quantitative DNase I footprinting to investigate the sequence preference and allosteric interactions of four peptides (representative footprinting is shown in Figure 7.1) to a complementary pair of 5′-$^{32}$P-labeled pBR322 duplexes[14] composed of a 158-mer duplex (upper strand, 5′-$^{32}$P-labeled) and a 135-mer duplex (lower strand, 5′-$^{32}$P-labeled). Circular dichroism was used to study the conformational and stoichiometric aspects of peptide binding, as well as to a 13-mer deoxyribonucleotide duplex. The thermodynamic basis of the allosteric interactions was investigated using isothermal titration calorimetry (ITC).

To evaluate the validity of our EEC hypothesis with reference to general biochemistry we searched the literature for DNA–protein and other DNA–small-ligand interactions which also show the EEC phenomenon, and

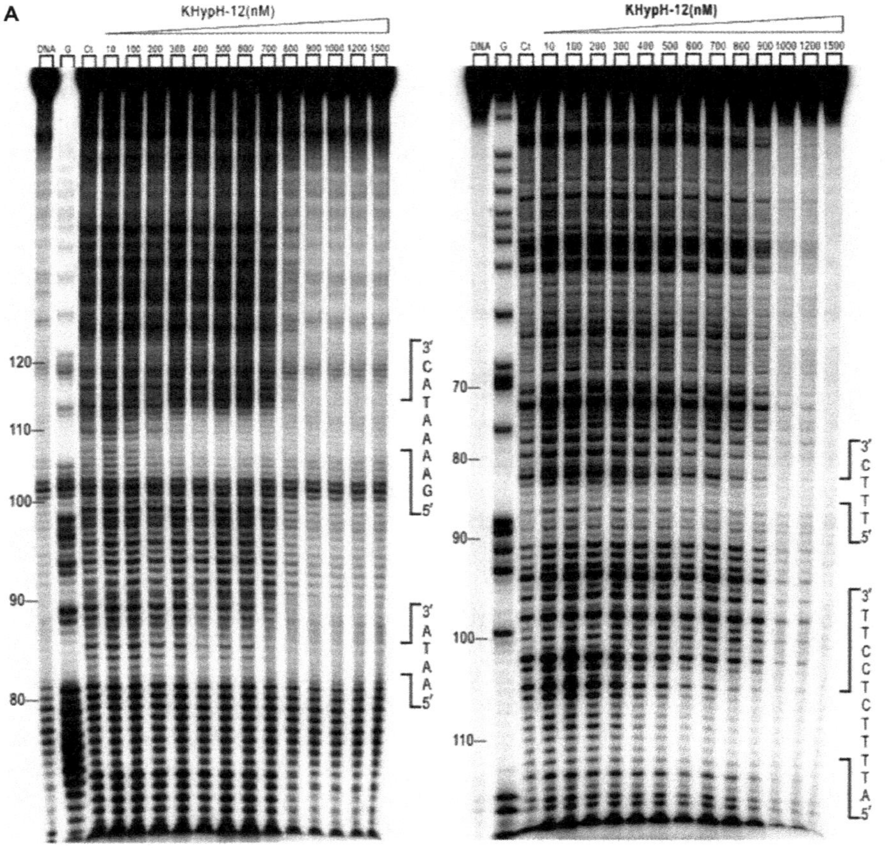


**Figure 7.1** (A) Autoradiographs showing DNase I footprinting of peptide KHypH-12 on DNA duplexes labelled at the 5′ end: 158-mer upper strand, left panel; 135-mer lower strand, right panel. Peptide KHypH-12 was equilibrated with the DNA in 5 mM sodium cacodylate buffer, pH 6.5 at 37 °C for 60 min before DNase 1 cleavage. G represents a Maxam–Gilbert guanine sequencing track and Ct indicates a DNase I digestion control lane. (B) Differential cleavage plots comparing the susceptibility of DNA fragments to DNase I cleavage after incubation with peptide KHypH-12 in cacodylate buffer at 37 °C for 60 min. The upper traces represent the differential cleavage plot for a given peptide bound to the 5′-[$^{32}$P]-labelled upper strand (158-mer) DNA fragment; the lower traces represent the corresponding plots for the 5′-[$^{32}$P]-labelled lower strand (135-mer) DNA fragment. Vertical dotted lines between DNA bases represent assignment of interstrand bidentate interactions where significant coincident H-bonding interactions occur between complementary bases in the two strands.

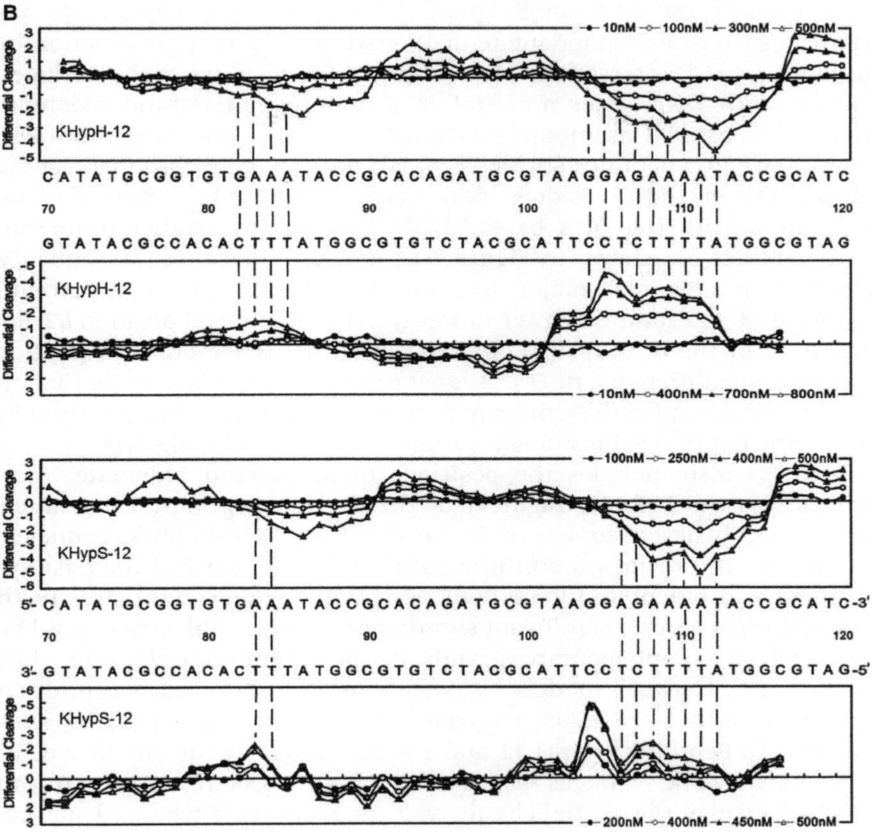


**Figure 7.1** Continued.

focused on the conservation of free energy change which is critical for maintaining enthalpy–entropy compensation.

### 7.1.1 Footprinting Analysis of DNA–Peptide Sequence-selective Interactions

Of the seven newly designed peptides, six were subjected to detailed footprinting analysis (KHypH-12, KHypS-12, PyWK-10, PySK-10, RHyM-10 and RHyH-10). The goal of footprinting studies is to verify our recent hypothesis that three different types of network-based allosteric communication occur when the designed synthetic peptides bind to multiple DNA recognition sites.[16–19] In addition, footprinting studies also allow us to gain some insights into the energetic basis for allosteric interaction of peptides incorporating the XP(Hyp)RK motifs with DNA. Hydrogen bonding between the peptide moieties and DNA bases is crucial to sequence-selective

DNA–protein[20] and DNA–small-ligand[16–19] recognition processes, and has been categorized as monodentate or interstrand bidentate interaction by Luscombe *et al.*[21] Interactions of the Arg and Lys residues of the designed peptides with DNA bases are vital in generating interstrand bidentate interactions as well as monodentate interactions.[20] The 'X' amino acid residues of these peptides such as Trp, His and Ser were chosen since the side chains of these residues may become involved in monodentate interactions with the DNA bases.[17–19,21] Footprinting studies (representative peptides are shown in Figure 7.1A and B) reveal that the sequence selectivity locates two major loci on the pBR322 fragment: around position 102–115 comprising the d(AAAA)–d(TTTT) site and position 82–90, consisting of the d(AAA)–d(TTT) site (Table 7.1). However, each peptide displays some difference in DNA sequence-selectivity. It is notable that all these peptides bind with significant positive cooperativity, characterised by Hill coefficient ($n_H$) values ranging from +1.6 to +9.3 (Table 7.1).

Table 7.1 also includes the positions of interstrand bidentate interactions with the PBR322 fragment by the six new peptides. These interactions are believed to be vitally involved in DNA–peptide sequence recognition. In previous footprinting studies[16–19] we marked the position of interstrand bidentate interactions assigned by connecting bases (with pecked lines) wherever significant simultaneous DNase I blockage could be discerned on complementary strands in the differential cleavage plots (Figure 7.1B). Network models[16–19] were constructed for each peptide in order to interpret the allosteric communication between binding sites (Figure 7.2). Peptides KHypH-12 and KHypS-12 display the circuit type of allosteric network, whereas peptides PyWK-10, PySK-10, RHyM-10 and RHyH-10 display the partial-circuit type. In the circuit type and partial-circuit type of interaction networks, strong intrastrand allosteric communications are conveyed by conformational changes between two peptide molecules bound to two sub-sites of a wide binding site (Figure 7.2). Conformational changes at particular DNA binding loci are readily transmitted to adjacent binding sites as well as to sites on the complementary strand, facilitating (positively cooperative) or impeding (negatively cooperative) the binding of further peptide molecules. Occupation of almost any binding locus by a functional portion of the ligand (for example, a peptide moiety which affords binding interactions) can be signalled to adjacent sub-binding sites and also to sites on the complementary strand *via* interstrand bidentate interactions. Results from footprinting studies confirm that allosteric communications may occasionally be transmitted to a more remote binding site on the same DNA strand, spanning as many as 12–16 base pairs. The frequent occurrence of efficient positive cooperative interaction networks in DNA–peptide binding suggest that these interaction networks represent nature's endeavour to minimize the binding energy required for corresponding interactions.

**Table 7.1** Sequence-selectivity and physicochemical parameters for binding of designed peptides to recognition sites in complementary upper (158-mer) and lower (135-mer) DNA strands at 37 °C determined by quantitative DNase I footprinting.[a]

| Ligand | Binding site position | Recognition sequence | $K_a$ | $n_H$ | Position of interstrand bidentate interactions | Type of interaction network |
|---|---|---|---|---|---|---|
| KHypH-12 | U83–89 | 5′-AAATACC-3′ | $3.2\times10^6$ | 1.6 | 83–85 | Circuit |
| | U104–115 | 5′-GGAGGAAAATAC-3′ | $3.7\times10^6$ | 1.5 | 104–115 | |
| | L85–82 | 5′-TTTC-3′ | $1.6\times10^6$ | 2.9 | | |
| | L112–102 | 5′-ATTTTCTCCTT-3′ | $2.6\times10^6$ | 2.9 | | |
| KHypS-12 | U83–90 | 5′-AAATACCG-3′ | $2.4\times10^6$ | 2.9 | 83–84 | Circuit |
| | U106–115 | 5′-AGAAAATACC-3′ | $3.9\times10^6$ | 3.6 | 106–112 | |
| | L84–82 | 5′-TTC-3′ | $2.3\times10^6$ | 1.8 | | |
| | L104–111 | 5′-TTTTCTCC-3′ | $4.3\times10^6$ | 1.9 | | |
| RHyM-10 | U103–112 | 5′-AGGAGAAAAT-3′ | $8.6\times10^5$ | 2.1 | 103–112 | Partial-circuit |
| | L113–103 | 5′-TATTTTCTCCT-3′ | $1.7\times10^6$ | 3.0 | | |
| RHyH-10 | U103–112 | 5′-AGGAGAAAAT-3′ | $1.7\times10^6$ | 3.8 | 105–112 | Partial-circuit |
| | U80–85 | 5′-GTGAAA-3′ | $1.6\times10^6$ | 2.6 | | |
| | L116–106 | 5′-CGGTATTTTCTC-3′ | $8.7\times10^5$ | 3.0 | | |
| PyWK-10 | U105–115 | 5′-GAGAAAATACC-3′ | $1.6\times10^6$ | 6.7 | 105–115 | Partial-circuit |
| | L115–103 | 5′-GGTATTTTCTCCT-3′ | $1.4\times10^6$ | 9.3 | | |
| PySK-10 | U104–113 | 5′-GGAGAAAATA-3′ | $1.3\times10^6$ | 5.3 | 104–113 | Partial-circuit |
| | L85–84 | 5′-TT-3′ | $9.9\times10^5$ | 7.4 | | |
| | L113–103 | 5′-TATTTTCTCCT-3′ | $1.2\times10^6$ | 1.8 | | |

[a] $K_a$ and $n_H$ are the apparent association constant and Hill coefficient determined from concentration-dependent DNase I footprinting studies, respectively. Binding sites on the upper and lower strands are identified by U and L respectively.

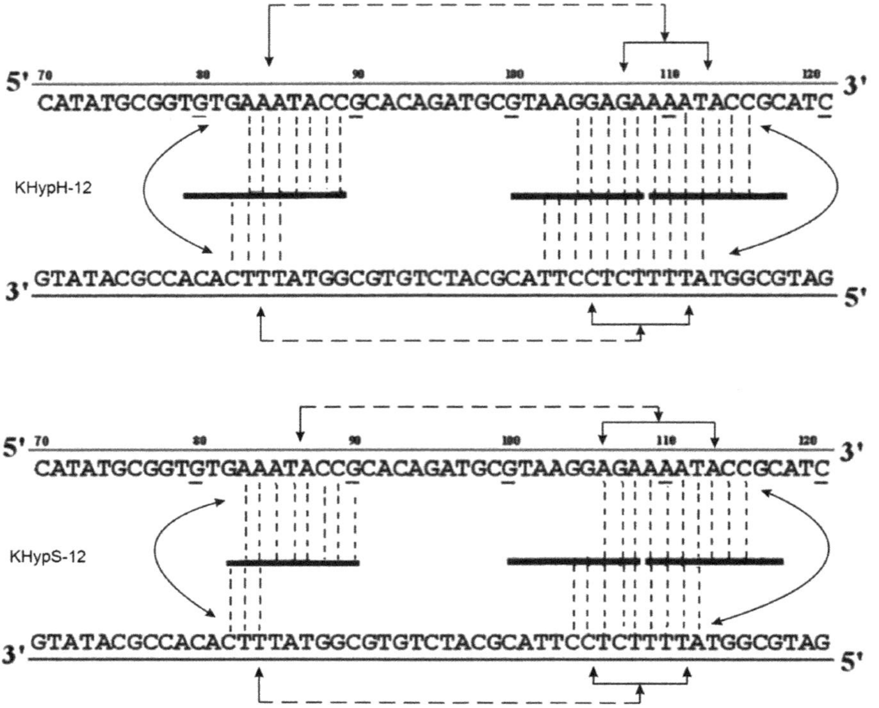


**Figure 7.2** Proposed allosteric interaction network models for cooperative binding of peptide KHypH-12 and KHypS-12 to pBR322 fragments based on quantitative footprinting experiments. The portion of the ligand binding to each DNA site/sub-site is represented by a thick horizontal line. Monodentate interactions and interstrand bidentate interactions are represented by vertical broken lines. The continuous horizontal arrow lines represent communication of allosteric interaction between DNA sub-binding sites. Broken horizontal arrow lines between neighbouring binding loci some 12–16 nucleotides apart represent moderate or weak cooperative communication.

### 7.1.2 Circular Dichroism Analysis of DNA–Peptide Interactions

Circular dichroism spectroscopy (CD) was used to investigate conformational changes in the helical structure when a peptide binds to DNA. For these experiments a 13-mer deoxyribonucleotide duplex was used in CD studies as previously described:[18,19] d(TAGGAGAAAATAC)–d(GTATTTTCTCCTA) (U4A–L4T). Representative CD spectra for peptides are shown in Figure 7.3A–E. Similar to other peptides, all nine new peptides produce a dose-dependent negative CD enhancement at around 250 nm as well as two strong positive CD enhancement bands at around 275 nm and 330 nm (Figure 7.3D). In general, in the difference spectra the negative band around 250 nm can be seen to be red-shifted to around 270 nm and two strong positive bands appear at around 285 nm and 330 nm. The induced positive band at around 330 nm can be used for plotting binding isotherms since it is peptide-concentration-dependent and free from interference by native DNA and peptide. The significant enhancement of CD around 330 nm by these peptides indicates that the principal interaction occurs in the minor groove of DNA.[22]

Unlike peptide PySK-10, peptides KHypH-12, PyWK-10, RHyM-10, RHyH-10, RHyE-10, and KHypS-12 induce significant ellipticity change, suggesting that drastic conformational changes in the DNA minor groove are induced by the binding of these peptides (Figure 7.3F and G).

### 7.1.3 Investigations of Enthalpy–Entropy Compensation Phenomena in General DNA–Ligand Interactions

For as much as four decades, considerable controversy and interest have surrounded apparent EEC phenomena,[23,24] with various theoretical interpretations being presented. Despite that, the physical basis of EEC has not been convincingly elucidated. In many of our studies of DNA–peptide interactions, the overwhelmingly large values of $\Delta H$ and $\Delta S$ measured by ITC for DNA–peptide interaction might offer a unique advantage for investigating the energetic basis of the EEC phenomenon.

According to Williams and colleagues,[25] the change in free energy of binding consists of a number of components.

$$\Delta G = \Delta G_{\mathrm{t+r}} + \Delta G_{\mathrm{r}} + \Delta G_{\mathrm{h}} + \Sigma\Delta G_{\mathrm{p}} + \Delta G_{\mathrm{conf.}} + \Delta G_{\mathrm{vdw}} \tag{7.1}$$

where i$\Delta G_{\mathrm{t+r}}$ refers to the unfavourable binding of a ligand to a receptor, losing its translational (t) and rotational (r) entropy considerably relative to the receptor. $\Delta G_{\mathrm{r}}$ refers to the energy cost of $n$ internal rotations. $\Delta G_{\mathrm{h}}$ represents the benefit of the net hydrophobic effect and $\Sigma\Delta G_{\mathrm{p}}$ expresses the benefit from formation of hydrogen bonding and ionic interactions. $\Delta G_{\mathrm{conf.}}$ is the total conformational strain energy generated upon binding, and $\Delta G_{\mathrm{vdw}}$ is the van der Waals energy difference between free and bound states.

Footprinting studies have shown that significant positive cooperativity can be observed in almost all DNA–peptide interactions (Table 7.1). Cooper and Dryden[26] suggested that positive cooperativity in ligand binding will induce a 'stiffening' or tightening effect in the protein/receptor, initiated by the loss of many internal vibration degrees of freedom. Williams and coworkers[25] also concluded that highly exothermic binding is associated with positive cooperativity with improved bonding within the receptor in its drug-bound state, accounting for the very unfavourable negative entropy change. Williams and colleagues[25] pointed out that positive cooperativity and negative cooperativity can coexist in the same system. Indeed, we observed that DNA binding of the peptide His-Pro-Arg-Lys-$(Py)_4$-His-Pro-Arg-Lys-$NH_2$ (PyPro-12)[17] involves concomitant positive and negative cooperative site binding at two loci. However, positive cooperativity is predominant for our peptide series incorporating the XP(Hyp)RK motifs. Positive cooperativity is likely to be associated with improved bonding within the DNA duplex upon peptide binding, manifested by multiple intrastrand monodentate interactions and interstrand bidentate interactions formed between the peptide moieties and the DNA bases. Ionic interactions between the positively charged Arg and Lys side chains of the peptides and DNA phosphates may restrict the conformational mobility of the DNA duplex and thus contribute

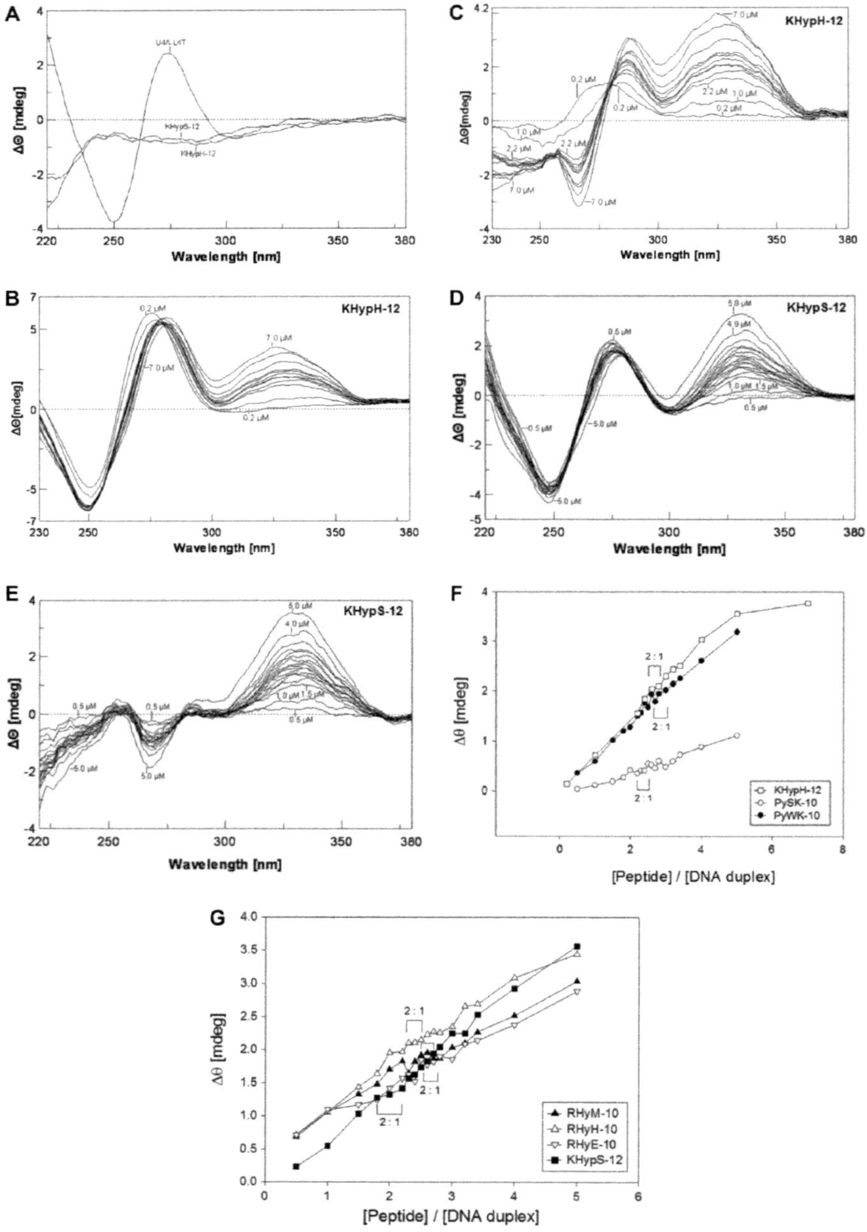

to the dramatic decrease in total entropy. Any loss in entropic energy must transform to exothermic energy, which in turn causes an increase in enthalpy. Moreover, high exothermic energy may promote allosteric adjustment of DNA conformation within the binding loci *via* hydrogen bonding and ionic interactions between DNA duplex and peptide, thereby augmenting the favourable energy components, $\Delta G_h$, $\Sigma\Delta G_p$, $\Delta G_{conf}$ and $\Delta G_{vdw}$, and facilitating positively cooperative binding of approaching peptide molecules. Williams and coworkers[25] also proposed that highly exothermic binding is associated with significant positive cooperativity. This suggestion is consistent with the results of our ITC experiments showing that very high $\Delta H$ values are generated by peptide binding to multiple DNA recognition sites with strong positive cooperativity.[17–19]

In DNA–peptide interactions, the overwhelmingly large magnitudes of $\Delta H$ and $\Delta S$ measured by ITC in previous studies as well as this study appear to offer a unique advantage for investigating the energetic basis of the compensation process. In a recent study employing a plot of $\Delta H$ *vs.* $T\Delta S$, the enthalpy–entropy compensation ratio $k_e$ (slope of the straight line) for 19 peptides containing XP(Hyp)RK or KXP(Hyp)R and Py motifs approaches unity (slope = 1.006), with the $y$-axis intercept equal to the average $\Delta G$ value: $-8.3$ kcal mol$^{-1}$ (Figure 7.4B).

$$\Delta H = k_e\, T\Delta S - 8.3 \tag{7.2}$$

When $k_e$ approaches unity,

$$\Delta H = -8.3 + T\Delta S \tag{7.3}$$

The above equation depicts the strict linear dependence of $\Delta H$ on $T\Delta S$ for the 19 peptide–DNA interactions studied for interpreting the enthalpy–entropy compensation phenomenon. It is apparent that the almost exact compensation of energy between the favourable negative $\Delta H$ and unfavourable negative $\Delta S$ energy components maintains the almost constant value of net binding free energy $\Delta G$. According to eqn (7.3), a negative $\Delta G$ of acceptable magnitude must be maintained for successful DNA–peptide binding. Any decrease of $\Delta G$ (less negative) would demand a lowering of

**Figure 7.3** (A) CD spectra of DNA duplex U4A-L4T alone and peptide KHypH-12 alone at 37 °C. (B) Titration of duplex U4A-L4T *versus* peptide KHypH-12 at peptide concentrations of 0.5, 1.0, 2.2, 2.4, 2.6, 2.8, 3.0, 3.2, 3.4, 4.0, 5.0, 7.0 μM at 37 °C. (C) Corresponding CD difference spectra with the contribution of free duplex and peptide KHypH-12 subtracted. (D) Titration of duplex U4A-4T *versus* peptide KHypS-12 with the same concentration as peptide KHypH-12. (E) Corresponding CD difference spectra with the contribution of free duplex and peptide KHypS-12 subtracted. (F, G). Plots of CD intensity at 322 nm *versus* [peptide]/[duplex] for the titration of duplex U4A-L4T with various peptides at 37 °C. The proposed stoichiometric binding ratios are as indicated, with binding <2 : 1 considered to be predominantly 1 : 1.

$\Delta H$ (less negative) and a concomitant decrease in $\Delta S$ (more negative)—both are unfavourable in DNA–ligand interactions.

For general application of eqn (7.3) to the EEC phenomenon for different DNA–ligand interactions,

$$\Delta H = k_e\ T\Delta S + \Delta G_c \tag{7.4}$$

$\Delta G_c$ is the free energy change of DNA–ligand interaction for a particular series of ligands which have structural or physicochemical similarity and exhibit the EEC phenomenon. The value of $\Delta G_c$ is always preserved even if conditions such as incubation temperature, ionic concentrations and pH of the reactions are varied.[19] $k_e$ is the slope of the straight line and $\Delta G_c$ is the intercept on the $y$-axis.

$$\Delta H - \Delta G_c = k_e T\Delta S \tag{7.5}$$

$$k_e = (\Delta H - \Delta G_c)/T\Delta S \tag{7.6}$$

where $k_e$ is the change in $(\Delta H - \Delta G_c)$ per unit change in $T\Delta S$.

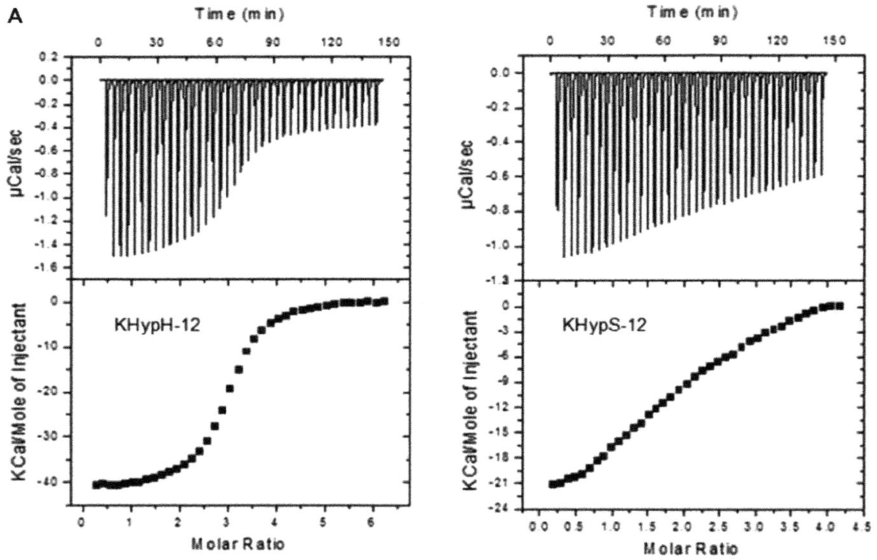


**Figure 7.4** (A) ITC curves for titration of peptides KHypH-12 and KHypS-12 into the U4A-L4T duplex at 25 °C. For each experiment the top panel represents the raw heat of binding generated with successive additions of peptide, and in the bottom panel the integrated heat is plotted *versus* peptide : DNA molar ratio. Data acquisition and analysis were performed using nonlinear least-squares fitting algorithm software (Microcal Origin 7.1). (B) Plot of enthalpy ($\Delta H$) *versus* entropy ($T\Delta S$) from ITC of 19 peptides added to the U4A-L4T duplex at 25 °C (298 K). (C) Three-dimensional plot of $\Delta H$, $T\Delta S$ and $\Delta G$ from titration of all 19 peptides into the U4A-L4T duplex. The short vertical projection plane (in green) corresponds to $\Delta G$ magnitudes sustained around $-8.2$ kcal mol$^{-1}$. The figure represents a view of the 'cube' from below.

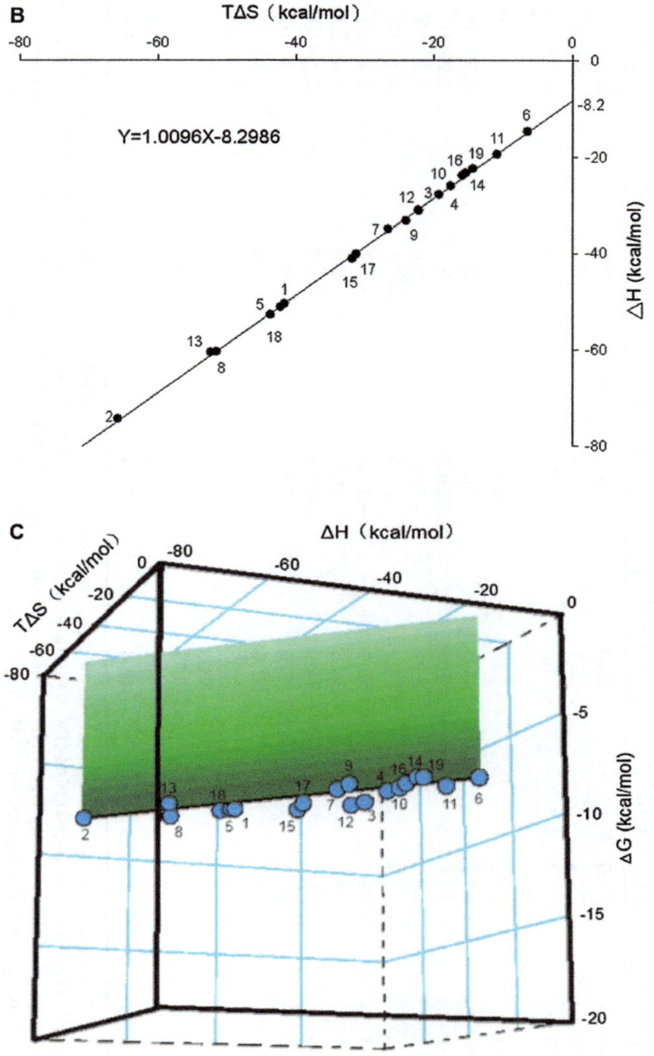


**Figure 7.4** Continued.

In this study we wish to further validate our recent interpretation that the energy compensation between $\Delta H$ and $T\Delta S$ operates synchronously to preserve EEC so as to maintain a favourable $\Delta G$ value within a constant or narrow range for successful complex formation between ligands and DNA. We employed CD and ITC to study the DNA interactions of seven new peptides incorporating the *N*-terminal KP(Hyp)XR, or C-terminal KRHyp-X-$NH_2$ motifs.

Table 7.2 shows that the binding free energy of 19 designed peptides containing XP(Hyp)RK, KPXR or KRHypX motifs to the U4A-L4T duplex lies within a fairly narrow range: −7.9 to −9.1 kcal $mol^{-1}$. To study the implicit EEC phenomenon, a plot of $\Delta H$ *versus* $T\Delta S$ for the 19 peptides titrated

**Table 7.2** Thermodynamic parameters from titration of peptides *versus* oligonucleotide duplex (U4A-L-4T) determined by isothermal titration calorimetry.[a]

| Entry | [peptide] (mM) | [DNA] (mM) | $\Delta H$ (kcal mol$^{-1}$) | $\Delta S$ (cal mol$^{-1}$) | $T\Delta S$ (kcal mol$^{-1}$) | $\Delta G$ (kcal mol$^{-1}$) | $K_a$ (M$^{-1}$) |
|---|---|---|---|---|---|---|---|
| 1 | HyM-10(0.15) | 0.0033 | −50.5 | −140 | −41.7 | −8.8 | $2.9\times10^6$ |
| 2 | HyQ-10(0.15) | 0.01 | −74.4 | −221 | −65.9 | −8.6 | $1.9\times10^6$ |
| 3 | HyS-10(0.15) | 0.006 | −31.2 | −74.5 | −22.2 | −9.0 | $4.2\times10^6$ |
| 4 | HyE-10(0.15) | 0.007 | −27.8 | −64.6 | −19.3 | −8.5 | $1.7\times10^6$ |
| 5 | HyH-10(0.05) | 0.0033 | −51.2 | −142 | −42.3 | −8.8 | $2.9\times10^6$ |
| 6 | PyHK-10(0.2) | 0.0066 | −14.7 | −21.8 | −6.5 | −8.2 | $9.8\times10^5$ |
| 7 | RY-12(0.15) | 0.01 | −35.0 | −89.7 | −26.7 | −8.2 | $1.0\times10^6$ |
| 8 | RPyY-12(0.05) | 0.0045 | −60.5 | −173 | −51.6 | −8.9 | $3.3\times10^6$ |
| 9 | PyHyp-12(0.05) | 0.004 | −33.2 | −80.9 | −24.1 | −9.1 | $4.4\times10^6$ |
| 10 | RHyp-12(0.15) | 0.005 | −26.0 | −59.1 | −17.6 | −8.4 | $1.4\times10^6$ |
| 11 | PyMK-10(0.15) | 0.005 | −19.4 | −36.5 | −10.9 | −8.5 | $1.8\times10^6$ |
| 12 | PyQK-10(0.15) | 0.006 | −31.0 | −74.9 | −22.3 | −8.7 | $2.2\times10^6$ |
| 13 | PyWK-10(0.15) | 0.007 | −60.6 | −176 | −52.4 | −8.2 | $1.0\times10^6$ |
| 14 | PySK-10(0.15) | 0.0085 | −23.3 | −51.6 | −15.4 | −7.9 | $6.3\times10^5$ |
| 15 | KHypH-12(0.15) | 0.005 | −41.2 | −107 | −31.9 | −9.1 | $5.0\times10^6$ |
| 16 | KHypS-12(0.15) | 0.0075 | −23.8 | −53.5 | −15.9 | −7.9 | $5.8\times10^5$ |
| 17 | RHyH-10(0.15) | 0.005 | −40.2 | −105 | −31.3 | −8.8 | $2.9\times10^6$ |
| 18 | RHyM-10(0.15) | 0.005 | −52.8 | −147 | −43.8 | −8.8 | $2.9\times10^6$ |
| 19 | RHyE-10(0.15) | 0.0035 | −22.4 | −48.4 | −14.4 | −7.9 | $6.7\times10^5$ |

[a]All ITC experiments were performed at 25 °C unless otherwise stated. The amino acid sequences of the peptides are: Met-Hyp-Arg-Lys-(Py)$_4$-Lys-Arg-NH$_2$ (HyM-10); Gln-Hyp-Arg-Lys-(Py)$_4$-Lys-Arg-NH$_2$ (HyQ-10); Ser-Hyp-Arg-Lys-(Py)$_4$-Lys-Arg-NH$_2$ (HyS-10); Glu-Hyp-Arg-Lys-(Py)$_4$-Lys-Arg-NH$_2$ (HyE-10); His-Hyp-Arg-Lys-(Py)$_4$-Lys-Arg-NH$_2$ (HyH-10); His-Pro-Arg-Lys- (Py)$_4$-Lys-Arg-NH$_2$ (PyHK-10); Tyr-Pro-Arg-Lys-(Py)$_4$-Lys-Arg-Pro-Tyr-NH$_2$ (RY-12); His-Hyp-Arg-Lys-(Py)$_4$-His-Hyp-Arg-Lys-NH$_2$ (PyHyp-12); His-Hyp-Arg-Lys-(Py)$_4$-Lys-Arg-Hyp-His-NH$_2$ (RHyp-12); His-Pro-Arg-Lys-(Py)$_4$-Lys-Arg-Pro-Tyr-NH$_2$ (RPyY-12); Lys-Pro-Met-Lys-(Py)$_4$-Lys-Arg-NH$_2$ (PyMK-10); Lys-Pro-Gln-Lys-(Py)$_4$-Lys-Arg-NH$_2$ (PyQK-10); Lys-Pro-Trp-Lys-(Py)$_4$-Lys-Arg-NH$_2$ (PyWK-10); Lys-Pro-Ser-Lys-(Py)$_4$-Lys-Arg-NH$_2$ (PySK-10); Lys-Hyp-His-Arg-(Py)$_4$-Arg-His-Hyp-Lys-NH$_2$ (KHypH-12); Lys-Hyp-Ser-Arg-(Py)$_4$-Arg-Ser-Hyp-Lys-NH$_2$ (KHypS-12); Arg-Lys-(Py)$_4$-Lys-Arg-Hyp-His-NH$_2$ (RHyH-10); Arg-Lys-(Py)$_4$-Lys-Arg-Hyp-Met-NH$_2$ (RHyM-10); Arg-Lys-(Py)$_4$-Lys-Arg-Hyp-Glu-NH$_2$ (RHyE-10).

against the U4A-L4T duplex affords the enthalpy–entropy compensation ratio $k_e$ (slope of the straight line) for the 19 peptides which approaches unity (slope = 1.01), with the *y* axis intercept equal to the average $\Delta G$ value: $-8.3$ kcal mol$^{-1}$ (Figure 7.4B). To underline the EEC phenomenon in the DNA–peptide interaction, it is instructive to plot a three-dimensional figure of $\Delta H$, $T\Delta S$ and $\Delta G$ for the 19 peptides titrated against the DNA duplex. The result is a near-straight line suspended in thermodynamic space, well demonstrating the linear enthalpy–entropy-free energy relationship (Figure 7.4C). The broad conclusion is that stringent constraints act in nature to preserve EEC so as to maintain favourable $\Delta G$ within a constant or narrow range for successful binding of these ligands to DNA.

In addition, to advance our understanding of the generality of applying eqn (7.4) to the universal EEC phenomenon, we searched the literature for DNA interactions with different series of small ligands (drugs) and large ligands (proteins) in relation to the EEC paradox. Although most DNA–ligand interactions show apparent EEC relationships, only a limited number of research studies give $\Delta H$ *versus* $T\Delta S$ with more than five data points. Therefore only representative DNA–ligand EEC relationships with adequate data points are shown in Table 7.3.

Various proteins show distinct EEC phenomena in binding to DNA. Allemann and colleagues reported the thermodynamics of transcription factor MASH-1 binding to DNA binding sites SP-1 and MCK-S in low-salt and high-salt MES buffer.[27] To verify our hypothesis about nature's thermodynamic trend to preserve EEC so as to maintain favourable $\Delta G$ within a constant or narrow range for successful binding of ligands to DNA, we plotted a three-dimensional figure of $\Delta H$, $T\Delta S$ and $\Delta G$ of SP-1 and MCK-S binding to MASH-1 (Figure 7.5B and C, Table 7.3). The result in each case is a nearly straight line, well demonstrating the preservation of $\Delta G$ values of $-7.9$ kcal mol$^{-1}$ and $-9.4$ kcal mol$^{-1}$, respectively. In addition, binding of bZIP domains of Jun-Fos heterodimer to dsDNA oligomers containing the wild type consensus motif TGACTCA and single nucleotide variants,[28] binding of Max p21 protein to E-box,[29] binding of Sso7d protein to DNA,[30] DNA binding by the c-Myb DNA-binding domain[31] and binding of the mammalian high mobility group protein HMGA2 to poly(dA–dT)2 and poly(dA)–poly(dT)[32] all show significant EEC relationships, with preservation of $\Delta G$ values. A three-dimensional plot of $\Delta H$, $T\Delta S$ and $\Delta G$ of bZIP domains for Jun-Fos heterodimer binding to dsDNA oligomers also reveals an almost straight line, with preservation of the $\Delta G$ value at $-6.8$ kcal mol$^{-1}$ (Figure 7.5C).

At the other end of the scale, the binding of small ligands or drugs to DNA can also display remarkable EEC phenomena, for example, 7-aminoactinomycin D,[33] ethidium,[34] netropsin,[34,35] and berberine,[36] as well as designed peptides (Table 7.3).[19] It is apparent that plots of $\Delta G$ *versus* $T\Delta S$ for different DNA-binding ligands with a wide range of molecular sizes titrated against DNA afford the net binding free energy that lies rather constantly within a certain range, showing distinctive EEC phenomenon.

**Table 7.3** Thermodynamic parameters of general DNA–ligand interactions.[a]

| DNA | Ligand | $\Delta G$ (kcal mol$^{-1}$) | $\Delta H$ range (kcal mol$^{-1}$) | $T\Delta S$ range (kcal mol$^{-1}$) | $k_e$ | Ref. |
|---|---|---|---|---|---|---|
| SP-1 oligonucleotide duplex in low-salt and high-salt buffer, pH 6.0 | Transcription factor MASH-1 | −7.9 | −28.6 to −39.8 | −20.0 to −30.3 | 1.21 | 27 |
| MCK-S oligonucleotide duplex in low-salt and high-salt buffer, pH 6.0 | Transcription factor MASH-1 | −9.4 | −32.6 to −39.4 | −24.0 to −38.8 | 1.06 | 27 |
| dsDNA containing the motif TGACTCA and single nucleotide variants | bZIP domains of Jun-Fos heterodimer | −6.8 | −14.0 to −42.9 | −6.2 to −29.5 | 1.04 | 28 |
| Homopolymeric DNA | Sso7d protein | −8.5 | +12.6 to +4.3 | −16.1 to −21.5 | 1.0 | 30 |
| C-Myb non-cognate DNA binding domain | R2R3 | −10.4 | −8.0 to −13.9 | +1.9 to −3.7 | 0.58 | 31 |
| C-Myb cognate DNA binding domain | R2R3 mutants | −11.5 | −9.6 to −12.5 | +0.41 to −1.4 | 0.70 | 31 |
| E-box | Max p21 and variants | −29.8 | −29.6 to −38.0 | −19.1 to −28.4 | 0.71 | 29 |
| $(AT)_n$, $(AA)_n$, $(AA)(AT)_n n = 2$–4 | 7-Amino-actinomycin D | −7.4 | −4.9 to −7.0 | +1.8 to +0.23 | −1.4 | 33 |
| -AATT containing ds-DNA construct, site 1 | Netropsin | −10.8 | −4.3 to −11.5 | +1 to −6.2 | 0.99 | 35 |
| -AATT containing ds-DNA construct, site 2 | Netropsin | −8.9 | −9.1 to −29.0 | +27.1 to −1.0 | 0.91 | 35 |
| CP,CT,EC,ML DNA, simple polynucleotides | Berberine | −5.8 | −1.1 to −32.5 | +1.6 to +9.7 | 0.95 | 36 |
| Chicken Erythrocyte Chromatin. | Netropsin,Ethidium | −6.8 | −7.4 to −11.4 | −0.4 to −4.9 | 1.12 | 34 |
| Oligonucleotide duplex U4A-L4T | 12 Designed peptides | −8.5 | −14.7 to −74.4 | −6.5 to −65.9 | 1.00 | 18 |
| Oligonucleotide duplex U4A-L4T | 19 Designed peptides | −8.3 | −14.7 to −74.4 | −6.5 to −65.9 | 1.01 | This study |
| Oligonucleotide duplex U4A-L4T | 4 Chlorambucil-peptide conjugates | −8.8 | −25.4 to −46.0 | −17.8 to −37.5 | 1.02 | 19 |

[a] $k_e$ is computed from published thermodynamic data (see references) using the equation: $k_e = (\Delta H - \Delta G_c)/T\Delta S$.

In DNA–ligand interactions, the entropy of association[37] can be expressed as:

$$\Delta S^{o}_{assoc} = \Delta S^{o}_{H} + \Delta S^{o}_{ions} + \Delta S^{o}_{rt} + \Delta S^{o}_{vib} + \Delta S^{o}_{conf} \quad (7.7)$$

$\Delta S^{o}_{assoc}$ is the standard entropy of binding, $\Delta S^{o}_{H}$ and $\Delta S^{o}_{ions}$ are $\Delta S$ arising from hydrophobic and polyelectrolyte effects, $\Delta S^{o}_{rt}$ reflects rotational and translational degrees of freedom on complex formation, $\Delta S^{o}_{vib}$ is the entropy change associated with the population of new low-frequency vibrational modes created by the complex formation and $\Delta S^{o}_{conf}$ is the conformational entropy change of complexation.

The strong negative entropy changes seen with the binding of all these designed peptides to DNA indicate that there are significant losses in rotational and translational motion of the DNA duplex as well as the peptide, suggesting that $\Delta S^{o}_{rt}$ is negative. The increase in entropy from solvent reorganisation $\Delta S^{o}_{H}$ and $\Delta S^{o}_{ions}$ favour DNA–ligand binding, whereas $\Delta S^{o}_{rt}$ is typically negative and opposed to binding. The sum of $\Delta S^{o}_{vib}$ and $\Delta S^{o}_{conf}$ equates to quasi-harmonic entropy change $\Delta S^{o}_{QH}$[37] that is also negative. When proteins or peptides bind to DNA, the coulombic interactions between the positively charged side chains of Arg and Lys residues and DNA phosphates lead to the release of bound counterions, affording favourable positive $\Delta S^{o}_{ions}$ and $\Delta S^{o}_{H}$. However, the significant decrease in total entropic energy upon binding indicates that $\Delta S^{o}_{conf}$ must be negative because of strong coulombic interactions between DNA phosphates and the positively charged ligand, simply restricting the conformational changes of the DNA–ligand complex.

Table 7.3 indicates that very often the binding of proteins to DNA is associated with significant negative $\Delta H$ and unfavourable negative $T\Delta S$ values. In receptor–ligand interactions, translational (t) and rotational (r) entropy are lost considerably relative to the receptor. According to the law of conservation of energy, any loss in entropic energy must transform to exothermic energy, hence causing an increase in enthalpy. For example, more than 100 nmol of cations can be released in DNA–protein interactions per nmole of protein whereas only 4 nmol of cations per nmole of our designed peptides can be released in DNA–peptide interactions, suggesting that DNA–peptide interactions are even more readily able to induce a decrease in total entropy. Salt-dependence CD experiments confirming that ionic interactions are crucial in DNA–peptide interactions have been reported.[19] Thus, ionic interactions between the positively charged Arg and Lys side chains of the peptides or conjugates and internucleotide phosphates, together with monodentate and interstrand bidentate interactions between the peptide moieties and DNA bases, furnish adequate energy to restrict the conformational mobility of the DNA duplex as well as the ligand and may account for the dramatic decrease in total entropy.

To eliminate EEC representational bias and to study the extent of enthalpy–entropy compensation in DNA–ligand interactions, we employed Williams' approach with differential plots[38] originally used for protein–ligand interactions, which diminishes any potential effect on the

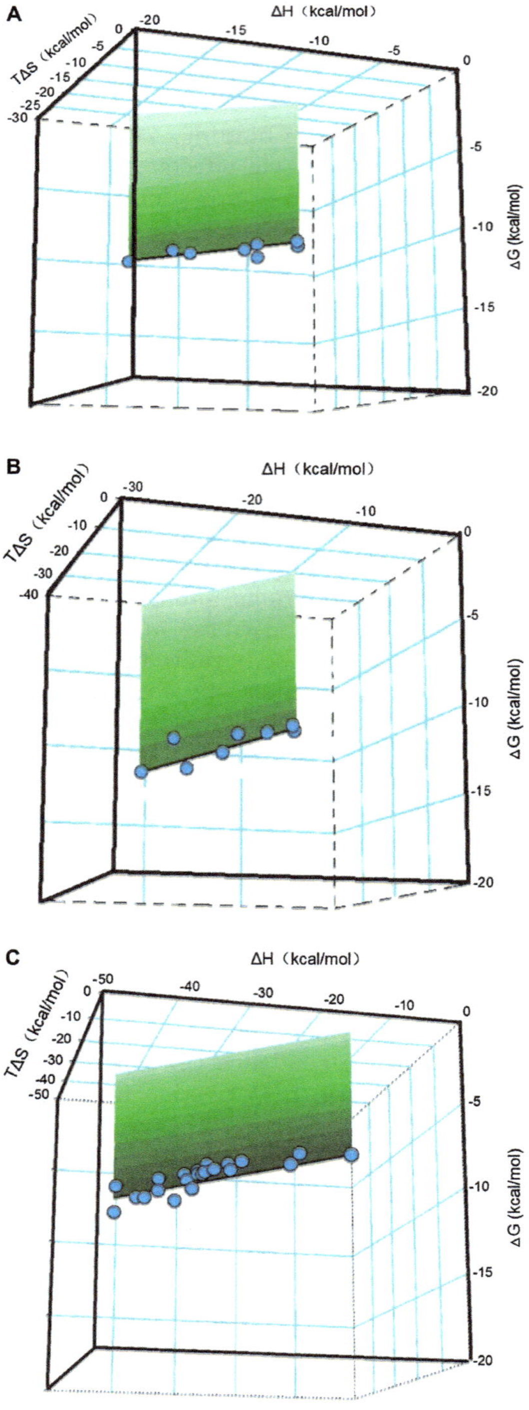
A
ΔH (kcal/mol)
TΔS (kcal/mol)
ΔG (kcal/mol)
-20
-15
-10
-5
0
-30
-25
-20
-15
-10
-5
0
-5
-10
-15
-20
B
ΔH (kcal/mol)
TΔS (kcal/mol)
ΔG (kcal/mol)
-30
-20
-10
0
-40
-30
-20
-10
0
-5
-10
-15
-20
C
ΔH (kcal/mol)
TΔS (kcal/mol)
ΔG (kcal/mol)
-50
-40
-30
-20
-10
0
-50
-40
-30
-20
-10
0
-5
-10
-15
-20

analysis from non-uniformity in the distribution of $\Delta G$ and $\Delta H$ measurements. Our application of Williams' differential-plots approach indicates that both $\Delta\Delta H$ *versus* $T\Delta\Delta S$ and $\Delta\Delta H$ *versus* $\Delta\Delta G$ plots show full compensation with ITC constraints for DNA–peptide interactions (Figure 7.6). In addition, most of the other ligands in Table 7.3 also show full compensation with ITC constraints in Williams' differential plots (data not shown).

After reviewing a number of DNA–protein and DNA–small-ligand interaction studies, it is possible to classify the EEC phenomenon in the DNA–ligand interaction field into four major categories *via* the gradient determinant $k_e$, where 'magnitude' of $\Delta H$ stands for absolute value (Table 7.3):

(A) $k_e = 1$ The magnitude of $\Delta H$ and that of $T\Delta S$ increase simultaneously
(B) $k_e > 1 > 0$ The magnitude of $\Delta H$ increases faster than that of $T\Delta S$
(C) $k_e < 1 > 0$ The magnitude of $T\Delta S$ increases faster than that of $\Delta H$
(D) When the sign of $k_e$ is negative, $\Delta H$ decreases as $T\Delta S$ increases.

In DNA–ligand interactions, $k_e$ signifies the type of energy compensation between $\Delta H$ and $T\Delta S$. In general, the EEC phenomenon occurs when a ligand interacts with different structurally related DNA species. Alternatively, the EEC phenomenon also arises when a DNA duplex interacts with different ligands sharing some structural resemblance, or with the same ligand under varying conditions of temperature, pH or ionic strength. In each of these cases, the energy compensation between $\Delta H$ and $T\Delta S$ may favour one of the four types of energy compensation shown above. In many DNA–ligand interactions, for example in DNA–protein, DNA–drug, and DNA–peptide interactions, $k_e$ equates to unity (type A) or near unity is more preferred (Table 7.3). However, other EEC gradient determinant types B,[27,34] C,[29,31] and D[33] can also be found in nature (Table 7.3).

Table 7.3 shows that most DNA–protein interactions are significantly exothermic, with $\Delta H$ values ranging between $-8.0$ and $-42.9$ kcal mol$^{-1}$. On the other hand, DNA–peptide interactions[19] are overwhelmingly enthalpy driven (Table 7.3), showing $\Delta H$ values between $-14.7$ and $-74.4$ kcal mol$^{-1}$ with most $\Delta H$ values exceeding $-26$ kcal mol$^{-1}$. Strikingly, total entropy change for proteins is mostly negative, with $T\Delta S$ values between $+1.9$ and $-38.8$ kcal mol$^{-1}$, suggesting that in general the sum of unfavourable negative entropic energy components $\Delta S^{o}_{rt}$, $\Delta S^{o}_{vib}$ and $\Delta S^{o}_{conf}$ exceeds that of

**Figure 7.5** Three-dimensional plots of $\Delta H$, $T\Delta S$ and $\Delta G$ from binding of transcription factor MASH-1to DNA binding sites SP-1 (A) and MCK-S (B). The short vertical projection plane (in green) corresponds to $\Delta G$ magnitudes sustained at around $-7.4$ kcal mol$^{-1}$ (SP-1) and $-9.4$ kcal mol$^{-1}$ (MCK-S). (C) Three-dimensional plot of $\Delta H$, $T\Delta S$ and $\Delta G$ of bZIP domains of Jun–Fos heterodimer binding to dsDNA oligonucleotides. The short vertical projection plane (in green) corresponds to $\Delta G$ magnitudes sustained at around $-6.8$ kcal mol$^{-1}$.

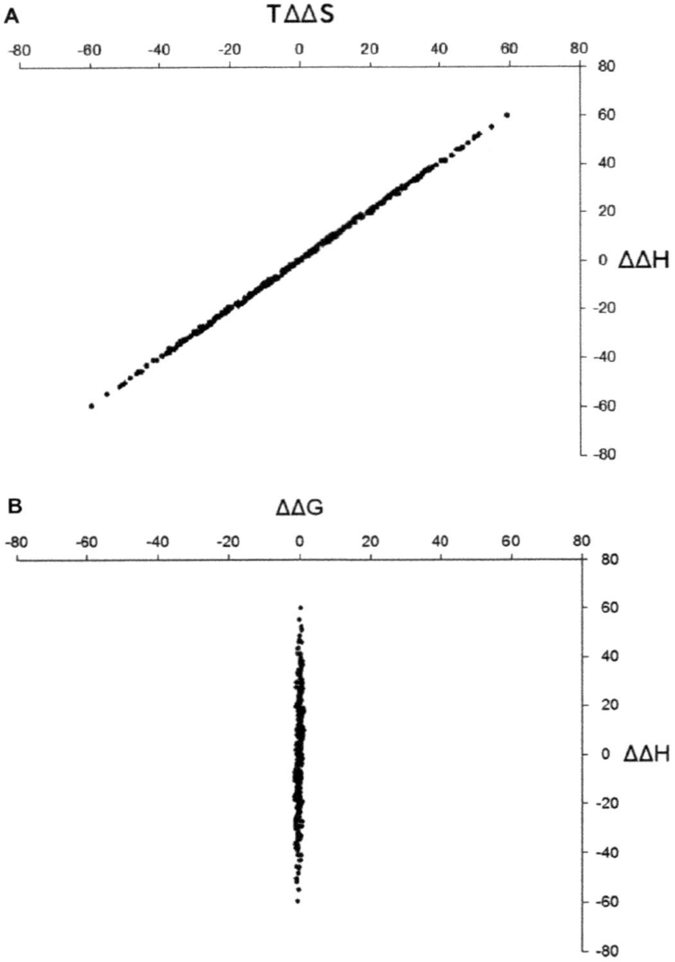


**Figure 7.6** Williams' differential-plots[38] for studying the extent of enthalpy–entropy compensation in DNA–peptide interactions for 19 designed peptides: A, Plot of $\Delta\Delta H$ *versus* $T\Delta\Delta S$; B, plot of $\Delta\Delta H$ *versus* $\Delta\Delta G$.

positive entropic components $\Delta S^{o}_{H}$ and $\Delta S^{o}_{ions}$. On binding to DNA, five different proteins with complex tertiary structures maintain $\Delta G$ values within a rather narrow range of $-6.8$ to $-11.5$ kcal mol$^{-1}$. Small-molecule drugs like 7-aminoactinomycin D, netropsin and berberine also sustain $\Delta G$ values ranging between $-5.8$ and $-10.8$ kcal mol$^{-1}$. On the other hand, the 19 medium sized peptides in this study maintain $\Delta G$ values falling within a narrow range of $-7.9$ to $-9.1$ kcal mol$^{-1}$, the average $\Delta G$ being $-8.3$ kcal mol$^{-1}$. It is noteworthy that most ligands shown in Table 7.3 have $\Delta G$ values varying over a relatively narrow range, from $-5.8$ to $-11.5$ kcal mol$^{-1}$, whereas the other thermodynamic parameters $\Delta H$ and $T\Delta S$ for these ligands yield values scattered over a fairly wide range.

## 7.2 Conclusions Regarding the EEC Phenomenon of General DNA–Ligand Interactions

Based upon the foregoing, we reason that in practically all circumstances the compensation of energy between $\Delta H$ and $T\Delta S$ in many DNA–ligand binding systems, ranging from very large ligands, such as proteins, to medium-sized ligands, such as peptides, and to small ligands, like drugs and other small molecules, provides a thermodynamic mechanism that maintains a favourable and almost constant value of negative $\Delta G_c$ that is vital for the binding process, and thus accounts for the observed EEC phenomenon. The maintenance of an almost constant value of negative $\Delta G_c$ is manifested by the gradient determinant $k_e$ which can be sub-divided into four categories. The enthalpy–entropy compensation effects associated with binding of the peptide Tyr-Pro-Arg-Lys-(Py)$_4$-Lys-Arg-Pro-Tyr-NH$_2$ (RY-12) to the U4A-L4T DNA duplex persist through various temperatures.[19] Similarly, DNA-binding ligands including proteins[27] and drugs[34] also show EEC phenomena despite variations in temperature or salt concentration. For our designed peptides[19] we have reasoned that high exothermic energy promotes allosteric adjustment of DNA conformation within and between binding loci by means of hydrogen bonding as well as ionic interactions between the DNA duplex and peptide, augmenting the favourable energy components, $\Delta G_h$, $\Sigma\Delta G_p$, $\Delta G_{conf}$, and $\Delta G_{vdw}$, and facilitating positively cooperative binding of approaching peptide molecules. This suggestion is consistent with the proposal of Williams and colleagues[25] that highly exothermic binding is generally associated with significant positive cooperativity, as verified by DNase I footprinting studies of positively cooperative peptide binding to multiple DNA recognition sites.[17–19] This communication of cooperativity between multiple binding loci furnishes the basis for DNA–peptide interaction networks proposed previously[18,19] as well as in this study.

We envisage that in a great variety of situations where DNA binding occurs, the EEC compensation mechanism represents nature's thermodynamic endeavour to promote successful DNA–ligand molecular recognition/interactions. The EEC phenomenon in general DNA–ligand interactions can be categorized into four major types as expressed by the gradient determinant $k_e$ of $\Delta H - T\Delta S$ energy plots. The maintenance of an almost constant negative value of $\Delta G_c$ is manifested by the gradient determinant $k_e$ which signifies the type of energy compensation between $\Delta H$ and $T\Delta S$. For any given category of DNA–ligand interaction, the chemical structure of the ligand and its DNA sequence-selectivity must underlie the type of energy compensation between $\Delta H$ and $T\Delta S$, in the process maintaining a favourable negative free energy change within a remarkably narrow range that assures successful DNA–ligand complex formation.

## Acknowledgements

This work was supported by grant NSC97-2113-M029-005 and funds from Tunghai Christian University and the Jeng Ching Ho Foundation. We thank

Prof. L. S. Kan and Prof. D. K. Chang, Academia Sinica, for helpful advice and access to ITC and CD facilities. We thank Dr Jonathan Huang for performing CD studies, and Mokai Liao, Y. C. Chen, and Y. C. Liao for some footprinting studies. We thank Prof. Bill Shen for helpful advice. This work is dedicated to the memory of Prof. Lin Ma.

## References

1. J. B. Chaires, K. B. Fox, J. E. Herrera, M. Britt and M. J. Waring, *Biochemistry*, 1987, **26**, 8227–8236.
2. J. D. Scott and R. M. Williams, *Chem. Rev.*, 2002, **102**, 1169–1730.
3. U. Galm, M. H. Hager, S. G. Van Lanen, J. Ju, J. S. Thorn and B. Shen, *Chem. Rev.*, 2005, **105**, 739–758.
4. D. L. Boger, J.-H. Chen and K. W. Saionz, *J. Am. Chem. Soc.*, 1996, **118**, 1629–1644.
5. C. Bailly and J. B. Chaires, *Bioconjugate Chem.*, 1998, **9**, 513–538, and references cited therein.
6. Y. H. Chen and J. W. Lown, *J. Am. Chem. Soc.*, 1994, **116**, 6995–7005.
7. W. L. Walker, E. M. Landaw, R. E. Dickerson and D. S. Goodsell, *Proc. Natl. Acad. Sci. U. S. A.*, 1997, **94**, 5634–5639.
8. A. Blasko and T. C. Bruice, *Proc. Natl. Acad. Sci. U. S. A.*, 1993, **90**, 10018–10022.
9. R. P. L. DeClairac, B. N. Geierstanger, M. Mrksich, P. B. Dervan and D. E. Wemmer, *J. Am. Chem. Soc.*, 1997, **119**, 7909–7916.
10. E. J. Fechter and P. B. Dervan, *J. Am. Chem. Soc.*, 2003, **125**, 8476–8485, and references cited therein.
11. K. L. Buchmueller, A. M. Staples, C. M. Howard, S. M. Horick, P. B. Uthe, N. Le Minh, K. K. Cox, B. Nguyen, K. A. O. Pacheco, W. D. Wilson and M. Lee, *J. Am. Chem. Soc.*, 2005, **127**, 742–750.
12. T. Brown, H. Mackay, M. Turlinton, A. Sutterfield, T. Smith, A. Sielaff, L. Westrate, C. Bruce, J. Kluza, C. O'Hare, B. Nguyen, W. D. Wilson, J. A. Hartley and M. Lee, *Bioorg. Med. Chem.*, 2008, **16**, 5266–5276.
13. M. Munde, M. A. Ismail, R. Arafa, P. Peixoto, C. J. Collar, I. Yang, L. Hu, M. H. David-Cordonnier, A. Lansiaux, C. Bailly, D. W. Boykin and W. D. Wilson, *J. Am. Chem. Soc.*, 2007, **129**, 13732–13743.
14. J. C. Chang, C. H. Yang, P. J. Chou, W. H. Yang, I. C. Chou, C. T. Lu, P. H. Lin, R. C. W. Hou, K. C. G. Jeng, C. C. Cheng and L. Sheh, *Bioorg. Med. Chem.*, 2004, **12**, 53–61.
15. C. H. Yang, W. F. Chen, M. C. Jong, B. J. Jong, J. C. Chang, M. J. Waring, L. Ma and L. Sheh, *J. Am. Chem. Soc.*, 2004, **126**, 8104–8105.
16. C. H. Yang, K. C. G. Jeng, W. H. Yang, Y. L. Chen, C. C. Hung, J. W. Lin, S. T. Chen, S. Richardson, C. R. H. Martin, M. J. Waring and L. Sheh, *ChemBioChem*, 2006, **7**, 1187–1196.
17. K. L. Kao, J. C. T. Huang, C. K. Yang, K. C. G. Jeng, J. C. Chang, W. C. Yao, S. C. Hsien, M. J. Waring, M. H. Chen, L. Ma and L. Sheh, *Bioorg. Med. Chem.*, 2010, **18**, 366–376; J. T. B. Huang, R. C. K. Yang, W. K. Hung,

M. J. Waring, L. Sheh in *Molecular Recognition*, ed. J. A. McEvoy, 2011, Nova Science Publishers Inc., New York, ISBN 978-1-61122-734-5; J. T. B. Huang, Y. C. Chen, J. C. Chang, K. C. G. Jeng, K. L. Kao, C. K. Yang, L. S. Kan, M. T. Wey, M. J. Waring, C. S. Chen, W. J. Chien and L. Sheh, *Bioorg. Med. Chem.*, 2010, **18**, 2575–2585.

18. R. C. K. Yang, J. T. B. Huang, S. C. Chien, R. Huang, K. C. G. Jeng, Y. C. Chen, M. Liao, J. R. Wu, W. K. Hung, C. C. Hung, Y. L. Chen, M. J. Waring and L. Sheh, *Org. Biomol. Chem.*, 2013, **11**, 48–61.
19. R. C. K. Yang, J. T. B. Huang, Y. L. Chen, C. C. Hung, M. Liao, W. C. Yao, C. H. Chen, C. C. Liou, M. J. Waring and L. Sheh, *Org. Biomol. Chem.*, 2014, **12**, 4890–4904.
20. M. Suzuki, *Nature*, 1990, **344**, 562–565.
21. N. M. Luscombe, R. A. Laskowski and J. M. Thornton, *Nucleic Acids Res.*, 2001, **29**, 2860–2874.
22. K. L. Buchmueller, A. M. Staples, C. M. Howard, S. M. Horick, P. B. Uthe, N. Le Minh, K. K. Cox, B. Nguyen, K. A. O. Pacheco, W. D. Wilson and M. Lee, *J. Am. Chem. Soc.*, 2005, **127**, 742–750.
23. E. B. Starikov and B. Norden, *J. Phys. Chem. B*, 2007, **111**, 14431–14435, and references quoted therein.
24. B. Breiten, M. R. Lockett, W. Sherman, S. Fujita, M. Al-Sayah, H. Lange, C. M. Bowers, A. Heroux, G. Krilov and M. Whitesides, *J. Am. Chem. Soc.*, 2013, **135**, 15579–15584.
25. D. H. Williams, E. Stephens, D. P. O'Brien and M. Zhou, *Angew. Chem., Int. Ed.*, 2004, **43**, 6596–6616, and references cited therein.
26. A. Cooper and D. T. F. Dryden, *Eur. Biophys. J.*, 1984, **11**, 103–109.
27. A. G. E. Kunne, M. Sieber, D. Meierhans and R. K. Allemann, *Biochemistry*, 1998, **37**, 4217–4223.
28. K. L. Seldeen, C. B. McDonald, B. J. Deegan and A. Farooq, *Biochemistry*, 2009, **48**, 1975–1983.
29. L. M. Andrejszki, S. Bjelic, J. F. Naud, P. Lavigne and I. Jelesarov, *Biochemistry*, 2007, **46**, 12427–12440.
30. T. Lundback, H. Hansson, S. Knapp, R. Ladenstein and T. Hard, *J. Mol. Biol.*, 1998, **276**, 775–786.
31. M. Thompson and N. W. Woodbury, *J. Mol. Biol.*, 1998, **276**, 571–590.
32. T. Cui, S. Wei, K. Brew and F. Leng, *J. Mol. Biol.*, 2005, **325**, 629–645.
33. X. Qu, J. Ren, P. V. Ricelli, A. S. Benight and J. B. Chaires, *Biochemistry*, 2003, **42**, 11960–11967.
34. A. Taquet, R. Labarbe and C. Houssier, *Biochemistry*, 1998, **37**, 9119–9126.
35. M. W. Freyer, R. Buscaglia, D. Cashman, S. Hyslop, W. D. Wilson, J. B. Chaires and E. A. Lewis, *Biophys. Chem.*, 2007, **126**, 186–196.
36. K. Bhadra, M. Maiti and G. S. Kumar, *Biochim. Biophys. Acta*, 2008, **178**, 1054–1061.
37. S. B. Dixit, D. Q. Andrews and D. L. Beveridge, *Biophys. J.*, 2005, **88**, 3147–3157.
38. T. S. G. Olsson, J. E. Ladbury, W. R. Pitt and M. A. Williams, *Protein Sci.*, 2011, **20**, 1607–1618.

CHAPTER 8

# *Structural Studies of DNA-binding Metal Complexes of Therapeutic Importance*

CHRISTINE J. CARDIN*[a] AND JAMES P. HALL[a,b]

[a] Department of Chemistry, University of Reading, Whiteknights, Reading RG6 6AD, UK; [b] Diamond Light Source, Harwell Science and Innovation Campus, Fermi Avenue, Didcot OX11 0DE, UK
*Email: c.j.cardin@reading.ac.uk

## 8.1 Introduction–Ruthenium Complexes as DNA Probes and DNA Damage Agents

Many years can elapse between the discovery of a compound of biomedical utility and the eventual identification of its target macromolecule. X-ray crystallographic studies are useful both to determine the structure of the compound itself and, usually much later, the compound, or its active metabolite, bound to the nucleic acid or protein target. In this chapter we examine the DNA-binding modes of ruthenium polypyridyl complexes, which are under development for a variety of biomedical applications but, by way of illustration, include some examples of related work where structural studies are instructive.

There has been a focus on investigating the use of group 8 metals, and in particular ruthenium, in the production of photoactive $d^6$ octahedral complexes. Metal complexes containing ruthenium and osmium have been intensively investigated because of their high kinetic inertness, due to their

Chemical Biology No. 7
DNA-targeting Molecules as Therapeutic Agents
Edited by Michael J. Waring

Published by the Royal Society of Chemistry, www.rsc.org

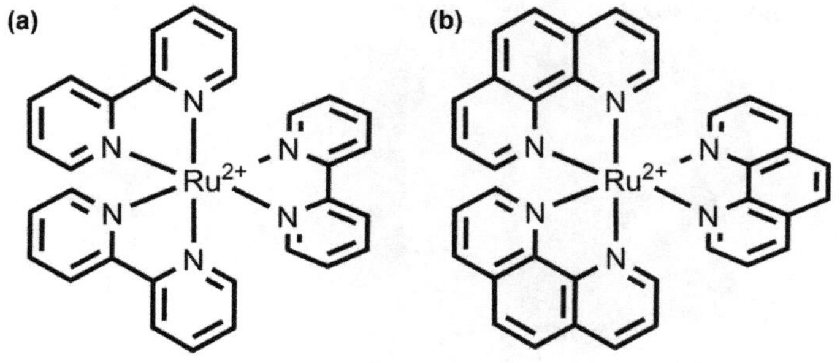


**Figure 8.1** (a) $[Ru(bpy)_3]^{2+}$ and (b) $[Ru(phen)_3]^{2+}$.

$4d^6$ electronic configuration meaning that they are both chemically and enantiomerically stable in solution and do not convert to the racemate over time, unlike those made with iron.

Polypyridyl ruthenium complexes have been investigated since the late 1950s with the first reported observation of fluorescence[1] from $[Ru(bpy)_3]^{2+}$ (bpy = 2,2-bipyridyl, Figure 8.1a). It was only at the time of the 1970s energy crisis that research into these compounds began in earnest as it was realised that they may be able to split $H_2O$ into $H_2$ and $O_2$ when irradiated by light.[2] While it soon became clear that the compounds would not be suitable for this purpose, there was still significant interest in their photophysical and luminescence properties.

## 8.2 The Versatility of Ruthenium Polypyridyl Complexes

This class of metal complexes have been extensively explored for a variety of applications. The DNA binding properties of $[Ru(bpy)_3]^{2+}$ and $[Ru(phen)_3]^{2+}$ (phen = 1,10-phenanthroline) (Figure 8.1b) were first studied in the 1980s,[3,4] and the likelihood of enantiomeric differences and range of binding modes were discussed at this time, without any structural confirmation available. While both compounds bound, the difference appeared to be that $[Ru(phen)_3]^{2+}$ could bind by intercalation (Figure 8.2a), with a phen group inserted into the DNA base pair stack. It subsequently became clear that the binding mode could be better described as semi-intercalation, or DNA kinking (Figure 8.2b,c), as determined by a range of biophysical measurements in solution and that this property, though striking, did not generate tight enough DNA binding to be a useful scaffold for potential therapeutic and sensing applications. Since this initial binding work, a number of groups have been trying to optimise Ru–polypyridyl complexes to produce compounds which combine useful photochemical or photophysical properties[2,5,6] with the ability to bind to DNA. Their luminescence behaviour arises from the ease with which one of the ruthenium *d* electrons can move to one of the ligands in the excited state, the so-called metal to ligand charge

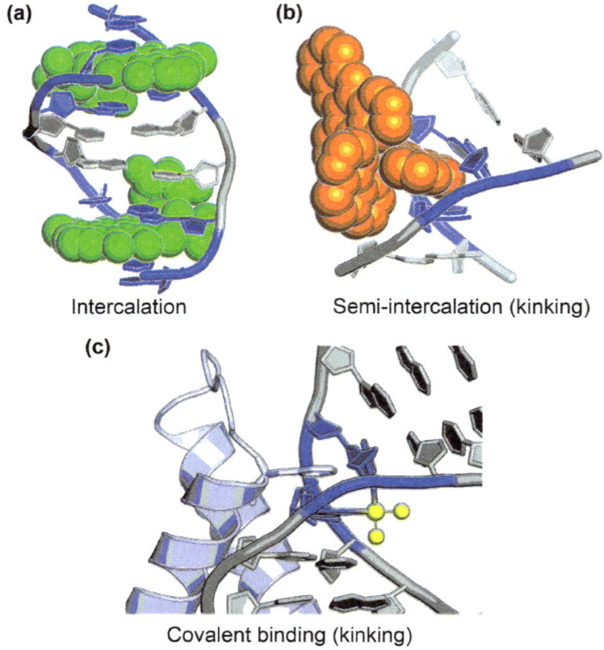


**Figure 8.2** (a) Intercalation into DNA by daunomycin (green) (PDB ID 1D11), DNA kinking by (b) a semi-intercalating Ru-polypyridyl complex (orange) (PDB ID 3QRN) and (c) a *cis*-platin (yellow) and the HMGB1 protein (light blue) (PDB ID 1CKT). The base pairs interacting with the complex are displayed in blue with the other bases in grey.

transfer (MLCT). In the excited state, Ru–polypyridyl complexes can act as luminescent DNA probes, induce indirect or direct DNA damage or, by photodissociation, yield therapeutically active species.[7]

X-ray crystallographic studies in this area have, so far, been confined to studies of the ground state of complexes, both unbound and bound to DNA. However, with the advances in time-resolved X-ray crystallography, it could be expected that structural studies of the excited-state geometry of such compounds could yield useful information, complementing the ultrafast spectroscopic studies that have been reported in solution.[8]

### 8.2.1 Early Spectroscopic Studies

The first ruthenium complexes to show tight binding, high luminescence, and intercalation as the binding mode were those containing ligands such as dppz (dipyrido[3,2-a;2′,3′-c]phenazine). The group of Barton reported the $[Ru(bpy)_2(dppz)]^{2+}$ (Figure 8.3a) complex in 1990,[9] laying the foundation for a large and constantly expanding area of investigation worldwide, this original paper is now approaching 1300 citations. This compound became known as the 'light-switch' compound, thanks to its property of luminescing

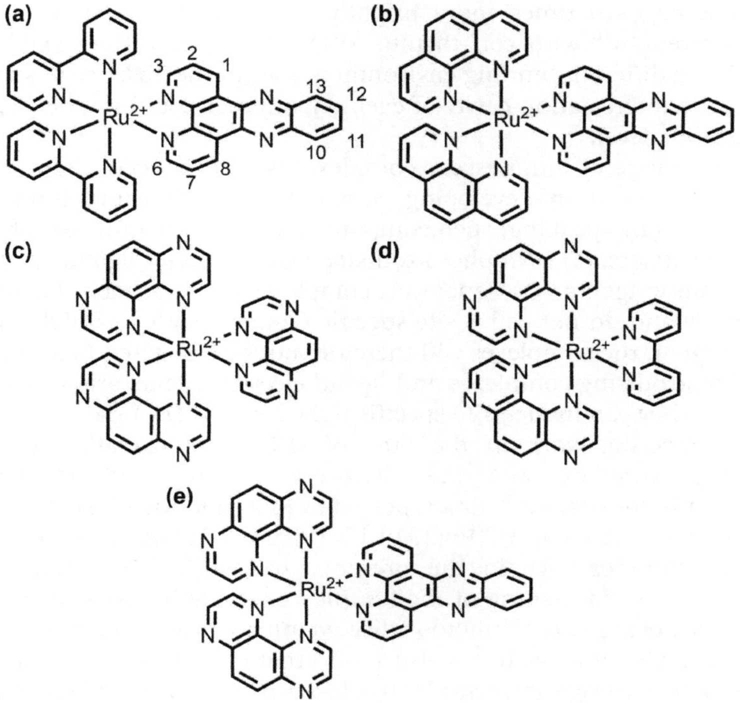


**Figure 8.3** (a) $[Ru(bpy)_2(dppz)]^{2+}$ with dppz numbering scheme, (b) $[Ru(phen)_2(dppz)]^{2+}$, (c) $[Ru(TAP)_3]^{2+}$, (d) $[Ru(TAP)_2(bpy)]^{2+}$ and (e) $[Ru(TAP)_2(dppz)]^{2+}$.

brightly (approximately $10^4$ increase in emission) when bound to DNA, or in an organic solvent such as acetonitrile, but not in aqueous solution or when in contact with hydrogen bonding species. The binding constant was also increased by a factor of $10^2$–$10^3$ compared with $[Ru(bpy)_3]^{2+}$, meaning that $[Ru(bpy)_2dppz]^{2+}$ had potential as a DNA probe. The theoretical basis for the luminescence has become an area of study in its own right,[10] with groups reporting that the observed emission is highly sensitive to the environment in which the complex is bound. It has subsequently been reported, from computational modelling studies, that the observed emission originates from a balance between bright $^3$MLCT states located on the ancillary ligands and dark $^3$MLCT and $^3$IL states on the dppz.[11,12] The presence of water–dppz hydrogen bonds provides a route of deactivation through the dark states and therefore one parameter which contributes to the observed emission is the accessibility of the dppz group to the solvent.[13] This can be affected by the orientation of the complex when bound as well as by intercalation depth. The observation of water quenching was key in explaining the luminescence differences between the two enantiomers of $[Ru(phen)_2(dppz)]^{2+}$ (Figure 8.3b), a light-switch complex with similar luminescence properties[14] to $[Ru(bpy)_2(dppz)]^{2+}$. The Δ enantiomer of the complex was found to luminesce 6–10 times more brightly than the Λ,[15] despite a similar binding strength[16] with calf thymus DNA. This observation could be explained by a different binding environment but, in the absence of structural studies, the intercalation cavity of each enantiomer could not be visualised at the atomic level.

Whilst a range of luminescent complexes were produced, there was also significant interest in developing compounds which could induce DNA damage. Broadly speaking, such compounds fit into one of three categories; (1) $^1O_2$ sensitisers, (2) DNA photooxidising agents and (3) ligand ejection covalent-binding agents. $^1O_2$-Generating complexes have not been characterised structurally and do not cause site-specific damage, as $^1O_2$ is highly prolific damage agent, the complexes will therefore not be discussed here. However, DNA photooxidising complexes and ligand ejection agents are of significant interest as they can induce site-specific light-induced DNA damage.

The pioneering work in the area of DNA photooxidation by a Ru–polypyridyl complex was the discovery of the TAP (1,4,5,8-tetra-azaphenanthrene) ligand,[17] first reported by Kelly and Kirsch-De Mesmaeker *et al.* in 1987. The complex $[Ru(TAP)_3]^{2+}$ (Figure 8.3c) was reported to have enhanced luminescence in the presence of poly[(A–T)] compared with poly[(G–C)]. The shortening of the luminescence lifetime was attributed to the complex being able to photooxidise guanine bases through a very rapid electron transfer process. It was also shown that the compound could induce single-strand cleavage into covalently-closed pBR233 DNA. Since then, the covalent adduct between guanosine monophosphate and a related complex, $[Ru(TAP)_2(bpy)]^{2+}$ (Figure 8.3d), has been isolated[18] and the mechanism of photooxidation of guanine by $[Ru(TAP)_2(dppz)]^{2+}$ (Figure 8.3e) has been proposed to be proton-coupled electron transfer, which was elucidated using ultrafast spectroscopic techniques.[19,20] The advantage of a complex which can cause direct DNA damage, instead of *via* a $^1O_2$ mechanism, is that the compound would still function in the hypoxic environment of large tumour masses. Such compounds could therefore have applications in both photo-activated chemotherapy (PACT) and photodynamic therapy (PDT).

## 8.3 PACT and PDT

On the face of it, it should be much easier to get a crystal structure of a new metal complex than of the same species bound to a DNA target molecule, but this is not always the case. Perhaps it is the tendency of these complexes to stack on themselves in solution, which is minimised if the correct stoichiometry of DNA is added, which inhibits facile crystallisation. Nevertheless, a range of structures have been determined, and it is appropriate to highlight some structural features which can be directly related to therapeutic potential.

Photoactivated chemotherapy is the light-activated dissociation of a ligand to generate a reactive ruthenium species capable of DNA binding in the manner of the structurally characterised, and still valuable, platinum drugs. The Glazer group have been very active in the area of PACT and have used *ortho* substitution by methyl groups on the dppz ligand to generate complexes which are unreactive in the dark and in the absence of DNA.[21] These compounds, of which an example, $[Ru(bpy)_2)3,6\text{-dimethyldppz}]^{2+}$ is illustrated in Figure 8.4a, undergo photoejection of the dppz ligand when irradiated in the presence of DNA *in vitro* to generate an active species which is presumed to react readily with the N7 atoms of guanine bases, perhaps to generate kinking of the duplex, as is known for the clinically useful platinum compounds. The structure of the compound shows the origins of the lability. The expected length of the Ru–N bond in polypyridyl complexes is 2.06 Å, whereas here the $Ru–N_{dppz}$ distance averages 2.115 Å. The dppz ligands in this class of compounds are typically somewhat nonplanar, both bound and unbound, but in this and related compounds, some extreme distortions, here a tilt of $\sim 20°$, are seen.

Photodynamic therapy (PDT) is based on the generation of a reactive species at the site of a tumour. A suitable sensitiser should be light activated in the therapeutic window (650–950 nm) and be non-toxic to normal tissue. The inertness of ruthenium complexes combined with their tunability has led to many studies in this area[22,23] and most commonly the active species is singlet oxygen, $^1O_2$. One currently used agent, Photofrin, is a porphyrin-based system, but is only slowly cleared from the body, and ruthenium polypyridyl complexes, with their endlessly tunable properties, are viewed as potential second-generation compounds. The parent complex $[Ru(bpy)_2dppz]^{2+}$ has $\lambda_{max}$ for the

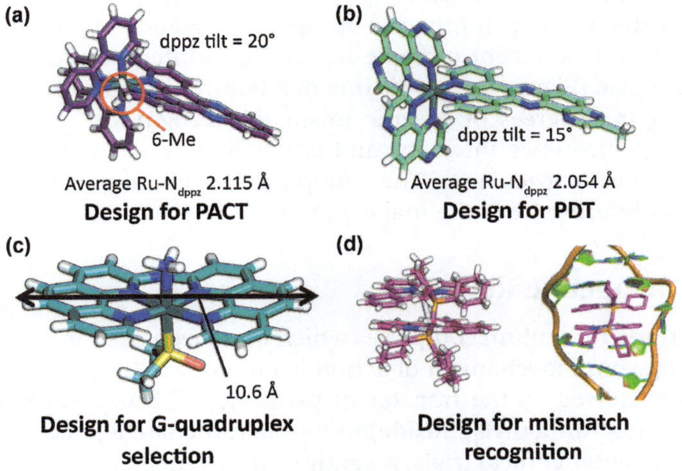


**Figure 8.4** Crystal structures of four metal complexes designed to bind to DNA or induce DNA damage. (a) $[Ru(bpy)_2(3,6\text{-dimethyldppz})]^{2+}$, (b) $[Ru(TAP)_2(3\text{-methylpyrazinodppz}]^{2+}$, (c) $[Ru(N,N\text{-bis-}(1,10\text{-phenanthrolin-2-yl})\text{-amine})(NH_3)(DMSO)]$, (d) a binuclear Pt-terpyridine complex able to bind into DNA C–C mismatches.
(d) has been adapted from S. K. Fung, T. Zou, B. Cao, T. Chen, W. P. To, C. Yang, C. N. Lok and C. M. Che, *Nat. Commun.*, 2016, 7, 10655, https://doi.org/10.1038/ncomms10655 (http://creativecommons.org/licenses/by/4.0/).

$^3$MLCT band at 460 nm, which is too energetic to fall in the 'skin transparency' window below 650 nm, therefore compound design has focused on modifying the complex so that light can be efficiently absorbed at longer wavelengths. Whilst the goal of modifying ruthenium complexes to be active in the transparency window has not yet been realised, a number of complexes have been produced which are phototoxic and can induce direct DNA damage, such as the complex illustrated in Figure 8.4b. In this example,[24] the Ru–N bond length has the normal value of 2.054 Å and the dppz ligand shows a typical curvature. The compound does not rely on singlet oxygen generation for its efficacy but rather causes direct DNA damage, important because many tumours are hypoxic. In order to activate complexes using the transparency window, groups have employed multi-photon excitation,[25] using two photons at 800 nm instead of one at 400 nm, to activate a complex. However, this approach requires that both photons are absorbed simultaneously by the complex and as such activation is only proportional to the square of the intensity, which can make this approach highly challenging.

The examples here, and most of the discussion in this chapter, are of complexes which specifically bind to duplex DNA. Stabilisation of a G-quadruplex structure in a useful way requires a molecular design which renders duplex binding structurally impossible. The Ru(II) complexes are normally of *cis* stereochemistry, but the *trans* complex (Figure 8.4c) has a planar tetra-chelation geometry with axial $NH_3$ ligands, preventing intercalation, and which recognises an antiparallel G-quadruplex.[26]

The recognition of DNA mismatches, covered in Chapter 6 of this volume, emphasises the necessity for the use of expansive ligands to prevent binding to duplex DNA. One recent example is strikingly different and worth mentioning here also (Figure 8.4d). It is one of a family of platinum compounds showing varying degrees of specific mismatch recognition in the form of enhanced luminescence intensity, and is remarkable for its recognition of CC mismatches particularly. The complex has chirality and has been modelled as binding from the major groove.[27]

## 8.3.1 Therapeutic Relevance

There are three ruthenium complexes which have been used in clinical trials (Figure 8.5), whose mechanism of action is not known, though it is thought they are transferred by the transferrin pathway.[28,29] NAMI-A, imidazolium trans-tetrachloro(dimethylsulfoxide)imidazole-ruthenate(III) was the first compound to enter clinical trials, recently terminated after the phase II trial, followed by trans-tetrachlorobis(1H-indazole)ruthenate(III). These agents are effective against metastasis and active against a range of tumours and are activated by reduction to ruthenium(II).

Another class of ruthenium compound with known antitumour properties are the organometallic 'piano-stool' compounds containing a labile anionic ligand. In this case the small-molecule structures were determined many years ago, but recently, binding to the nucleosome was shown by X-ray

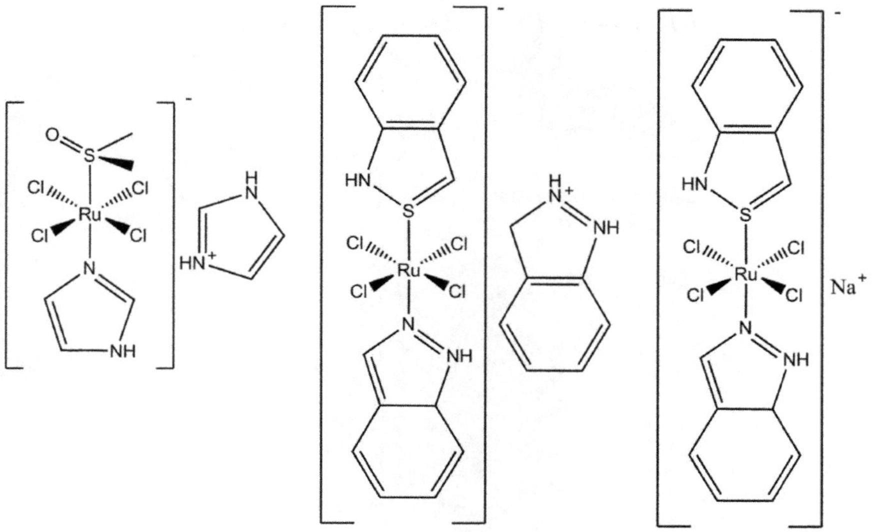

NAMI-A KP1019 (N)KP1339

**Figure 8.5** Three ruthenium complexes which have entered into clinical trials.

crystallography. Two closely related structures were determined,[30] one showing the cation [($\eta^6$-*p*-cymene)-chloro-1,2-ethylenediamineruthenium(II)]$^+$ covalently bound to the N7 atom of a C–G base pair of the nucleosome (Figure 8.6). The closely related adamantyl derivative was bound to one of the histone proteins. These are the only examples of therapeutically active ruthenium compounds bound to macromolecular targets currently in the PDB.

## 8.4 Intercalation by Ruthenium–dppz Complexes

Until 2010 there was little structural information to show how Ru–polypyridyl complexes could bind to DNA. However, since then a number of structures have been published which show how Ru–dppz complexes can bind to DNA with multiple binding modes. There is no question that the predominant binding mode is intercalation, but the nature of the precise orientation or interaction of the complex can be changed by seemingly small changes to the DNA or environment. Here X-ray crystallographic studies can provide great insight, as structures can be obtained to atomic resolution, showing the precise 3D arrangement of atoms with picometer precision. From such studies detailed nucleic acid parameters,[31] used to describe DNA deformation or conformational change, can also be obtained. The primary binding force for intercalation is the stacking interaction between the planar aromatic ligand, here dppz, and the base pairs of the intercalation cavity but the electrostatic attraction between the positively charged complex and negatively charged DNA backbone is also highly significant.

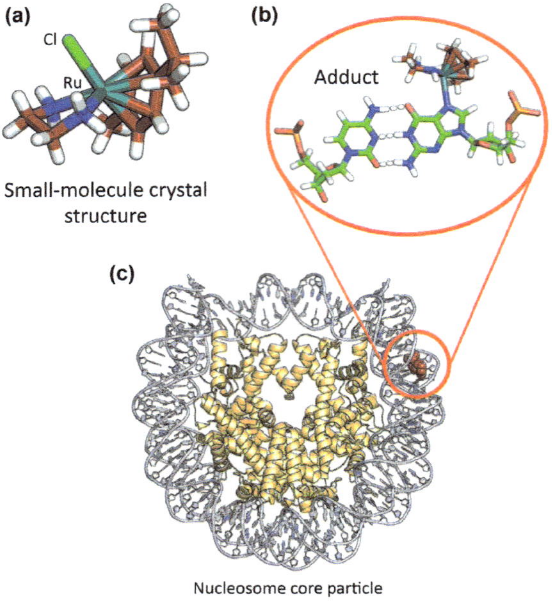


**Figure 8.6** (a) [($\eta^6$-*p*-cymene)-chloro-1,2-ethylenediamineruthenium(II)].Cl. (b) The complex can form an adduct with guanine. Atoms are coloured according to type with the complex carbon atoms in brick red, nitrogen in blue, ruthenium in teal, and complex chlorine in green. DNA carbon atoms are illustrated in green with phosphorus in orange, and oxygen in red. (c) The adduct has been observed in a crystal structure of the nucleosome core particle. DNA is drawn in white with protein in orange and adduct in red. Nucleosome drawn from PDB ID 5CP6.

The first X-ray structural study of DNA-bound ruthenium dppz complexes, that of $[Ru(TAP)_2dppz]^{2+}$ bound to $d(TCGGCGCCGA)_2$, was reported in 2011, and since then a series of other structures have been published by the Cardin and Barton groups (Table 8.1).

One feature which has emerged is that crystal growth is nearly always enantiospecific with respect to the metal complex, which should not be surprising given the strong chirality of DNA. In some combinations, crystals of the DNA with a particular enantiomer form from a racemic mixture, with the other enantiomer simply remaining in solution, but in others crystallisation only, or more readily, occurs from a solution of one enantiomer. The enormous improvements in technology, at Diamond Light Source as well as other synchrotrons, mean that usable data can then be obtained from a crystal of dimensions as small as $10\times10\times10$ μm, with the X-ray phases obtained routinely from the anomalous scattering of the ruthenium atoms in the crystal. A dataset can be measured in 90 seconds and a structure refined and deposited in a few days. What can take much longer, as well as finding a suitable crystal, is the interpretation of the result in terms of the solution measurements or other available information, and for patterns to emerge.

**Table 8.1** Reported X-ray structures of monomeric Ru complexes bound to oligodeoxynucleotides by intercalation or insertion.

| Complex | Sequence | PDB |
|---|---|---|
| Λ-$[Ru(TAP)_2dppz]^{2+}$ | TCGGCGCCGA | 3QRN |
| | TCGGTACCGA | 3UYB |
| | TCGGCGCCGA | 4LTG |
| | TCGGCGCCIA | 4QI0 |
| | TCIGCGCCGA | 4RE7 |
| | CCGGATCCGG | 4YMC |
| | TTGGCGCCAA | 5ET2 |
| | TCGGCICCGA | 5IU5 |
| | CCGGATCCGG | 4E7Y |
| | TCGGTACCGA | 4LY2 |
| | (Br-C)GGC | 5LFW |
| Λ-$[Ru(TAP)_2(11\text{-CN-dppz})]^{2+}$ | TCGGCGCCGA | 5NBE |
| Λ-$[Ru(TAP)_2(11\text{-Cl-dppz})]^{2+}$ | TCGGCGCCGA | 4III |
| Λ-$[Ru(TAP)_2(11\text{-Me-dppz})]^{2+}$ | TCGGCGCCGA | 4X18 |
| Λ-$[Ru(TAP)_2(10\text{-Me-dppz})]^{2+}$ | TCGGCGCCGA | 4MJ9 |
| Λ-$[Ru(TAP)_2(10,12\text{-Me}_2\text{-dppz})]^{2+}$ | TCGGCGCCGA | 4X1A |
| Λ-$[Ru(phen)_2dppz]^{2+}$ | CCGGTACCGG | 3U38 |
| | CCGGATCCGG | 4E7Y |
| | TCGGTACCGA | 4LY2 |
| | (Br-C)GGC/GCCG | 5LFW |
| Δ-$[Ru(phen)_2dppz]^{2+}$ | TCGGCGCCGA | 5JEU |
| | TCGGCGCCGA | 5JEV |
| Δ-$[Ru(bpy)_2dppz]^{2+}$ | CGGAAATTACCG | 5LFS |
| Λ and Δ-$[Ru(phen)_2(dppz)]^{2+}$ | ATGCAT | 4JD8 |

Table 8.1 illustrates that, due to relative ease of crystallisation, we know more about the binding modes in crystals of lambda complexes, and for this series are able to make detailed comparisons and see such patterns. As will be discussed below, these structures have revealed much about the nature of the intercalation site, but in addition have provided evidence for a range of binding modes of the metal complexes, showing some site preference (sequence dependence) alongside a very clear enantiomer specificity of the binding mode in some cases.

### 8.4.1 B-DNA Duplexes—Intercalation Geometries and Sequence Specificities

Planar aromatic cations, containing $Ru^{2+}$, of suitable dimensions to fit a DNA cavity, normally bind by intercalation, and for these cations all the X-ray evidence is that intercalation occurs from the minor groove side, placing the dppz group between two base pairs and the ruthenium atom between the negatively charged phosphate groups in a characteristically angled mode of binding. This is unlike the classical perpendicular (daunomycin)[32] or parallel (proflavine[33] and acridines[34]) intercalation modes (Figure 8.7), which in the case of the original fibre diffraction studies led to descriptions in terms of helical parameters and the use of the term 'depth of intercalation' to

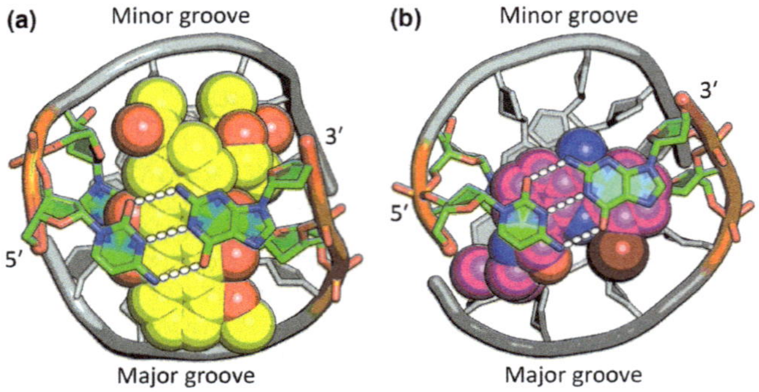


**Figure 8.7** (a) Daunomycin intercalates into the minor groove with the long axis of the chromophore perpendicular to the base hydrogen bonds. Daunomycin carbon atoms are drawn in yellow (PDB ID 1D11). (b) In contrast, an acridine-4-carboxamide derivative binds into the major groove with its long axis nearly parallel to the base hydrogen bonds (PDB ID 367D). Acridine carbon atoms are illustrated in pink. All other atoms are coloured according to type with nitrogen in blue, carbon in green, phosphorus in orange and bromine in brown.

describe the relationship between the chromophore and the helix axis of the fibre.

This term is less helpful for defining angled intercalation, and a better way for a detailed analysis (beyond the scope of this chapter) may be the use of ring centroids to relate the dppz and purine or pyrimidine stackings, which emphasises the extent of overlap rather than the difficult to define orientation angle. The structures indicate, in the case of the lambda enantiomer, that the orientation is determined by the hydrophobic stacking of aromatic surface of the ancillary ligand on the side of the deoxyribose sugar ring, which is typically in the B-DNA conformation. DNA binding is from the minor groove, and is therefore affected by the presence of the projecting 2-amino group when guanine is present. There are 10 possible DNA steps and therefore our knowledge of sequence dependence is currently incomplete. In addition, an extremely limited range of ancillary ligands have been studied to date (bpy, phen and TAP), as shown in Table 8.1. Changing the ancillary ligand can make a big difference to the ease of crystallisation, and there can be effects on the crystal packing and stability. In the cases we have studied so far, the dppz orientation is more dependent on base sequence and on dppz substitution than on the ancillary ligand.

### 8.4.2 Lambda Enantiomer

The lambda enantiomer of $[Ru(LL)_2dppz]^{2+}$ (LL = bpy, phen, TAP) has been crystallised in a number of DNA sequences and DNA steps. It can adopt two different intercalative binding modes—canted (angled) or symmetrical. The angled mode has been observed at a number of DNA steps, including CC/GG, TC/GC and TT/AA (Figure 8.8a–c). At an unsymmetrical step, such as

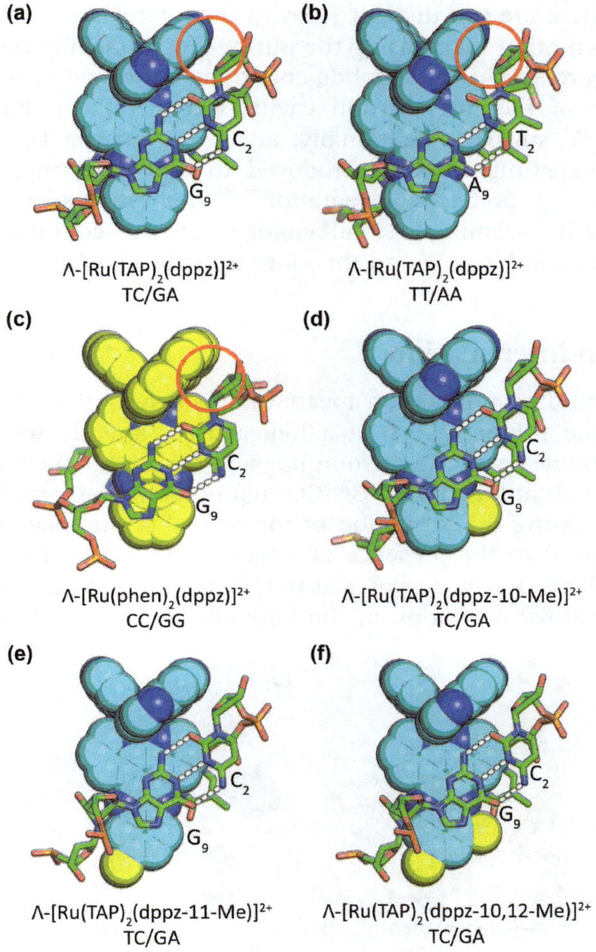


**Figure 8.8** Λ-[Ru(TAP)2(dppz)]$^{2+}$ (carbon atoms in cyan) can intercalate with an angled geometry into a (a) TC/GA and (b) TT/AA step. (c) This binding is consistent with that observed for Λ-[Ru(phen)2(dppz)]$^{2+}$ (carbon atoms in yellow) in CC/GG sites. Circled in red is a close contact between the phen/TAP group and the sugar ring. (d–f) Introduction of methyl substitutions into Λ-[Ru(TAP)2(dppz)]$^{2+}$ causes the complex to bind with the methyl groups (yellow) angled in one direction. DNA atoms are coloured according to type with nitrogen in blue, carbon in green, oxygen in red and phosphorus in orange.

CaCb/GxGy, the dppz group of the complex intercalates into the step with the long axis of the chromophore at an angle of approximately 55° with respect to the Ca–Gx hydrogen bonds and 70° to the Cb–Gy. This canted mode is characterised by a low helical twist, optimising the base stacking onto the dppz chromophore, and rotated such that the TAP ring contacts the sugar ring of (here) cytosine Cb. In principle there are two orientations of the canting, as there are two distinct sides to the intercalation cavity, with the preferred orientation maximising the purine–dppz stacking interactions, as shown in Figure 8.8a–c. The lambda enantiomer always contacts the 3′ side of the sugar of the intercalation cavity, whereas the delta enantiomer contacts the 5′ sugar.[35] Additionally, at an asymmetric DNA step, when asymmetric substitutions are introduced to the dppz ring, the complex can bind with a preferred orientation[36,37] (Figure 8.8d–f). This shows that introducing seemingly small changes on the complex can have a profound effect on binding and therefore must be carefully considered.

### 8.4.3 Semi-intercalation

The angled binding orientation means that the second ancillary ligand is exposed in the minor groove of a helical DNA strand, and is therefore available to interact with the surroundings. In the crystal, the interaction can take the form of kinking of the GG/CC step of a symmetry-related duplex,[38] possibly facilitating the formation of the overall crystal packing. The kink can be stabilised by the presence of a metal cation coordinated to the N7 position on the two guanine bases at that step (Figure 8.9a,b), although this is not essential for formation of the kink site (Figure 8.9c). However, this

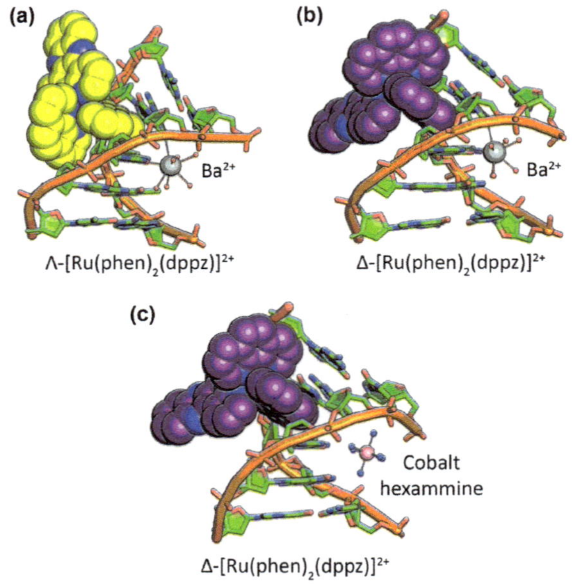


**Figure 8.9** (a–b) Both Λ (yellow) and Δ-$[Ru(phen)_2(dppz)]^{2+}$ (purple) can semi-intercalate into DNA inducing a kink. This may be stabilised by a bound $Ba^{2+}$ cation but the same binding mode has been observed with (c) cobalt hexammine instead of barium. Atoms are coloured according to type with nitrogen in blue, carbon in green, oxygen in red, phosphorus in orange and cobalt in pink.

binding mode is not merely an artefact of crystallization and was proposed by Lincoln and Nordèn in 1998 on the basis of linear dichroism and emission anisotropy data.[39] A reasonable assumption is therefore that the weak binding observed to natural DNA by $[Ru(phen)_3]^{2+}$ is therefore semi-intercalation and not intercalation. Semi-intercalation has a large effect on the overall DNA structure and induces an approximately 50° kink into each semi-intercalation site. There can be at least two sites per DNA duplex which, combining with bending at the central step, can give a DNA 10 base pair duplex an overall bend of 120°. This can be further increased by dehydrating the duplex and can yield an overall curvature of 150°.[40] This curvature highlights how remarkably flexible the DNA duplex can be.

### 8.4.4 Symmetrical Intercalation

A second intercalative binding mode, termed symmetrical intercalation, can also be adopted. This is, so far, specific for the TA/TA[41] step (and not, for example, the AT/AT step), and is characterised by higher twist angle (40°) and the packing of the ancillary phen ligands against, in this case, both, not just one, of the sugars (Figure 8.10a). This binding preference has not only been observed in the crystal but has also been validated in solution[42] for $\Lambda$-$[Ru(TAP)_2(dppz)]^{2+}$. The TA/TA step is distinguished from the remaining

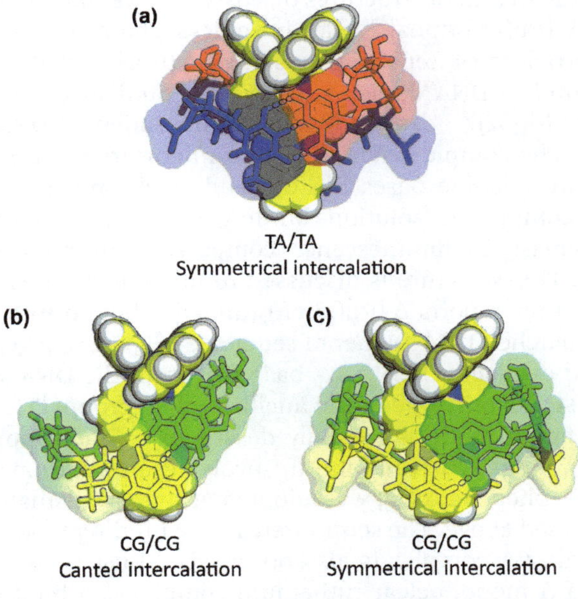


**Figure 8.10** (a) $\Lambda$-$[Ru(phen)_2(dppz)]^{2+}$ (carbon atoms in yellow) binds symmetrically into TA/TA steps. (b) The CG/CG canted intercalation site. (c) A model for a CG/CG symmetrical intercalation site, in which the intercalation depth is reduced. DNA bases are coloured according to type with T in blue, A in red, C in yellow and G in green.

nine base steps by having the lowest stacking energy[43] and so far has only been observed for this enantiomer. In the crystal structures this mode is distinguished by its occurrence on twofold axes in the crystal lattice coinciding with the central unique step of the duplex. The CG/CG step could be expected to give a symmetrical intercalation site. However, perhaps counterintuitively, canted intercalation is also the preferred binding mode in this sequence context (Figure 8.10b,c). The binding site gives the angled intercalation mode with a twist of 28° with the tetrameric (5Br-C)GGC/GCCG duplex.[44] In this instance, the crystal packing is completely different and, because kinking of DNA is not involved, is also obtained when the ancillary ligand is bpy. The angled intercalation mode is preferred because of the steric effect of the two guanine 2-amino groups projecting into the minor groove of the duplex. However, if the complex were to bind symmetrically, it would have a shallower intercalation depth. This would reduce the solvent accessibility of the dppz nitrogen atoms and may lead to a longer luminescence lifetime. In the literature, multiple emission lifetimes have been observed for $\Lambda$-$[Ru(phen)_2(dppz)]^{2+}$ even when bound to DNA homopolymers. It is therefore possible that canted intercalation is preferred at CG/CG steps with symmetrical intercalation being a minor component.

### 8.4.5 Delta Enantiomer

In contrast to the wealth of structures discussed above, the binding the delta enantiomer of $[Ru(LL)_2dppz]^{2+}$ has been less characterised. At time of writing only two studies, reporting crystal structures of delta mononuclear complexes bound to DNA duplexes, have been published. The first shows how $\Delta$-$[Ru(bpy)_2(dppz)]^{2+}$ can bind to an adenine–adenine DNA mismatch[45] (Figure 8.11a). The complex enters into the minor groove at the mismatch site and the two adenine bases flip out and stack onto the ancillary bpy groups of the complex. In solution, binding at the mismatch site causes a significant increase in luminescence compared with binding into well-matched DNA. This structure is discussed in more detail in Chapter 15.

The second study reports $\Delta$-$[Ru(phen)_2(dppz)]^{2+}$ bound into the terminal step of a well-matched DNA 10-mer of sequence $d(T_1C_2G_3G_4C_5G_6C_7C_8G_9A_{10})$. The compound displaces the $T_1:A_{10}$ base pair from the DNA stack and $A_{10}$ flips over and stacks onto one of the ancillary phen groups[46] (Figure 8.11b). Whilst this binding mode is formally described as end-capping, and not intercalation, a similar binding motif, involving an additional H-bond as well as adenine–phen stacking, was adopted at the AA mismatch site in the structure discussed above. The semi-intercalative binding mode observed for the corresponding $\Lambda$ complex is also observed. Whilst no other structures showing how a $\Delta$ mononuclear ruthenium complex can bind to DNA have been reported, there is one structure of a $\Delta\Delta$ dinuclear complex bound to a DNA duplex. The complex, $\Delta,\Delta$-$[\mu\text{-}(11,11'\text{-bidppz})(phen)_4Ru_2]^{4+}$, is able to thread through a DNA duplex at a TA well-matched step[47] (Figure 8.11c). This causes both the bases to flip out and the adenine stacks onto a phen group

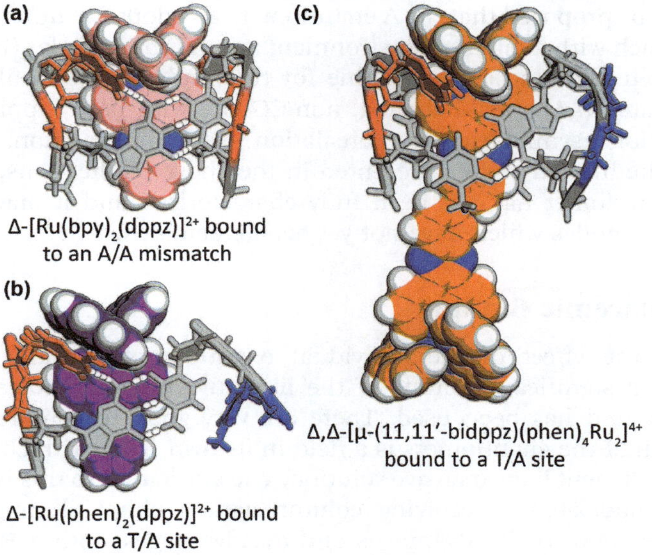


**Figure 8.11** (a) Δ-$[Ru(bpy)_2(dppz)]^{2+}$ (pink) bound to an A–A mismatch site, with both adenine bases flipped out and stacked on the bpy ligand groups from the complex (PDB ID 4E1U). (b) Δ-$[Ru(phen)_2(dppz)]^{2+}$ (purple) binds into $d(TCGGCGCCGA)_2$ at the terminal step, with $T_1$ and $A_{10}$ flipped out. $A_{10}$ stacks onto an ancillary phen group (PDB ID 5JEU). (c) A similar binding motif crystallised in the threading of Δ,Δ-$[\mu\text{-}(11,11'\text{-bidppz})(phen)_4Ru_2]^{4+}$ (orange) at the TA/TA step of $d(CGTACG)_2$ (PDB ID 5GQJ). Both a thymine and adenine base are flipped out, with the adenine stacking on an ancillary phen group. In all three structures, a hydrogen bond is formed between the flipped adenine and a $G(NH_2)$ on the 5′ side of the binding site, although this not adjacent to the threading cavity with Δ,Δ-$[\mu\text{-}(11,11'\text{-bidppz})(phen)_4Ru_2]^{4+}$. In this figure, adenine bases are drawn in red with thymine bases in blue. The atoms in the complex are coloured according to type with nitrogen in deep blue and hydrogen in white. Other DNA bases are drawn in grey. Hydrogen bonds between the bases are drawn as dashed lines in white. Figure adapted from J. P. Hall, P. M. Keane, H. Beer, K. Buchner, G. Winter, T. Sorensen, D. J. Cardin, J. Brazier and C. J. Cardin, *Nucleic Acids Res.*, 2016, **44**, 9472–9482, https://doi.org/10.1093/nar/gkw753 (http://creativecommons.org/licenses/by/4.0/)

on the complex. If the three binding sites discussed above are drawn side-by-side they look remarkably similar and all three involve the formation of a hydrogen bond between the flipped out *syn* adenine and a guanine base.

It has previously been reported that the luminescence lifetime of Δ-$[Ru(phen)_2(dppz)]^{2+}$ is usually longer than for the Λ form. It has also been shown that the observed luminescence from the complex, when bound to DNA, increases as a function of A–T content of the sequence.[48,49] It is therefore plausible that this binding mode, base flipping combined with stacking to an ancillary ligand, occurs in solution with A–T-rich DNA. It can therefore be proposed that the Δ enantiomer can adopt five different binding modes, each with a different environment around the complex (Figure 8.12). The expected luminescence lifetime for the binding sites would therefore be mismatch (AA) > well-matched, non-CG site with base flipping ≥ canted intercalation > symmetrical intercalation > semi-intercalation. However, beyond the limited results presented in the above publications, binding of the Δ enantiomer has not been fully characterised and it may bind with additional modes which have not yet been elucidated.

### 8.4.6 Racemic Binding

Studying the effect of the individual enantiomers certainly has merit. However, a significant volume of the literature reports studies where the *rac* compound has been used. There are very good reasons for this. The separation of the enantiomers is a field in its own right, with chiral HPLC[50] being an efficient but expensive solution. The alternative to this is separation on a Sephadex-based resolving column with a chiral eluting salt. Both techniques have their advantages and disadvantages, with a HPLC set up being efficient but expensive. In contrast the resolving column is comparatively inexpensive but the enantiomers must then be separated from the chiral resolving salt, which can lead to a significant reduction in yield.

To date, there is only one structural study looking at how a *racemic* Ru–polypyridyl complex can bind to a DNA duplex. The structure shows how both Λ and Δ-$[Ru(phen)_2(dppz)]^{2+}$ can bind to a short DNA duplex (Figure 8.13a) of the sequence $d(A_1T_2G_3C_4A_5T_6)$. In the structure, the Δ enantiomer binds into a TG/CA step[51] and both enantiomers are bound in the minor groove, consistent with the previously reported structures. This enantiomer can bind at a more acute angle than usually observed for canted intercalation, as there is no direct contact between the 3′-sugar and the ancillary phen ligand, which is inevitable for lambda enantiomers. The 5′-sugar, here attached to thymine, only contacts the phen ligand at H4′, and the alignment of sugar and phen rings is no longer approximately parallel. The phosphate backbone is opened out by the *anti* conformation of the $\gamma$ torsion angle (Figure 8.13b), creating an asymmetric cavity seen in several of the structures described in this chapter. In contrast, a phen group from the Λ enantiomer stacks onto the 3′ sugar on base $A_5$ and adopts a binding angle more consistent with canted intercalation. Again, the phosphate backbone on one side of the intercalation cavity has adopted the *anti* conformation. There is a single, highly ordered water layer between the phen groups for both complexes.

Due to the differences in intercalation geometry of the two enantiomers, each complex has a different dppz solvent accessibility. The one dppz nitrogen atom on the Δ enantiomer is occluded from the solvent but, for the Λ enantiomer, both dppz nitrogen atoms are accessible (Figure 8.13c). It has previously been reported that the solvent accessibility can affect the the observed luminescence. When the same sequence and complex were examined in solution, it was found that the order of emission was Δ > *rac* > Λ

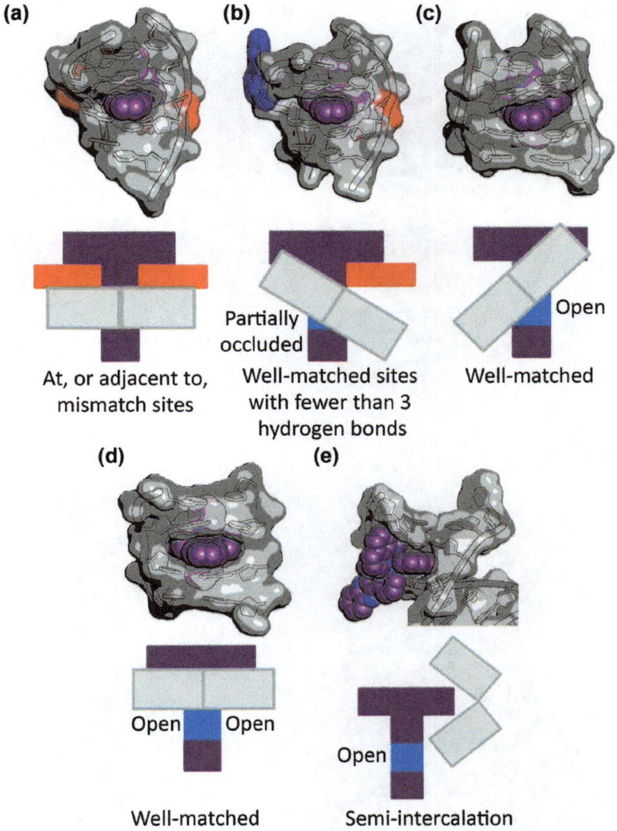


**Figure 8.12** Five possible binding modes for Δ-$[Ru(phen)_2(dppz)]^{2+}$ to DNA. (a) The complex binds at, or adjacent to, a mismatch site, with flanking adenine bases stacking on ancillary ligands, reducing intercalation depth. (b) Insertion into well-matched T/A sites. The purine flips out and is able to π-stack onto an ancillary ligand. The pyrimidine also flips out but does not stack. This partially, but not completely, occludes one dppz nitrogen. (c) Canted (angled) intercalation into a well-matched base pair leaves one dppz nitrogen atom entirely exposed to solvent. (d) Model for symmetrical intercalation by a Δ enantiomer at a 5′-AT/AT-3′ step, exposing both dppz nitrogen atoms to solvent. (e) Semi-intercalation by an ancillary ligand into the DNA duplex, exposing both phenazine nitrogen atoms to solvent. DNA is drawn in grey as a solvent-accessible surface with a radius of 1.5 Å. The carbon atoms of the complex are in purple with nitrogen atoms in blue. In the schematic diagram, the complex is drawn in purple with dppz nitrogen atoms as blue. DNA bases are represented by grey blocks with flanking adenine bases as red rectangles.
Figure adapted from J. P. Hall, P. M. Keane, H. Beer, K. Buchner, G. Winter, T. Sorensen, D. J. Cardin, J. Brazier and C. J. Cardin, *Nucleic Acids Res.*, 2016, **44**, 9472–9482, https://doi.org/10.1093/nar/gkw753 (http://creativecommons.org/licenses/by/4.0/).

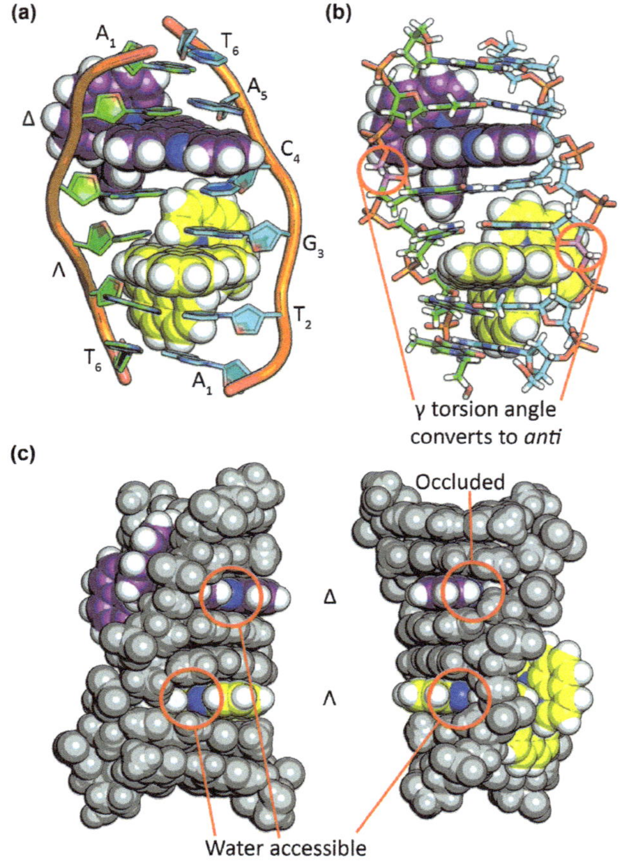


**Figure 8.13** (a,b) Λ (yellow) and Δ-[Ru(phen)$_2$(dppz)]$^{2+}$ (purple) simultaneously bound to d(ATGCAT)$_2$. (c) When the complexes bind, $\gamma$ (pink) increases opening up the backbone at the intercalation sites. (c) Differences in the binding geometry mean that both dppz nitrogen atoms on the Λ are accessible to solvent whereas one of the Δ dppz nitrogen atoms is occluded. DNA atoms are coloured according to chain with chain A carbon atoms in green and chain B in cyan. Other atoms are coloured according to type with oxygen in red, phosphorus in orange, nitrogen in blue and hydrogen in white.

and that the binding ratio of the rac complex–DNA duplex was 2 : 1 *i.e.* one Λ and one Δ per DNA duplex. This indicates that the binding observed in the structure could also be present in solution.

## 8.5 Binding of Ru-polypyridyl Complexes to DNA G-quadruplexes

G-quadruplexes are four-stranded, higher-order nucleic acid structures (Figure 8.14a–c), formed by G-rich DNA and RNA sequences.[52] They have been shown to play an important role in gene expression,[52] regulation,[53] cell

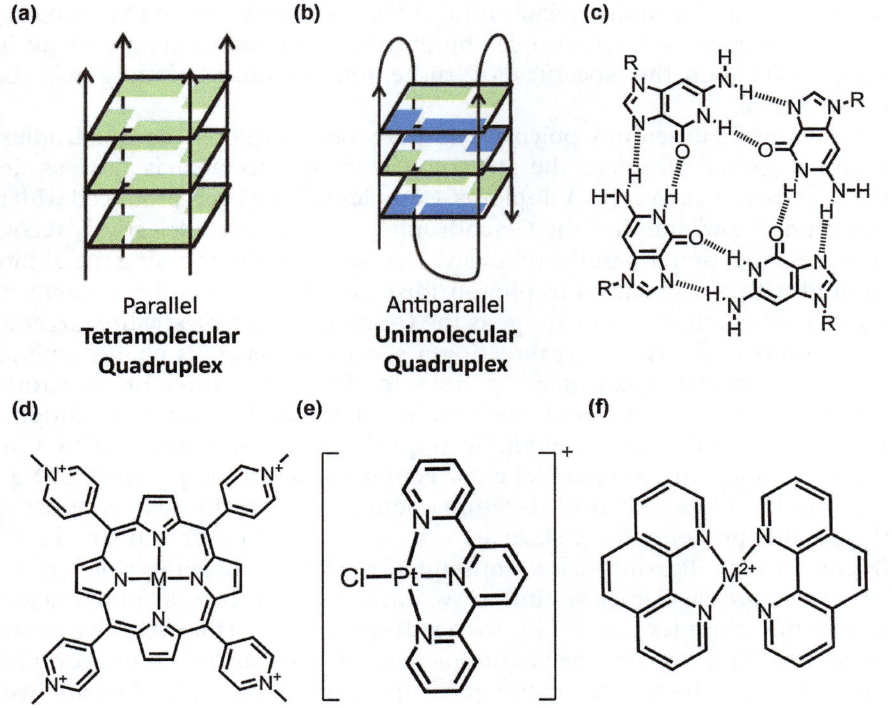


**Figure 8.14** (a) The parallel tetramolecular quadruplex and (b) a unimolecular antiparallel basket quadruplex. *Syn* guanine bases are displayed in blue with *anti* in green. (c) The guanine base tetrad in the DNA G-quadruplex. A number of G-quadruplex binding agents have been characterised including (d) metalloporphyrins (e) platinum terpyridines and (f) metal phenanthroline complexes.

replication[54] and human disease[55] and have been visualised in human cells.[56] Sequences able to form G-quadruplexes can be found in the telomeric region of the human genome[57] and in promoter regions.[58] Long 3' overhangs, approximately 300–500 repeats of the sequence $d(TTAGGG)_n$, regulate cell longevity through systematic cleavage in each cycle of cell replication.[59] However, in approximately 85% of cancer cells, the enzyme telomerase is activated[60] which can extend the length of the telomere repeat sequences, preventing apoptosis. Targeting the G-quadruplex, by small-molecule binders, is an area of intense research activity due to stabilisation of the structure being an effective method of inducing apoptosis in cancer cells.[61] A wide variety of compounds have been produced which can bind to G-quadruplexes including metallo-porphyrins[62] (Figure 8.14d), Pt-terpyridines[63] (Figure 8.14e), acridines,[64] distamycin and derivatives, naphthalene-based compounds[65] and Cu(II)-phenanthroline derivatives[66] (Figure 8.14f). All of these classes of compounds possess extended planar aromatic systems which are able to π-stack with the G-quartets in the quadruplex, and ligands typically bind at the interface between the G-quartet core and loop region. Quadruplex binding is discussed in greater detail in Chapter 12 but the specifics of ruthenium complex binding will be covered here.

Octahedral ruthenium polypyridyls have been explored as quadruplex-binding agents, although the structural work in this area is far less developed than that with DNA duplexes. Complexes have been produced which can bind to and stabilise the G-quadruplex, and can even selectively recognise a parallel or antiparallel topology. The benefits of combining the ability to bind with the potential to photooxidise guanine are readily apparent in the case of quadruplex targeting, as the G-quartets are typically the target of such compounds. However, targeting a specific topology is challenging as the unimolecular quadruplex is able to adopt 26 different structural forms which are dependent on both sequence and external conditions. For example, the human telomeric G-quadruplex, d(A(GGGTTA)$_3$GGG), in solution adopts an anti-parallel chair conformation in the presence of $Na^+$ but with $K^+$ adopts a hybrid structure (Figure 8.15a,b). This is in contrast to the parallel propeller form observed with $K^+$ in the X-ray crystal structure[67] (Figure 8.15c). Possibly as a consequence of this flexibility, no crystal structures are available, at time of writing, showing how a mononuclear ruthenium complex can bind to quadruplex DNA. However, two NMR structures, showing how the enantiomers of a binuclear ruthenium complex can bind to the human telomeric quadruplex, have been published and will be discussed.

## 8.5.1 Quadruplex Binding—Mononuclear Complexes

Mononuclear G-quadruplex-targeting Ru–polypyridyl complexes have been found to interact with the DNA G-quadruplex and, given the correct synthetic

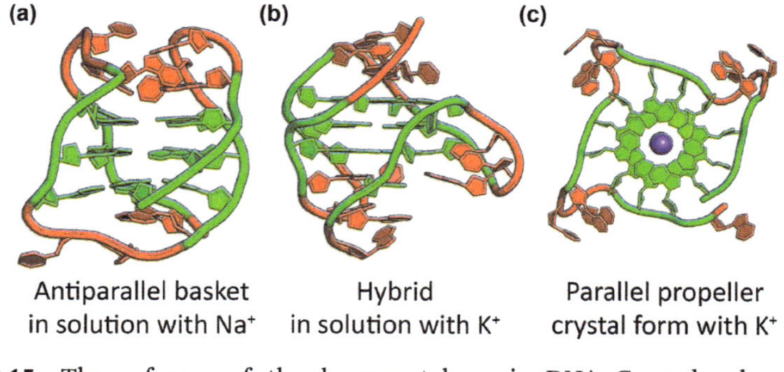


**Figure 8.15** Three forms of the human telomeric DNA G-quadruplex forming sequence. (a) The antiparallel basket form adopted in solution in the presence of $Na^+$. (b) The hybrid quadruplex adopted in solution with $K^+$. (c) The propeller conformation adopted when crystallized with $K^+$. Guanine bases are illustrated in green with the loop regions in red. A $K^+$ cation is shown in purple.

modification, can target the quadruplex specifically in preference to other DNA structures. One strategy which is often employed when designing a compound which targets G-quadruplex DNA is to produce a compound with a planar aromatic ligand approximately the same size and shape as a G-quartet, and is therefore too large to bind to a DNA duplex by intercalation. This has been used to design an octahedral ruthenium complex which is able to selectively bind to G-quadruplexes, by using the bpa ligand (Figure 8.16a) to produce a complex which contains a planar aromatic group in the *x*–*y* plane, with groups which could interact with the central metal ion in the G-quadruplex in the *z* direction. This complex was found to stabilise the human telomeric G-quadruplex whilst reducing the binding affinity toward to duplex DNA.[26]

A range of other ruthenium polypyridyls, more closely related to the $[Ru(L)_2(dppz)]^{2+}$ family, have also been produced. Due to the difficulty in separating the enantiomers of such compounds, it is still unusual to find studies where the binding of the two enantiomers have been directly compared. However, Shi *et al.* studied the binding of the enantiomers of one such complex, $[Ru(bpy)_2(dppz\text{-}idzo)]^{2+}$ (idzo = 10,11-imidazolone) (Figure 8.16b), in the presence of several G-quadruplex-forming sequences. They report that the Λ enantiomer is able to bind with a higher binding stoichiometry and yields greater luminescence emission than the Δ with both the $Na^+$ and $K^+$ forms of the human telomeric G-quadruplex.[68] In the same study, modelling

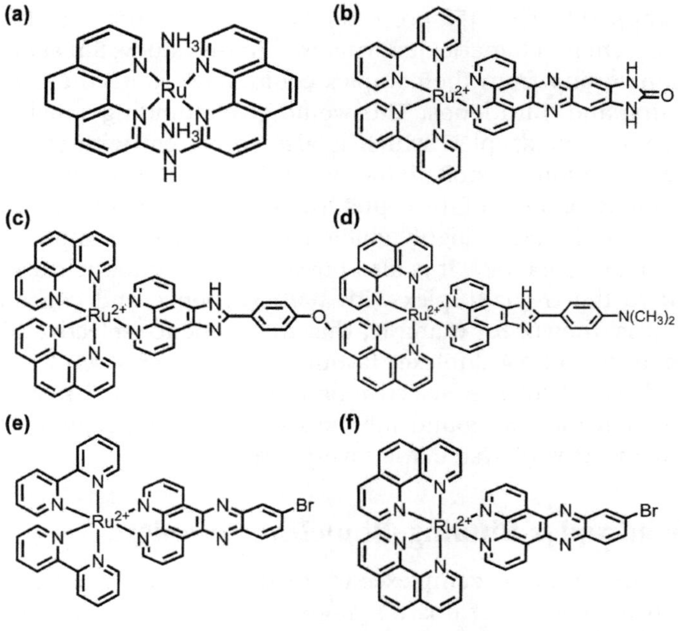


**Figure 8.16** Selected mononuclear ruthenium polypyridyl complexes which have been shown to bind to the DNA G-quadruplex.

indicated that a mixture of groove binding and G-tetrad stacking could occur with the Λ enantiomer, whereas the Δ would only cap the guanine core.

A related complex, $[Ru(phen)_2(p\text{-MOPIP})]^{2+}$ (p-MOPIP = 2-(4-methoxyphenyl)-imidazo[4,5f]phenanthroline) (Figure 8.16c), has also been examined in both solution and through cell studies. In solution, the binding behaviour of the two enantiomers was shown to be different. Whilst both enantiomers could bind to the $Na^+$ form of the human telomeric quadruplex, the Λ was shown to induce greater stabilisation, bind more strongly and luminesce more brightly. Intriguingly, this chiral selectivity has not just been observed in solution. The Λ enantiomer of the compound was found to inhibit telomerase, and therefore induce cell death, to a greater extent than the Δ in a human gastric cancer cell line.[69] The issue of chirality is also important in cell localisation, with the G-quadruplex targeting compound Λ-$[Ru(phen)_2(p\text{-DMNP})]^{2+}$ (*p*-DMNP = 2-(4-N,N-dimethyl aminophenyl)imidazo[4,5-f] [1,10]phenanthroline) (Figure 8.16d) localising in the nucleus of HepG2 cells and Δ-$[Ru(phen)_2(p\text{-DMNP})]^{2+}$ in the cytoplasm.[70]

Ru-polypyridyl complexes can also be used as luminescent DNA probes which can, given the correct synthetic modifications, emit more brightly when bound to quadruplex DNA than duplex,[71] an example of which is $[Ru(bpy)_2(dppz\text{-}11\text{-}Br)]^{2+}$ (Figure 8.16e). There are of course questions about how mononuclear ruthenium complexes can bind to quadruplexes, especially given how underdeveloped the structural work is in this area. A number of studies have attempted to address this issue by reporting computationally derived models. One such study reports a model for how $[Ru(phen)_2(dppz\text{-}11\text{-}Br)]^{2+}$ (Figure 8.16f) can bind to both the $Na^+$ and $K^+$ forms of the human telomeric quadruplex.[72] In this study, Shi *et al.* reported that the dppz group from the complex could stack, onto the G-tetrad, between the first and third loops. This would give a binding stoichiometry of one complex per quadruplex, which is also consistent with their reported solution results. Binding stoichiometry can be affected by changes to both the complex and topology of the G-quadruplex into which the compound has bound and it is therefore highly challenging to design compounds which target a particular topology. It is also possible that mononuclear complexes could bind to the G-quadruplex with both sequence and conformational selectivity. It is known, for example, that the sequence-selective binding of such compounds to DNA duplexes favours 5′-TA-3′ over 5′-AT-3′[41,42] in both solution and crystal studies. Selective binding should therefore not be ruled out, especially if the compound interacts with the loop region. This is an area certainly worthy of systematic investigation.

### 8.5.2 Quadruplex Binding–Binuclear Complexes

The binding of binuclear complexes to quadruplex DNA has been studied, although, in the context of unambiguous structural work, it is also a relatively underdeveloped area. The binding of such compounds to quadruplex DNA has been studied by a number of spectroscopic methods, showing that certain complexes can bind to and stabilise the G-quadruplex, with a key objective being to visualise such structures *in vivo*. For example, $[Ru_2(obip)(bpy)_4]^{4+}$ (Figure 8.17a) has been shown to stabilise both the $Na^+$ and $K^+$ form of the human telomeric G-quadruplex and is therefore able to bind to both an anti-parallel and mixed DNA conformation.[73] The stabilising effect of binding is strong and can even fold the quadruplex in the absence of metal cations. Similar stabilisation effects were observed for a number of other complexes[74] which were also able to fold single-stranded DNA into a G-quadruplex in low-metal conditions. Both this set of compounds and $[Ru_2(obip)(bpy)_4]^{4+}$ (obip = 2-(2-pyridyl)imidazo[4,5-f][1,10]-phenanthroline) possess a large, and relatively inflexible, linking ligand which can $\pi$-stack with the end of the G-tetrad. However, another binuclear complex contains a long and highly flexible non-cyclic 18-crown-6 linker[75] (Figure 8.17b). The flexibility in this linking group is similar to that in the

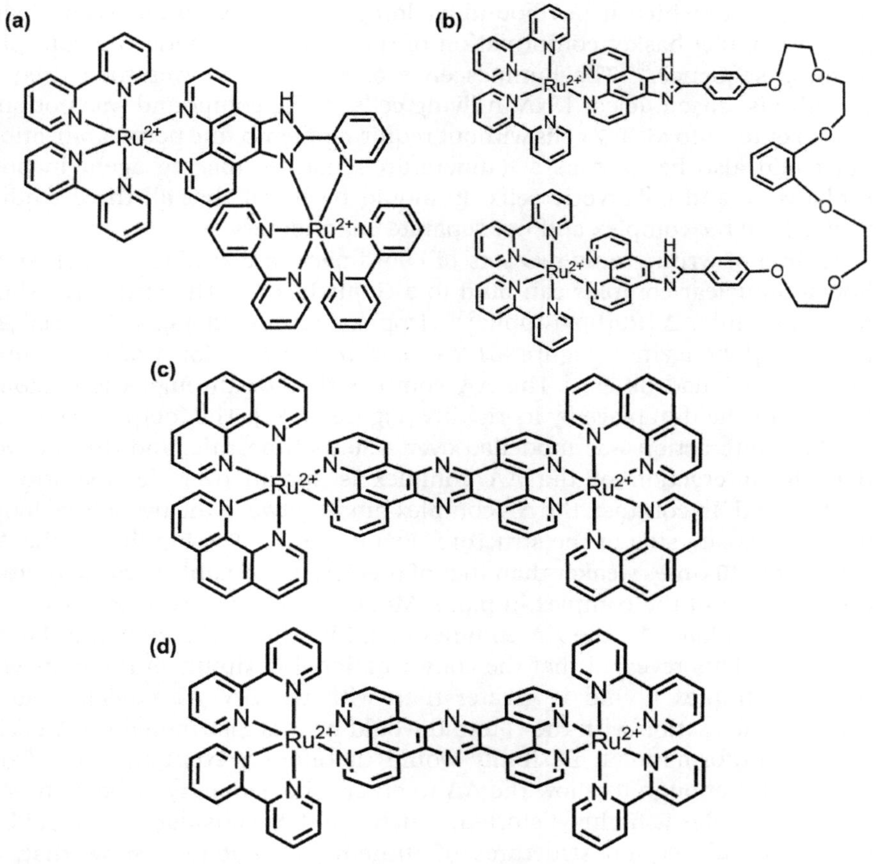


**Figure 8.17** Selected binuclear ruthenium polypyridyl complexes which have been shown to bind to the DNA G-quadruplex.

bis-intercalating compounds developed to bind to DNA duplexes, enabling such compounds to either crosslink adjacent duplexes[76] or intercalate into two sites.[77] However, adjacent crosslinking with G-quadruplexes would be unlikely and it is therefore possible that either the compound binds on one side of the G-tetrad or that the two ends of the compound bind on opposite sides of the quadruplex core. All of the binuclear complexes referred to so far show a substantial increase in luminescence when bound to DNA G-quadruplexes compared with duplex DNA, showing that the compounds can act as structure-sensitive DNA probes.

Two related binuclear ruthenium complexes have been found to bind to the DNA G-quadruplex, one of which has been structurally investigated. The complex,[78] $[\{Ru(phen)_2\}_2tpphz]^{4+}$ (Figure 8.17c), was initially found to bind to both duplex and quadruplex DNA. However, the binding strength to the DNA G-quadruplex was higher and emission from the complex when bound was 2.5 times higher than when bound to a DNA duplex.[79] It was subsequently found that emission of the complex was sensitive to the topology of the quadruplex to which it has bound, yielding the highest luminescence with the antiparallel basket conformation of the human telomeric G-quadruplex forming sequence.[80] The luminescence of the complex was then used to directly visualise nuclear DNA in living cells.[81] The compound was not only able to enter into MCF-7 cells without requiring membrane permeabilization, but could also be used as a temperature-sensitive imaging agent in both prokaryotic and eukaryotic cells. It should be noted that all three studies utilised the *rac* complex and not separate enantiomers.

At time of writing only two sets of coordinates are available which show how a binuclear complex can bind to a G-quadruplex. The structures show how ΛΛ and ΔΔ-$[Ru(bpy)_2tpphz]^{4+}$ (tpphz = tetrapyrido[3,2-a:2′,3′-c:3″,2″-h:2‴,3‴-j]phenazine) (Figure 8.17d), bind to the $Na^+$ form of the human telomeric G-quadruplex.[82] The ΛΛ complex threads through the diagonal loop and is held in place by its rigidity (Figure 8.18a). The footprint of the π-stacking interaction was modelled over a 50 ns timescale, and this showed that the interaction of the ΛΛ complex is spread over the majority of the G-tetrad. In contrast, the ΔΔ complex binds between the two lateral loops on the opposite side of the structure (Figure 8.18b). The binding of the ΔΔ complex is 40 times weaker than that of the ΛΛ, presumably because there is no loop to hold the complex in place. Modelling studies were performed to try and elucidate why the ΛΛ complex could bind into the loop and the ΔΔ could not. This revealed that the conformational flexibility of the loop with the ΛΛ complex bound is greater than with the ΔΔ. As a result, the π-stacking interaction with the G-tetrad would be less distributed for ΔΔ. The lack of conformational flexibility would therefore prevent the loop from opening far enough to allow the ΔΔ to enter. The geometry of both the ΛΛ and ΔΔ complex is highly distorted, which could be considered unusual but small-molecule crystal structures of ruthenium complexes show that, in some cases, the dppz group can be highly deformed away from the planar structure usually assumed for conjugated aromatic systems. This study

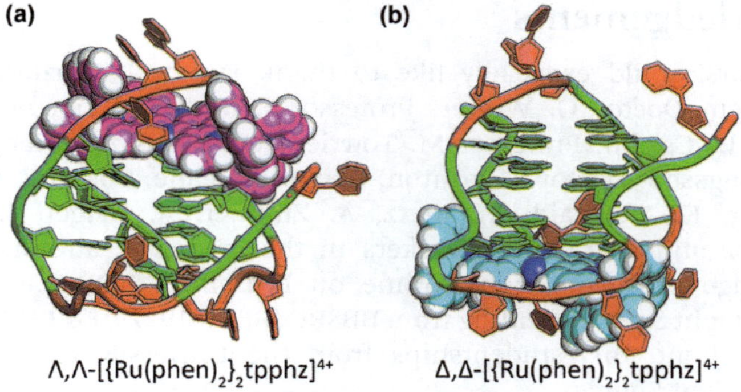


**Figure 8.18** (a) ΛΛ-$[\{Ru(bpy)_2\}_2tpphz]^{4+}$ (pink) binds into the diagonal loop of the $Na^+$ form human telomeric G-quadruplex. (b) ΔΔ-$[\{Ru(bpy)_2\}_2tpphz]^{4+}$ adopts a $\pi$-stacking interaction with the G-tetrad and is located between the lateral loops, on the opposite side of the G-quadruplex core to the lateral loop. Guanine bases are drawn in green with loop regions in red. Other atoms are coloured according to type with hydrogen in white and nitrogen in blue. The images are drawn from entries 2MCO and 2MCC, in the Protein Data Bank, for the ΛΛ- and ΔΔ-containing structures respectively.

provided the first, and only, structural view of how a binuclear complex could bind to a G-quadruplex.

## 8.6 Summary and Future Outlook

In this chapter, we have reviewed the major structural studies which show how ruthenium polypyridyl complexes can bind to DNA. Such compounds show therapeutic potential, with three compounds reaching the clinical trial stage of the drug development pathway. Complexes have been developed which can bind to DNA through a variety of binding modes and show promise as potential PACT and PDT agents. It has been shown, by both us and the Barton group that the binding modes adopted by the Λ and Δ enantiomers can significantly differ, which can affect both the luminescence and photooxidising potential of the complexes. The next step is to ensure that the link between structure and biological activity is fully explored. An area which requires additional focus and effort is to ensure that studies are performed using the enantiomers of any complexes in addition to the racemic mixture. Of particular interest is how Ru–polypyridyl complexes can bind to DNA G-quadruplexes, an area which has not yet been fully exploited. It should also be noted that there is no structural evidence to show how a ruthenium complex can bind to RNA, despite a number of complexes being shown to only reach the cytoplasm in cell-localisation studies. The area of Ru-polypyridyl development is an exciting and active area and it can be expected that many advancements will be made in the future.

## Acknowledgements

The authors would especially like to thank Doctor J. Brazier, Professor T. Sorensen, Doctor G. Winter, Professor J. M. Kelly, Doctor S. Quinn, Professor D. Cardin, Professor M. Towrie, Doctor I. Sazanovich, Professor T. Gunnlaugsson, Doctor F. Poynton, Doctor P. Keane, Doctor K. O'Sullivan, S. Gurung, K. McQuaid, M. Merz, A. Zimmer, A. Geiger, J. Fröhlich, C. Schwarz and all other co-workers in the area. The authors gratefully acknowledge provision of beamtime on beamlines I02, I03 and I04 at Diamond Light Source, funding from BBSRC (BB/K019279/1 & BB/M004635/1) and three joint PhD studentships from the University of Reading and Diamond Light Source.

## References

1. J. P. Paris and W. W. Brandt, *J. Am. Chem. Soc.*, 1959, **81**, 5001–5002.
2. J. G. Vos and J. M. Kelly, *Dalton Trans.*, 2006, 4869–4883.
3. J. K. Barton, *Science*, 1986, **233**, 727–734.
4. N. J. Turro, J. K. Barton and D. A. Tomalia, *Acc. Chem. Res.*, 1991, **24**, 332–340.
5. N. A. Smith and P. J. Sadler, *Philos. Trans. R. Soc., A*, 2013, **371**, 20120519.
6. G. Li, L. Sun, L. Ji and H. Chao, *Dalton Trans.*, 2016, **45**, 13261–13276.
7. E. C. Glazer, *Isr. J. Chem.*, 2013, **53**, 391–400.
8. C. J. Cardin, J. M. Kelly and S. J. Quinn, *Chem. Sci.*, 2017, DOI: DOI:10.1039/C7SC01070B.
9. A. E. Friedman, J. C. Chambron, J. P. Sauvage, N. J. Turro and J. K. Barton, *J. Am. Chem. Soc.*, 1990, **112**, 4960–4962.
10. F. E. Poynton, J. P. Hall, P. M. Keane, C. Schwarz, I. V. Sazanovich, M. Towrie, T. Gunnlaugsson, C. J. Cardin, D. J. Cardin, S. J. Quinn, C. Long and J. M. Kelly, *Chem. Sci.*, 2016, **7**, 3075–3084.
11. A. Chantzis, T. Véry, C. Daniel, A. Monari and X. Assfeld, *Chem. Phys. Lett.*, 2013, **578**, 133–137.
12. T. Véry, D. Ambrosek, M. Otsuka, C. Gourlaouen, X. Assfeld, A. Monari and C. Daniel, *Chem. - Eur. J.*, 2014, **20**, 12901–12909.
13. J. Olofsson, B. Önfelt and P. Lincoln, *J. Phys. Chem. A*, 2004, **108**, 4391–4398.
14. R. M. Hartshorn and J. K. Barton, *J. Am. Chem. Soc.*, 1992, **114**, 5919–5925.
15. C. Hiort, P. Lincoln and B. Nordén, *J. Am. Chem. Soc.*, 1993, **115**, 3448–3454.
16. I. Haq, P. Lincoln, D. Suh, B. Nordén, B. Z. Chowdhry and J. B. Chaires, *J. Am. Chem. Soc.*, 1995, **117**, 4788–4796.
17. J. M. Kelly, D. J. McConnell, C. OhUigin, A. B. Tossi, A. K.-D. Mesmaeker, A. Masschelein and J. Nasielski, *J. Chem. Soc., Chem. Commun.*, 1987, **24**, 1821–1823.
18. L. Jacquet, R. J. H. Davies, A. Kirsch-De Mesmaeker and J. M. Kelly, *J. Am. Chem. Soc.*, 1997, **119**, 11763–11768.

19. I. Ortmans, B. Elias, J. M. Kelly, C. Moucheron and A. Kirsch-DeMesmaeker, *Dalton Trans.*, 2004, 668–676.
20. B. Elias, C. Creely, G. W. Doorley, M. M. Feeney, C. Moucheron, A. Kirsch-DeMesmaeker, J. Dyer, D. C. Grills, M. W. George, P. Matousek, A. W. Parker, M. Towrie and J. M. Kelly, *Chem. – Eur. J.*, 2008, **14**, 369–375.
21. E. Wachter, B. S. Howerton, E. C. Hall, S. Parkin and E. C. Glazer, *Chem. Commun.*, 2014, **50**, 311–313.
22. C. Mari, V. Pierroz, S. Ferrari and G. Gasser, *Chem. Sci.*, 2015, **6**, 2660–2686.
23. J. K. White, R. H. Schmehl and C. Turro, *Inorg. Chim. Acta*, 2017, **454**, 7–20.
24. B. C. Poulsen, S. Estalayo-Adrián, S. Blasco, S. A. Bright, J. M. Kelly, D. C. Williams and T. Gunnlaugsson, *Dalton Trans.*, 2016, **45**, 18208–18220.
25. H. Huang, B. Yu, P. Zhang, J. Huang, Y. Chen, G. Gasser, L. Ji and H. Chao, *Angew. Chemie Int. Ed.*, 2015, **54**, 14049–14052.
26. L. He, X. Chen, Z. Meng, J. Wang, K. Tian, T. Li and F. Shao, *Chem. Commun.*, 2016, **52**, 8095–8098.
27. S. K. Fung, T. Zou, B. Cao, T. Chen, W. P. To, C. Yang, C. N. Lok and C. M. Che, *Nat. Commun.*, 2016, **7**, 10655.
28. K. D. Mjos and C. Orvig, *Chem. Rev.*, 2014, **114**, 4540–4563.
29. P. Zhang and P. J. Sadler, *Eur. J. Inorg. Chem.*, 2017, **2017**, 1541–1548.
30. Z. Adhireksan, G. E. Davey, P. Campomanes, M. Groessl, C. M. Clavel, H. Yu, A. A. Nazarov, C. H. F. Yeo, W. H. Ang, P. Dröge, U. Rothlisberger, P. J. Dyson and C. A. Davey, *Nat. Commun.*, 2014, **5**, 3462.
31. X.-J. Lu and W. K. Olson, *Nat. Protoc.*, 2008, **3**, 1213–1227.
32. A. H. J. Wang, G. Ughetto, G. J. Quigley and A. Rich, *Biochemistry*, 1987, **26**, 1152–1163.
33. S. Neidle, A. Achari, G. L. Taylor, H. M. Berman, H. L. Carrell, J. P. Glusker and W. C. Stallings, *Nature*, 1977, **269**, 304–307.
34. A. K. Todd, A. Adams, J. H. Thorpe, W. A. Denny, L. P. Wakelin and C. J. Cardin, *J. Med. Chem.*, 1999, **42**, 536–540.
35. P. Keane, J. Hall, F. Poynton, B. Poulsen, S. Gurung, I. Clark, I. Sazanovich, M. Towrie, T. Gunnlaugsson, S. Quinn, C. Cardin and J. M. Kelly, *Chem. – Eur. J.*, 2017, DOI: 10.1002/chem.201701447.
36. J. P. Hall, H. Beer, K. Buchner, D. J. Cardin and C. J. Cardin, *Philos. Trans. R. Soc., A*, 2013, **371**, 20120525.
37. J. P. Hall, H. Beer, K. Buchner, D. J. Cardin and C. J. Cardin, *Organometallics*, 2015, **34**, 2481–2486.
38. J. P. Hall, K. O'Sullivan, A. Naseer, J. A. Smith, J. M. Kelly and C. J. Cardin, *Proc. Natl. Acad. Sci. U. S. A.*, 2011, **108**, 17610–17614.
39. P. Lincoln and B. Nordén, *J. Phys. Chem. B*, 1998, **102**, 9583–9594.
40. J. P. Hall, J. Sanchez-Weatherby, C. Alberti, C. H. Quimper, K. O'Sullivan, J. A. Brazier, G. Winter, T. Sorensen, J. M. Kelly, D. J. Cardin and C. J. Cardin, *J. Am. Chem. Soc.*, 2014, **136**, 17505–17512.

41. H. Niyazi, J. P. Hall, K. O'Sullivan, G. Winter, T. Sorensen, J. M. Kelly and C. J. Cardin, *Nat. Chem.*, 2012, **4**, 621–628.
42. P. M. Keane, F. E. Poynton, J. P. Hall, I. V. Sazanovich, M. Towrie, T. Gunnlaugsson, S. J. Quinn, C. J. Cardin and J. M. Kelly, *Angew. Chem.*, 2015, **127**, 8484–8488.
43. E. Protozanova, P. Yakovchuk and M. D. Frank-Kamenetskii, *J. Mol. Biol.*, 2004, **342**, 775–785.
44. J. P. Hall, S. P. Gurung, J. Henle, P. Poidl, J. Andersson, P. Lincoln, G. Winter, T. Sorensen, D. J. Cardin, J. A. Brazier and C. J. Cardin, *Chem. – Eur. J.*, 2017, **23**, 4981–4985.
45. H. Song, J. T. Kaiser and J. K. Barton, *Nat. Chem.*, 2012, **4**, 615–620.
46. J. P. Hall, P. M. Keane, H. Beer, K. Buchner, G. Winter, T. Sorensen, D. J. Cardin, J. Brazier and C. J. Cardin, *Nucleic Acids Res.*, 2016, **44**, 9472–9482.
47. D. R. Boer, L. Wu, P. Lincoln and M. Coll, *Angew. Chem. Int. Ed.*, 2014, **53**, 1949–1952.
48. A. W. McKinley, J. Andersson, P. Lincoln and E. M. Tuite, *Chem. – Eur. J.*, 2012, **18**, 15142–15150.
49. A. W. McKinley, P. Lincoln and E. M. Tuite, *Dalton Trans.*, 2013, **42**, 4081–4090.
50. Y. Shu, Z. S. Breitbach, M. K. Dissanayake, S. Perera, J. M. Aslan, N. Alatrash, F. M. MacDonnell and D. W. Armstrong, *Chirality*, 2015, **27**, 64–70.
51. J. P. Hall, D. Cook, S. R. Morte, P. McIntyre, K. Buchner, H. Beer, D. J. Cardin, J. A. Brazier, G. Winter, J. M. Kelly and C. J. Cardin, *J. Am. Chem. Soc.*, 2013, **135**, 12652–12659.
52. H. J. Lipps and D. Rhodes, *Trends Cell Biol.*, 2009, **19**, 414–422.
53. S. Kendrick and L. H. Hurley, *Pure Appl. Chem.*, 2010, **82**, 1609–1621.
54. A.-L. Valton, V. Hassan-Zadeh, I. Lema, N. Boggetto, P. Alberti, C. Saintome, J. F. Riou and M.-N. Prioleau, *EMBO J.*, 2014, **33**, 732–746.
55. Y. Wu and R. M. Brosh, *FEBS J.*, 2010, **277**, 3470–3488.
56. G. Biffi, D. Tannahill, J. McCafferty and S. Balasubramanian, *Nat. Chem.*, 2013, **5**, 182–186.
57. K. Paeschke, T. Simonsson, J. Postberg, D. Rhodes and H. J. Lipps, *Nat. Struct. Mol. Biol.*, 2005, **12**, 847–854.
58. B. Onel, C. Lin and D. Yang, *Sci. China: Chem.*, 2014, **57**, 1605–1614.
59. V. L. Makarov, Y. Hirose and J. P. Langmore, *Cell*, 1997, **88**, 657–666.
60. N. Kim, M. Piatyszek, K. Prowse, C. Harley, M. West, P. Ho, G. Coviello, W. Wright, S. Weinrich and J. Shay, *Science*, 1994, **266**, 2011–2015.
61. M. Düchler, *J. Drug Targeting*, 2012, **20**, 389–400.
62. Q. Cao, Y. Li, E. Freisinger, P. Z. Qin, R. K. O. Sigel and Z.-W. Mao, *Inorg. Chem. Front.*, 2017, **4**, 10–32.
63. D. L. Ang, B. W. J. Harper, L. Cubo, O. Mendoza, R. Vilar and J. Aldrich-Wright, *Chem. – Eur. J.*, 2016, **22**, 2317–2325.
64. S. M. Haider, S. Neidle and G. N. Parkinson, *Biochimie*, 2011, **93**, 1239–1251.

65. S. Neidle, *Curr. Opin. Struct. Biol.*, 2009, **19**, 239–250.
66. K. Duskova, S. Sierra, M.-S. Arias-Pérez and L. Gude, *Bioorg. Med. Chem.*, 2016, **24**, 33–41.
67. G. N. Parkinson, M. P. H. Lee and S. Neidle, *Nature*, 2002, **417**, 876–880.
68. S. Shi, J.-H. Xu, X. Gao, H.-L. Huang and T.-M. Yao, *Chem. – Eur. J.*, 2015, **21**, 11435–11445.
69. X. Zhang, Z. Huang, S. Wu, R. Lin, J. Liu and N. Su, *Inorg. Chem. Commun.*, 2016, **72**, 1–6.
70. D. Sun, Y. Liu, Q. Yu, D. Liu, Y. Zhou and J. Liu, *J. Inorg. Biochem.*, 2015, **150**, 90–99.
71. E. Wachter, D. Moyá, S. Parkin and E. C. Glazer, *Chem. – Eur. J.*, 2016, **22**, 550–559.
72. S. Shi, H. L. Huang, X. Gao, J. L. Yao, C. Y. Lv, J. Zhao, W. L. Sun, T. M. Yao and L. N. Ji, *J. Inorg. Biochem.*, 2013, **121**, 19–27.
73. S. Shi, J. Liu, T. Yao, X. Geng, L. Jiang, Q. Yang, L. Cheng and L. Ji, *Inorg. Chem.*, 2008, **47**, 2910–2912.
74. L. Xu, X. Chen, J. Wu, J. Wang, L. Ji and H. Chao, *Chem. – Eur. J.*, 2015, **21**, 4008–4020.
75. L. Xu, D. Zhang, J. Huang, M. Deng, M. Zhang and X. Zhou, *Chem. Commun.*, 2010, **46**, 743–745.
76. N. H. Hopcroft, A. L. Brogden, M. Searcey and C. J. Cardin, *Nucleic Acids Res.*, 2006, **34**, 6663–6672.
77. A. Wang, G. Ughetto, G. Quigley, T. Hakoshima, G. van der Marel, J. van Boom and A. Rich, *Science*, 1984, **225**, 1115–1121.
78. F. M. MacDonnell and S. Bodige, *Inorg. Chem.*, 1996, **35**, 5758–5759.
79. C. Rajput, R. Rutkaite, L. Swanson, I. Haq and J. A. Thomas, *Chem. – Eur. J.*, 2006, **12**, 4611–4619.
80. T. Wilson, M. P. Williamson and J. A. Thomas, *Org. Biomol. Chem.*, 2010, **8**, 2617–2621.
81. M. R. Gill, J. Garcia-Lara, S. J. Foster, C. Smythe, G. Battaglia and J. A. Thomas, *Nat. Chem.*, 2009, **1**, 662–667.
82. T. Wilson, P. J. Costa, V. Félix, M. P. Williamson and J. A. Thomas, *J. Med. Chem.*, 2013, **56**, 8674–8683.

CHAPTER 9

# Therapeutic Potential of DNA Gene Targeting using Peptide Nucleic Acid (PNA)

PETER E. NIELSEN

University of Copenhagen, Department of Cellular and Molecular Medicine, The Panum Institute, Blegdamsvej 3c, DK-2200 Copenhagen N, Denmark
Email: ptrn@sund.ku.dk

## 9.1 Introduction

Peptide nucleic acids (PNA) were introduced in 1991 as a new class of pseudo peptide DNA mimics (Figure 9.1). PNA was originally conceived as a ligand that could recognize double-stranded DNA sequence specifically in the major groove, mimicking the DNA recognition of triplex-forming oligonucleotides.[1,2] Surprisingly, however, it was discovered that homopyrimidine PNA oligomers are able to invade the DNA helix by forming very stable triplex invasion complexes in which a $PNA_2$–DNA triplex is formed on the sequence-complementary purine DNA strand while the pyrimidine strand is extruded as a single stranded loop (P-loop)[1,3,4] (Figure 9.2).

It was also immediately realized that PNA is a very potent structural mimic of DNA and RNA, capable of forming both PNA–DNA and PNA–RNA (as well as PNA–PNA) duplexes which are generally of higher stability and are endowed with greater sequence discrimination than the corresponding DNA–DNA or DNA–RNA duplexes.[2,5] Subsequent studies have also revealed that more complex structures, such as triplexes, quadruplexes and hairpins can

Chemical Biology No. 7
DNA-targeting Molecules as Therapeutic Agents
Edited by Michael J. Waring

Published by the Royal Society of Chemistry, www.rsc.org

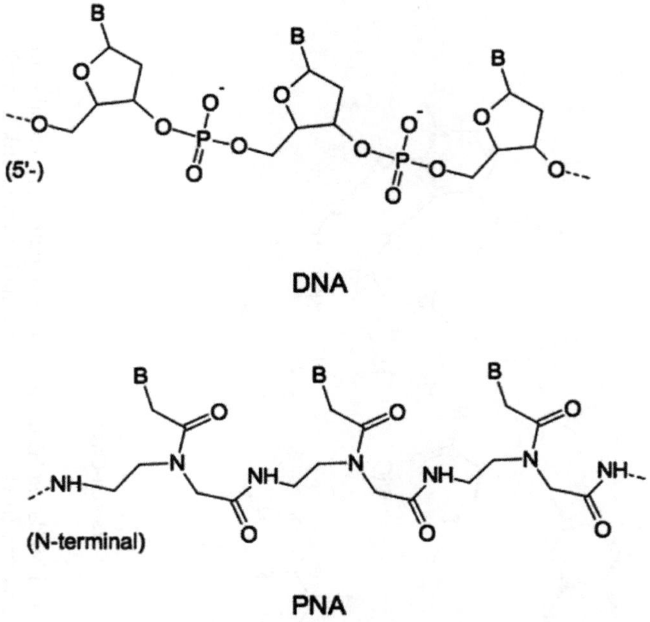


**Figure 9.1** Chemical structures of the original aeg (aminoethylglycine) PNA compared with DNA.
Adapted from *Mol. Biotechnol.*, **26**, 2004, 233, P. E. Nielsen, (© 2004 Humana Press Inc.). With permission of Springer.

be adopted by PNA oligomers of appropriate sequence.[6–10] The properties of PNA have inspired and attracted the attention of many disciplines of science ranging from organic and physical chemistry, through molecular biology to genetic diagnostics, nanotechnology and drug discovery[11–14]—even so far as to include aspects of the origin of life.[15,16]

This chapter will focus on the efforts and progress made towards exploiting PNA for DNA-targeted effects on gene expression in drug discovery approaches.

## 9.2 Duplex DNA Recognition *In Vitro*

Four distinct modes of recognizing duplex DNA have been discovered for PNA oligomers (Figure 9.3) depending upon the target and PNA sequence as well as the conditions. Triplex invasion is (so far) restricted to homopurine DNA targets, and is thermodynamically and kinetically stabilized by the formation of a Watson–Crick–Hoogsteen $PNA_2$–DNA triplex.[3,4] Simple duplex invasion relies on Watson–Crick base pairing and in principle suffers no sequence constraints, but under most conditions these complexes are not very stable *in vitro* for simple aegPNAs. However, binding has been demonstrated with homopurine PNA oligomers which form exceptionally stable PNA–DNA duplexes,[17] and also for γ-substituted PNAs (Figure 9.4), which

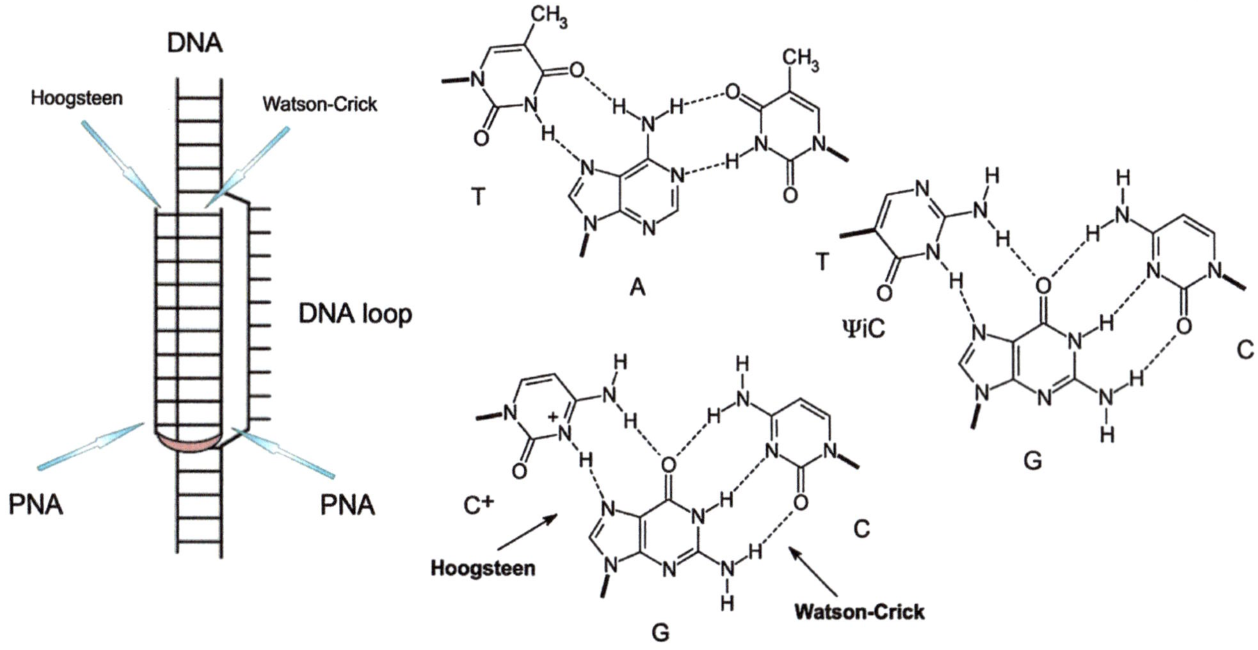


**Figure 9.2** Triplex invasion by homopyrimidine PNA oligomers. One PNA strand binds *via* Watson–Crick base pairing (preferably in the antiparallel orientation), while the other binds *via* Hoogsteen base pairing (preferably in the parallel orientation). It is usually advantageous to connect the two PNA strands covalently *via* a flexible linker (red connection) into a bis-PNA, and to substitute all cytosines in the Hoogsteen strand with pseudoisocytosines (ΨiC), which do not require low pH for protonation at $N^3$. Adapted from *Mol. Biotechnol.*, **26**, 2004, 233, P. E. Nielsen, 

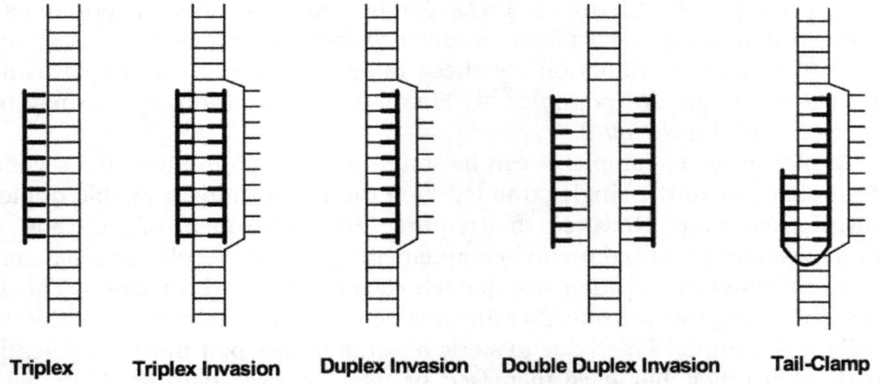


**Figure 9.3** Five different types of PNA–dsDNA complexes. DNA is schematically drawn as a ladder, and the PNA oligomers are in bold.
Adapted from *Mol. Biotechnol.*, **26**, 2004, 233, P. E. Nielsen, (© 2004 Humana Press Inc.). With permission of Springer.

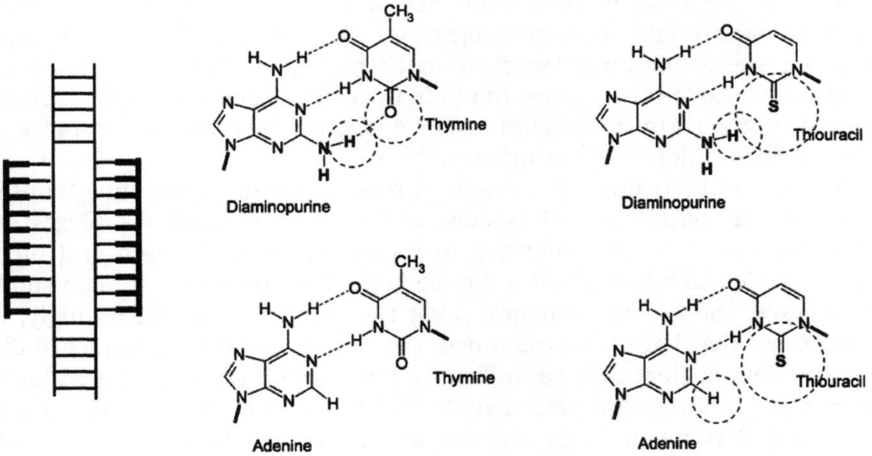


**Figure 9.4** Double duplex invasion of pseudo-complementary PNAs. In order to obtain efficient binding the target (and thus the PNAs) should contain at least 50% AT, though no other sequence constraints apply, and in the PNA oligomers all adenines and thymines are substituted with 2,6-diaminopurine or 2-thiouracil respectively. Pairing between these base analogues is very unstable due to steric hindrance. Therefore the two sequence-complementary PNAs will not be able to bind to each other, but they bind to their complementary DNA sequences very well.
Adapted from *Mol. Biotechnol.*, **26**, 2004, 233, P. E. Nielsen, (© 2004 Humana Press Inc.). With permission of Springer.

exhibit increased PNA–DNA duplex stability relative to unmodified PNA.[18] The efficiency of γ-PNA double duplex invasion is also enhanced by acridine conjugation[18] or by using helix-stabilizing modified nucleobases such as the G-clamp.[19] Furthermore, PNAs conjugated to cationic peptides can bind to double-stranded targets under negative super helical stress of the DNA.[20] Finally, combinations of these binding modes, for example using PNA 'tail-clamps' are possible.[21,22] However, the situation *in vivo* may be more complex (*vide infra*).

Duplex invasion complexes can be stabilized through binding of a second PNA oligomer to the single stranded loop thereby forming a double duplex invasion complex. However, this requires the employment of sterically or otherwise compromised pseudo-complementary PNA molecules because such a pair of PNAs would otherwise quench each other by hybridization. This is possible using the adenine–thymine analogues diaminopurine and 2-thiouracil (or –thymine), which due to steric hindrance base pair most inefficiently with each other, but have increased or retained base pairing ability with thymine and adenine, respectively[23] (Figure 9.4). Most interestingly, pseudo-complementarity can also be accomplished through backbone modifications illustrated by the pyrrolidinyl PNAs (Figure 9.5). In this case pyrrolidinyl PNAs hybridize very poorly with each other, but very well with DNA.[24]

Finally, conventional triplex binding of homopyrimidine PNAs in the major groove of the DNA helix at sequence-complementary homopurine DNA targets may also take place by Hoogsteen base pairing, and such PNA–$DNA_2$ triplexes are generally more stable than the analogous DNA–$DNA_2$ triplexes.[25]

More special DNA structures found in the genome such as G-quadruplexes and H-DNA *via* triplex formation can be targeted by PNAs that stabilize or disrupt these folded DNA structures.[26,27]

All types of helix invasion complexes require opening ('breathing') of the DNA double helix, and the kinetics of formation of such complexes are therefore very slow and sensitive to increasing ionic strength and other factors which stabilize the DNA double helix. Thus in order to obtain stable complexes, the dissociation rate must also be very slow. Accordingly, in most cases that have been examined, the binding, including sequence discrimination, is kinetically controlled—a phenomenon that can be exploited to control binding and selectivity.[4,28,29] In contrast, conventional triplex formation has much faster kinetics and is much less sensitive to ionic strength.[6,25] This mode of binding has not yet been extensively investigated.

Although PNA binding does not show the almost all-or-none orientation dependence observed with DNA oligonucleotides, Watson–Crick duplexes are most stable in the antiparallel configuration (amino-end of the PNA

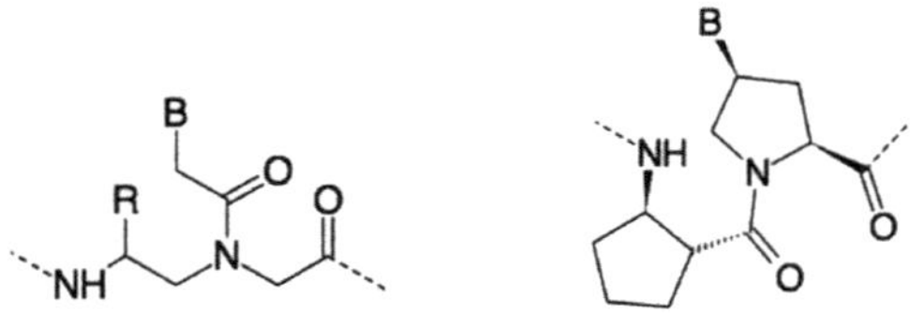


**Figure 9.5** Chemical structures of $\gamma$-substituted (R) PNA (left) and pyrrolidinyl-PNA (right).

facing the 3′-end of the oligonucleotide), whereas the parallel orientation is preferred for Hoogsteen binding to DNA.[30] Based on these observations, bis-PNAs (or PNA clamps) for triplex invasion have been constructed in which the Watson–Crick and Hoogsteen domains are connected *via* a flexible chemical linker.[30] This allows advantageous replacement of cytosine by pseudoisocytosine in the Hoogsteen domain which virtually eliminates the requirement for an acidic pH otherwise needed to ensure $N^3$-protonation of cytosine[30] (Figure 9.2).

## 9.3 PNA Conjugates

A variety of other DNA-interactive ligands have been conjugated to (bis) PNAs in order to improve or to modify DNA recognition and binding. The rate of binding can be significantly improved by making cationic PNAs,[31] for example *via* lysine ‘conjugation’, which results in a rate enhancement of almost 10-fold for each charge (up to +4). This effect is ascribed to an increase in local PNA concentration proximal to the DNA thereby increasing the probability of the PNA ‘catching’ a dynamic DNA opening.

An analogous effect is obtained using PNAs conjugated to the DNA intercalator 9-aminoacridine (Figure 9.6), but additionally the binding of such conjugates to dsDNA is much less sensitive to inhibition by elevated ionic strength. In fact bis-PNA–acridine conjugates bind to relaxed duplex DNA targets at submicromolar concentration in buffer containing 150 mM $K^+$, 2 mM $Mg^{++}$.[32]

Although it has been demonstrated that decameric (bis)PNAs bind to their sequence-complementary target with very high kinetically controlled discrimination,[28,29] it has (not surprisingly) been observed that this kinetic discrimination is dramatically reduced with increasing target length.

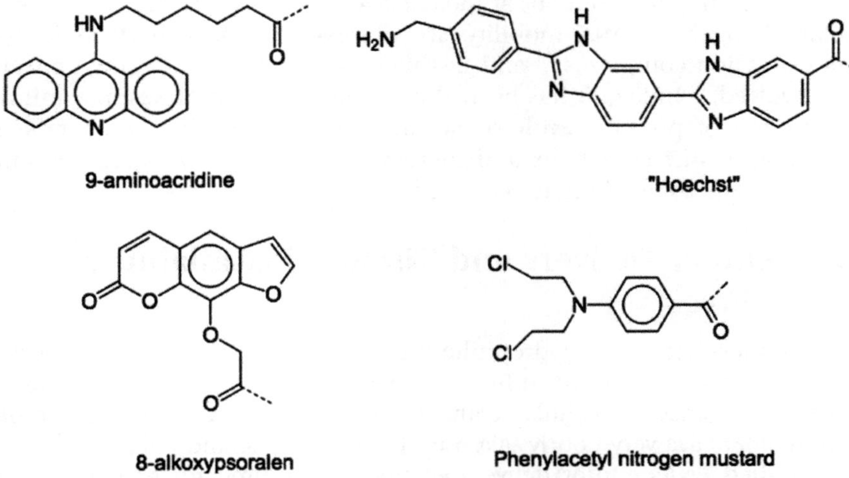


**Figure 9.6** Ligands used for conjugation to PNAs.

Therefore, 15–16-mer targets which would be required for uniqueness in the human genome cannot be selectively targeted by triplex-invading PNA oligomers. However, the kinetic control can be exploited using PNAs conjugated to another DNA-recognizing ligand that relies on thermodynamically controlled recognition with a fast on-rate, thereby poising the PNA in close proximity to its recognition site on the DNA. This principle has been exemplified using a bis-PNA conjugated to a minor groove-binding Hoechst analogue.[33] It was demonstrated that this PNA was able to recognize a PNA target proximal to an AT-rich Hoechst binding site up to 30 times more effectively than an 'isolated' PNA target. Therefore, by using other DNA ligands, such as hairpin polyamides, that by themselves exhibit exquisite sequence-specificity[34] PNA conjugates that effectively recognize longer DNA sequences should be feasible.[35]

PNA conjugates may also be exploited in order to direct covalent modification to specific DNA targets as exemplified by conjugates attached to nitrogen mustard,[36] chlorambucil or psoralen.[37] With such conjugates DNA cross-linking can specifically be introduced proximal to a PNA target.

## 9.4 Effect of PNA Binding on DNA Structure

Obviously PNA invasion complexes have dramatic effects on the DNA structure by the mere fact that the helix is opened. However, quite dramatic decreases in the mobility of PNA–dsDNA complexes upon gel electrophoretic analysis [electrophoretic mobility shift assays (EMSA)] have indicated that the overall properties of the DNA helix such as its flexibility and/or (directional) bending can also be affected. More detailed studies of this phenomenon have shown that triplex invasion unwinds the DNA helix by 12° per base, essentially reflecting the larger pitch of a $PNA_2$–DNA triplex (*ca.* 16 bases per turn) compared with B-DNA (10.5 bp). It also introduces an additional unwinding of about 60° at the dsDNA–$PNA_2$–DNA triplex junctions.[38] Similar changes in DNA mobility are observed for PNA–dsDNA double duplex invasion complexes,[23] and in this case directional bending appears to be involved.[39] Indeed it has been shown that such complexes may mimic the function of proteins evolved by nature to induce kinks or bends in the otherwise stiff DNA helix and thereby bring distant DNA regions into proximity so as to facilitate functional interaction.[39]

## 9.5 Cellular Delivery and Tissue Bioavailability *In Vivo*

PNA oligomers are large, hydrophilic molecules and—like oligonucleotides and most peptides—do not diffuse passively through the lipid membrane. Furthermore, since no cellular transporters appear to be present, simple PNA oligomers are very poorly taken up by prokaryotic and eukaryotic cells, and modified PNAs and/or delivery agents are required for effective penetration into cells.[40] With cells in culture, these methods include cationic liposome-assisted delivery of PNA–DNA hybrids[41] in which the negatively charged oligonucleotide functions as a carrier of the uncharged PNA. Alternatively, association of PNA with the liposomes may be accomplished through conjugation to cholesterol or cholic acid[42] or other lipophilic entities such as polyaromatics, of which 9-aminoacridine[43] and a Hoechst analogue[44] (Figure 9.6) have proved particularly effective. Finally, PNAs with a phosphonate amino acid peptide tail exhibit cellular activities rivalling those of small interfering RNA (siRNA) when delivered by cationic liposomes.[45]

A variety of cationic 'cell penetrating peptides' (CPPs) have also been successfully employed.[40,46] However, several studies have shown that most of these peptides are taken up by endocytotic pathways,[47,48] and therefore the PNA conjugates have to escape from the endosomal compartment in order to reach their target within the nucleus (or cytoplasm). Endosomal release is triggered by a variety of viral peptides,[49] which could be conjugated to PNA, but which unfortunately are rather large (20-mers or larger). Alternatively, a variety of auxiliary agents such as chloroquine, $Ca^{2+}$ ions[50] and photosensitizing dyes[51] have been found to very significantly augment the biological effects of PNA by triggering endosomal release. Thus sufficiently effective methods are available for studies of biological effects of PNA in cell culture.

PNAs also exhibit very poor tissue bioavailabity *in vivo* (studied in mice, rats and baboons) due to very fast renal excretion ($t_{1/2}$ of 30 minutes)[52,53] and consequently good activity in animals requires the use of chemically modified PNAs. Simple conjugation to peptides such as oligolysine[54] resulted in improved bioavailability, as has conjugation to receptor-targeted ligands such as oligo *N*-acetylgalactosamine,[52] which has high affinity for the hepatic asialoglycoprotein receptor (ASGP-R). Nanoparticle formulations have also been used for successful *in vivo* administration (*vide infra*). However, significantly improved and toxicologically safe methods for *in vivo* administration as well as efficient tissue delivery of PNA oligomers are needed before these molecules can be effectively used for therapeutic development.

## 9.6 Cellular Gene Targeting

Specific binding of PNA to duplex DNA targets *in vivo* or *ex vivo* in cells in culture has so far not been directly demonstrated, but has been inferred from biological effects. Using homopyrimidine PNAs Glazer and co-workers have found that PNAs delivered to cells in culture can produce a twofold to threefold increase in mutation rates within or proximal to the PNA target, indicating that PNA triplex invasion complexes are, at least to some extent, mutagenic, probably by activating the repair machinery.[55]

Most surprisingly Cutrona *et al.* have reported that treatment of cancer cells with mixed purine–pyrimidine sequence PNAs targeted to the sense strand of the c-myc (and mycn) gene results in a decreased level of c-myc mRNA transcript as well as cell death by apoptosis.[56,57] The authors ascribe these effects to anti-gene binding of the PNA to the DNA of the c-myc gene, but have no evidence supporting this assumption. Indeed the accumulated knowledge from experiments *in vitro* indicates that complexes between mixed purine–pyrimidine sequence PNAs and double-stranded DNA are neither thermodynamically nor kinetically stable. However, such complexes may form at DNA ends[58] or with cationic peptide–PNA conjugates directed to negatively supercoiled targets,[20] which, in general, can increase PNA binding by up to two orders of magnitude.[59] Thus the reported cellular effects of these anti-myc PNAs are not mechanistically understood at present, but transcription bubble facilitation or protein stabilization as discussed below DNA could play a role. However, also possible regulatory RNA targets functioning in trans such as sense RNA transcripts, micro RNAs or long non-coding (lnc) RNAs should be considered.

It is well established that PNA triplex invasion complexes can arrest elongating RNA polymerases *in vitro*, in particular when positioned on the template DNA strand.[60–62] Most interestingly, Corey *et al.* demonstrated that mixed-sequence PNA oligomers targeted to the open transcription initiation complex of RNA polymerase can cause specific inhibition of transcription.[63] They ascribed the observed effect to binding of the PNA to the DNA loop in the open complex. This interpretation is consistent with earlier results reported by Sigman *et al.* who showed that oligonucleotides can specifically bind to isolated *Escherichia coli* RNA polymerase open promoter complexes and thereby block transcription initiation.[64–66] Moreover, the results are compatible with earlier findings showing that active transcription catalyzes PNA invasion.[67]

Analogously to the facilitation of PNA helix invasion by biological processes resulting in DNA unwinding or helix opening, invasion complexes once formed may be stabilized by proteins recognizing and binding single-stranded DNA, as illustrated by *in vitro* stabilization of PNA duplex invasion complexes by SSB (single strand binding) protein[68,69] (Figure 9.7).

## 9.7 Activation of Gene Transcription

The PNA helix invasion P-loop is reminiscent of an open transcription initiation complex with RNA polymerase, and indeed it was found that *E. coli* as well as mammalian RNA polymerase recognizes this DNA structure as a starting point for transcription.[70] There is even evidence that transcription in cells may be activated by PNA triplex invasion complexes.[71] Thus PNA targets may function as artificial 'promoters' where cognate PNA oligomers can act after the manner of transcription factors. Following a more traditional approach, conjugates between PNA oligomers and peptide transcriptional activator domains have been shown to activate transcription *in vitro*.[72,73] These approaches have not yet been pursued further.

## 9.8 Gene-targeted Repair

The possibility of being able to correct genes in somatic cells by sequence-targeted repair has great appeal in basic science and not least in drug discovery. Although results of initial reports exploiting DNA–RNA chimeras for the purpose could not be reproduced, it was subsequently found that

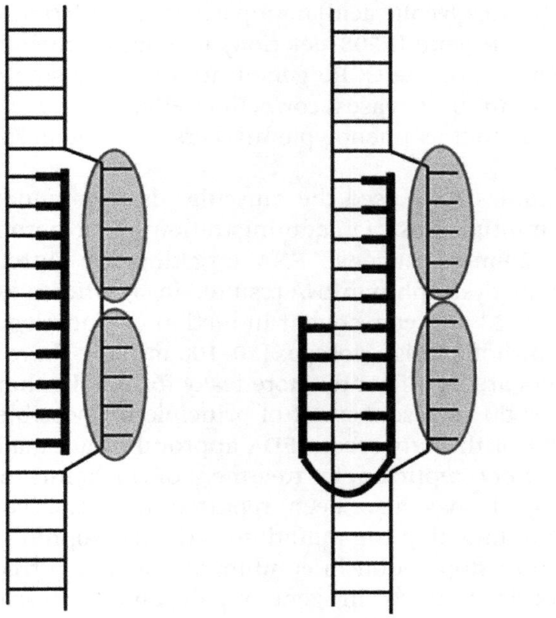

**Figure 9.7** Schematic drawing illustrating SSB (single strand binding) protein stabilization of PNA–dsDNA invasion complexes.

activation of homologous recombination through site-specific introduction of double-strand DNA breaks by a genetically expressed engineered nuclease in the presence of a co-expressed or otherwise delivered donor DNA of the correcting sequence can result in up to 20% gene repair in mammalian cells.[74] It is noteworthy that DNA adducts formed by PNA–psoralen[75] or PNA–nitrogen mustard[76] conjugates would induce double-strand breaks upon action by the cellular repair machinery, and it may well be that PNA conjugates of these types may analogously be employed to catalyze gene-targeted repair.[37,77] Even more encouragingly, co-administration of a triplex forming bis-PNA clamp targeting a site proximal to the gene to be corrected in conjunction with a sequence-correcting donor oligo-deoxyribonucleotide can induce low levels (a few percent) of gene repair of the targeted gene,[78] and, most interestingly, this PNA-directed gene repair concept can be translated to *in vivo* applications in mice.

## 9.9 *In Vivo* Gene Targeting and Repair by PNA Oligomers

Within the past few years two studies have reported biological, clinically relevant effects of PNA-targeted gene correction in mice by homologous recombination gene repair using a co-administered donor deoxy-oligonucleotide. In both cases triplex-forming tail clamp PNAs[79,80] administered in a

PLGA [poly(lactic-glycolic acid] nanoparticle formulation were used to target either the CFTR gene (F508 deletion) in lung epithelium cells (treatment of cystic fibrosis) or the CCR5 gene (introducing a stop codon) in hematopoietic cells. In both cases correction effects were verified by genome sequencing as well as phenotypic markers, and the absolute efficiency was about 0.5%.

Several studies have used the muscular dystrophy mdx mouse model to show that multiple-dose *i.v.* administration of an unmodified or peptide-conjugated 20-mer antisense PNA targeting the intron22–exon23 splice junction of the dystrophin mRNA resulted in significant exon skipping of the mutated exon 23 and consequent induction of (functional) dystrophin synthesis.[81–85] Although high dosages (10–100 mg $kg^{-1}$) have to be used and the muscle fibers are significantly more leaky for uptake than healthy muscles, these studies do represent proof of principle for possible PNA therapy, not least in light of the very recent FDA approval of an analogous morpholino (PMO) oligomer (Eteplirsen) for treatment of Duchenne muscular dystrophy. Unexpectedly, it has also been reported that *i.m.* administration of a PNA oligomer targeting the mutation in the dystrophin gene that gives rise to a premature stop codon in combination with a correcting donor deoxy-oligonucleotide resulted in gene repair and expression of full length dystrophin.[86] This would very surprisingly imply that even a simple duplex-forming PNA may induce site-directed homologous recombination repair, but this conclusion requires additional experimental support.

Effective and tissue-specific availability of PNA upon *in vivo* administration is clearly a major challenge for eventually obtaining potent gene-targeting PNA drugs. Some recent progress has made extensive use of nanoparticles (see above),[79,80] and in an anticancer drug discovery context, PNA conjugation to a pH-driven tumor-targeting peptide (pHLIP) has proved effective in PNA targeting of a regulatory miRNA (miR-155) and a lncRNA.[87,88]

PNA-targeted gene repair has also been investigated in an *ex vivo* therapeutic modality in which stem cells were treated in culture *ex vivo* and subsequently reintroduced into the host animal.[89,90] This approach has the advantage of exploiting clonal expansion of the corrected cells. However, due to the, at least presently, very low efficiency (at best a few percent) of the PNA-enhanced gene correction it is difficult to see how it can compare with the newly developed effective CRISPR technology.[91]

Thus this exciting new area of therapeutic gene-correction modalities deserves close attention, although many and severe challenges in terms of efficacy, specificity, delivery and administration as well as toxicity (not least due to-off target mutation problems) need to be addressed before safe gene-correction drug regimes can be developed.

## 9.10 Therapeutic Prospects

Clearly sequence-specific gene targeting with PNA oligomers at the level of DNA offers great opportunities due to the variety and versatility associated with different PNA–dsDNA binding modes as well as the availability of a wide selection of chemical modifications of PNA oligomers. Nonetheless, much still needs to be learned about DNA targeting in cells and its consequences for gene expression. Also, cellular delivery and not least bioavailability of these large molecules and conjugates *in vivo* is still very much a challenge that demands attention. However, the reward, if successful, will be the possibility of manipulating gene function at the sequence level, not only in animals but also in humans; perhaps even with organ- and cell-specificity. PNAs are now recognized as highly promising agents for the discovery and development of novel tools for biological research as well as specifically designed gene-targeted drugs in molecular medicine. Although CRISPR technology is at the moment 'stealing the stage', and well may offer advantages for *ex vivo* cell correction therapy, systemic therapy will be challenged by much the same obstacles that are so far prohibiting conventional gene therapy from a major breakthrough. By contrast the 'two oligonucleotide'-based modalities represent conventional drug therapies analogous to, albeit more complex than, antisense therapy, which within the past couple of years have yielded major breakthroughs in terms of several newly approved drugs. Thus the coming years should be very exciting.

# References

1. P. E. Nielsen, M. Egholm, R. H. Berg and O. Buchardt, Sequence-selective recognition of DNA by strand displacement with a thymine-substituted polyamide, *Science*, 1991, **254**, 1497–1500.
2. M. Egholm, O. Buchardt, L. Christensen, C. Behrens, S. M. Freier and D. A. Driver, *et al.*, PNA hybridizes to complementary oligonucleotides obeying the Watson-Crick hydrogen-bonding rules, *Nature*, 1993, **365**, 566–568.
3. P. E. Nielsen, M. Egholm and O. Buchardt, Evidence for (PNA)2/DNA triplex structure upon binding of PNA to dsDNA by strand displacement, *J. Mol. Recognit.*, 1994, **7**, 165–170.
4. V. V. Demidov, E. Protozanova, K. I. Izvolsky, C. Price, P. E. Nielsen and M. D. Frank-Kamenetskii, Kinetics and mechanism of the DNA double helix invasion by pseudocomplementary peptide nucleic acids, *Proc. Natl. Acad. Sci. U. S. A.*, 2002, **99**, 5953–5958.
5. P. Wittung, P. E. Nielsen, O. Buchardt, M. Egholm and B. Nordén, DNA-like double helix formed by peptide nucleic acid, *Nature*, 1994, **368**, 561–563.
6. P. Wittung, P. Nielsen and B. Nordén, Extended DNA-recognition repertoire of peptide nucleic acid (PNA): PNA-dsDNA triplex formed with cytosine-rich homopyrimidine PNA, *Biochemistry*, 1997, **36**, 7973–7979.
7. S. A. Kushon, J. P. Jordan, J. L. Seifert, H. Nielsen, P. E. Nielsen and B. A. Armitage, Effect of Secondary Structure on the Thermodynamics and Kinetics of PNA Hybridization to DNA Hairpins, *J. Am. Chem. Soc.*, 2001, **123**, 10805–10813.

8. B. Datta, C. Schmitt and B. A. Armitage, Formation of a PNA2-DNA2 Hybrid Quadruplex, *J. Am. Chem. Soc.*, 2003, **125**, 4111–4118.
9. Y. Krishnan-Ghosh, E. Stephens and S. Balasubramanian, A PNA(4) quadruplex, *J. Am. Chem. Soc.*, 2004, **126**, 5944–5945.
10. B. Petersson, B. B. Nielsen, H. Rasmussen, I. K. Larsen, M. Gajhede and P. E. Nielsen, *et al.*, Crystal Structure of a Partly Self-Complementary Peptide Nucleic Acid (PNA) Oligomer Showing a Duplex-Triplex Network, *J. Am. Chem. Soc.*, 2005, **127**, 1424–1430.
11. H. Stender, PNA FISH: an intelligent stain for rapid diagnosis of infectious diseases, *Expert Rev. Mol. Diagn.*, 2003, **3**, 649–655.
12. G. L. Igloi, Single-nucleotide polymorphism detection using peptide nucleic acids, *Expert Rev. Mol. Diagn.*, 2003, **3**, 17–26.
13. V. L. Marin, S. Roy and B. A. Armitage, Recent advances in the development of peptide nucleic acid as a gene-targeted drug, *Expert Opin. Biol. Ther.*, 2004, **4**, 337–348.
14. O. Brandt and J. D. Hoheisel, Peptide nucleic acids on microarrays and other biosensors 50, *Trends Biotechnol.*, 2004, **22**, 617–622.
15. P. E. Nielsen, Peptide nucleic acid (PNA): a model structure for the primordial genetic material?, *Origins Life Evol. Biospheres*, 1993, **23**, 323–327.
16. A. Singhal and P. E. Nielsen, Cross-catalytic peptide nucleic acid (PNA) replication based on templated ligation, *Org. Biomol. Chem.*, 2014, **12**, 6901–6907.
17. P. E. Nielsen and L. Christensen, Strand displacement binding of a duplex-forming homopurine PNA to a homopyrimidine duplex DNA target, *J. Am. Chem. Soc.*, 1996, **118**, 2287–2288.
18. S. Rapireddy, G. He, S. Roy, B. A. Armitage and D. H. Ly, Strand invasion of mixed-sequence B-DNA by acridine-linked, gamma-peptide nucleic acid (gamma-PNA), *J. Am. Chem. Soc.*, 2007, **129**, 15596–15600.
19. S. Rapireddy, R. Bahal and D. H. Ly, Strand invasion of mixed-sequence, double helical B-DNA by γPNAs containing G-clamp nucleobases under physiological conditions, *Biochemistry*, 2011, **50**, 3913–3918.
20. X. Zhang, T. Ishihara and D. R. Corey, Strand invasion by mixed base PNAs and a PNA-peptide chimera, *Nucleic Acids Res.*, 2000, **28**, 3332–3338.
21. T. Bentin, H. J. Larsen and P. E. Nielsen, Combined triplex/duplex invasion of double-stranded DNA by "tail-clamp" peptide nucleic acid, *Biochemistry*, 2003, **42**, 13987–13995.
22. K. Kaihatsu, R. H. Shah, X. Zhao and D. R. Corey, Extending Recognition by Peptide Nucleic Acids (PNAs): Binding to Duplex DNA and Inhibition of Transcription by Tail-Clamp PNA-Peptide Conjugates, *Biochemistry*, 2003, **42**, 13996–14003.
23. J. Lohse, O. Dahl and P. E. Nielsen, Double duplex invasion by peptide nucleic acid: a general principle for sequence-specific targeting of double-stranded DNA, *Proc. Natl. Acad. Sci. U. S. A.*, 1999, **96**, 11804–11808.
24. P. R. Bohlander, T. Vilaivan and H. A. Wagenknecht, Strand displacement and duplex invasion into double-stranded DNA by pyrrolidinyl peptide nucleic acids, *Org. Biomol. Chem.*, 2015, **13**, 9223–9230.

25. M. E. Hansen, T. Bentin and P. E. Nielsen, High-affinity triplex targeting of double stranded DNA using chemically modified peptide nucleic acid oligomers, *Nucleic Acids Res.*, 2009, **37**, 4498–4507.
26. I. G. Panyutin, M. I. Onyshchenko, E. A. Englund, D. H. Appella and R. D. Neumann, . Targeting DNA G-quadruplex structures with peptide nucleic acids, *Curr. Pharm. Des.*, 2012, **18**, 1984–1991.
27. H. Bergquist, C. S. Rocha, R. Alvarez-Asencio, C. H. Nguyen, M. W. Rutland, C. I. Smith, L. Good, P. E. Nielsen and R. Zain, Disruption of Higher Order DNA Structures in Friedreich's Ataxia (GAA)n Repeats by PNA or LNA Targeting, *PLoS One*, 2016, e0165788.
28. H. Kuhn, V. V. Demidov, M. D. Frank-Kamenetskii and P. E. Nielsen, Kinetic sequence discrimination of cationic bis-PNAs upon targeting of double-stranded DNA, *Nucleic Acids Res.*, 1998, **26**, 582–587.
29. H. Kuhn, V. V. Demidov, P. E. Nielsen and M. D. Frank-Kamenetskii, An Experimental Study of Mechanism and Specificity of Peptide Nucleic Acid (PNA) Binding to Duplex DNA, *J. Mol. Biol.*, 1999, **286**, 1337–1345.
30. M. Egholm, L. Christensen, K. L. Dueholm, O. Buchardt, J. Coull and P. E. Nielsen, Efficient pH-independent sequence-specific DNA binding by pseudoisocytosine-containing bis-PNA, *Nucleic Acids Res.*, 1995, **23**, 217–222.
31. M. C. Griffith, L. M. Risen, M. J. Greig, E. A. Lesnik, K. G. Sprankle and R. H. Griffey, *et al.*, Single and Bis Peptide Nucleic Acids as Triplexing Agents: Binding and Stoichiometry, *J. Am. Chem. Soc.*, 1995, **117**, 831–832.
32. T. Bentin and P. E. Nielsen, Superior Duplex DNA Strand Invasion by Acridine Conjugated Peptide Nucleic Acids, *J. Am. Chem. Soc.*, 2003, **125**, 6378–6379.
33. P. E. Nielsen, K. Frederiksen and C. Behrens, Extended Target Sequence Specificity of PNA-Minor-Groove Binder Conjugates, *ChemBioChem*, 2005, **6**, 66–68.
34. P. B. Dervan and B. S. Edelson, Recognition of the DNA minor groove by pyrrole-imidazole polyamides, *Curr. Opin. Struct. Biol.*, 2003, **13**, 284–299.
35. W. Kameshima, T. Ishizuka, M. Minoshima, M. Yamamoto, H. Sugiyama, Y. Xu, and M. Komiyama. 2013. Conjugation of Peptide Nucleic Acid with a Pyrrole/Imidazole Polyamide to Specifically Recognize and Cleave DNA. *Angew. Chem. Int. Ed. Engl.* **52**, 13681–13684.
36. Z. V. Zhilina, A. J. Ziemba, P. E. Nielsen and S. W. Ebbinghaus, PNA-Nitrogen Mustard Conjugates Are Effective Suppressors of HER-2/neu and Biological Tools for Recognition of PNA/DNA Interactions, *Bioconjugate Chem.*, 2006, **17**, 214–222.
37. K. H. Kim, P. E. Nielsen and P. M. Glazer, Site-Specific Gene Modification by PNAs Conjugated to Psoralen, *Biochemistry*, 2006, **45**, 314–323.
38. J. H. Kim, K. H. Kim, N. E. Møllegaard, P. E. Nielsen and H. S. Koo, Helical periodicity of $(PNA)_2(DNA)$ triplexes in strand displacement complexes, *Nucleic Acids Res.*, 1999, **27**, 2842–2847.

39. E. Protozanova, V. V. Demidov, V. Soldatenkov, S. Chasovskikh and M. D. Frank-Kamenetskii, Tailoring the activity of restriction endonuclease PleI by PNA-induced DNA looping, *EMBO Rep.*, 2002, **3**, 956–961.
40. U. Koppelhus and P. E. Nielsen, Cellular delivery of peptide nucleic acid (PNA), *Adv. Drug Delivery Rev.*, 2003, **55**, 267–280.
41. S. E. Hamilton, C. G. Simmons, I. S. Kathiriya and D. R. Corey, Cellular delivery of peptide nucleic acids and inhibition of human telomerase, *Chem. Biol.*, 1999, **6**, 343–351.
42. T. Shiraishi and P. E. Nielsen, Nanomolar cellular antisense activity of peptide nucleic acid (PNA) cholic acid ("umbrella") and cholesterol conjugates delivered by cationic lipids, *Bioconjugate Chem.*, 2012, **23**, 196–202.
43. T. Shiraishi and P. E. Nielsen, Down-regulation of MDM2 and activation of p53 in human cancer cells by antisense 9-aminoacridine-PNA (peptide nucleic acid) conjugates, *Nucleic Acids Res.*, 2004, **32**, 4893–4902.
44. T. Shiraishi, N. Nadia Bendifallah and P. E. Nielsen, Cellular Delivery of Polyheteroaromate-Peptide Nucleic Acid (PNA) Conjugates Mediated by Cationic Lipids, *Bioconjugate Chem.*, 2006, **17**, 189–194.
45. T. Shiraishi, R. Hamzavi and P. E. Nielsen, Subnanomolar antisense activity of phosphonate-peptide nucleic acid (PNA) conjugates delivered by cationic lipids to HeLa cells, *Nucleic Acids Res.*, 2008, **36**, 4424–4432.
46. K. Kilk and U. Langel, Cellular delivery of peptide nucleic acid by cell-penetrating peptides, *Methods Mol. Biol.*, 2005, **298**, 131–141.
47. U. Koppelhus, S. K. Awasthi, V. Zachar, H. U. Holst, P. Ebbesen and P. E. Nielsen, Cell-dependent differential cellular uptake of PNA, peptides, and PNA-peptide conjugates, *Antisense Nucleic Acid Drug Dev.*, 2002, **12**, 51–63.
48. M. Fotin-Mleczek, R. Fischer and R. Brock, Endocytosis and cationic cell-penetrating peptides–a merger of concepts and methods, *Curr. Pharm. Des.*, 2005, **11**, 3613–3628.
49. J. S. Wadia, R. V. Stan and S. F. Dowdy, Transducible TAT-HA fusogenic peptide enhances escape of TAT-fusion proteins after lipid raft macropinocytosis, *Nat. Med.*, 2004, **10**, 310–315.
50. T. Shiraishi, S. Pankratova and P. E. Nielsen, Calcium ions effectively enhance the effect of antisense Peptide nucleic acids conjugated to cationic tat and oligoarginine peptides, *Chem. Biol.*, 2005, **12**, 923–929.
51. T. Shiraishi and P. E. Nielsen, Photochemically enhanced cellular delivery of cell penetrating peptide-PNA conjugates, *FEBS Lett.*, 2006, **580**, 1451–1456.
52. R. Hamzavi, F. Dolle, B. Tavitian, O. Dahl and P. E. Nielsen, Modulation of the Pharmacokinetic Properties of PNA: Preparation of Galactosyl, Mannosyl, Fucosyl, N-Acetylgalactosaminyl, and N-Acetylglucosaminyl Derivatives of Aminoethylglycine Peptide Nucleic Acid Monomers and Their Incorporation into PNA Oligomers, *Bioconjugate Chem.*, 2003, **14**, 941–954.
53. L. C. Boffa, G. Cutrona, M. Cilli, M. R. Mariani, S. Matis, M. Pastorino, G. Damonte, E. Millo, S. Roncella and M. Ferrarini, Therapeutically

Promising PNA Complementary to a Regulatory Sequence for c-myc: Pharmacokinetics in an Animal Model of Human Burkitt's Lymphoma, *Oligonucleotides*, 2005, **15**, 85–93.
54. P. Sazani, F. Gemignani, S.-H. Kang, M. A. Maier, M. Manoharan and M. Persmark, *et al.*, Systemically delivered antisense oligomers upregulate gene expression in mouse tissues, *Nat. Biotechnol.*, 2002, **20**, 1228–1233.
55. A. F. Faruqi, M. Egholm and P. M. Glazer, Peptide nucleic acid-targeted mutagenesis of a chromosomal gene in mouse cells, *Proc. Natl. Acad. Sci. U. S. A.*, 1998, **95**, 1398–1403.
56. G. Cutrona, E. M. Carpaneto, M. Ulivi, S. Roncella, O. Landt and M. Ferrarini, *et al.*, Effects in live cells of a c-myc anti-gene PNA linked to a nuclear localization signal, *Nat. Biotechnol.*, 2000, **18**, 300–303.
57. R. Tonelli, S. Purgato, C. Camerin, R. Fronza, F. Bologna and S. Alboresi, *et al.*, Anti-gene peptide nucleic acid specifically inhibits MYCN expression in human neuroblastoma cells leading to cell growth inhibition and apoptosis, *Mol. Cancer Ther.*, 2005, **4**, 779–786.
58. I. V. Smolina, V. V. Demidov, V. A. Soldatenkov and S. G. Chasovskikh, Frank-Kamenetskii MD: End invasion of peptide nucleic acids (PNAs) with mixed-base composition into linear DNA duplexes, *Nucleic Acids Res.*, 2005, **33**, e146.
59. T. Bentin and P. E. Nielsen, Enhanced Peptide Nucleic Acid Binding to Supercoiled DNA: Possible Implications for DNA "Breathing" Dynamics, *Biochemistry*, 1996, **35**, 8863–8869.
60. N. J. Peffer, J. C. Hanvey, J. E. Bisi, S. A. Thomson, C. F. Hassman and S. A. Noble, *et al.*, Strand-invasion of duplex DNA by peptide nucleic acid oligomers, *Proc. Natl. Acad. Sci. U. S. A.*, 1993, **90**, 10648–10652.
61. P. E. Nielsen, M. Egholm and O. Buchardt, Sequence-specific transcription arrest by peptide nucleic acid bound to the DNA template strand, *Gene*, 1994, **149**, 139–145.
62. E. Guffanti, R. Corradini, S. Ottonello and G. Dieci, Functional dissection of RNA polymerase III termination using a peptide nucleic acid as a transcriptional roadblock, *J. Biol. Chem.*, 2004, **279**, 20708–20716.
63. B. A. Janowski, K. Kaihatsu, K. E. Huffman, J. C. Schwartz, R. Ram and D. Hardy, *et al.*, Inhibiting transcription of chromosomal DNA with antigene peptide nucleic acids, *Nat. Chem. Biol.*, 2005, **1**, 210–215.
64. L. Milne, Y. Xu, D. M. Perrin and D. S. Sigman, An approach to gene-specific transcription inhibition using oligonucleotides complementary to the template strand of the open complex, *Proc. Natl. Acad. Sci. U. S. A.*, 2000, **97**, 3136–3141.
65. L. Milne, D. M. Perrin and D. S. Sigman, Oligoribonucleotide-based gene-specific transcription inhibitors that target the open complex, *Methods*, 2001, **23**, 160–168.
66. J. T. Hwang, F. E. Baltasar, D. L. Cole, D. S. Sigman, C. H. Chen and M. M. Greenberg, Transcription inhibition using modified pentanucleotides, *Bioorg. Med. Chem.*, 2003, **11**, 2321–2328.

67. H. J. Larsen and P. E. Nielsen, Transcription-mediated binding of peptide nucleic acid (PNA) to double-stranded DNA: sequence-specific suicide transcription, *Nucleic Acids Res.*, 1996, **24**, 458–463.
68. T. Ishizuka, K. Otani, J. Sumaoka and M. Komiyama, Strand invasion of conventional PNA to arbitrary sequence in DNA assisted by single-stranded DNA binding protein, *Chem. Commun.*, 2009, 1225–1227.
69. T. Ishizuka, T. Tedeschi, R. Corradini, M. Komiyama, S. Sforza and R. Marchelli, SSB-Assisted Duplex Invasion of Preorganized PNA into Double-Stranded DNA, *ChemBioChem*, 2009, **10**, 2607–2612.
70. N. E. Møllegaard, O. Buchardt, M. Egholm and P. E. Nielsen, Peptide nucleic acid-DNA strand displacement loops as artificial transcription promoters, *Proc. Natl. Acad. Sci. U. S. A.*, 1994, **91**, 3892–3895.
71. G. Wang, X. Xu, B. Pace, D. A. Dean, P. M. Glazer and P. Chan, *et al.*, Peptide nucleic acid (PNA) binding-mediated induction of human γ-globin gene expression, *Nucleic Acids Res.*, 1999, **27**, 2806–2813.
72. B. Liu, Y. Han, D. R. Corey and T. Kodadek, Toward Synthetic Transcription Activators: Recruitment of Transcription Factors to DNA by a PNA-Peptide Chimera, *J. Am. Chem. Soc.*, 2002, **124**, 1838–1839.
73. B. Liu, Y. Han, A. Ferdous, D. R. Corey and T. Kodadek, Transcription Activation by a PNA-Peptide Chimera in a Mammalian Cell Extract, *Chem. Biol.*, 2003, **10**, 909–916.
74. F. D. Urnov, J. C. Miller, Y. L. Lee, C. M. Beausejour, J. M. Rock, S. Augustus, A. C. Jamieson, M. H. Porteus, P. D. Gregory and M. C. Holmes, Highly efficient endogenous human gene correction using designed zinc-finger nucleases, *Nature*, 2005, **435**, 646–651.
75. K. H. Kim, P. E. Nielsen and P. M. Glazer, Site-Specific Gene Modification by PNAs Conjugated to Psoralen, *Biochemistry*, 2006, **45**, 314–323.
76. Z. V. Zhilina, A. J. Ziemba, P. E. Nielsen and S. W. Ebbinghaus, PNA-Nitrogen Mustard Conjugates Are Effective Suppressors of HER-2/neu and Biological Tools for Recognition of PNA/DNA Interactions, *Bioconjugate Chem.*, 2006, **17**, 214–222.
77. H. Birkedal and P. E. Nielsen, Targeted gene correction using psoralen, chlorambucil and camptothecin conjugates of triplex forming peptide nucleic acid (PNA), *Artif. DNA PNA XNA*, 2011, **2**, 23–32.
78. N. A. McNeer, J. Y. Chin, E. B. Schleifman, R. J. Fields, P. M. Glazer and W. M. Saltzman, Nanoparticles Deliver Triplex-forming PNAs for Site-specific Genomic Recombination in CD34(+) Human Hematopoietic Progenitors, *Mol. Ther.*, 2011, **19**, 172–180.
79. N. A. McNeer, E. B. Schleifman, A. Cuthbert, M. Brehm, A. Jackson, C. Cheng, K. Anandalingam, P. Kumar, L. D. Shultz, D. L. Greiner, S. W. Mark and P. M. Glazer, Systemic delivery of triplex-forming PNA and donor DNA by nanoparticles mediates site-specific genome editing of human hematopoietic cells in vivo, *Gene Ther.*, 2012, **20**, 658–669.
80. N. A. McNeer, K. Anandalingam, R. J. Fields, C. Caputo, S. Kopic, A. Gupta, E. Quijano, L. Polikoff, Y. Kong, R. Bahal, J. P. Geibel,

P. M. Glazer, W. M. Saltzman and M. E. Egan, Nanoparticles that deliver triplex-forming peptide nucleic acid molecules correct F508del CFTR in airway epithelium, *Nat. Commun.*, 2015, **6**, 6952.
81. H. Yin, Q. Lu and M. Wood, Effective Exon Skipping and Restoration of Dystrophin Expression by Peptide Nucleic Acid Antisense Oligonucleotides in mdx Mice, *Mol. Ther.*, 2008, **16**, 38–45.
82. H. Yin, C. Betts, A. F. Saleh, G. D. Ivanova, H. Lee, Y. Seow, D. Kim, M. J. Gait and M. J. A. Wood, Optimization of Peptide Nucleic Acid Antisense Oligonucleotides for Local and Systemic Dystrophin Splice Correction in the mdx Mouse, *Mol. Ther.*, 2010, **18**, 819–827.
83. X. Gao, X. Shen, X. Dong, N. Ran, G. Han, L. Cao, B. Gu and H. Yin, Peptide Nucleic Acid Promotes Systemic Dystrophin Expression and Functional Rescue in Dystrophin-deficient mdx Mice, *Mol. Ther. – Nucleic Acids*, 2015, **4**, e255.
84. X. Gao, X. Shen, X. Dong, N. Ran, G. Han, L. Cao, B. Gu and H. Yin, Peptide Nucleic Acid Promotes Systemic Dystrophin Expression and Functional Rescue in Dystrophin-deficient mdx Mice, *Mol. Ther. – Nucleic Acids*, 2015, **4**, e255.
85. C. Brolin, T. Shiraishi, P. Hojman, T. O. Krag, P. E. Nielsen and J. Gehl, Electroporation Enhanced Effect of Dystrophin Splice Switching PNA Oligomers in Normal and Dystrophic Muscle, *Mol. Ther. – Nucleic Acids*, 2015, **4**, e267.
86. R. Kayali, F. Bury, M. Ballard and C. Bertoni, Site-directed gene repair of the dystrophin gene mediated by PNA-ssODNs, *Hum. Gene Ther.*, 2010, **19**, 3266–3281.
87. C. J. Cheng, R. Bahal, I. A. Babar, Z. Pincus, F. Barrera, C. Liu, A. Svoronos, D. T. Braddock, P. M. Glazer, D. M. Engelman, W. M. Saltzman and F. J. Slack, MicroRNA silencing for cancer therapy targeted to the tumour microenvironment, *Nature*, 2014, **518**, 107–110.
88. A. R. Özeş, Y. Wang, X. Zong, F. Fang, J. Pilrose and K. P. Nephew, Therapeutic targeting using tumor specific peptides inhibits long non-coding RNA HOTAIR activity in ovarian and breast cancer, *Sci. Rep.*, 2017, **7**, 894.
89. E. B. Schleifman, N. A. McNeer, A. Jackson, J. Yamtich, M. A. Brehm, L. D. Shultz, D. L. Greiner, P. Kumar, W. M. Saltzman and P. M. Glazer, Site-specific Genome Editing in PBMCs With PLGA Nanoparticle-delivered PNAs Confers HIV-1 Resistance in Humanized Mice, *Mol. Ther. – Nucleic Acids*, 2013, **2**, e135.
90. F. Nik-Ahd and C. Bertoni, Ex vivo gene editing of the dystrophin gene in muscle stem cells using peptide nucleic acid single stranded oligodeoxynucleotides (PNA-SsODNs) induces stable expression of dystrophin in a mouse model for Duchenne muscular dystrophy, *Stem Cells*, 2014, **32**, 1817–1830.
91. T. I. Cornu, C. Mussolino and T. Cathomen, Refining strategies to translate genome editing to the clinic, *Nat. Med.*, 2017, **23**, 415–423.

CHAPTER 10

# *Sequence-selective Interactions of Actinomycin D with DNA: Discovery of a Thermodynamic Switch*

JASON S. HUDSON,[a] AARON LUCIUS,[a] MICHAEL J. WARING[b] AND DAVID E. GRAVES*[a]

[a] Department of Chemistry, University of Alabama at Birmingham, 901 14th Street, Birmingham, AL 35294, USA; [b] Department of Biochemistry, Cambridge University, Cambridge CB2 1GA, England, UK
*Email: dgraves1978@gmail.com

## 10.1 Summary

The interactions of actinomycin D with nucleic acids have served as a paradigm for DNA base sequence selectivity for the past half century through its demonstrated preference for binding to the d(GpC) sequence.[1–4] In addition to the d(GpC) intercalation step, our laboratory has demonstrated that the 5′-flanking base adjacent to the d(GpC) intercalation site plays a critical role in directing the thermodynamic mechanism(s) associated with complex formation. Isothermal titration calorimetry (ITC) and UV–visible spectroscopy were used to discern the thermodynamic and kinetic properties associated with drug–DNA complex formation. These studies revealed significant thermodynamic and kinetic changes resulting from single base modifications at the 5′-flanking base adjacent to the d(GpC) intercalation

Chemical Biology No. 7
DNA-targeting Molecules as Therapeutic Agents
Edited by Michael J. Waring

Published by the Royal Society of Chemistry, www.rsc.org

step.[5] Historically, the interaction of actinomycin with native DNA has been characterized as an entropy-driven process with observed binding enthalpy approximately 0 to $-1$ kcal mol$^{-1}$.[6] However, the interaction of actinomycin D with a deoxyoligonucleotide duplex containing the -TGCA- intercalation site is characterized by a high binding affinity and large favourable binding entropy ($-5.9$ kcal mol$^{-1}$). However, substitution of the flanking 5′-T with 5′-C results in complex formation that is characterized by a smaller binding enthalpy ($-3.4$ kcal mol$^{-1}$) revealing that the 5′ flanking base adjacent to the GpC intercalation site can significantly modulate the binding enthalpy.[5] Using a photoreactive analogue of actinomycin D, (7-azido-actinomycin D),[7] we were able to demonstrate that the binding of actinomycin D to DNA was more promiscuous than previously reported with additional strong intercalation sites occurring at "atypical" base sequences including the d(GpG) and d(GpGpG) steps.[8] Utilizing biophysical thermodynamic and kinetic experimental techniques, the driving forces behind the unique thermodynamic properties associated with complex formation have been probed. Our results reveal that the DNA base sequence both at the site of intercalation as well as the bases surrounding the intercalation site effectively acts as a "switch" that governs the thermodynamic mechanism for complex formation. Changing the sequence within the intercalation site by addition of one base, 5′-TGGT-3′ to 5′-TGGGT-3′ results in a 6 kcal mol$^{-1}$ difference in binding enthalpy.[9] Both the 5′-TGGT-3′ and 5′-TGGGT-3′ sequences exhibit similar Gibbs free energies ($\Delta G$); hence, these results suggest large enthalpy/entropy compensation in the thermodynamic binding mechanisms of actinomycin D complex formation for these two sequences. Using circular dichroism spectrometry, the mode of binding of actinomycin D to both sequences was confirmed to be intercalation. To examine the possibility of a kinetic correlation with the binding thermodynamics, SDS-induced dissociation kinetics and stopped-flow association kinetic experiments were conducted. The association kinetic profiles revealed significant differences, suggesting a complex association process which may be evidence for a higher order of redistribution (*i.e.* shuffling) between potential binding sites in the 5′-TGGT-3′ sequence compared to the 5′-TGGGT-3′ sequence. Our findings demonstrate a unique thermodynamic mechanism for binding of actinomycin D to the 5′-TGGT-3′ and 5′-TGGGT-3′ motif in double-stranded DNA that has both thermodynamic and kinetic implications. Hence, subtle changes to the DNA intercalation sequence or bases adjacent to the intercalation step result in significant changes to the thermodynamic binding mechanism for the ligand.

## 10.2 Introduction

The interactions of actinomycin D with nucleic acids have garnered considerable attention from the biophysical community over more than half a century. During the past 60 years, this anticancer antibiotic has served as a paradigm for DNA base-sequence-selective binding agents that has led to considerable insights into the structural and bioenergetic properties associated with ligand–DNA interactions.[1–4] As early as 1965, Gellert and coworkers reported that actinomycin D binds preferentially to the 5′-GpC-3′ sequence.[6] Subsequently, high resolution NMR and X-ray diffraction studies provided details of the structural nature of the actinomycin D–DNA complex.[10–16] Key features thought to drive this selectivity for the d(GpC) sequence include intercalation of the phenoxazone ring between the 5′-G and 3′-C resulting in stabilization of the complex through stacking interactions together with formation of specific hydrogen bonds between the threonine carbonyl residues of the drug and the 2-amino group of guanine as well as the threonine amide-NH and guanine-N3 within the intercalation site. These hydrogen bonds are thought to be stabilized due to their shielding from water by the two cyclic pentapeptide chains of actinomycin D (structure shown in Figure 10.1) that reside in the minor groove of the DNA helix above and below the intercalation site.[17–19]

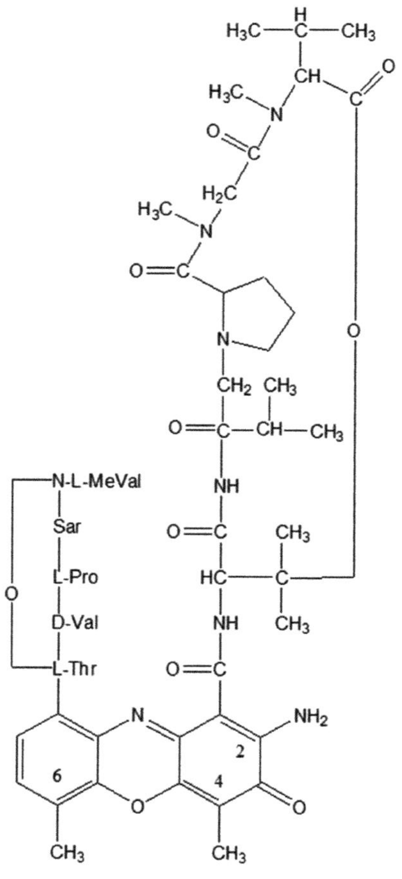


**Figure 10.1** Structure of actinomycin D showing the numbering scheme of the chromophore.

The ability of actinomycin D to interact with DNA sequences other than the classic 5′-GpC-3′ sequence gained interest with observations reported by Rill and coworkers in 1994 using a photoaffinity analog of actinomycin D to footprint binding sites at very low bound-drug densities.[8] Their studies revealed novel atypical high-affinity actinomycin D binding sites consisting of the 5′-T(G)$_n$T-3′ sequence motif where n equals 2–4 guanines.[20] Subsequently, Graves and coworkers provided a more detailed biophysical study examining the binding of actinomycin D to a series of deoxyoligonucleotide (11-mer) duplexes containing a central -T(G)$_n$T- binding site and reported that the interactions of actinomycin with the -TGGGT- and -TGGT- sequences were similar to those with the duplex containing the traditional -XGCX- binding site.[9] Further studies by this group revealed that the base sequence of the actinomycin D binding site along with the bases that flank the binding site were responsible for directing the thermodynamic mechanism(s) associated with complex formation.[5,9]

The unusual and highly complex dissociation kinetics of actinomycin D from heterogeneous DNAs have been the subject of intense examination since initial reports in 1971 by Crothers and later by Bittman and Blau in 1975, then Brown and Shafer in 1987.[21–24] The role that the DNA sequence plays in modulating the kinetic properties of actinomycin binding is well documented.[24,25] These unique dissociative properties led Waring and Fox propose the ‘shuffling hypothesis’ in 1984, wherein actinomycin D would initially interact with DNA bases in a relatively nonspecific manner and subsequently shuffle along the DNA helix to locate preferred base sequences for the formation of more stable ligand–DNA complexes.[26,27] Utilizing 7-azidoactinomycin D (a photoaffinity analog of actinomycin D), Waring and coworkers were able to demonstrate by direct analysis that the ‘shuffling model’ was indeed correct.[28] Substantial change in binding densities of actinomycin D between low- and high-affinity sites as a function of time were observed and confirmed the lateral diffusion (or shuffling) of actinomycin D along the DNA lattice, relying on the availability and formation of specific hydrogen bond contacts, van der Waals interactions and hydrophobic contacts between the ligand and DNA, all of which translates into the thermodynamic profile of sequence recognition.[28,29] Arising from these studies were observations of promiscuity in the ‘sequence-specific’ nature of actinomycin–DNA interactions and, of particular interest, the interaction of this antibiotic with the 5′-T(G)nT-3′ sequence motif where *n* equals 2 or 3 guanines.

Recently, we have used more sensitive biophysical methods, including isothermal titration calorimetry (ITC), to measure directly the binding enthalpy of actinomycin D to deoxyoligonucleotide duplexes containing the central core sequences 5′-TGGT-3′ and 5′-TGGT-3′ to probe the nature of their markedly different actinomycin D binding enthalpies. Confirmation of these results was followed by a more detailed biophysical study of the mechanistic possibilities that could give rise to a thermodynamic switch operative between the two deoxyoligonucleotide duplexes.

The primary focus of the work described here is to provide mechanistic insights into the energetics and kinetic linkages that influence the binding of actinomycin D to 13-mer duplex deoxyoligonucleotides, each containing a single atypical 5′-(T(G)$_n$T-3′ actinomycin binding site. These studies demonstrate that actinomycin D has high affinity for sites other than the classic dGpC sequences and the thermodynamic and kinetic mechanism(s) that drive complex formation are highly dependent on the base sequence both at and surrounding the intercalation site.

## 10.3 DNA Sequence Dictates Binding Energetics

### 10.3.1 The Energetic Mechanism Is Sequence-dependent

Initial experiments were carried out to determine the enthalpy of binding of actinomycin D to 13-mer oligodeoxynucleotide duplexes each containing a single 5′-TGGT-3′ or 5′-TGGGT-3′ site using the excess site ITC method described by Chaires and coworkers.[30–32] This method assumes that at each injection all ligand is bound due to the excess DNA concentration, and the enthalpy of binding can be calculated as the sum of the heat of binding minus the heat of dilution. Representative ITC traces for the ITC experiments are shown in Figure 10.2 and the resulting energetic parameters are provided in Table 10.1.

Injection of actinomycin D into the 13-mer duplex DNA containing a single 5′-TGGT-3′ ligand binding site results in the observation of an endothermic heat for complex formation and reveals an unfavorable binding enthalpy of approximately $+2.5$ kcal mol$^{-1}$. In contrast, titration of actinomycin D into the 13-mer duplex DNA containing a single 5′-TGGGT-3′ ligand binding site results in an exothermic heat of formation and favorable binding enthalpy of $-3.4$ kcal mol$^{-1}$.

Hence this calorimetric study demonstrated that a single base change within the actinomycin D binding site (5′-TGGT-3′ to 5′-TGGGT-3′) resulted in a change in both the magnitude and sign of the binding enthalpy by approximately 6 kcal mol$^{-1}$ and the thermodynamic binding mechanism changes from entropy driven to an enthalpy driven process.

The free energies ($\Delta G^\circ$) for the binding of actinomycin D to the 5′-TGGT-3′ and 5′-TGGGT-3′ sequences are relatively similar ($-7.6 \pm 0.2$ kcal mol$^{-1}$ and $-8.0 \pm 0.1$ kcal mol$^{-1}$, respectively).[8] Consequently these data reveal that the formation of the actinomycin D–DNA complex must be highly dependent on the composition of the base sequences that constitute the binding site and that subtle changes to those sequences may serve as a ‘thermodynamic switch’ (*i.e.* enthalpy-driven *versus* entropy-driven) in directing the thermodynamic mechanism that drives complex formation.[5,8,33–36]

### 10.3.2 The Mode of Binding is Intercalation

Of fundamental importance was to determine whether the interaction of actinomycin D with the atypical 5′-T(G)$_n$T-3′ was intercalative in nature.

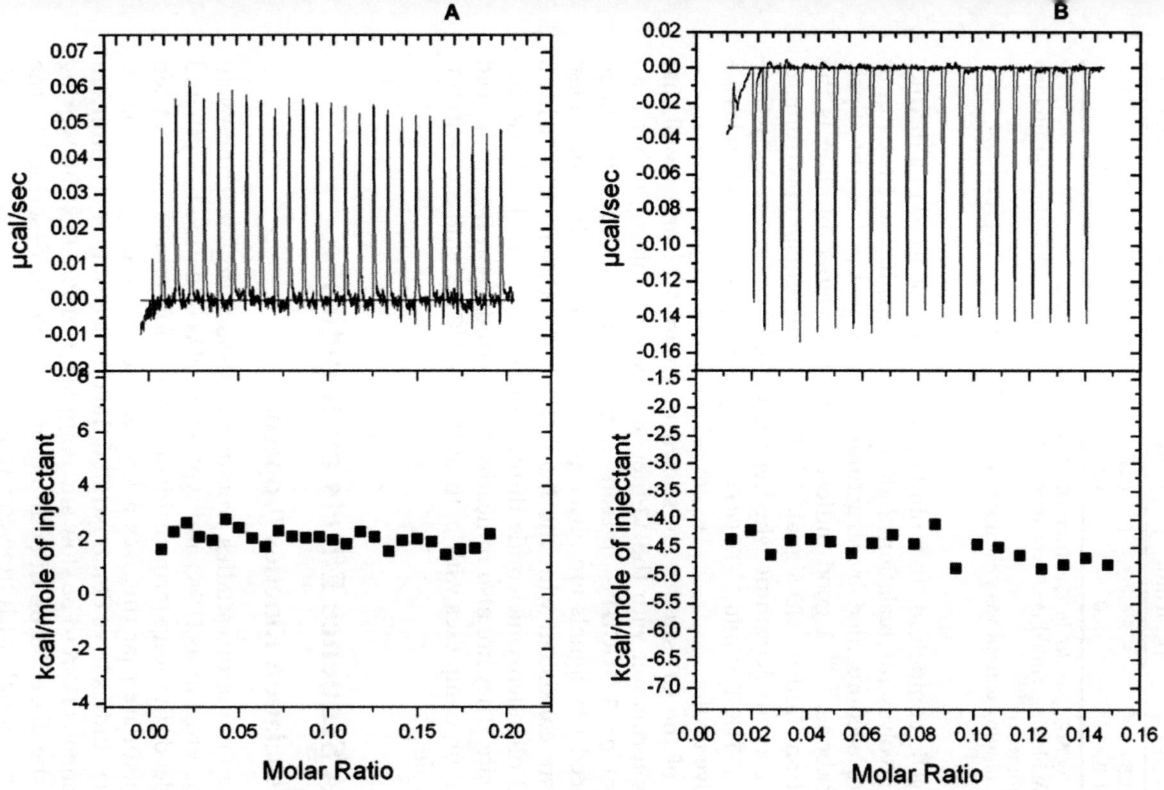


**Figure 10.2** Isothermal titration calorimetry (ITC) of the interaction of actinomycin D with (A) the 13-mer DNA of base sequence 5′-ACT **ATG GT**T ATC A-3′/5′-T GAT AAC CAT AGT-3′ (referred to as 5′-TGGT-3′) or (B) the 13-mer DNA of base sequence, 5′-ACT **ATG GGT** ATC A-3′/5′-T GAT ACC CAT AGT-3′ (referred to as 5′-TGGGT-3′) using the 'Model-Free' ITC protocol of Chaires and coworkers.[30–32] The model-free method allows each injection to evolve the maximum enthalpy of binding due the large excess of binding sites present for each injection. The sample cell was filled to capacity, ~1.8 ml, with duplex oligodeoxynucleotide (100 μM) at 25 °C. Actinomycin D (100 μM) was then injected in 20 aliquots of 10 μl each into the DNA. The enthalpy was derived from the area under each peak and averaged over the 20 injections. The net enthalpy of binding was determined by subtracting the heat of dilution of actinomycin D (~ −0.5 kcal mol$^{-1}$) from each value.

**Table 10.1** Thermodynamic parameters for the interaction of actinomycin D with 13-mer duplexes containing the single binding sites 5′-TGGT-3′ and 5′-TGGGT-3′.

| Sequence | $K_{int}$ [a] ($M^{-1}$) | $\Delta G^{o}_{obs}$ [a] (kcal mol$^{-1}$) | $n$ (duplex)[a] | $\Delta H^{o}_{obs}$ [b] (kcal mol$^{-1}$) | $\Delta S^{o}_{obs}$ [c] (cal mol$^{-1}$) |
|---|---|---|---|---|---|
| 5′-TGGT-3′ | $2.5\times10^{5}$ | −7.6 | 0.9 | +2.5 | 33 |
| 5′-TGGGT-3′ | $4.0\times10^{5}$ | −8.0 | 0.9 | −3.5 | 14 |

[a] $\Delta G^{o}_{obs}$ calculated by $\Delta G^{o}_{obs} = -RT \ln K_{obs}$ from DNA binding isotherms.
[b] Changes in DNA binding enthalpies were obtained from 'model free' isothermal titration calorimetry experiments.[34,35]
[c] Changes in DNA binding entropy were calculated from $\Delta S^{o}_{obs} = (\Delta H^{o}_{obs} - \Delta G^{o}_{obs})/T$.

Small heterocyclic compounds that bind to DNA typically do so in a distinct manner that involves intercalation and/or binding within the DNA minor groove and are accompanied by structural changes to the DNA base-pair stack and backbone.[37,38] A good indicator for intercalative binding is the observation of an induced CD signal.[39–41] Circular dichroism spectrometry was therefore used to determine whether the complexes formed by actinomycin D with 5′-TGGT-3′ and 5′-TGGGT-3′ duplexes appeared to be classically intercalative or not.

The results of the CD studies are shown in Figure 10.3 and reveal that upon complex formation with the 13-mer duplexes containing either a 5′-TGGT-3′ (blue) or 5′-TGGGT-3′ (green) actinomycin binding site, large negative induced CD signals are observed at 440 nm and 380 nm. These observations are consistent with the interaction of actinomycin D with a similarly sized oligodeoxynucleotide duplex containing a single 5′-TGCA-3′ drug binding site. They are also indicative that actinomycin D does indeed form intercalation complexes with both of the 5′-TGGT-3′ and 5′-TGGGT-3′ containing duplexes.

## 10.4 DNA Sequence Effects on Kinetics

### 10.4.1 Dissociation Kinetics Properties

There have been numerous studies probing the dissociation of actinomycin D from both heterogenous DNAs and synthetic DNAs of defined length and sequence. Indeed, the long-term residency of actinomycin D on DNA led Crothers to postulate that this was a key feature of the drug's biological activity.[2] These studies have demonstrated the dissociation of actinomycin from heterogeneous DNA to be slow and complex. In 1984, Fox and Waring reported that the dissociation of actinomycin D could be described by the sum of two exponentials with time constants of 86 and 1800 s. In contrast, the dissociation of actinomycin D from poly(dGdC) could be described by a single exponential with a time constant of 1100 s.[26] The $k_{off}$ for the dissociation of actinomycin D from our 13-mer duplexes, each containing

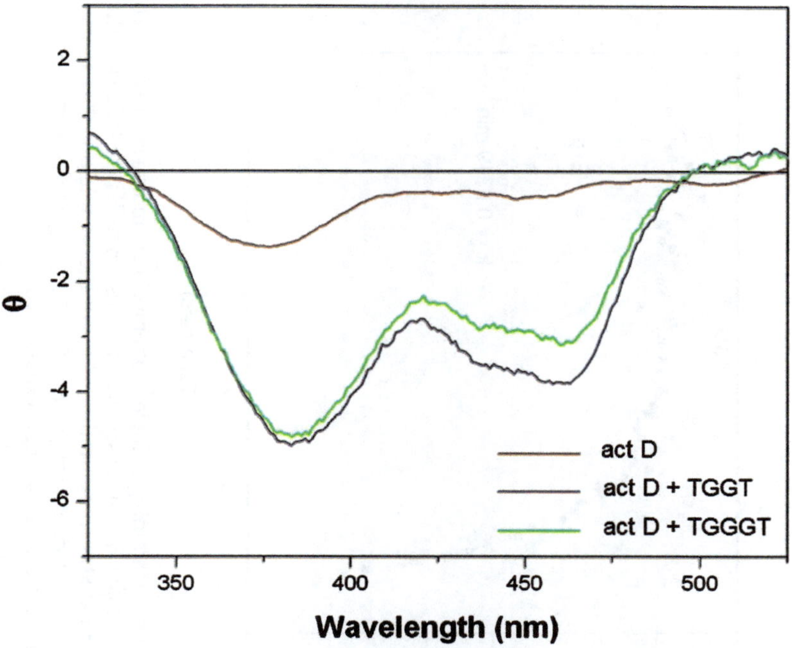


**Figure 10.3** Induced CD spectra of actinomycin D complexed with duplex DNAs containing a single actinomycin D binding site; 5′-TGGT-3′ (black) or 5′-TGGGT-3′ (green). The spectrum for actinomycin D with no DNA is shown in red. All experiments were conducted at 25 °C in 1 cm pathlength rectangular cells. The data were collected from 300 to 550 nm every 1 nm with a bandwidth of 3 nm. Final concentrations of DNA and actinomycin D were 100 micromolar (duplex) and 10 micromolar drug, respectively, to ensure complete binding of the ligand. The large negative induced CD signals observed at 440 and 380 nm are consistent with intercalation of the actinomycin D chromophore into the duplex DNA.

either a 5′-TGGT-3′ or a 5′-TGGGT-3′ binding site, are shown in Figure 10.4. The dissociation rate constants, calculated in accordance with a single exponential model, resulted in values of 1.57 $min^{-1}$ for the 5′-TGGT-3′ duplex and 0.872 $min^{-1}$ for the 5′-TGGGT-3′ duplex, reflecting dissociation half-life-lives of 26.5 and 47.8 s, respectively.

These values are in sharp contrast to the markedly longer half-life reported by Chen and coworkers 1996 of 16–18 min for the dissociation of actinomycin D from sequences consisting of optimal dGpC binding sites.[25] The half-life for the 5′-TGGGT-3′ duplex was twice as long as that for the 5′-TGGT-3′ duplex and is reflective of the differences in binding affinities between the two sequences (5′-TGGT-3′ duplex of $K_a = 1.50 \times 10^6$ $M^{-1}$ and 5′-TGGGT-3′ duplex of $K_a = 0.77 \times 10^6$ $M^{-1}$).

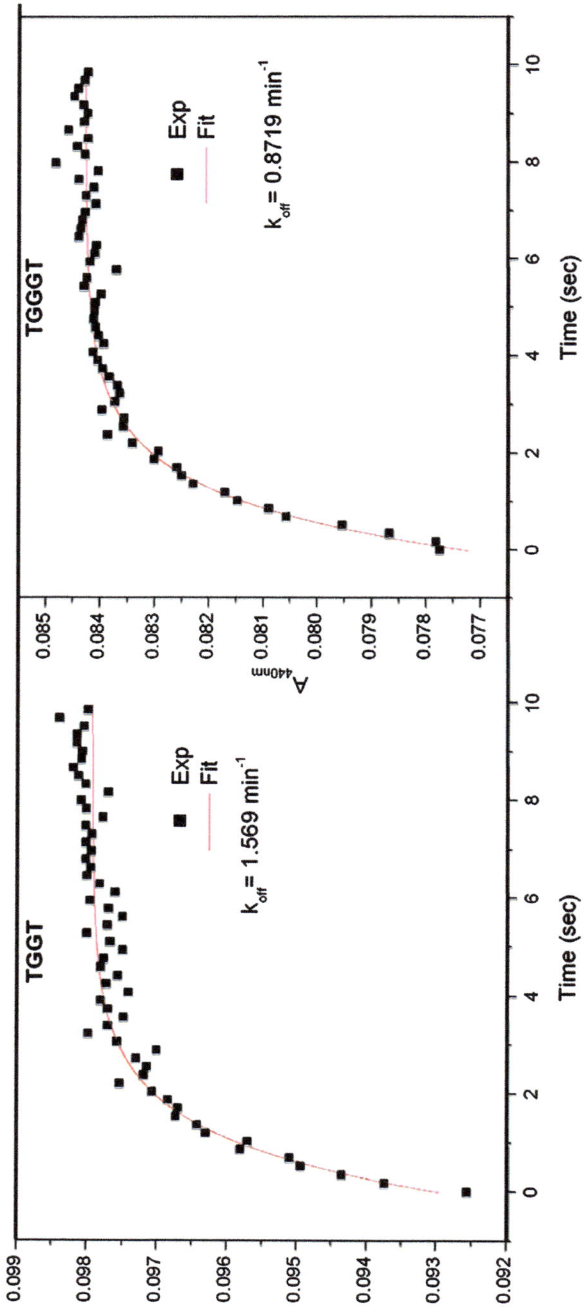


**Figure 10.4** SDS-induced dissociation of actinomycin D from 13-mer duplexes containing a single actinomycin D binding site (left panel) 5′-TGGT-3′ or (right panel) 5′-TGGGT-3′. Data were recorded at 440 nm at 10 second intervals immediately following the addition of a 20% stock solution of SDS sufficient to yield a final SDS concentration of 1% in the sample cell. The 13-mer duplexes and actinomycin D solutions were pre-equilibrated at a 1 : 1 duplex:drug ratio. The resulting curves were analyzed using a first-order rate equation and rate constants were well described by a single exponential.

### 10.4.2 Association Kinetics Controlled by DNA Sequence

In an effort to discern potential linkages between the thermodynamic binding mechanisms and the association of actinomycin D with the 5′-TGGT-3′ and 5′-TGGGT-3′ 13-mer duplexes, stopped-flow kinetics was applied to observe the association reaction. Representative histograms are shown in Figure 10.5 for the association of the fluorescent 7-amino analog of actinomycin D with the 13-mer duplexes containing 5′-TGGT-3′ and 5′-TGGGT-3′.

These kinetic traces reveal the association process to be double-exponential having a relatively fast component ($A_1$) and a much slower component ($A_2$). For the interaction of 7-aminoactinomycin D with the 13-mer duplex containing the 5′-TGGT-3′ binding site, the $k_1$ and $k_2$ are 6.8 and 0.27 $s^{-1}$, respectively. This corresponds to an initial association ($A_1$) half-life of 0.1 s and subsequent equilibrium half-life ($A_2$) of 25.3 s. These values are markedly faster than the association kinetics of actinomycin D binding to the more typical dGpC binding sequence.[43–45] Similarly, the interaction of 7-aminoactinomycin D with the 13-mer duplex containing the 5′-TGGGT-3′ binding site is characterized by a $k_1$ and $k_2$ of 3.02 $s^{-1}$ and 0.00032 $s^{-1}$, respectively, corresponding to an initial association half-life ($A_1$) of 0.23 s (twice the rate observed for the 5′-TGGT-3′ site) and subsequent equilibrium half-life ($A_2$) of 2166 s, that is similar to the rates observed for the more traditional 5′-XGCY-3′ actinomycin D binding site[26,42–45] (Table 10.2).

### 10.4.3 Linkage of Energetics and Kinetics to the Shuffling Model

The signal observed for the slower association step ($A_2$) is indicative that the formation of the actinomycin D–DNA complex is not a simple bimolecular association. The second association step is not as clearly defined by the equation given in the Figure 10.5 caption and is most probably attributable to the limiting concentrations of DNA. A plot of the amplitude of the kinetics trace as a function of increasing DNA concentration is used to determine whether the long association process is a consequence of shuffling between potential binding sites or a consequence of conformational changes in the DNA backbone to accommodate the binding of actinomycin D. The sequential conformational theory predicts that the amplitudes should remain effectively constant over the range of DNA concentrations. The results of the treatment of the data in this manner can be observed in Figure 10.6.[26]

As seen in Figure 10.6A, the association of 7-aminoactinomycin D with the 13-mer duplex containing the 5′-TGGT-3′ binding site shows no systematic response with respect to the DNA concentration. In contrast, the association of 7-aminoactinomycin D with the 13-mer duplex containing the 5′-TGGGT-3′ binding site in Figure 10.6B reveals a systematic increase in amplitude as a function of increasing DNA concentration. These data indicate that the 5′-TGGGT-3′ binding site provides the 7-aminoactinomycin D additional

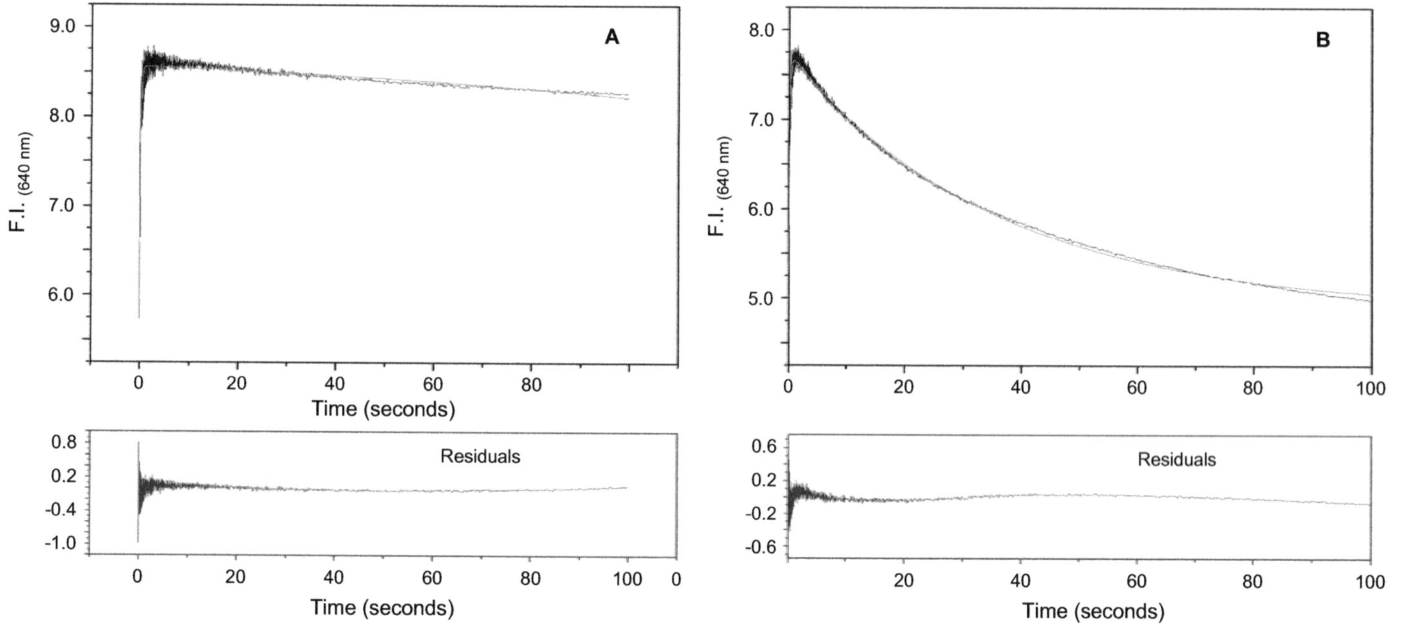


**Figure 10.5** Stopped-flow kinetic traces showing the binding of 7-aminoactinomycin D to the 13-mer duplex containing a single 5′-TGGT-3′ binding site (A) or a 5′-TGGGT-3′ binding site (B). Data are shown in black and the solid red line is the best fit using a single-exponential fit. The panels below are the residuals plots from the calculated fit. The 7-aminoactinomycin D fluorophore was excited at 540 nm and the fluorescence emission was monitored at all wavelengths greater than 645 nm using a long-pass filter. Concentrations of 13-mer DNA duplexes ranged from 0.2 mM to 20 mM (duplex) while 7-aminoactinomycin D was held constant at 0.2 μM. Toward the higher concentration range, the 13-mer DNA duplex concentration remained relatively constant as the reaction approximates pseudo-first-order kinetics. Experimental method and treatment of the data were as described by Fox and Waring.[26] The kinetic traces were analyzed using Micromath Scientist (Micromath, Inc.) and described by a two-exponential equation: $F_t = A_1^{(1-\exp(-k_{obs1}\times t))} + A_2^{(1-\exp(-k_{obs2}\times t))} + b$ where $F_t$ is the total fluorescence, $A_1$ and $A_2$ are the amplitudes of the fast step and slow step, respectively, and $k_{obs1}$ and $k_{obs2}$ are the time constants of the decay.[26]

**Table 10.2** Kinetic parameters for the association and dissociation of actinomycin D measured with 13-mer duplexes containing single binding sites 5′-TGGT-3′ and 5′-TGGGT-3′.

| Sequence | $k_1(A_1)^a$ ($s^{-1}$) | $k_2(A_2)^a$ ($s^{-1}$) | $T_{1/2}(A_1)$ (s) | $T_{1/2}(A_2)$ (s) | $k_{diss}{}^a$ ($min^{-1}$) | $T_{1/2(diss)}$ (s) |
|---|---|---|---|---|---|---|
| 5′-TGGT-3′ | 6.8 | 0.27 | 0.10 | 25.3 | 1.57 | 26.5 |
| 5′-TGGGT-3′ | 3.0 | 0.00032 | 0.23 | 2166 | 0.87 | 47.8 |

[a]Rate constants determined by fitting single-exponential curves to the data, collected at 20 °C in BPES.

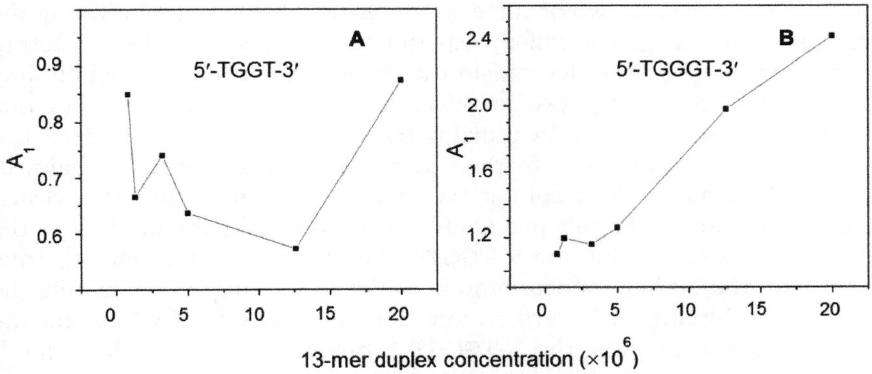


**Figure 10.6** Effect of DNA concentration on the time constants ($A_1$) obtained by stopped-flow association kinetics. (A) A plot of the $A_1$ obtained for the 13-mer duplex with the 5′-TGGT-3′ actinomycin D binding site. There is no discernible systematic variation. In contrast, (B) shows a plot of $A_1$ for the interaction of actinomycin D with the 13-mer duplex containing the 5′-TGGGT-3′ binding site. A systematic increase in the amplitude ($A_1$) is observed a function of increasing duplex concentration.

binding sites with the capacity to shuffle between them, allowing optimization of binding energies and conformations. In contrast, the 5′-TGGT-3′ binding site provides a single site for actinomycin complex formation.

## 10.5 Discussion

The goal of the work presented here was to probe the sequence-selective interaction(s) of actinomycin D with nucleic acids. Although once thought of as the gold-standard for sequence-selective (*i.e.* dGpC) DNA binding agents, more recently there have been numerous examples described of the promiscuity of actinomycin D binding to nucleic acids at sites other than DNA duplexes containing dGpC: single-stranded DNAs, G-quadruplex DNAs and others.[1] Through the use of time-resolved photoaffinity labeling, Waring and coworkers clearly demonstrated that actinomycin D binding is a highly complex process involving initial weak binding of the drug to the DNA, sampling numerous sequences along the DNA lattice until a high-affinity sequence is located. Once at the high-affinity site, actinomycin D shows significantly slower dissociation from that preferred binding site.[28,29] These studies, testing and demonstrating the shuffling model are clear indicators of sequence-specific ligand–DNA interactions. The apparently critical first-step criterion was revealed to be the exclusion of actinomycin from DNA sequences having high AT content, indicating that the initial event in the sequence-recognition process requires recognition of a global structural feature such as minor groove geometry.[28,29]

The focus of the work presented here was to revisit the observation of preferences in actinomycin D binding to the ‘atypical’ 5′-T(G)nT-3′ sequence motifs where *n* equals two or three guanines. In addition to confirming the previous work through calorimetry experiments we chose to make an effort to characterize the binding mechanism that gives rise to such unusual binding thermodynamics between two identical sequences that differ by only one guanine residue. We began by utilizing the highly sensitive ‘model-free’ isothermal titration calorimetry method that directly measures the enthalpy of complex formation. The resulting ITC traces shown in Figure 10.2 clearly demonstrates an exothermic process for the binding of actinomycin D to the 13-mer duplex containing the 5′-TGGT-3′ binding site and an endothermic process for the binding of actinomycin to the 13-mer duplex containing the 5′-TGGGT-3′ binding site. Furthermore, the enthalpy value (ΔHobs) for the 13-mer duplex containing the 5′-TGGGT-3′ binding site of $-3.4$ kcal mol$^{-1}$, and the 13-mer duplex containing the 5′-TGGT-3′ binding site sequence of $+2.5$ kcal mol$^{-1}$, reveal a ΔΔHobs of almost 6 kcal mol$^{-1}$ in binding energy, comparing the two actinomycin binding sites. This is a very significant shift in binding enthalpy, based on a simple change of the binding site from 5′-TGGT-3′ to 5′-TGGGT-3′. Even more striking is the shift in the thermodynamic binding mechanism from an entropy-driven to an enthalpy-driven process.

The question of why and how such a change in thermodynamic binding mechanisms could arise between two very similar sequences led to additional experiments in an effort to provide further insights into this base-sequence-directed thermodynamic switch. Circular dichroism spectrometry is an extremely sensitive technique that is commonly used to identify structural changes arising in macromolecules. In addition, for some DNA-binding ligands, circular dichroism spectrometry can provide insight into structural features of the ligand–DNA complex. Actinomycin D has a relatively symmetrical structure and alone yields no significant CD spectrum; however, upon intercalation of its planar phenoxazine ring between adjacent base pairs in duplex DNA, the actinomycin is placed in an asymmetric environment, resulting in an induced CD signal. This phenomenon was previously identified and reported by Krugh and Young in 1977.[33] As observed in Figure 10.3, similar induced CD signals are observed at 375 and 460 nm for both of the 13-mer duplexes containing the 5′-TGGT-3′ and 5′-TGGGT-3′ binding sites, indicating that the modes of binding of actinomycin D to these ‘atypical’ or non-GpC binding sequences are both intercalative in nature and similar in geometry.

The kinetics of the interaction of actinomycin D with 13-mer duplexes containing the 5′-TGGT-3′ and 5′-TGGGT-3′ binding sites were investigated with the aim of determining whether a kinetic component might be linked with the observed differences in thermodynamic binding mechanisms. Investigations into the kinetic properties associated with the binding of actinomycin D to various DNA sequences have been extensively reported.[22–26] The association kinetics of this ligand–DNA interaction can be complex, and this complexity has been attributed both to the heterogeneity of the sequence as well as the nature of bases flanking the intercalation site.[25,26] Depending on the base sequence, it may take up to five exponentials to fully describe the association process.[22,23] In contrast, analyses of the dissociation kinetics of actinomycin D are less complex and can in many cases be described by a single exponential decay.[42] Accordingly, SDS dissociation experiments were conducted comparing the release of 7-aminoactinomycin D from the 13-mer duplexes containing the 5′-TGGT-3′ and 5′-TGGGT-3′ binding sites. As shown in Figure 10.4, the values of $k_{off}$ for the sequences were calculated by a single exponential and resulted in $k_{off}$ values of 1.57 $min^{-1}$ for the 13-mer duplex containing the 5′-TGGT-3′ binding site. For comparison, the dissociation of actinomycin D from the 13-mer duplex containing the 5′-TGGGT-3′ binding site led to an apparent dissociation constant of 0.87 $min^{-1}$. This corresponds to dissociation half-lives of 26.5 s for the 5′-TGGT-3′ binding site and 48 s for the 5′-TGGGT-3′ binding site. The dissociation half-lives for the 5′-T(G)nT-3′ binding sites are markedly shorter than those observed for the dissociation of actinomycin D from 5′-XGCY-3′ binding sites.[42–45] These values indicate that the longer residence time of the ligand on the 5′-TGGGT-3′ binding site may be attributable to the additional potential binding sites within the 5′-TGGGT-3′ sequence as compared with the 5′-TGGT-3′ sequence, which is also reflected in the differences in binding affinities between the two sequences (TGGGT $K_a = 1.50 \times 10^6\ M^{-1}$ (duplex) and TGGT $K_a = 0.77 \times 10^6\ M^{-1}$ (duplex).

The association kinetic traces for the TGGGT and TGGT sequences as determined by stopped-flow fluorescence experiments are shown in Figure 10.5. In 1984, Fox and Waring described the kinetic evidence for the redistribution of actinomycin D between potential binding sites and referred to this redistribution along the DNA lattice as shuffling.[26,27] Under pseudo-first-order conditions, 13-mer duplexes containing either the 5′-TGGT-3′ or the 5′-TGGGT-3′ actinomycin D binding sites were subjected to a kinetic analysis and demonstrated strikingly different association properties. As illustrated in Figure 10.5, the overall reactions forming the ligand–DNA complexes are slow, taking up to 100 s to reach equilibrium. The amplitudes that are associated with the fast ($A_1$) and slow ($A_2$) steps vary significantly between these two sequences. Exponential curves fitted to these data (as reflected by the continuous red line) along with their resulting residuals are shown in Figure 10.5. Fox and Waring provided similar spectra for the association of actinomycin D and polydGpC.[26,42] They noted that the kinetic contributions may arise from two sources; a conformational change of the

DNA lattice to accommodate the binding of the ligand or a redistribution of the actinomycin D to more favorable binding sequences along the DNA lattice; hence the name, shuffling.[26,27] The DNA sequences containing 5′-TGGT-3′ and 5′-TGGGT-3′ binding sites may contain multiple binding sites within the central core sequence, each having different binding affinities. In the case of the 5′-$TG_1G_2G_3T$-3′ sequence, actinomycin D may intercalate between $G_1$ and $G_2$ or between $G_2$ and $G_3$. In contrast, for 5′-$TG_1G_2T$-3′, only one favorable intercalation site is available to actinomycin D, that between $G_1$ and $G_2$. In the case of actinomycin D binding to oligodeoxynucleotides containing the 5′-XGCY-3′ binding sites, the preferred 5′ flanking base appeared to be T with A or C less favorable.[28] Based on the promiscuity demonstrated by the interactions of actinomycin D with non-dGpC sequences, actinomycin D might also bind, albeit weakly, to dTpG or dGpT steps in both sequences. The possibility exists that these low-affinity binding sites could allow for a statistical opportunity for actinomycin D to bind initially to these sites and then dissociate so as to bind to adjacent higher-affinity sites. According to Fox and Waring,[26–28] a shuffling mechanism *versus* a conformational change in the DNA lattice can be investigated by plotting the initial fast step of the association reaction ($A_1$) *versus* increasing DNA concentration and probing for systematic trends. If conformational changes give rise to these kinetic traces, then the relationship between $A_1$ with increasing DNA concentrations should remain constant and reveal no systematic variation. A plot of these data are shown in Figure 10.6. As observed for the 5′-TGGGT-3′ binding site, we observe a systematic increase in the amplitude of the fast step as a function of increasing DNA concentrations. This response is not consistent with the 5′-TGGT-3′ binding site wherein the amplitude appears to vary in no discernible pattern with increasing DNA concentration. These data indicate the possibility of a higher-order shuffling of the actinomycin D, with ligand sampling among the multiple potential binding sites of the 5′-TGGGT-3′ binding site in comparison to the more limited sampling afforded by the 5′-TGGT-3′ binding site.

## 10.6 Summary

The focal point of this chapter has revolved around the effects of base sequence on actinomycin D–DNA interactions. Of particular interest are the results of experiments to examine the kinetic and energetic properties that drive these base sequence effects using a couple of 'atypical' (non-dGpC) containing 13-mer duplex oligodeoxynucleotides, to which actinomycin D has been demonstrated to bind with high affinity. Comparing the thermodynamic properties associated with complex formation between the 5′-TGGT-3′ and 5′-TGGGT-3′ binding sites reveals the complex formed with the 5′-TGGT-3′ site to be entropy driven, with a $\Delta H$ equal to $+3.5$ kcal mol$^{-1}$, while the drug–DNA complex formed by actinomycin binding to the 5′-TGGGT-3′ site is enthalpically driven with a $\Delta H$ of $-2.5$ kcal mol$^{-1}$. The subtle addition of a single G within the actinomycin D binding site results in a large change of 6 kcal $mol^{-1}$, shifting the binding mechanism from an entropy-driven to an enthalpy-driven process. These changes are also reflected in the association and dissociation kinetics describing these drug–DNA interactions and are compatible with the shuffling model proposed by Waring and Fox in 1986.[26] According to their hypothesis, actinomycin initially interacts with DNA bases in a relatively non-specific manner and subsequently 'shuffles' along the DNA lattice probing the energetic landscape for preferred base sequences, resulting in the formation of more stable ligand–DNA complexes. Using a photoreactive analogue of actinomycin (7-azidoactinomycin D), Waring and coworkers were able to establish that the shuffling hypothesis was correct.[28,29] The changes in binding densities of actinomycin D between low- and high-affinity sites as a function of time were observed and confirmed the lateral diffusion (or shuffling) of actinomycin D along the DNA, relying on the availability and formation of specific hydrogen bond contacts, van der Waals contacts and hydrophobic interactions between the ligand and DNA, all of which translates into the kinetic and thermodynamic profile of sequence recognition. The association kinetics studies reported here demonstrate that complex formation between actinomycin D and the 13-mer duplex containing the binding site sequence 5′-TGGT-3′ has an association half-life of ~25.3 s. In contrast, the association half-life for the interaction of the drug with the 5′-TGGGT-3′ sequence is 100 times longer, 2165 s, indicating that with the added propensity of additional available binding sites, the actinomycin redistributes between those sites so as to allow optimization of the formation of specific hydrogen bond contacts, van der Waals contacts and hydrophobic interactions between the ligand and DNA. Clearly, the addition of a single G within the intercalation site core has profound energetic and kinetic consequences.

# References


1. D. E. Graves 2006. Actinomycin D: Sixty years of progress in characterizing a sequence-selective DNA-binding agent, in *Sequence-Specific DNA Binding Agents* (2006), ed. M. Waring, RSC Publishing, Cambridge, UK, pp. 109–129.
2. W. Muller and D. M. Crothers, Studies of the binding of actinomycin and related compounds to DNA, *J. Mol. Biol.*, 1968, **35**, 251–290.
3. H. M. Sobell, How actinomycin binds to DNA, *Sci. Am.*, 1974, **231**, 82–91.
4. M. J. Waring, DNA modification and cancer, *Annu. Rev. Biochem.*, 1981, **50**, 159–192.
5. S. A. Bailey, D. E. Graves, R. Rill and G. Marsch, Influence of DNA base sequence on the binding energetics of actinomycin D, *Biochemistry*, 1993, **32**, 5881–5887.
6. M. Gellert, C. E. Smith, D. Neville and G. Felsenfeld, Actinomycin binding to DNA: Mechanism and Specificity, *J. Mol. Biol.*, 1965, **11**, 445–457.

7. R. L. Rill, G. A. Marsch and D. E. Graves, 7-Azidoactinomycin D: A photoaffinity probe of the sequence specificity of DNA binding by actinomycin D, *J. Biomol. Struct. Dyn.*, 1989, **7**, 591–605.
8. S. A. Bailey, D. E. Graves and R. Rill, Binding of actinomycin D to the T(G)nT motif of double-stranded DNA: determination of the guanine requirement in nonclassical, non-GpC binding sites, *Biochemistry*, 1994, **33**, 11493–11500.
9. T. R. Krugh and Y. C. Chen, Actinomycin D-deoxynucleotide complexes as models for the actinomycin D-DNA complex. The use of nuclear magnetic resonance to determine the stoichiometry and the geometry of the complexes, *Biochemistry*, 1975, **14**, 4912–4922.
10. T. R. Krugh, Association of actinomycin D and deoxyribonucleotides as a model for binding of the drug to DNA, *Proc. Natl. Acad. Sci. U. S. A.*, 1972, **69**, 1911–1914.
11. D. J. Patel, Peptide antibiotic-deoxyoligonucleotide interactions. Nuclear magnetic resonance investigations of complex formation between actinomycin D and d-ApTpGpCpApT in aqueous solution, *Biochemistry*, 1974, **13**, 2396–2402.
12. E. V. Scott, G. Zon, L. G. Marzilli and W. D. Wilson, 2D NMR investigation of the binding of the anticancer drug actinomycin D to duplexed dATGCGCAT: conformational features of the unique 2:1 adduct, *Biochemistry*, 1988, **27**, 7940–7951.
13. H. M. Sobell, S. C. Jain, T. D. Sakore and C. E. Nordman, Stereochemistry of actinomycin-d DNA binding, *Nat. New Biol.*, 1971, **231**, 200–205.
14. S. Kamitori and F. Takusagawa, Crystal structure of the 2 : 1 complex between d(GAAGCTTC) and the anticancer drug actinomycin D, *J. Mol. Biol.*, 1992, **225**, 445–456.
15. S. Kamitori and F. Takusagawa, Multiple Binding Modes of Anticancer Drug Actinomycin D: X-ray, Molecular Modeling, and Spectroscopic Studies of d(GAAGCTTC)$_2$-Actinomycin D Complexes and Its Host DNA, *J. Am. Chem. Soc.*, 1994, **116**, 4154–4165.
16. C. Bailly and M. J. Waring, Transferring the purine 2-amino group from guanines to adenines in DNA changes the sequence-specific binding of antibiotics, *Nucleic Acids Res.*, 1995, **23**, 885–892.
17. M. J. Waring and C. Bailly, The purine 2-amino group as a critical recognition element for binding of small molecules to DNA, *Gene*, 1994, **149**, 69–79.
18. C. Bailly, C. Marchand and M. J. Waring, New binding sites for antitumor antibiotics created by relocating the purine 2-amino group in DNA, *J. Am. Chem. Soc.*, 1993, **115**, 3784–3785.
19. R. M. Wadkins and D. E. Graves, 7-Azidoactinomycin D: a novel probe for examining actinomycin D-DNA interactions, *J. Biol. Chem.*, 1989, **5**, 7262–7266.
20. R. L. Rill, G. A. Marsch and D. E. Graves, 7-Azidoactinomycin D: A photoaffinity probe of the sequence specificity of DNA binding by actinomycin D, *J. Biomol. Struct. Dyn.*, 1989, **7**, 591–605.

21. G. A. Marsch, D. E. Graves and R. L. Rill, Photoaffinity approaches to determining the sequence selectivities of DNA–small molecule interactions: actinomycin D and ethidium, *Nucleic Acids Res.*, 1995, **23**, 1252–1259.
22. D. M. Crothers, Kinetics of binding drugs to DNA., *Prog. Mol. Subcell. Biol.*, 1971, **2**, 10–20.
23. R. Bittman and L. Blau, Stopped-flow kinetic studies of actinomycin binding to DNAs, *Biochemistry*, 1975, **14**, 2138–2145.
24. S. C. Brown and R. H. Shafer, Kinetic studies of actinomycin D binding to mono-, oligo-, and polynucleotides, *Biochemistry*, 1987, **26**, 277–282.
25. C. Liu and F. M. Chen, Actinomycin D binds strongly and dissociates slowly at the dGpdC site with flanking T/T mismatches, *Biochemistry*, 1996, **35**, 16346–16353.
26. K. R. Fox and M. J. Waring, Kinetic evidence for redistribution of actinomycin molecules between potential DNA-binding sites, *Eur. J. Biochem.*, 1984, **145**, 579–586.
27. K. R. Fox and M. J. Waring, Footprinting reveals that nogalamycin and actinomycin shuffle between DNA binding sites, *Nucleic Acids Res.*, 1986, **14**, 2001–2014.
28. C. Bailly, D. E. Graves, G. Ridge and M. J. Waring, Use of a photoactive derivative of actinomycin to investigate shuffling between binding sites on DNA, *Biochemistry*, 1994, **33**, 8736–8745.
29. G. S. Ridge, C. Bailly, D. E. Graves and M. J. Waring, Daunomycin modifies the sequence-selective recognition of DNA by actinomycin, *Nucleic Acids Res.*, 1994, **22**, 5241–5246.
30. J. Ren, T. C. Jenkins and J. B. Chaires, Energetics of intercalation reactions, *Biochemistry*, 2000, **39**, 8439–8447.
31. I. Haq, B. Z. Chowdhry and T. C. Jenkins, Calorimetric techniques in the study of high-order DNA-drug interactions, *Methods Enzymol.*, 2001, **340**, 109–149.
32. J. B. Chaires, A thermodynamic signature for drug-DNA binding mode., *Arch. Biochem. Biophys.*, 2006, **453**, 26–31.
33. T. R. Krugh and M. A. Young, Daunorubicin and adriamycin facilitate actinomycin D binding to poly(dA-dT)-poly(dA-dT), *Nature*, 1977, **269**, 627–628.
34. L. A. Marky, J. G. Snyder, D. P. Remeta and K. J. Breslauer, Thermodynamics of drug-DNA interactions, *J. Biomol. Struct. Dyn.*, 1983, **1**, 487–507.
35. K. J. Breslauer, D. P. Remeta, W. Y. Chou, R. Ferrante, J. Curry, D. Zaunczkowski, J. G. Snyder and L. A. Marky, Enthalpy-entropy compensations in drug-DNA binding studies., *Proc. Natl. Acad. Sci. U. S. A.*, 1987, **84**, 8922–8926.
36. X. Qu, J. Ren, P. V. Riccelli, A. S. Benight and J. B. Chaires, Enthalpy/entropy compensation: influence of DNA flanking sequence on the binding of 7-aminoactinomycin D to its primary binding site in short DNA duplexes, *Biochemistry*, 2003, **42**, 11960–11967.

37. J. B. Chaires, A thermodynamic signature for drug-DNA binding mode, *Arch. Biochem. Biophys.*, 2006, **453**, 26–31.
38. M. J. Waring, Drugs which affect the structure and function of DNA, *Nature*, 1968, **219**, 1320–1325.
39. K. R. Fox and M. J. Waring, DNA structural variations produced by actinomycin and distamycin as revealed by DNAase I footprinting, *Nucleic Acids Res.*, 1984, **12**, 9271–9285.
40. R. Lyng, T. Hard and B. Norden, Induced CD of DNA intercalators: Electric dipole allowed transitions, *Biopolymers*, 1987, **26**, 1327–1345.
41. P. V. Scaria and R. H. Shafer, Binding of ethidium bromide to a DNA triple helix. Evidence for intercalation, *J. Biol. Chem.*, 1991, **266**, 5417–5423.
42. M. C. Fletcher and K. R. Fox, Dissociation Kinetics of Actinomycin D from Individual GpC sites in DNA, *Eur. J. Biochem.*, 1996, **237**, 164–170.
43. F.-M. Chen, Binding specificities of actinomycin D to non-self-complementary -XGCY- tetranucleotide sequences, *Biochemistry*, 1992, **31**, 6223–6228.
44. F.-M. Chen, Kinetic and Equilibrium binding studies of actinomycin D with some dTGCA- containing dodecamers, *Biochemistry*, 1998, **27**, 1843–1848.
45. F.-M. Chen, Observation of an anomalously slow association kinetics in the binding of actinomycin D to d(CATGGCCATG), *Biochemistry*, 1990, **29**, 7684–7690.

CHAPTER 11

# Molecular Modelling Approaches for Assessing Quadruplex–Small Molecule Interactions

MARTYNA KUTA,[a] BARIRA ISLAM,[b] SHOZEB HAIDER[b] AND STEPHEN NEIDLE*[b]

[a] Laboratory of Quantum Chemistry, Faculty of Chemistry, Adam Mickiewicz University, Umultowska Street 89b, Poznań 61-614, Poland; [b] UCL School of Pharmacy, University College London, 29-39 Brunswick Square, London WC1N 1AX, UK
*Email: s.neidle@ucl.ac.uk

## 11.1 Introduction

The primary focus and target of DNA-binding small molecules and therapeutic agents has traditionally been double-helical DNA, *i.e.* the overwhelming component of the human and other genomes. The past 20 years has seen the emergence of an entirely new category of DNA target, which is based on the concept of higher-order four-stranded structures that can only be formed by certain DNA sequences. This gives these structures inherent selectivity compared with the rest of the genome, a feature that, it is increasingly realised, can be exploited for therapeutic gain. These quadruplex structures have been extensively reviewed elsewhere and the interested reader is referred to a wide range of articles elsewhere describing their

Chemical Biology No. 7
DNA-targeting Molecules as Therapeutic Agents
Edited by Michael J. Waring

Published by the Royal Society of Chemistry, www.rsc.org

structures, biophysics and biology, some of which are referenced in this chapter: a brief outline of the major features of quadruplex architecture is given in the following sections.

### 11.1.1 A Brief Overview of Quadruplexes

The concept that four-stranded nucleic acid structures can exist, originated with seminal fibre-diffraction studies on fibres formed from guanosine monophosphate gels.[1] The fibre diffraction patterns obtained by Gellert and colleagues were consistent with a right-handed helix whose dimensions suggested the involvement of four 'strands'—these could not be formally continuous since they were formed from discrete guanosine monophosphate units. They also suggested that the strands were held together by bases hydrogen-bonded in a way analogous to Watson–Crick base pairs but with the crucial difference that the arrangement involves four bases, and that all are by definition, guanines. The guanine (G)-quartet (sometimes termed a tetrad), involves the so-called Hoogsteen and Watson–Crick edges of a guanine base, with each participating in four hydrogen bonds, thus making eight in total for a single quartet (Figure 11.1a). Subsequent fibre-diffraction and biophysical studies on synthetic polynucleotides confirmed and extended these findings, demonstrating the unequivocal existence of four-stranded, right-handed polynucleotide helices.[2–4]

As well as polynucleotide four-stranded structures, analogous arrangements, termed quadruplexes, can be formed by certain guanine-rich oligonucleotide sequences.[5–7] These are discrete four-stranded mini-helical nucleic acid structures that can form from repeating runs of guanine-rich DNA or RNA sequences, typically $G_{2-5}\ X_n\ G_{2-5}\ X_o\ G_{2-5}\ X_p\ G_{2-5}$, where $X_{n-p}$ are intervening general sequences. Such sequences were first identified in telomeric DNA, at the ends of eukaryotic chromosomes.[6–8] The guanine tracts do not need to be of identical length. Quadruplexes can be formed by discrete sequences, or can be formed within duplex or single-stranded DNA or RNA. All quadruplexes are constructed around a core of two or more stacked G-quartets, highly stable arrangements of Hoogsteen hydrogen-bonded guanine bases. Since a G-quartet comprises four bases rather than the two in a Watson–Crick base pair, quadruplexes necessarily contain four nucleotide strands. These may be continuous, in which case the quadruplexes are unimolecular (the most significant in terms of biological prevalence and importance), consist of two strands (bimolecular) or consist of four (tetramolecular). A quadruplex with three separate strands is also possible, although appears to be of lesser biological relevance.

Several G-quartets can stack on top of one another and typically form the stem of a quadruplex. Cations are located in the central helical cavity and coordinate with carbonyl oxygen from guanines (Figure 11.1a). The cation type has been suggested to affect quadruplex topology, especially for the flexible ones derived from human telomeric DNA sequences. The overall topology of a quadruplex can be described as the order and mutual

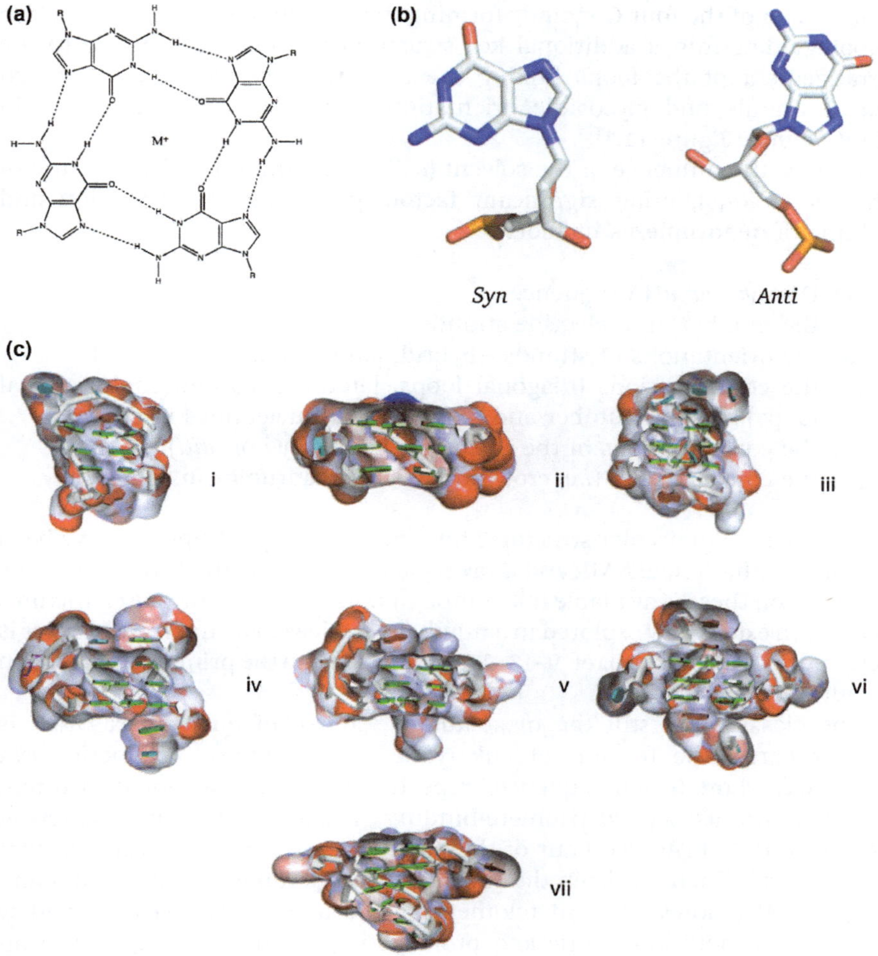


**Figure 11.1** (a) A G-quartet with a mono-cation in the central channel. (b) The *syn/anti* conformations of guanine nucleotides. (c) Molecular structures of various quadruplexes, determined by crystallographic or NMR methods. In each instance the quadruplex is shown in cartoon representation, with guanine bases coloured green, together with part of the solvent-accessible surface (coloured in charge representation). Each surface has been cut away to reveal the quadruplex fold and loops (i) A $Na^+$ form of a human intramolecular telomeric quadruplex, from NMR studies (PDB id 143D) (ii) A $K^+$ form of a human intramolecular telomeric quadruplex, from crystallographic studies (PDB id 1KF1) (iii) A (3 + 1) hybrid form of a human intramolecular telomeric quadruplex, from NMR studies (PDB id 2GKU) (iv) A quadruplex formed from a promoter sequence in the human VEGF gene, from NMR studies (PDB id 2M27) (v) A quadruplex formed from a promoter sequence in the human k-RAS gene, from NMR studies (PDB id 5I2V) (vi) A quadruplex formed from a promoter sequence in the human c-MYC gene, from NMR studies (PDB id 1XAV) (vii) A quadruplex formed from a promoter sequence in the human c-KIT gene, from crystallographic studies (PDB id 4WO3).

orientation of the four G-strands forming the G-stem of the quadruplex. The topology determines additional key structural features. It is linked to the arrangement of the loops linking the G-strands, groove widths between the G-strands and glycosidic orientation (-*syn* or -*anti*) of guanines in the G-stems (Figure 11.1b).

Besides the influence of the solvent (salt concentration and the nature of the metal ions), other significant factors governing the formation and folding of quadruplexes include:

(1) DNA *versus* RNA sequences
(2) the number of nucleotide strands
(3) the orientations of strands—hybrid, parallel and anti-parallel
(4) the conformations (diagonal loops, lateral loops and chain-reversal loops) and the number and nature of the nucleotides in the loop
(5) the conformations of the glycosidic bonds (*syn* or *anti*)
(6) the concentration (*i.e.* crowding) of the quadruplex in solution

Atomic-level molecular structures for a number of quadruplexes have been obtained by high-field NMR and X-ray crystallographic methods (Figure 11.1c), and data on these is available in the Protein Data Bank. Quadruplex structures have been extensively explored in a number of reviews. The interested reader is referred, in particular, to ref. 9–13, 23, 24, as well as to the primary literature on quadruplex structures.

The classic, and still the most-studied, source of quadruplex DNAs is in telomeres, the termini of eukaryotic chromosomes. Telomeric DNA comprises short G-rich sequences repeated up to several thousand times, together with associated telomere-binding proteins. In humans the repeat sequence is 5′-TTAGGG. Four of these repeats can form an intramolecular quadruplex, which has been the subject of a large number of structural and biophysical studies. Beyond telomeres, quadruplex sequences are widely prevalent in both eukaryotic and prokaryotic genomes, although most attention has focussed on the human genome. The number of such sequences has been variously estimated on the basis of bioinformatics, and more recently by next-generation sequencing, as between *ca.* 350 000[14,15] and 700 000.[16] However experimental trapping of actual quadruplex structures in human chromatin has found a much lower prevalence, of *ca.* 10 000.[17] Quadruplex-forming sequences and structures are over-represented in several regions of the genome, in particular in promoter regions of cancer-related genes[17,18] and in untranslated regions.[19] The prevalence in promoter regions in particular, has led to the development of therapeutic strategies targeting these regions in genes such as c-MYC, c-KIT and RAS.[20] The approach has been given added credence by the direct visualisation of quadruplexes in human cancer cells using quadruplex-specific antibodies,[21,22] validating quadruplexes as targets in human cancer.

A large number of small molecules have been shown to bind to quadruplexes—these enhance quadruplex stability and thus the particular feature of quadruplexes, which is that they can interrupt normal cellular processes such as transcription, translation and replication, as well as DNA damage repair, in the absence of mechanisms to unwind them. A major current impetus of the quadruplex field is thus to develop small molecules with enhanced affinity and selectivity to promote the ability of quadruplexes to inhibit the progression of these cellular processes. This chapter focusses on computational and molecular modelling approaches to understanding and guiding the design of present and future quadruplex-binding small molecules. The reader is referred to a number of reviews for more detailed discussions on individual quadruplex-binding small molecules and their associated biology.[23–33]

## 11.2 G-quadruplex Stabilising Ligands

Ligand-induced stabilization of quadruplex formation, especially in telomeres, gene promoters and 5′ and 3′ untranslated RNA regions, have been much studied as a way of enhancing these potentially attractive targets for potential therapies, especially against human cancers.[34–36] A wide variety of small molecules have been either identified by library screening or designed on the basis of their ability to interact with the planar G-quartet group of a quadruplex, in order to stabilize these quadruplex structures. Rather few such compounds have been designed to date on the basis of molecular structural considerations—exceptions include BRACO-19 and several naphthalene diimides.

The diverse nature of the topology of quadruplexes (see for example Figure 11.1c) provides potential advantages for small molecules to interact in a specific way with a particular quadruplex, compared with duplex or other nucleic acid structures.[32] The structure–activity relationships of small molecules binding to quadruplex structures have been widely reported. Those compounds (the overwhelming majority) contain an aromatic chromophore, which can stack on the G-quartet through $\pi$–$\pi$ interactions.[32] Compounds with side chains of different optimal lengths can interact in the grooves of a quadruplex. The presence of positively charged atoms in the side chains increases quadruplex affinity and (often) also biological activity.[36] The ligand can interact with an external exposed G-quartet, loops or/and grooves exploiting these structural features (Figure 11.2).

The increasing use of structural methods (crystallography and NMR) for native quadruplexes and their ligand complexes has provided a firm basis for the rise of modelling and computational approaches, which are used for both structure prediction and the rationalisation of biophysical and biological data. This chapter focusses on the computational methods used to study quadruplex–ligand interactions with examples from a number of well-studied quadruplex-binding compounds: the acridine derivative BRACO-19,[37,38] the naphthalene diimide derivative MM41,[39] the acridine derivative RHPS4,[40] the tetrasubstituted porphyrin derivative TMPyP4,[41–43] BMVC,[44,45]

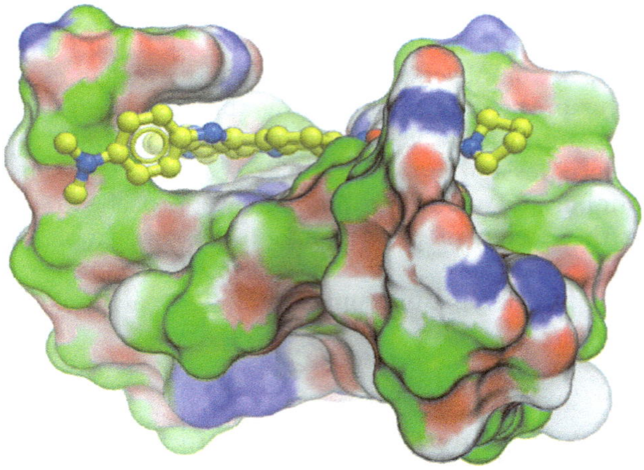

**Figure 11.2** Results from a simulation showing the acridine compound BRACO-19 (yellow) and its interactions with a human telomeric quadruplex. The BRACO-19 side-chains are shown fitting into the quadruplex grooves.

MMQ1, the anticancer drug daunomycin,[38,46] the alkaloid berberine,[47,48] PHEN-DC3[49] and the natural product telomestatin[50–52] (Figure 11.3).

BRACO-19 is a 3,6,9-trisubstituted acridine derivative, designed to stabilize telomeric quadruplex DNA structures.[53] It is, to our knowledge, the first rationally designed quadruplex-targeted telomerase inhibitor[54] and exhibits significant antitumor activity that has been shown to be caused by this action.[55] MMQ1, a quinacridine-based ligand was first described as an anticancer agent in 2007.[56] The G-quadruplex stabilising properties of the MMQ1 series were also confirmed by Hou *et al.*[38] and Gabelica *et al.*[57] MM41 is a tetra-substituted naphthalene-diimide derivative shown to be[58] a potent stabilizer of human gene promoter DNA quadruplexes and inhibits the growth of human cancer cells *in vitro* and *in vivo*. Significant *in vivo* antitumour activity against the MIA-PaCa-2 pancreatic cancer xenograft model has been reported.[39] TmPyP4 is a tetracationic porphyrin derivative which has been extensively studied by Hurley and co-workers.[59–63] It has high affinity for quadruplexes, inhibits telomerase activity and downregulates c-MYC oncogene expression *via* binding to a quadruplex-forming promoter region of the c-MYC gene.[64] TmPyP4 has a preference for binding to parallel-stranded quadruplexes and can convert antiparallel to parallel forms of G-quadruplexes.[65] This ligand does not bind selectively to quadruplex structures and also has significant affinity for duplex DNA. 3,6-Bis(1-methyl-4-vinylpyridinium)carbazole diiodide (BMVC) was one of the first selective quadruplex ligands to be described.[66] It is a diphenylcarbazole derivative; it stabilizes telomeric quadruplexes and exhibits remarkably potent inhibition of the telomerase enzyme complex ($IC_{50}$ of 0.05 μM).[67–70] Phen-DC3 is a phenanthroline analogue[71] which has a perfect geometric match with a G-quartet resulting in high affinity and selectivity to human telomeric

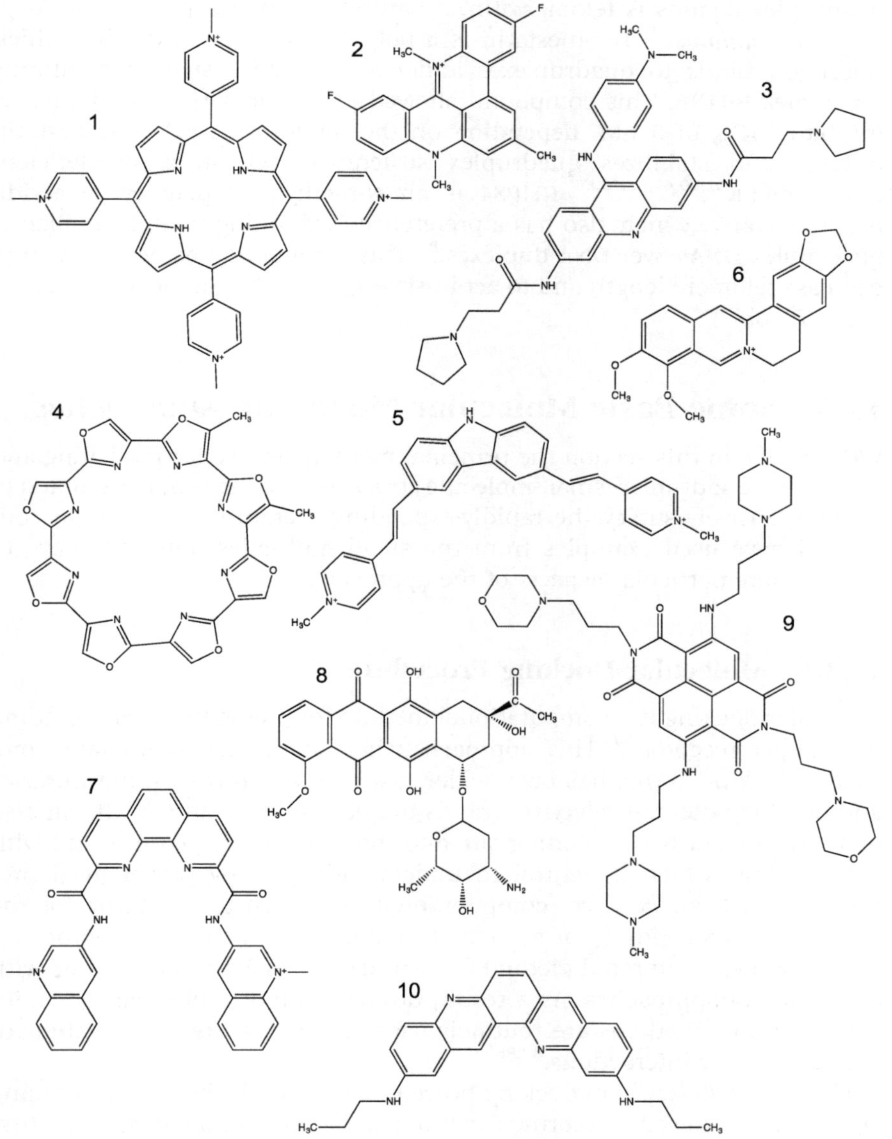


**Figure 11.3** Structures of quadruplex-binding compounds discussed in this chapter 1, TmPyP4; 2, RHPS4; 3, BRACO-19; 4, Telomestatin; 5, BMVC; 6, Berberine; 7, Phen-DC3; 8, Daunomycin; 9, MM41; 10, MMQ1.

quadruplexes.[72,73] Daunomycin has high affinity to duplex B-DNA but also binds to quadruplexes: it forms a complex with a parallel tetramolecular quadruplex, crystallizing as a trimer onto a terminal quartet.[74] Berberine, an alkaloid with anti-bacterial activity[75] has been shown to bind quadruplex DNAs and inhibits telomere elongation.[76,77] One of the more unusual quadruplex ligands is telomestatin, a natural product isolated from *Streptomyces annulatus*.[78] Telomestatin is a polycyclic oxazole derivative, which selectively binds to quadruplexes and does not have significant affinity for duplex B-DNA. This compound showed potent activity as a telomerase inhibitor ($IC_{50}$ of 5 nM, depending on the mode of assay). Telomestatin induces and stabilizes quadruplex structures even in a salt-deficient environment.[51,59,61,62,79,80] RHPS4 is an N-methylated pentacyclic acridinium derivative which also has a preference for binding to and stabilizing, quadruplex DNAs over DNA duplexes.[81] This compound has been shown to decrease telomere length and to act in synergy with the classical anti-cancer agent Taxol.[82,83]

## 11.3 Some Basic Molecular Modelling Approaches

We describe in this section the principal methods in use for modelling and simulating quadruplex–small-molecule structures. We have not attempted to comprehensively survey the rapidly-expanding literature on this topic, but instead have used examples from the small molecules outlined above to demonstrate particular aspects of the approaches.

### 11.3.1 Molecular Docking Procedures

Molecular docking is a computational method for predicting ligand binding to a target receptor.[84] This approach, when automated (automated molecular docking, AMD) has been widely used to screen compound libraries and develop potential selective ligands to quadruplexes.[35,37,39,85] It can also be used to identify the binding site and interactions of bound ligand with the receptor, which is central to understanding many biochemical processes.[86] Such an *in silico* (computational) approach to screening for the interactions and affinity of an actual or potential ligand to receptor is a relatively cheap and rapid method for ligand optimisation as compared with experimental approaches. As a result, docking and/or molecular dynamics computational methods are routinely used as the first step in the study of ligand–receptor interactions.[87,88]

There are two stages in docking procedures—initially the use of sampling algorithms, followed by scoring functions to define ranking order. The first stage involves prediction of the position and orientation of the ligand within the binding site. The second stage enables estimates of the binding energies/affinities.[89] In an ideal scenario, experimentally-determined binding modes should be reproduced by using appropriate sampling algorithms and the scoring function should also rank the binding mode as among the highest generated conformations. In reality this is rarely the case, but the approach has most value in enabling useful new hit compounds to be generated, for subsequent experimental validation and lead compound generation.

### 11.3.2 Classical Molecular Dynamics Simulations

X-ray and NMR studies have provided detailed atomic insight into DNA quadruplex structures, and to a much lesser extent, into RNA quadruplex structure. Molecular dynamics (MD) simulations combined with other computational methods can be used to supplement these experimental approaches. Molecular dynamics simulation consists of step-by-step solutions of the classical equations of motion, which for the force on a simple atomistic system may be represented as $F = ma$. The force on an individual atom is contributed by its bonded and non-bonded interactions. Bonded terms correspond to contributions from atoms that are covalently bonded, including parameters of bond length, bond angle, out-of-plane distortions and torsional terms. Non-bonded terms describe contributions to potential energy resulting from interactions between atoms, which are not covalently bound—van der Waals interactions, electrostatic interactions and hydrogen bonds. The initial position of the atoms is an initial distribution of velocities and the acceleration is determined by the gradient of the potential energy function. A trajectory is calculated by tracing the course of position vectors as a function of time. The determination of a suitable potential function is extremely important for an MD simulation. Lennard-Jones potentials are most commonly used, which represent the interatomic behaviour of atoms so that they attract each other when they are at long distance while they repel at short interatomic distances. MD simulations are usually carried out in canonical (NVT (constant number of particles, volume, and temperature)) or isobaric–isothermal (NPT (constant number of particles, pressure, and temperature)) ensembles that are generated if the number of atoms, volume, pressure and temperature are kept constant during the simulation. Explicit solvated systems are generated with the solute being immersed in a solvent box that is replicated in all directions to satisfy periodic boundary conditions.

## 11.4 Force Fields for Quadruplexes

In molecular mechanics and dynamics, force fields are used to describe the molecular behaviour of the system using empirically derived, quantum-mechanical or experimentally derived parameters. The total energy of the molecule is calculated by taking into account all-atom contributions such as Columbic, polarization, dispersion and repulsive energies. To describe the intramolecular interactions of any system, appropriate parameterizations are needed, as mentioned above. The ensemble of descriptors of the intramolecular and intermolecular behaviour of a system is termed the 'force-field'. While many force fields are available for simulating proteins, there are a very limited number that can accurately reproduce experimental observations for nucleic acids. This is further limited for reliable simulations of quadruplexes. Various polarization effects that have not been included in force field limit the representation of a quadruplex in the MD simulations. In 2007 Orozco *et al.* published the parmbsc0 force field. This is a refinement of the AMBER parm99 force field with particular emphasis on the correct representation of the $\alpha/\beta$ concerted rotation in nucleic acids. Their work was based on an earlier study,[90] which found very large $\alpha/\gamma$ transitions to the *gauche*$^{+}$, *trans* geometry (away from the $g^{-}$, $g^{+}$ state), which introduced severe distortions in DNA in 50 ns MD trajectories.[91] The parmbsc0 force field improved the representation of the $\alpha/\beta$ conformational space, which was poorly represented even in 50 ns long B-DNA MD simulations with the earlier parm94/94 and parm99 force fields.[90] The parametrization was based on quantum calculations using B3LYP and LMP2 methods and validated by CCSD(t)/CBS calculations. This newer force-field formalism can be used to study canonical B-DNAs over long MD time-scales and in parallel can successfully recognize and 'repair' large structural errors while still preserving the essential flexibility of the duplex. This new force field has been successfully used for telomeric quadruplex simulations, reproducing the general overall features of telomeric parallel and antiparallel quadruplexes.[91] However, complete agreement of quadruplex backbone torsion angles in simulations with the experimentally observed values has still not been achieved.[92–95] The other advantage of this improved force field is that it can capture fast conformational transition in duplex DNAs and can be used for RNA and RNA–DNA hybrids. While the force fields can adequately represent guanine stems, loop conformations show some deviation from experimental structures due to multiple force-field issues.[93] These issues could not be resolved even with the improved parmbsc1 version of the force field, with modifications to the sugar puckers, the $\chi$ glyosidic torsion angles, and the $\varepsilon/\zeta$ torsion angles.[95,96] Similar force-field improvements in parmbsc0 have been introduced, leading to the development of the OL15 force field. This includes the parmbsc0 force field with refinements in $\chi$,[97] $\varepsilon/\zeta$[98] and $\beta$ torsion angles.[99] A subsequent study of loop conformations in a simulated quadruplex showed deviations from the experimental structure in all these force field versions.[92,115] It has to be realised that these recent modifications for DNA in the AMBER, bsc1 and OL15 force-fields, were originally developed to improve the accuracy of molecular dynamics (MD) simulations of double-stranded DNA (hopefully including four-stranded DNA structures) on relatively long time-scales.[100] Clearly these major force-field improvements have improved our ability to reliably simulate quadruplexes: as computer power continues to increase, the ability to undertake yet more extended simulations at biophysical and biological time-scales will further test the robustness of these force-fields.

### 11.4.1 Long-range Electrostatic Interactions

Molecular dynamics simulation is an iterative process in which the impact of all contributing factors on inter- and intramolecular energies has to be taken into account. Long-range electrostatic interactions have a significant impact on the stability of quadruplexes. It would be ineffective to sum all the non-bonded interactions in an MD simulation set up. Therefore, spherical cut-off procedures have been introduced.[101] However, such methods have been demonstrated to cause errors and destabilize trajectories, leading to collapse of quadruplex structure due to the dislodging of cations from the quadruplex stem.[101] The introduction of the particle-mesh Ewald (PME) truncation method eliminates this fundamental problem and produces stable DNA simulations.[102,103] One caveat is that the PME method is more time-demanding and can over-stabilize simulated systems,[104] but even so this is the current method of choice for simulating quadruplex structures.

### 11.4.2 Base Stacking and Backbone Descriptions

The AMBER force-field parameters for nucleic acids provide descriptions of interactions between the bases. There is no quantum mechanical operator or direct experiment to determine atomic charges on the bases.[105] Instead 'arbitrary' charges can be assigned by fitting to molecular electronic potential (MEP, ESP) charges. The AMBER force field derives the charges so as to reproduce the electrostatic potential (ESP) around the molecules.[106] Due to their partial $sp^3$ hybridization[105] the amino substituent groups of nucleic acid bases tend to be slightly non-polar.[107] This effect stabilizes bifurcated H-bonds, close amino group contacts, non-planar G/A base pairs and some other specific interactions.[93,95,105] The force fields assume purely $sp^2$ amino hybridisation for ring nitrogen atoms. This should be sufficient for most interactions, as primary H-bonds stabilize the $sp^2$ electronic structure. However, parametrization of a force field that allows neither out-of-plane H-bonds nor amino–acceptor interactions to occur is not straightforward. Also, the backbone of a nucleic acid is difficult to deal with due to the high flexibility of the structures and highly negative charge on the phosphate groups. This is not adequately handled by the non-polarizable atom–atom pair-additive force field. A variable correction of the $\alpha/\gamma$ DNA backbone profile (parmbsc0)[91] has been introduced, which has been further improved in the parmbsc1 and OL15 force fields.

### 11.4.3 Molecular Docking and Dynamic Simulations of DNA and RNA Quadruplex–Ligand Complexes—Some Examples

The BRACO series of compounds was the first to be rationally designed as quadruplex DNA stabilisers and telomerase inhibitors.[53] Their selectivity and potent telomerase inhibitory activity was subsequently demonstrated experimentally. Manual docking was used in the design process to position ligand between the diagonal $T_2A$ loop and the top G-quartet of the only human telomeric quadruplex structure then available, the NMR structure in sodium solution of an intramolecular human telomeric DNA quadruplex (PDB id 143D). The Cornell *et al.*[108] force field was used with the AMBER simulation program,[109] which produced 1000 ps of MD simulation. It was indicated in this study that BRACO-19, compared with four other ligands examined, possesses substituents on all three side chains, which facilitate optimal non-bonded interactions with three out of four quadruplex grooves. The crystal structure of a human telomeric quadruplex complex[110] with BRACO-19 (PDB id: 3CE5), which validated these predictions, has been in turn MD-simulated[38] to better understand the stacking interactions between BRACO-19 and the external G-quartet. Simulations have also been performed on several other quadruplex complexes having experimentally-determined molecular structures—with daunomycin (PDB id 1L1H),[74] RHPS4 (PDB id: 1NZM),[111] MMQ1 (PDB id: 2JWQ),[56] and TMPyP4 (PDB id: 2A5R).[112] These 100 ns MD simulations showed that once a ligand is stably bounded (preferably on the surface of the external G-quartet), the stability of Hoogsteen hydrogen bonds in the G-quartet is also increased, contributing to the overall quadruplex-complex stability. Observed hydrogen bonds between ligands and receptors can further increase the binding affinity of ligands and quadruplexes, and enhance ligand selectivity. MD simulations with the parmbsc0 force field supplemented with $\chi_{OL4}$ modification have been recently used[39] to study the interactions between the tetra-substituted naphthalene diimide compound MM41 and quadruplexes formed in gene promoter sequences. These MD simulations have demonstrated that the favourable binding mode of MM41 to one of the BCL-2 quadruplexes is closely analogous to that observed in the crystal structures of the MM41 complex with a human telomeric quadruplex.[113] Clearly there are differences in detailed interactions, but this result gives confidence in the concept that modelling on a 'generic' and well-studied quadruplex structure, such as the human telomeric parallel intramolecular quadruplex, can provide useful insights relevant to other quadruplex targets (especially when their 3-D structures are unknown). In particular overall shape, size and nature of ligand skeletons and substituents can be optimised in this way, and then fine-tuned.

A comparative study[114] has been made of interactions of TMPyP4 with several different human telomeric quadruplex DNA topologies, namely, the parallel propeller-type (PP), the antiparallel basket-type (AB), and a mixed hybrid-type (MH) G-quadruplex DNA. The AutoDock 4.0 program[115] was used with the Lamarckian genetic algorithm to perform blind and selective docking of the TMPyP4 ligand. The docking procedure effectively identified the most potent binding sites. Subsequent simulations were performed with the parmbsc0 force field and showed that these docked complexes were stable during a 50 ns simulation. The results also showed that the ligand binds most efficiently with a parallel propeller quadruplex topology due to the enhanced effect of the favourable $\pi$–$\pi$ interaction of the G-quartets with the core aromatic moieties of the ligand.

Telomestatin interactions with the human Tel-22 quadruplex have been evaluated.[52] Telomestatin was docked manually over the external G-quartet, on both 5′ and 3′ ends of the quadruplex core.[50,116] All systems were simulated with the ff99 force field in $K^+$ solution, with the PME method applied to treat long-range electrostatics. The 5 ns simulation of the complex with an anti-parallel chair-type quadruplex revealed that one of the central $K^+$ ions exchanged with a water molecule. Simulation of telomestatin with an anti-parallel basket-type quadruplex highlighted that the one central $K^+$ ion shifted to the interstitial space between bound telomestatin and the subsequent quartet, whereas for the models with mixed hybrid type of quadruplex no movement of ions in the central channel was observed.[52] This indicated that $K^+$ ions have a significant influence on stability *via* interactions between telomestatin and several conformations of telomeric quadruplexes.

Combined experimental and computational experiments[48] have been used to investigate the interaction between the berberine ligand and a hybrid-type human telomeric quadruplex structure. The optimized berberine molecule was docked to the 26-mer mixed-hybrid structure (PDB id 2HY9), using the SUFLEDOCK module in the SYBYL molecular modelling system. MD simulations were then performed on two complexes selected on the basis of their energy ranking. Simulation indicated that the berberine molecule forms stable interactions with the external G-quartet by stacking, with its positively-charged N7 atom positioned directly above the centre of the G-quartet. This behaviour was observed in both simulations, which indicated that this interaction could have physical reality.

As well as end-stacking, groove binding has been observed to be the ligand binding mode for the classic duplex groove-binding compound distamycin-A,[117] on binding to a tetramolecular parallel DNA quadruplex.[118,119] Two contrasting models have been suggested for this complex. Distamycin-A can either groove-bind or stack on a terminal G-quartet, as well as possibly adopting a mixed groove/G-quartet mode.[120] The specificity, affinity and binding modes of potential quadruplex groove binders ligands have also been explored by virtual screening docking calculations.[121] All ligands were docked using the AutoDock4 program,[115] to a tetramolecular parallel DNA quadruplex (PDB id: 1S45). The Lamarckian genetic algorithm local search (GALS) method was used to calculate energies for the ligand complexes and a low-frequency local search based on the method of Solis and Wets[122] was applied. The docking results were broadly in accord with NMR experimental screening and indicated that appropriate groove-bound compounds can bind with high affinity to at least this quadruplex.[123]

#### 11.4.3.1 *Targeting RNA Quadruplexes*

A principal focus to date for small-molecule design studies has been telomeric DNA quadruplexes. However, the knowledge that telomeric DNA can be transcribed into RNA (TERRA) sequences, also with quadruplex-forming ability, has made this design concept more complex, since inadvertent stabilization of TERRA quadruplexes could potentially increase telomerase activity (by reducing the amount of hTR (human telomerase RNA component) template-RNA which is maintained in duplex form).

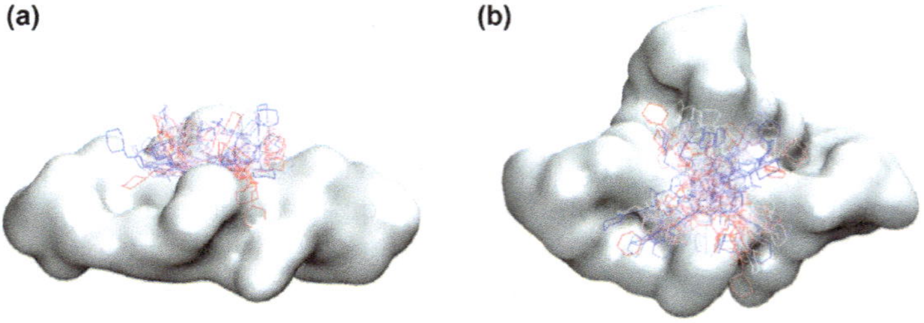


**Figure 11.4** Side and top view illustrating how tetrasubstituted naphthalene diimide ligands may scan the terminal quartet and loops to maximise quadruplex interactions during molecular dynamics simulations.

The interactions between BRACO-19 and naphthalene diimide derivatives and DNA and RNA human telomeric quadruplexes have been systematically compared by molecular modelling[104] (Figure 11.4). The native 22-mer DNA parallel-stranded human telomeric quadruplex was used (PDB id: 1KF1) in this study and an RNA 22-mer telomeric quadruplex was generated from the DNA structure by direct addition of 2′-OH groups. The protocol for all complexes was the same. This work has revealed several factors differentiating DNA from RNA quadruplexes. Firstly, RNA complexes are more stable than the analogous DNA ones due to the presence of the 2′-OH groups in the former. MD simulations indicate that the OH group imparts rigidity to the nucleotide backbone by interacting with the O5′ groups and water molecules in the grooves. The presence of additional 2′-OH groups in the RNA quadruplex constricts the space available to the ligand side-chains to interact with the loops, by reducing the depth and the width of the UUA loops. Thirdly, just one naphthalene diimide ligand forms a strong complex with a telomeric RNA G4 monomer. The simulations also showed that substitution of one side-chain group (–$NMe_2$ for an –OH) increased RNA quadruplex–ligand affinity by some 15-fold. The most important conclusion from this is that it is plausible to rationally design appropriate small-molecule ligands capable of discriminating between DNA and RNA quadruplexes,[104] or of binding selectively to other types of RNA quadruplex, for example those encoded in untranslated regions of a gene.

The results from the MD simulations[124] also explain the differences in interactions between naphthalene-derivative ligands and a DNA telomeric quadruplex compared with an RNA one. A notable finding was that from three ligands which have high affinity to the human telomeric quadruplex DNA just one binds to the equivalent RNA quadruplex with similar affinity. MD simulations showed that the presence of the 2′-OH group in the RNA quadruplex makes multiple interactions within the TTA loops, thereby reducing the depth and width of loops when compared with the equivalent DNA quadruplex structure. These findings indicated that the RNA telomeric quadruplex is less amenable to binding ligands with side-chains terminating in bulky and/or inflexible functional groups, as are present in the two ligands with reduced affinity to the RNA quadruplex.

## 11.5 Enhanced Sampling Methods

### 11.5.1 Simulated Annealing Algorithms

Simulated annealing (SA) is a local search algorithm (meta-heuristic) for solving unconstrained and bound-constrained optimization problems.[125] This method carries out random optimization based on a Monte Carlo iterative strategy. The name of the algorithm is derived from the physical process of heating a material and then slowly lowering the temperature to decrease defects, used, for example, in crystallization. During the changes of temperature a SA algorithm searches randomly for the solutions by using the metropolis criterion and escapes from local minima through the acceptance of 'bad solutions' with a certain probability.[126] By accepting 'bad solutions', the algorithm avoids being trapped in local minima, and is able to explore globally for a greater number of possible solutions. The process is repetitive while the temperature descends gradually until the global optimization solution is found. Due to the effectiveness of non-linear combinatorial optimization methods, simulated annealing algorithms are used in many application fields.[127]

Simulated annealing has been used to determine the optimum ligand orientation and key molecular interactions[128] of telomestatin with a human telomeric intramolecular quadruplex. The approach used simulated annealing (SA) docking to study the binding interactions with an anti-parallel topology for this quadruplex. The results indicated that that one or two molecules of telomestatin can bind to many different binding sites on a quadruplex (*via* loops or external G-quartets), although the comparison of binding energies of the complexes (1 : 1 and 1 : 2) revealed that two bound telomestatin molecules are more favourable. The SA docking also reveals that this telomestatin–quadruplex complex undergoes a conformational rearrangement that leads to significant changes in the relative position, orientation and potential energy of both the telomestatin and the quadruplex.[50]

The novel triazatruxene derivative azatrux binds to the parallel form of the human telomeric quadruplex under molecular crowding conditions.[129] A simulated annealing docking approach has been used to find the most probable conformation for this ligand. SA docking identified ten possible conformations of the ligand bound on the external quartet face, following which the positions and orientations of the docked ligand were analysed in order to decide if a particular position was feasible. This protocol was repeated until a stable conformation with optimal G-quartet stacking was identified.[129] A simulated annealing approach has also been used in order to find reasonable positions for the research compound GQR bound to a quadruplex.[130]

### 11.5.2 Principal Component Analysis

The Principal Component Analysis (PCA) method is a statistical tool which transforms a number of possibly correlated variables into a smaller number of uncorrelated variables termed principal components. In a simulated system not all motions are important. PCA separates large-amplitude motions from irrelevant fluctuations. The translation and rotation of the structure through the trajectory is removed by changes in the simulated structure, which is translated to the geometrical centre of the molecule by least-square fit superimposition onto a reference structure.[131] The appropriate configurational space is then constructed using a simple linear transformation in Cartesian coordinate space to generate a $3N\times3N$ covariance matrix. The matrices are summed and averaged over the entire trajectory. The resulting matrix is then diagonalised, generating a set of eigenvectors that gives a vectoral description of each component of the motion by indicating the direction of the motion. Each eigenvector describing the motion has a corresponding eigenvalue that represent the energetic contribution of that particular component to the motion. The eigenvalue is the average square displacement of the structure in the direction of the eigenvector. Projection of a trajectory on a particular eigenvector highlights the time-dependent motions that the component performs in the particular vibrational mode. The time-average of the projections shows the contribution of components of the atomic vibrations to this mode of concerted motions.[132] The eigenvalues are placed in descending order where the first eigenvectors and eigenvalue describes the largest internal motion of the structure. The eigenvalues decline sharply, highlighting the possibility of separating the dynamics into a small essential space and a relatively large space containing only small atomic fluctuations. On average only about 5% of eigenvectors are necessary to describe 90% of the total dynamics. Although PCA is a convenient method for visualising trajectories, its limitations should also be taken into consideration when interpreting results.[133] PCA is most suited to analysing trajectories of system that undergo transitional changes instead of trajectories that highlight thermal fluctuations to flexible molecules.[133]

PCA has been applied to study the molecular dynamics of the planar perylene derivative Tel03, binding to the dimeric quadruplex formed by the sequence d(GGGTGGGTGGGTGGGT).[134] The largest directional motion in the free dimer occurred on the loop regions. The motions of loops were clearly reduced in the dimer–Tel03 complexes. It was suggested that binding of the ligand to the quadruplex dimer is able to stabilize the model by reducing the motion of the loops. Furthermore, for the 1 : 1 complexes, the flexibility of the individual quadruplex unit that does not interact with the ligand directly was also clearly reduced, without interference from the other unit. A general conclusion is that using PCA reveals that the dominant motions in the free dimer occur mostly in the loop regions, and that the presence of ligand reduces loop motion.[135]

### 11.5.3 Free-energy Calculations

Calculating a reliable and accurate value for the free energy of a small molecule binding to a quadruplex continues to be a difficult challenge. More physically meaningful and easier to calculate is to take into account two separate states A and B, where both are separate components of the complex system under consideration. There are several methods available, some of which have been used to study quadruplex–ligand interactions.

#### *11.5.3.1 The MM/PBSA Method*

The Molecular Mechanics/Poisson–Boltzmann Surface Area (MM/PBSA) approach is a computational method that combines molecular mechanics energy and implicit solvation models used for energy calculations. The objective of the MM/PBSA methods is to calculate the difference in free energy of two states, usually unbound (ligand and receptor) and bound (complex).[136] The binding free energy is a sum of internal energy (bond, angle and torsion/dihedral energies), electrostatic energy, van der Waals energy, sum of electrostatic solvation energy (polar contribution) and the non-electrostatic solvation component (nonpolar contribution) minus conformational entropy change. The PB model is used to calculate polar contribution while the solvent accessible surface area (SASA) is used to estimate the non-polar energy component. The MM/PBSA method is commonly used to calculate the free energy of binding in protein–ligand or protein–peptide complexes. This approach to estimate the free energy of binding has also been extensively used for small molecule–quadruplex complexes.[137]

MM/PBSA has been used to calculate the free energy of binding between a dimeric model with end to end positioning of an intra-molecular parallel-stranded quadruplex and the N,N′-bis-(2-(dimethylamino)ethyl)-3,4,9,10-perylenetetracarboxylic acid diimide ligand Tel03.[134] Five $K^+$ ions in the central channel of the quadruplex were included in the calculations. In order to calculate the electrostatic contribution to the solvation free energy the PBSA program[138] was used, while to estimate the nonpolar contribution to the solvation free energy the MOLSURF program was employed.[139] The resulting calculated free energy of binding indicated that Tel03 can efficiently bind to the dimer through end stacking. In a quadruplex model complexed with two ligands (1 : 2), one at each end, ligand stacked with the 3′ terminal thymine was indicated to be the most favourable binding mode.[134]

A number of tri- and tetra-substituted naphthalene diimides have been shown experimentally to be exceptional quadruplex binding agents.[85] These experimental results are in accord with MM/PBSA calculations of the binding free energy. For example one of these tetrasubstituted compounds was shown both experimentally and computationally to have greater binding affinity than BRACO-19 for a human telomeric quadruplex.[85] MM/PBSA has also been used to compare the energies for various topologies of the human telomeric quadruplex.[140] However variations in energy within a simulation have been a major limitation hindering the making of any more general conclusions from such a study.[140]

#### 11.5.3.2 Thermodynamic Integration

The thermodynamic integration (TI) method for energy calculation was originally proposed by Kirkwood.[141] It is one of the commonly used methods for calculating binding energies, giving accurate results compared with experiment even though it is computationally expensive.[142] When the two discrete states (*e.g.* initial state A and final state B) with potential energy $U_A$ and $U_B$ are dependent in a different way from the spatial coordinates then one can use TI methods to compare the difference in free energy between them. The paths between two states are defined as a thermodynamic cycle, which is used to estimate the energy difference. An example of a thermodynamic cycle is shown in Figure 11.5.

The binding of TMPyP4 and Phen-DC3 (Figure 11.3) to a human telomeric DNA (Tel22) quadruplex has been characterized thermodynamically and structurally.[143] The NACCESS program was used to calculate SASAs (solvent accessible polar, $\Delta A_P$, and non-polar, $\Delta A_N$, surface area). The values calculated as the differences between SASA of the complex and the summed SASA of the ligand-free Tel22 and the free (unbound) ligand are correlated with the process of formation of ligand–quadruplex complexes.[143] The results of the global thermodynamics analysis also revealed that the process of binding complexes involves a large number of adjustable parameters in order to fully define quadruplex–ligand complex thermodynamics in this way. For example, it is necessary to estimate the driving forces for the conversion between intermediate and folded quadruplex separately. This then can provide more reliable thermodynamic fingerprints of ligand binding to human telomeric DNA quadruplexes.

### 11.5.4 Umbrella Sampling

Umbrella sampling as a general algorithm, was developed in 1977 by G. M. Torrie and J. P. Valleau[144] and provides data on the free energy along a

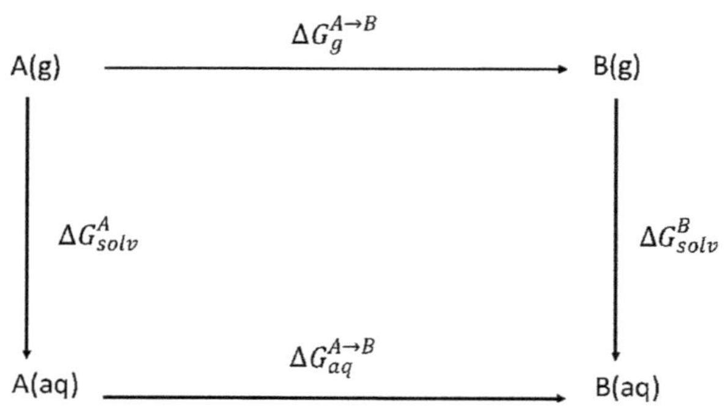


**Figure 11.5** An example of a thermodynamic cycle.

reaction coordinate. The method is based on the assumption that bias potential along a reaction coordinate drives a system from state A to B (*e.g.*, reactant and product). The bias potential can have any functional form. If there are any intermediate steps a series of windows needs to be considered, at each of which a molecular dynamics simulation has to be performed. In general, it is preferable to sample many windows for shorter times than fewer windows for longer.[145] The bias potential of the windows is an additional energy term that depends only on the reaction coordinate. The effect of the bias potential to connect energetically-separated regions in phase space gives rise to the name umbrella sampling. The free energy of the system projected into an order parameter can be obtained by histogram analysis.[146]

Steered molecular dynamics (SMD) simulations together with umbrella sampling have been used to investigate the molecular mechanism and kinetics of a ligand related to BMVC (Figure 11.3), in which the nitrogen atom in the pyridinium group is replaced with a carbon atom, during dissociation processes leading to native basket, propeller and hybrid human telomeric quadruplex structures. Umbrella sampling was used together with SMD due to the difficulty of producing unfavourable states using conventional MD simulation protocols. SMD was used to computationally remove the ligand from its binding site, following which the initial coordinates for umbrella sampling calculations were generated. The use of umbrella sampling provides an improved understanding of the pathway and the mechanism of how the BMVC derivative leaves (and enters) the external G-quartet binding site and forms a stable complex. Two possible pathways were obtained for the ligand dissociating from the ligand-bound quadruplex: the direct pathway, in which the ligand escapes from the end-stacked binding site to the bulk state, and an indirect pathway, which utilises surface sliding. The protocol of combined docking, molecular dynamic simulations, PMF and umbrella sampling gave results which were consistent with experimental findings. This protocol can be applied to other quadruplex ligands to assess their robustness and rank a series of derivatives with similar potencies.[45]

### 11.5.5 Markov State Models

Markov state modelling (MSM) has been widely used for studying protein dynamics and protein–ligand binding.[147–149] MSM provides a statistical framework for efficient sampling and a human-readable view of the key aspects of functional dynamics from data analysis.[150] We have recently used data from 10 μs (30 μs in total) long MD simulations of a parallel-stranded quadruplex to construct MSM models (Figure 11.6).[92] Our aim was to construct a coarse-grained model to understand conformational transitions of propeller loops in a parallel-stranded telomeric quadruplex. The MSM was in broad agreement with clustering analyses of the simulations. Transition path theory[151] of MSM suggests that although the loops of quadruplexes are flexible, they require microsecond time scales for transitions from one state

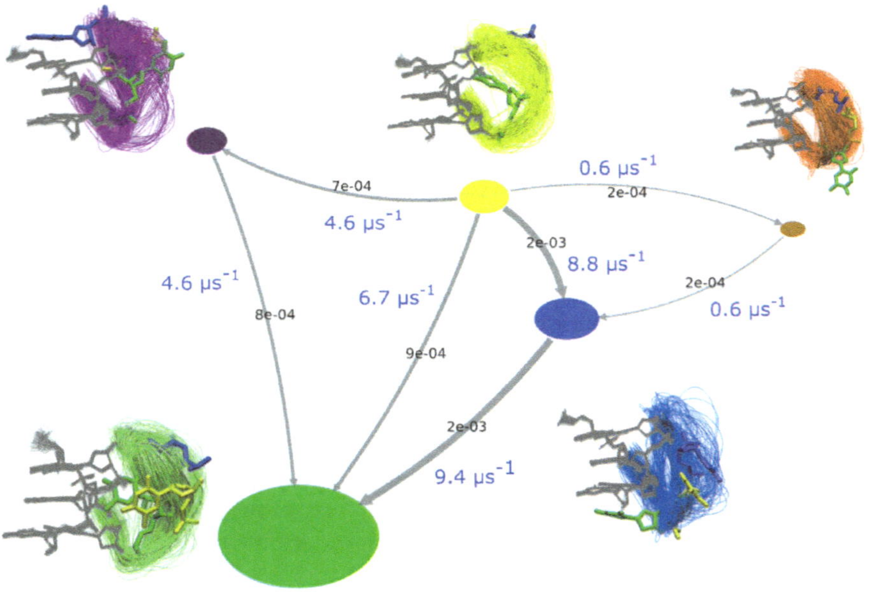


**Figure 11.6** Conformational searching using Markov State Models (MSM), showing Transition Path Theory (TPT) analysis of the loop dynamics of three 10 μs long simulations of human telomeric quadruplex TTA propeller loops, constructed at a 250 ns MSM lag time. The backbone torsion angles ($\alpha$, $\beta$, $\gamma$, $\delta$, $\varepsilon$, $\zeta$, $\chi$) as sin–cos pairs of TTA propeller loops were used as a feature for MSM. MSM was coarse-grained into five macrostates. The crystal structure- and the dominant state are shown in yellow and green. The area of the circle representing the macrostate is proportional to its population. Transition fluxes are shown as net probability (in black) and as rates (probability per time unit) in $\mu s^{-1}$ (in blue). The G-containing strands connected by loops are shown as grey lines. The ensemble of loop backbone geometries contained in each set is shown by displaying overlays of the most probable structures of the set (opaque lines) on top of samples of the entire set (transparent lines) to show both intra-state conformational variability and inter-state conformational differences. The most probable position of loop nucleotides in each state is shown in liquorice representation; the first and second thymine and adenine of loops are shown in green, yellow and blue, respectively.

to another. Markov state models provide an alternate approach for identifying the kinetically relevant states and intermediates of interconversion between them. Therefore, they could be further employed to study quadruplex–ligand interaction and energetics.

## 11.6 Conclusions

Several more recent theoretical studies on DNA quadruplex–small-molecule interactions have been reported that have utilised a number of the concepts outlined in this chapter and have demonstrated their utility in using theory to explain experimental findings and predict new directions to take. Examples include:

1. Molecular dynamics simulations on RHPS4 with three distinct human telomeric quadruplex folds[40]
2. Ion specificity studied by free-energy simulations[152]
3. RNA *vs.* DNA quadruplex specificity for the ligand pyridostatin using molecular dynamics and docking approaches[153]
4. Molecular dynamics simulation of telomestatin binding to a telomeric quadruplex[154]
5. Analysis of the protonation state of naphthalene diimide ligands bound to telomeric quadruplexes using molecular dynamics[155]

It is notable that the overwhelming majority of computational quadruplex studies have been performed on human telomeric quadruplexes, even though structural data is available in the Protein Data Bank on a number of DNA promoter quadruplexes (notably c-MYC, c-KIT and k-RAS—see Figure 1C). We look forward to future computational exploitation of these and other quadruplexes with diverse 3-D structures, in order to find selective ligands for particular disease processes. A further challenge is to simulate and model quadruplexes embedded in duplex sequences, which will provide more realistic models of quadruplexes within their genomic environments.

## Acknowledgements

SH thanks University College London for the award of an Excellence Fellowship.

## References

1. M. Gellert, M. N. Lipsett and D. R. Davies, Helix formation by guanylic acid, *Proc. Natl. Acad. Sci. U. S. A.*, 1962, **48**, 2013–2018.
2. S. Arnott, R. Chandrasekaran and C. M. Marttila, Structures for polyinosinic acid and polyguanylic acid, *Biochem. J.*, 1974, **141**, 537–543.
3. S. B. Zimmerman, G. H. Cohen and D. R. Davies, X-ray fiber diffraction and model-building study of polyguanylic acid and polyinosinic acid, *J. Mol. Biol.*, 1975, **92**, 181–192.
4. F. B. Howard and H. T. Miles, Poly(inosinic acid) helixes: essential chelation of alkali metal ions in the axial channel, *Biochemistry*, 1982, **21**, 6736–6745.
5. E. Henderson, C. C. Hardin, S. K. Walk, I. Tinoco and E. H. Blackburn, Telomeric DNA oligonucleotides form novel intramolecular structures containing guanine-guanine base pairs, *Cell*, 1987, **51**, 899–908.
6. W. I. Sundquist and A. Klug, Telomeric DNA dimerizes by formation of guanine tetrads between hairpin loops, *Nature*, 1989, **342**, 825–829.

7. D. Sen and W. Gilbert, A sodium-potassium switch in the formation of four-stranded G4-DNA, *Nature*, 1990, **344**, 410–414.
8. J. Meyne, R. L. Ratliff and R. K. Moyzis, Conservation of the human telomere sequence (TTAGGG)n among vertebrates, *Proc. Natl. Acad. Sci. U. S. A.*, 1989, **86**, 7049–7053.
9. J. T. Davis, G-quartets 40 years later: from 5′-GMP to molecular biology and supramolecular chemistry, *Angew. Chem.*, 2004, **43**, 668–698.
10. S. Burge, G. N. Parkinson, P. Hazel, A. K. Todd and S. Neidle, Quadruplex DNA: sequence, topology and structure, *Nucleic Acids Res.*, 2006, **34**, 5402–5415.
11. D. J. Patel, A. T. Phan and V. Kuryavyi, Human telomere, oncogenic promoter and 5′-UTR G-quadruplexes: diverse higher order DNA and RNA targets for cancer therapeutics, *Nucleic Acids Res.*, 2007, **35**, 7429–7455.
12. S. Neidle and S. Balasubramanian, in *Fundamentals of Quadruplex Structures*, ed. S. Neidle and S. Balasubramanian, The Royal Society of Chemistry, 2006.
13. J. B. Chaires and D. Graves, in *Quadruplex Nucleic Acids*, ed. J. B. Chaires and D. Graves, Springer-Verlag, Berlin Heidelberg, 2013.
14. A. K. Todd, M. Johnston and S. Neidle, Highly prevalent putative quadruplex sequence motifs in human DNA, *Nucleic Acids Res.*, 2005, **33**, 2901–2907.
15. J. L. Huppert and S. Balasubramanian, Prevalence of quadruplexes in the human genome, *Nucleic Acids Res.*, 2005, **33**, 2908–2916.
16. V. S. Chambers, G. Marsico, J. M. Boutell, M. Di Antonio, G. P. Smith and S. Balasubramanian, High-throughput sequencing of DNA G-quadruplex structures in the human genome, *Nat. Biotechnol.*, 2015, **33**, 877–881.
17. R. Hansel-Hertsch, D. Beraldi, S. V. Lensing, G. Marsico, K. Zyner, A. Parry, M. Di Antonio, J. Pike, H. Kimura, M. Narita, D. Tannahill and S. Balasubramanian, G-quadruplex structures mark human regulatory chromatin, *Nat. Genet.*, 2016, **48**, 1267–1272.
18. J. L. Huppert and S. Balasubramanian, G-quadruplexes in promoters throughout the human genome, *Nucleic Acids Res.*, 2007, **35**, 406–413.
19. P. Murat and S. Balasubramanian, Existence and consequences of G-quadruplex structures in DNA, *Curr. Opin. Genet. Dev.*, 2014, **25**, 22–29.
20. S. Balasubramanian, L. H. Hurley and S. Neidle, Targeting G-quadruplexes in gene promoters: a novel anticancer strategy?, *Nat. Rev. Drug Discovery*, 2011, **10**, 261–275.
21. G. Biffi, D. Tannahill, J. McCafferty and S. Balasubramanian, Quantitative visualization of DNA G-quadruplex structures in human cells, *Nat. Chem.*, 2013, **5**, 182–186.
22. A. Henderson, Y. Wu, Y. C. Huang, E. A. Chavez, J. Platt, F. B. Johnson, R. M. Brosh, D. Sen and P. M. Lansdorp, Detection of G-quadruplex DNA in mammalian cells, *Nucleic Acids Res.*, 2014, **42**, 860–869.

23. S. Neidle, The structures of quadruplex nucleic acids and their drug complexes, *Curr. Opin. Struct. Biol.*, 2009, **19**, 239–250.
24. M. Adrian, B. Heddi and A. T. Phan, NMR spectroscopy of G-quadruplexes, *Methods*, 2012, **57**, 11–24.
25. D. Monchaud and M. P. Teulade-Fichou, A hitchhiker's guide to G-quadruplex ligands, *Org. Biomol. Chem.*, 2008, **6**, 627–636.
26. S. A. Ohnmacht and S. Neidle, Small-molecule quadruplex-targeted drug discovery., *Bioorg. Med. Chem. Lett.*, 2014, **24**, 2602–2612.
27. S. Muller and R. Rodriguez, G-quadruplex interacting small molecules and drugs: from bench toward bedside, *Expert Rev. Clin. Pharmacol.*, 2014, **7**, 663–679.
28. S. L. Cree and M. A. Kennedy, Relevance of G-quadruplex structures to pharmacogenetics, *Front. Pharmacol.*, 2014, **5**, 160.
29. C. Sissi and M. Palumbo, Telomeric G-quadruplex architecture and interactions with potential drugs, *Curr. Pharm. Des.*, 2014, **20**, 6489–6509.
30. V. Sekaran, J. Soares and M. B. Jarstfer, Telomere maintenance as a target for drug discovery, *J. Med. Chem.*, 2014, **57**, 521–538.
31. M. K. Islam, P. J. Jackson, K. M. Rahman and D. E. Thurston, Recent advances in targeting the telomeric G-quadruplex DNA sequence with small molecules as a strategy for anticancer therapies, *Future Med. Chem.*, 2016, **8**, 1259–1290.
32. S. Neidle, Quadruplex Nucleic Acids as Novel Therapeutic Targets, *J. Med. Chem.*, 2016, **59**, 5987–6011.
33. S. Neidle, Quadruplex nucleic acids as targets for anticancer therapeutics, *Nat. Rev. Chem.*, 2017, **1**, 0041.
34. G. W. Collie, R. Promontorio, S. M. Hampel, M. Micco, S. Neidle and G. N. Parkinson, Structural Basis for Telomeric G-Quadruplex Targeting by Naphthalene Diimide Ligands,*J. Am. Chem. Soc.*, 2012, **134**, 2723–2731.
35. S. M. Haider, S. Neidle and G. N. Parkinson, A structural analysis of G-quadruplex/ligand interactions, *Biochimie*, 2011, **93**, 1239–1251.
36. S. Neidle, A Personal History of Quadruplex-Small Molecule Targeting, *Chem. Rec.*, 2015, **15**, 691–710.
37. S. Sparapani, S. M. Haider, F. Doria, M. Gunaratnam and S. Neidle, Rational Design of Acridine-Based Ligands with Selectivity for Human Telomeric Quadruplexes, *J. Am. Chem. Soc.*, 2010, **132**, 12263–12272.
38. J.-Q. Hou, S.-B. Chen, J.-H. Tan, T.-M. Ou, H.-B. Luo, D. Li, J. Xu, L.-Q. Gu and Z.-S. Huang, New insights into the structures of ligand-quadruplex complexes from molecular dynamics simulations, *J. Phys. Chem. B*, 2010, **114**, 15301–15310.
39. S. A. Ohnmacht, C. Marchetti, M. Gunaratnam, R. J. Besser, S. M. Haider, G. DiVita, H. L. Lowe, M. Mellinas-Gomez, S. Diocou, M. Robson, J. Šponer, B. Islam, R. Barbara Pedley, J. A. Hartley and S. Neidle, A G-quadruplex-binding compound showing anti-tumour activity in an in vivo model for pancreatic cancer, *Sci. Rep.*, 2015, **5**, 11385.
40. K. Mulholland, F. Siddiquei and C. Wu, Binding modes and pathway of RHPS4 to human telomeric G-quadruplex and duplex DNA probed by

all-atom molecular dynamics simulations with explicit solvent, *Phys. Chem. Chem. Phys.*, 2017, **19**, 18685–18694.
41. H. Han, D. R. Langley, A. Rangan and L. H. Hurley, Selective Interactions of Cationic Porphyrins with G-Quadruplex Structures, *J. Am. Chem. Soc.*, 2001, **123**, 8902–8913.
42. D. J. Cashman, R. Buscaglia, M. W. Freyer, J. Dettler, L. H. Hurley and E. A. Lewis, Molecular modeling and biophysical analysis of the c-MYC NHE-III1 silencer element, *J. Mol. Model.*, 2007, **14**, 93–101.
43. C. Marchetti, A. Minarini, V. Tumiatti, F. Moraca, L. Parrotta, S. Alcaro, R. Rigo, C. Sissi, M. Gunaratnam, S. A. Ohnmacht, S. Neidle and A. Milelli, Macrocyclic naphthalene diimides as G-quadruplex binders, *Bioorg. Med. Chem.*, 2015, **23**, 3819–3830.
44. S.-Y. Sheu, C.-H. Huang, J.-K. Zhou and D.-Y. Yang, Relative stability of G-quadruplex structures: Interactions between the human Bcl2 promoter region and derivatives of carbazole and diphenylamine, *Biopolymers*, 2014, **101**, 1038–1050.
45. J.-K. Zhou, D.-Y. Yang and S.-Y. Sheu, The molecular mechanism of ligand unbinding from the human telomeric G-quadruplex by steered molecular dynamics and umbrella sampling simulations, *Phys. Chem. Chem. Phys.*, 2015, **17**, 12857–12869.
46. M. Wilhelm, A. Mukherjee, B. Bouvier, K. Zakrzewska, J. T. Hynes and R. Lavery, Multistep Drug Intercalation: Molecular Dynamics and Free Energy Studies of the Binding of Daunomycin to DNA, *J. Am. Chem. Soc.*, 2012, **134**, 8588–8596.
47. M. Ebrahimi and T. Khayamian, Interactions of G-quadruplex DNA binding site with berberine derivatives and construct a structure-based QSAR using docking descriptors, *Med. Chem. Res.*, 2013, **23**, 1327–1339.
48. A. Arora, C. Balasubramanian, N. Kumar, S. Agrawal, R. P. Ojha and S. Maiti, Binding of berberine to human telomeric quadruplex – spectroscopic, calorimetric and molecular modeling studies, *FEBS J.*, 2008, **275**, 3971–3983.
49. T. Wilson, P. J. Costa, V. Félix, M. P. Williamson and J. A. Thomas, Structural Studies on Dinuclear Ruthenium(II) Complexes That Bind Diastereoselectively to an Antiparallel Folded Human Telomere Sequence, *J. Med. Chem.*, 2013, **56**, 8674–8683.
50. M.-Y. Kim, H. Vankayalapati, K. Shin-ya, K. Wierzba and L. H. Hurley, Telomestatin, a Potent Telomerase Inhibitor That Interacts Quite Specifically with the Human Telomeric Intramolecular G-Quadruplex, *J. Am. Chem. Soc.*, 2002, **124**, 2098–2099.
51. W. J. Chung, B. Heddi, M. Tera, K. Iida, K. Nagasawa and A. T. Phan, Solution Structure of an Intramolecular (3 + 1) Human Telomeric G-Quadruplex Bound to a Telomestatin Derivative, *J. Am. Chem. Soc.*, 2013, **135**, 13495–13501.
52. S. Agrawal, R. P. Ojha and S. Maiti, Energetics of the human Tel-22 quadruplex-telomestatin interaction: a molecular dynamics study, *J. Phys. Chem. B*, 2008, **112**, 6828–6836.

53. M. Read, R. J. Harrison, B. Romagnoli, F. A. Tanious, S. H. Gowan, A. P. Reszka, W. D. Wilson, L. R. Kelland and S. Neidle, Structure-based design of selective and potent G quadruplex-mediated telomerase inhibitors, *PNAS*, 2001, **98**, 4844–4849.
54. R. J. Harrison, J. Cuesta, G. Chessari, M. A. Read, S. K. Basra, A. P. Reszka, J. Morrell, S. M. Gowan, C. M. Incles, F. A. Tanious, W. D. Wilson, L. R. Kelland and S. Neidle, Trisubstituted acridine derivatives as potent and selective telomerase inhibitors, *J. Med. Chem.*, 2003, **46**, 4463–4476.
55. S. M. Gowan, J. R. Harrison, L. Patterson, M. Valenti, M. A. Read, S. Neidle and L. R. Kelland, A G-Quadruplex-Interactive Potent Small-Molecule Inhibitor of Telomerase Exhibiting in Vitro and in Vivo Antitumor Activity, *Mol. Pharmacol.*, 2002, **61**, 1154–1162.
56. C. Hounsou, L. Guittat, D. Monchaud, M. Jourdan, N. Saettel, J.-L. Mergny and M.-P. Teulade-Fichou, G-Quadruplex Recognition by Quinacridines: a SAR, NMR, and Biological Study, *ChemMedChem*, 2007, **2**, 655–666.
57. V. Gabelica, E. S. Baker, M.-P. Teulade-Fichou, E. De Pauw and M. T. Bowers, Stabilization and structure of telomeric and c-myc region intramolecular G-quadruplexes: the role of central cations and small planar ligands, *J. Am. Chem. Soc.*, 2007, **129**, 895–904.
58. M. Micco, G. W. Collie, A. G. Dale, S. A. Ohnmacht, I. Pazitna, M. Gunaratnam, A. P. Reszka and S. Neidle, Structure-based design and evaluation of naphthalene diimide G-quadruplex ligands as telomere targeting agents in pancreatic cancer cells, *J. Med. Chem.*, 2013, **56**, 2959–2974.
59. D. Sun, K. Guo, J. J. Rusche and L. H. Hurley, Facilitation of a structural transition in the polypurine/polypyrimidine tract within the proximal promoter region of the human VEGF gene by the presence of potassium and G-quadruplex-interactive agents, *Nucleic Acids Res.*, 2005, **33**, 6070–6080.
60. A. Siddiqui-Jain, C. L. Grand, D. J. Bearss and L. H. Hurley, Direct evidence for a G-quadruplex in a promoter region and its targeting with a small molecule to repress c-MYC transcription, *Proc. Natl. Acad. Sci. U. S. A.*, 2002, **99**, 11593–11598.
61. T. S. Dexheimer, D. Sun and L. H. Hurley, Deconvoluting the structural and drug-recognition complexity of the G-quadruplex-forming region upstream of the bcl-2 P1 promoter, *J. Am. Chem. Soc.*, 2006, **128**, 5404–5415.
62. K. Guo, A. Pourpak, K. Beetz-Rogers, V. Gokhale, D. Sun and L. H. Hurley, Formation of pseudosymmetrical G-quadruplex and i-motif structures in the proximal promoter region of the RET oncogene, *J. Am. Chem. Soc.*, 2007, **129**, 10220–10228.
63. J. Dai, M. Carver, L. H. Hurley and D. Yang, Solution structure of a 2 : 1 quindoline-c-MYC G-quadruplex: insights into G-quadruplex-interactive small molecule drug design, *J. Am. Chem. Soc.*, 2011, **133**, 17673–17680.

64. H. R. Nasiri, N. M. Bell, K. I. E. McLuckie, J. Husby, C. Abell, S. Neidle and S. Balasubramanian, Targeting a c-MYC G-quadruplex DNA with a fragment library, *Chem. Commun.*, 2014, **50**, 1704–1707.
65. H. Zhang, X. Xiao, P. Wang, S. Pang, F. Qu, X. Ai and J. Zhang, Conformational conversion of DNA G-quadruplex induced by a cationic porphyrin, *Spectrochim. Acta, Part A*, 2009, **74**, 243–247.
66. C.-C. Chang, J.-Y. Wu, C.-W. Chien, W.-S. Wu, H. Liu, C.-C. Kang, L.-J. Yu and T.-C. Chang, A fluorescent carbazole derivative: high sensitivity for quadruplex DNA, *Anal. Chem.*, 2003, **75**, 6177–6183.
67. C.-C. Chang, I. C. Kuo, I. F. Ling, C.-T. Chen, H.-C. Chen, P.-J. Lou, J.-J. Lin and T.-C. Chang, Detection of quadruplex DNA structures in human telomeres by a fluorescent carbazole derivative, *Anal. Chem.*, 2004, **76**, 4490–4494.
68. C.-C. Chang, C.-W. Chien, Y.-H. Lin, C.-C. Kang and T.-C. Chang, Investigation of spectral conversion of d(TTAGGG)$_4$ and d(TTAGGG)$_{13}$ upon potassium titration by a G-quadruplex recognizer BMVC molecule, *Nucleic Acids Res.*, 2007, **35**, 2846–2860.
69. T.-C. Chang and C.-C. Chang, Detection of G-Quadruplexes in Cells and Investigation of G-Quadruplex Structure of d($T_2AG_3$)$_4$ in $K+$ Solution by a Carbazole Derivative: BMVC, in *G-Quadruplex DNA*, ed. P. Baumann, Humana Press, 2010, pp. 183–206.
70. D.-Y. Yang, T.-C. Chang and S.-Y. Sheu, Interaction between human telomere and a carbazole derivative: a molecular dynamics simulation of a quadruplex stabilizer and telomerase inhibitor, *J. Phys. Chem. A*, 2007, **111**, 9224–9232.
71. A. De Cian, E. DeLemos, J.-L. Mergny, M.-P. Teulade-Fichou and D. Monchaud, Highly Efficient G-Quadruplex Recognition by Bisquinolinium Compounds, *J. Am. Chem. Soc.*, 2007, **129**, 1856–1857.
72. W. J. Chung, B. Heddi, F. Hamon, M.-P. Teulade-Fichou and A. T. Phan, Solution Structure of a G-quadruplex Bound to the Bisquinolinium Compound Phen-DC3, *Angew. Chem., Int. Ed.*, 2014, **53**, 999–1002.
73. A. Piazza, J.-B. Boulé, J. Lopes, K. Mingo, E. Largy, M.-P. Teulade-Fichou and A. Nicolas, Genetic instability triggered by G-quadruplex interacting Phen-DC compounds in Saccharomyces cerevisiae, *Nucleic Acids Res.*, 2010, **38**, 4337–4348.
74. G. R. Clark, P. D. Pytel, C. J. Squire and S. Neidle, Structure of the first parallel DNA quadruplex-drug complex, *J. Am. Chem. Soc.*, 2003, **125**, 4066–4067.
75. A. K. Ghosh, F. K. Bhattacharyya and D. K. Ghosh, Leishmania donovani: Amastigote inhibition and mode of actior of berberine, *Exp. Parasitol.*, 1985, **60**, 404–413.
76. I. Naasani, H. Seimiya, T. Yamori and T. Tsuruo, FJ5002: A Potent Telomerase Inhibitor Identified by Exploiting the Disease-oriented Screening Program with COMPARE Analysis, *Cancer Res.*, 1999, **59**, 4004–4011.

77. X. Shi and J. B. Chaires, Sequence- and structural-selective nucleic acid binding revealed by the melting of mixtures, *Nucleic Acids Res.*, 2006, **34**, e14.
78. K. Shin-ya, K. Wierzba, K. Matsuo, T. Ohtani, Y. Yamada, K. Furihata, Y. Hayakawa and H. Seto, Telomestatin, a novel telomerase inhibitor from Streptomyces anulatus, *J. Am. Chem. Soc.*, 2001, **123**, 1262–1263.
79. P. S. Shirude, E. R. Gillies, S. Ladame, F. Godde, K. Shin-Ya, I. Huc and S. Balasubramanian, Macrocyclic and helical oligoamides as a new class of G-quadruplex ligands, *J. Am. Chem. Soc.*, 2007, **129**, 11890–11891.
80. M. Tera, H. Ishizuka, M. Takagi, M. Suganuma, K. Shin-ya and K. Nagasawa, Macrocyclic Hexaoxazoles as Sequence- and Mode-Selective G-Quadruplex Binders, *Angew. Chem., Int. Ed.*, 2008, **47**, 5557–5560.
81. J. Stanslas, D. J. Hagan, M. J. Ellis, C. Turner, J. Carmichael, W. Ward, T. R. Hammonds and M. F. G. Stevens, Antitumor Polycyclic Acridines. 7. Synthesis and Biological Properties of DNA Affinic Tetra- and Pentacyclic Acridines, *J. Med. Chem.*, 2000, **43**, 1563–1572.
82. E. Salvati, C. Leonetti, A. Rizzo, M. Scarsella, M. Mottolese, R. Galati, I. Sperduti, M. F. G. Stevens, M. D'Incalci, M. Blasco, G. Chiorino, S. Bauwens, B. Horard, E. Gilson, A. Stoppacciaro, G. Zupi and A. Biroccio, Telomere damage induced by the G-quadruplex ligand RHPS4 has an antitumor effect, *J. Clin. Invest.*, 2007, **117**, 3236–3247.
83. J. C. Cookson, F. Dai, V. Smith, R. A. Heald, C. A. Laughton, M. F. G. Stevens and A. M. Burger, Pharmacodynamics of the G-Quadruplex-Stabilizing Telomerase Inhibitor 3,11-Difluoro-6,8,13-trimethyl-8H-quino[4,3,2-kl]acridinium methosulfate (RHPS4) in Vitro: Activity in Human Tumor Cells Correlates with Telomere Length and Can Be Enhanced, or Antagonized, with Cytotoxic Agents, *Mol. Pharmacol.*, 2005, **68**, 1551–1558.
84. F. Lopez-Vallejo, T. Caulfield, K. Martinez-Mayorga, M. A. Giulianotti, A. Nefzi, R. A. Houghten and J. L. Medina-Franco, Integrating Virtual Screening and Combinatorial Chemistry for Accelerated Drug Discovery, *Comb. Chem. High throughput Screening*, 2011, **14**, 475–487.
85. F. Cuenca, O. Greciano, M. Gunaratnam, S. Haider, D. Munnur, R. Nanjunda, W. D. Wilson and S. Neidle, Tri- and tetra-substituted naphthalene diimides as potent G-quadruplex ligands, *Bioorg. Med. Chem. Lett.*, 2008, **18**, 1668–1673.
86. C. S. D. Magalhães, H. J. C. Barbosa and L. E. Dardenne, A genetic algorithm for the ligand-protein docking problem, *Genet. Mol. Biol.*, 2004, **27**, 605–610.
87. J. Hughes, S. Rees, S. Kalindjian and K. Philpott, Principles of early drug discovery: principles of early drug discovery, *Br. J. Pharmacol.*, 2011, **162**, 1239–1249.
88. J. D. Durrant and J. A. McCammon, Molecular dynamics simulations and drug discovery, *BMC Biol.*, 2011, **9**, 71.

89. X.-Y. Meng, H.-X. Zhang, M. Mezei and M. Cui, Molecular Docking: A powerful approach for structure-based drug discovery, *Curr. Comput.-Aided Drug Des.*, 2011, **7**, 146–157.
90. P. Várnai and K. Zakrzewska, DNA and its counterions: a molecular dynamics study, *Nucleic Acids Res.*, 2004, **32**, 4269–4280.
91. A. Pérez, I. Marchán, D. Svozil, J. Sponer, T. E. Cheatham, C. A. Laughton and M. Orozco, Refinement of the AMBER Force Field for Nucleic Acids: Improving the Description of $\alpha/\gamma$ Conformers, *Biophys. J.*, 2007, **92**, 3817–3829.
92. B. Islam, P. Stadlbauer, A. Gil-Ley, G. Pérez-Hernández, S. Haider, S. Neidle, G. Bussi, P. Banas, M. Otyepka and J. Sponer, Exploring the Dynamics of Propeller Loops in Human Telomeric DNA Quadruplexes Using Atomistic Simulations, *J. Chem. Theory Comput.*, 2017, **13**, 2458–2480.
93. B. Islam, P. Stadlbauer, M. Krepl, J. Koca, S. Neidle, S. Haider and J. Sponer, Extended molecular dynamics of a c-kit promoter quadruplex, *Nucleic Acids Res.*, 2015, **43**, 8673–8693.
94. B. Islam, M. Sgobba, C. Laughton, M. Orozco, J. Sponer, S. Neidle and S. Haider, Conformational dynamics of the human propeller telomeric DNA quadruplex on a microsecond time scale, *Nucleic Acids Res.*, 2013, **41**, 2723–2735.
95. J. Šponer, G. Bussi, P. Stadlbauer, P. Kührová, P. Banáš, B. Islam, S. Haider, S. Neidle and M. Otyepka, Folding of guanine quadruplex molecules–funnel-like mechanism or kinetic partitioning? An overview from MD simulation studies, *Biochim. Biophys. Acta, Gen. Subj.*, 2017, **1861**, 1246–1263.
96. I. Ivani, P. D. Dans, A. Noy, A. Pérez, I. Faustino, A. Hospital, J. Walther, P. Andrio, R. Goñi, A. Balaceanu, G. Portella, F. Battistini, J. L. Gelpí, C. González, M. Vendruscolo, C. A. Laughton, S. A. Harris, D. A. Case and M. Orozco, Parmbsc1: a refined force field for DNA simulations, *Nat. Methods*, 2016, **13**, 55–58.
97. M. Krepl, M. Zgarbová, P. Stadlbauer, M. Otyepka, P. Banáš, J. Koča, T. E. Cheatham, P. Jurečka and J. Sponer, Reference simulations of noncanonical nucleic acids with different $\chi$ variants of the AMBER force field: quadruplex DNA, quadruplex RNA and Z-DNA, *J. Chem. Theory Comput.*, 2012, **8**, 2506–2520.
98. M. Zgarbová, F. J. Luque, J. Sponer, T. E. Cheatham, M. Otyepka and P. Jurečka, Toward Improved Description of DNA Backbone: Revisiting Epsilon and Zeta Torsion Force Field Parameters, *J. Chem. Theory Comput.*, 2013, **9**, 2339–2354.
99. M. Zgarbova, J. Sponer, M. Otyepka, T. E. Cheatham, R. Galindo-Murillo and P. Jurecka, Refinement of the Sugar-Phosphate Backbone Torsion Beta for AMBER Force Fields Improves the Description of Z- and B-DNA, *J. Chem. Theory Comput.*, 2015, **11**, 5723–5736.
100. R. Galindo-Murillo, J. C. Robertson, M. Zgarbová, J. Šponer, M. Otyepka, P. Jurečka and T. E. Cheatham, Assessing the Current State

of Amber Force Field Modifications for DNA, *J. Chem. Theory Comput.*, 2016.
101. W. S. Ross and C. C. Hardin, Ion-Induced Stabilization of the G-DNA Quadruplex: Free Energy Perturbation Studies, *J. Am. Chem. Soc.*, 1994, **116**, 6070–6080.
102. D. M. York, T. A. Darden and L. G. Pedersen, The effect of long-range electrostatic interactions in simulations of macromolecular crystals: A comparison of the Ewald and truncated list methods, *J. Chem. Phys.*, 1993, **99**, 8345.
103. T. E. I. Cheatham, J. L. Miller, T. Fox, T. A. Darden and P. A. Kollman, Molecular Dynamics Simulations on Solvated Biomolecular Systems: The Particle Mesh Ewald Method Leads to Stable Trajectories of DNA, RNA, and Proteins, *J. Am. Chem. Soc.*, 1995, **117**, 4193–4194.
104. G. Collie, A. P. Reszka, S. M. Haider, V. Gabelica, G. N. Parkinson and S. Neidle, Selectivity in small molecule binding to human telomeric RNA and DNA quadruplexes, *Chem. Commun.*, 2009, 7482.
105. J. Šponer, J. Leszczynski and P. Hobza, Electronic properties, hydrogen bonding, stacking, and cation binding of DNA and RNA bases, *Biopolymers*, 2001, **61**, 3–31.
106. C. I. Bayly, P. Cieplak, W. Cornell and P. A. Kollman, A well-behaved electrostatic potential based method using charge restraints for deriving atomic charges: the RESP model, *J. Phys. Chem.*, 1993, **97**, 10269–10280.
107. B. Luisi, M. Orozco, J. Sponer, F. J. Luque and Z. Shakked, On the potential role of the amino nitrogen atom as a hydrogen bond acceptor in macromolecules, *J. Mol. Biol.*, 1998, **279**, 1123–1136.
108. W. D. Cornell, P. Cieplak, C. I. Bayly, I. R. Gould, K. M. Merz, D. M. Ferguson, D. C. Spellmeyer, T. Fox, J. W. Caldwell and P. A. Kollman, A Second Generation Force Field for the Simulation of Proteins, Nucleic Acids, and Organic Molecules, *J. Am. Chem. Soc.*, 1995, **117**, 5179–5197.
109. D. A. Pearlman, D. A. Case, J. W. Caldwell, W. S. Ross, T. E. Cheatham, S. DeBolt, D. Ferguson, G. Seibel and P. Kollman, AMBER, a package of computer programs for applying molecular mechanics, normal mode analysis, molecular dynamics and free energy calculations to simulate the structural and energetic properties of molecules, *Comput. Phys. Commun.*, 1995, **91**, 1–41.
110. N. H. Campbell, G. N. Parkinson, A. P. Reszka and S. Neidle, Structural basis of DNA quadruplex recognition by an acridine drug, *J. Am. Chem. Soc.*, 2008, **130**, 6722–6724.
111. E. Gavathiotis, R. A. Heald, M. F. G. Stevens and M. S. Searle, Drug Recognition and Stabilisation of the Parallel-stranded DNA Quadruplex d(TTAGGGT)4 Containing the Human Telomeric Repeat, *J. Mol. Biol.*, 2003, **334**, 25–36.
112. A. T. Phan, V. Kuryavyi, H. Y. Gaw and D. J. Patel, Small-molecule interaction with a five-guanine-tract G-quadruplex structure from the human MYC promoter, *Nat. Chem. Biol.*, 2005, **1**, 167–173.

113. G. Prato, S. Silvent, S. Saka, M. Lamberto and D. Kosenkov, Thermodynamics of Binding of Di- and Tetrasubstituted Naphthalene Diimide Ligands to DNA G-Quadruplex, *J. Phys. Chem. B*, 2015, **119**, 3335–3347.
114. A. Ali, M. Bansal and S. Bhattacharya, Ligand 5,10,15,20-tetra(N-methyl-4-pyridyl)porphine (TMPyP4) prefers the parallel propeller-type human telomeric G-quadruplex DNA over its other polymorphs, *J. Phys. Chem. B*, 2015, **119**, 5–14.
115. G. M. Morris, D. S. Goodsell, R. S. Halliday, R. Huey, W. E. Hart, R. K. Belew and A. J. Olson, Automated docking using a Lamarckian genetic algorithm and an empirical binding free energy function, *J. Comput. Chem.*, 1998, **19**, 1639–1662.
116. E. M. Rezler, J. Seenisamy, S. Bashyam, M.-Y. Kim, E. White, W. D. Wilson and L. H. Hurley, Telomestatin and Diseleno Sapphyrin Bind Selectively to Two Different Forms of the Human Telomeric G-Quadruplex Structure, *J. Am. Chem. Soc.*, 2005, **127**, 9439–9447.
117. B. Pagano, A. Virno, C. A. Mattia, L. Mayol, A. Randazzo and C. Giancola, Targeting DNA quadruplexes with distamycin A and its derivatives: An ITC and NMR study, *Biochimie*, 2008, **90**, 1224–1232.
118. M. J. Cocco, L. A. Hanakahi, M. D. Huber and N. Maizels, Specific interactions of distamycin with G-quadruplex DNA, *Nucleic Acids Res.*, 2003, **31**, 2944–2951.
119. A. Randazzo, A. Galeone and L. Mayol, 1H-NMR study of the interaction of distamycin A and netropsin with the parallel stranded tetraplex $[d(TGGGGT)]_4$, *Chem. Commun.*, 2001, 1030–1031.
120. L. Martino, A. Virno, B. Pagano, A. Virgilio, S. Di Micco, A. Galeone, C. Giancola, G. Bifulco, L. Mayol and A. Randazzo, Structural and Thermodynamic Studies of the Interaction of Distamycin A with the Parallel Quadruplex Structure [d(TGGGGT)]4, *J. Am. Chem. Soc.*, 2007, **129**, 16048–16056.
121. S. Cosconati, L. Marinelli, R. Trotta, A. Virno, L. Mayol, E. Novellino, A. J. Olson and A. Randazzo, Tandem application of virtual screening and NMR experiments in the discovery of brand new DNA quadruplex groove binders, *J. Am. Chem. Soc.*, 2009, **131**, 16336–16337.
122. F. J. Solis and R. J.-B. Wets, Minimization by Random Search Techniques, *Math. Oper. Res.*, 1981, **6**, 19–30.
123. R. Trotta, S. De Tito, I. Lauri, V. La Pietra, L. Marinelli, S. Cosconati, L. Martino, M. R. Conte, L. Mayol, E. Novellino and A. Randazzo, A more detailed picture of the interactions between virtual screening-derived hits and the DNA G-quadruplex: NMR, molecular modelling and ITC studies, *Biochimie*, 2011, **93**, 1280–1287.
124. G. W. Collie, S. M. Haider, S. Neidle and G. N. Parkinson, A crystallographic and modelling study of a human telomeric RNA (TERRA) quadruplex, *Nucleic Acids Res.*, 2010, **38**, 5569–5580.
125. P. J. van Laarhoven, in *Simulated Annealing*, ed. P. J. V. Laarhoven and E. H. Aarts, Springer, Netherlands, 1987, vol. 37, pp. 7–15.

126. P. Lv, L. Yuan and J. Zhang, Cloud theory-based simulated annealing algorithm and application, *Eng. Appl. Artif. Intell.*, 2009, **22**, 742–749.
127. A. S. Aspbury and R. M. Gibson, Long-range visibility of greater sage grouse leks: a GIS-based analysis, *Anim. Behav.*, 2004, **67**, 1127–1132.
128. W. Duan, A. Rangan, H. Vankayalapati, M.-Y. Kim, Q. Zeng, D. Sun, H. Han, O. Y. Fedoroff, D. Nishioka, S. Y. Rha, E. Izbicka, D. D. V. Hoff and L. H. Hurley, Design and Synthesis of Fluoroquinophenoxazines That Interact with Human Telomeric G-Quadruplexes and Their Biological Effects1, *Mol. Cancer Ther.*, 2001, **1**, 103–120.
129. L. Petraccone, I. Fotticchia, A. Cummaro, B. Pagano, L. Ginnari-Satriani, S. Haider, A. Randazzo, E. Novellino, S. Neidle and C. Giancola, The triazatruxene derivative azatrux binds to the parallel form of the human telomeric G-quadruplex under molecular crowding conditions: Biophysical and molecular modeling studies, *Biochimie*, 2011, **93**, 1318–1327.
130. L. Zhang, J. C. Er, K. K. Ghosh, W. J. Chung, J. Yoo, W. Xu, W. Zhao, A. T. Phan and Y.-T. Chang, Discovery of a Structural-Element Specific G-Quadruplex "Light-Up" Probe, *Sci. Rep.*, 2014, 4.
131. A. Amadei, A. B. Linssen, B. L. de Groot, D. M. van Aalten and H. J. Berendsen, An efficient method for sampling the essential subspace of proteins, *J. Biomol. Struct. Dyn.*, 1996, **13**, 615–625.
132. D. M. van Aalten, A. Amadei, A. B. Linssen, V. G. Eijsink, G. Vriend and H. J. Berendsen, The essential dynamics of thermolysin: confirmation of the hinge-bending motion and comparison of simulations in vacuum and water, *Proteins*, 1995, **22**, 45–54.
133. K. Réblová, Z. Střelcová, P. Kulhánek, I. Beššeová, D. H. Mathews, K. Van Nostrand, I. Yildirim, D. H. Turner and J. Šponer, An RNA molecular switch: Intrinsic flexibility of 23S rRNA helices 40 and 68 5′-UAA/5′-GAN internal loops studied by molecular dynamics methods, *J. Chem. Theory Comput.*, 2010, **6**, 910–929.
134. M.-H. Li, Q. Luo, X.-G. Xue and Z.-S. Li, Molecular dynamics studies of the 3D structure and planar ligand binding of a quadruplex dimer, *J. Mol. Model.*, 2011, **17**, 515–526.
135. J. Bhat and S. Chatterjee, Skeleton selectivity in complexation of chelerythrine and chelerythrine-like natural plant alkaloids with the G-quadruplex formed at the promoter of c-MYC oncogene: in silico exploration, *RSC Adv.*, 2016, **6**, 36667–36680.
136. F. Fogolari, A. Brigo and H. Molinari, Protocol for MM/PBSA Molecular Dynamics Simulations of Proteins, *Biophys. J.*, 2003, **85**, 159–166.
137. T. Hou, J. Wang, Y. Li and W. Wang, Assessing the performance of the MM/PBSA and MM/GBSA methods. 1. The accuracy of binding free energy calculations based on molecular dynamics simulations, *J. Chem. Inf. Model.*, 2011, **51**, 69–82.

138. R. Luo, L. David and M. K. Gilson, Accelerated Poisson-Boltzmann calculations for static and dynamic systems, *J. Comput. Chem.*, 2002, **23**, 1244–1253.
139. D. Sitkoff, K. A. Sharp and B. Honig, Accurate Calculation of Hydration Free Energies Using Macroscopic Solvent Models, *J. Phys. Chem.*, 1994, **98**, 1978–1988.
140. B. Islam, P. Stadlbauer, S. Neidle, S. Haider and J. Sponer, Can We Execute Reliable MM-PBSA Free Energy Computations of Relative Stabilities of Different Guanine Quadruplex Folds?, *J. Phys. Chem. B*, 2016, **120**, 2899–2912.
141. J. G. Kirkwood, Statistical Mechanics of Fluid Mixtures, *J. Chem. Phys.*, 1935, **3**, 300.
142. S. A. Martins, S. F. Sousa, M. J. Ramos and P. A. Fernandes, Prediction of Solvation Free Energies with Thermodynamic Integration Using the General Amber Force Field, *J. Chem. Theory Comput.*, 2014, **10**, 3570–3577.
143. M. Bončina, Č. Podlipnik, I. Piantanida, J. Eilmes, M.-P. Teulade-Fichou, G. Vesnaver and J. Lah, Thermodynamic fingerprints of ligand binding to human telomeric G-quadruplexes, *Nucleic Acids Res.*, 2015, **43**, 10376–10386.
144. G. M. Torrie and J.-P. Valleau, Nonphysical sampling distributions in Monte Carlo free-energy estimation: Umbrella sampling, *J. Comput. Phys.*, 1977, **23**, 187–199.
145. T. C. Beutler and W. F. v. Gunsteren, The computation of a potential of mean force: Choice of the biasing potential in the umbrella sampling technique, *J. Chem. Phys.*, 1994, **100**, 1492–1497.
146. Y.-K. Wang, D.-Q. Wei, R.-X. Gu, H.-M. Fan and J. Ulmschneider, Applications of rare event dynamics on the free energy calculations for membrane protein systems, *Can. J. Chem.*, 2013, **91**, 769–774.
147. D. Shukla, C. X. Hernández, J. K. Weber and V. S. Pande, Markov State Models Provide Insights into Dynamic Modulation of Protein Function, *Acc. Chem. Res.*, 2015, **48**, 414–422.
148. N. Plattner and F. Noe, Protein conformational plasticity and complex ligand-binding kinetics explored by atomistic simulations and Markov models, *Nat. Commun.*, 2015, **6**, 7653.
149. F. Noé, C. Schütte, E. Vanden-Eijnden, L. Reich and T. R. Weikl, Constructing the equilibrium ensemble of folding pathways from short off-equilibrium simulations, *Proc. Natl. Acad. Sci. U. S. A.*, 2009, **106**, 19011–19016.
150. V. S. Pande, K. Beauchamp and G. R. Bowman, Everything you wanted to know about Markov State Models but were afraid to ask, *Methods*, 2010, **52**, 99–105.
151. F. Noé and S. Fischer, Transition networks for modeling the kinetics of conformational change in macromolecules, *Curr. Opin. Struct. Biol.*, 2008, **18**, 154–162.

152. T. Siebenmorgen and M. Zacharias, Origin of Ion Specificity of Telomeric DNA G-Quadruplexes Investigated by Free-Energy Simulations, *Biophys. J.*, 2017, **112**, 2280–2290.
153. R. Rocca, C. Talarico, F. Moraca, G. Costa, I. Romeo, F. Ortuso, S. Alcaro and A. Artese, Molecular recognition of a carboxy pyridostatin toward G-quadruplex structures: Why does it prefer RNA?, *Chem. Biol. Drug Des.*, 2017.
154. K. Mulholland and C. Wu, Binding of Telomestatin to a Telomeric G-Quadruplex DNA Probed by All-Atom Molecular Dynamics Simulations with Explicit Solvent, *J. Chem. Inf. Model.*, 2016, **56**, 2093–2102.
155. A. Spinello, G. Barone and J. Grunenberg, Molecular recognition of naphthalene diimide ligands by telomeric quadruplex-DNA: the importance of the protonation state and mediated hydrogen bonds, *Phys. Chem. Chem. Phys.*, 2016, **18**, 2871–2877.

CHAPTER 12

# Molecular Recognition of DNA by Py–Im Polyamides: From Discovery to Oncology

PETER B. DERVAN,* ALEXIS A. KURMIS AND PAUL B. FINN

Division of Chemistry and Chemical Engineering, California Institute of Technology, Pasadena, CA 91125, USA
*Email: dervan@caltech.edu

## 12.1 Introduction—DNA-targeted Therapeutics

DNA-targeted therapeutics have been a mainstay in oncology due to their efficacy and broad applicability.[1–5] As such, understanding the physical and biological characteristics of these molecules has been an area of intense research. Pyrrole–imidazole (Py–Im) polyamides evolved over the course of 20 years from the natural product distamycin to a class of programmable DNA-binding oligomers with high sequence specificity and affinity. The natural product distamycin contains three N-methylpyrrole (Py) amino acids and binds DNA at A–T tracts four to five base pairs (bp) in size.[6,7] Structural studies of distamycin–DNA complexes revealed that the crescent-shaped molecule binds in the minor groove in both 1 : 1 and 2 : 1 ligand:DNA stoichiometries.[8–11] Both the shape complementarity and the specific hydrogen-bonding with the edges of A–T base pairs account for its affinity and specificity. The development of unbiased methods, such as footprinting and affinity cleavage, allowed new heterocycle amino acids to be screened for sequence specificity.[12–17] A set of heterocycle ring pairs that bind a large number of different DNA sequences in a programmable fashion was created

Chemical Biology No. 7
DNA-targeting Molecules as Therapeutic Agents
Edited by Michael J. Waring

Published by the Royal Society of Chemistry, www.rsc.org

and characterised.[18] Successive iterations advanced from small synthetic molecules modelled on distamycin to the modern hairpin polyamide structure. A set of aromatic amino acids can be combined as modular pairs in the minor groove of DNA to bind a large repertoire of DNA sequences with affinity and specificity similar to those of DNA-binding proteins. In this chapter we will summarise briefly highlights of the evolution of the design of small molecules for sequence-specific recognition of DNA. The main goal will be an update regarding biological properties since our last reviews 2001–2005.[18–20] This chapter will focus on research from the Dervan laboratory at Caltech.

## 12.2 Pairing Rules in the Minor Groove

In a formal sense, the edges of the four Watson–Crick base pairs can be differentiated on the minor groove floor by the specific positions of hydrogen bond donors and acceptors, by differences in shape, and by electronic potential surfaces. Footprinting/affinity cleaving experiments revealed that the imidazole (Im)-containing polyamide ImPyPy bound to the five bp sequence 5′-WGWCW-3′ (where W = A or T) in an antiparallel 2 : 1 polyamide–DNA complex.[21,22] The data were consistent with an Im–Py pair targeting G–C, Py–Im targeting C–G and Py–Py degenerate for T–A and A–T. NMR studies and crystal structures confirmed that this specificity is due to hydrogen bonding between the imidazole nitrogen of the polyamide and the exocyclic 2-amino group of guanine on the minor groove edge of a G–C base pair.[23–25] DNA binding affinity can be attributed, in part, to the hydrophobic interaction of the polyamide with the deep, narrow DNA minor groove; biophysical experiments showed no significant interaction with a much wider, shallower RNA minor groove.[26–28] The Im–Py pair has been explored by extensive studies including analyses of binding in hundreds of different sequence contexts.[29–33] The unanticipated discovery that an unsymmetrical ring pair Im–Py was key to distinguishing the edge of a Watson–Crick base pair (G–C *vs.* C–G) set us to dream of conquering the missing half of the recognition puzzle, selecting T–A from A–T, by inventing a new ring pair.

In contrast to G–C, the A–T base pair is fairly symmetrical; both adenine and thymine present hydrogen bond acceptors on the floor of the minor groove. A Py–Py pairing does not discriminate between A–T and T–A base pairs.[10,11] There exists, however, a small asymmetric cleft between the thymine $O_2$ and adenine $C_2$, and the $N_3$ of adenine presents only one lone pair while the $O_2$ of thymine presents two lone pairs capable of hydrogen bonding. Guided by high-resolution crystallographic data and our simple bump and hole model, 3-hydroxypyrrole (Hp) was designed to specifically bind a T–A base pair when paired with Py.[34] Crystallographic studies on polyamides containing the Hp–Py pair have shown that specificity arises from a combination of hydrogen bonding between the hydroxyl and the thymine $O_2$, together with shape-selective recognition of the asymmetric cleft as predicted.[35,36]

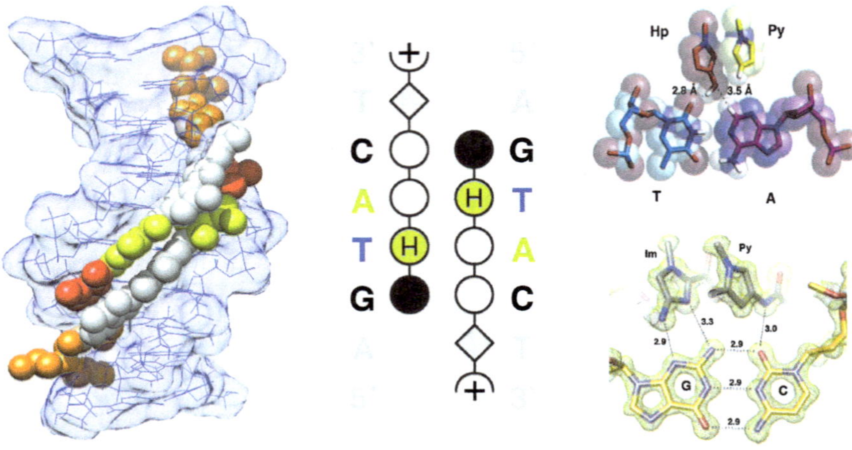


**Figure 12.1** (Left) X-ray structure of $(ImHpPyPy)_2 \cdot 5'$-CC**AGTACT**GG-3′. The aromatic residues of the dimer are coloured: Hp is yellow, Py is white, Im is red. The β-alanine (β) and dimethylaminopropylamine (Dp) are shown in orange. (Middle) A schematic with the aromatic residues of the ImHpPyPy polyamide indicated by filled circles and β by the diamonds. (Right) From the X-ray structure, space-filling model of the Hp–Py pair and Im–Py interacting with the T–A base pair and the G–C base pair, respectively.

Together, these three aromatic amino acids—Py, Im and Hp—can be combined as four unsymmetrical ring pairs to recognise and distinguish each of the four Watson–Crick base pairs (Figure 12.1).[35–37] Footprinting and affinity cleavage studies have demonstrated that Im–Py is specific for G–C and Hp–Py for T–A.[21,34,38] Although these interactions can be conveniently described as pairing rules, those rules should be considered as guidelines. For example, the pairing code Im–Py for G–C is agnostic regarding sequence context. Undoubtedly, the sequence-dependent microstructure of DNA (minor groove width, curvature, tilt and twist of the base pairs) would be expected to play a role regarding energetics of binding, some sequences being more ideal than others.

## 12.3 The Hairpin Structure

In principle, crescent-shaped dimers of multi-ring Py–Im polyamides could bind antiparallel to each other in perfect register with complete overlap.[39] Alternatively, antiparallel oligomers in the minor groove might adopt slipped dimer structures, affording some ambiguity in predicting the sequence targeted.[40,41] In order to align ring pairs in a predictable sense, covalent linkage of the carboxy and amino termini with an aliphatic tether should afford a hairpin structure where unsymmetrical aromatic ring pairs in an oligomer are unambiguously paired when folded in the minor groove at precise target base-pair assignment.[42–47] Turn units of varying alkyl length have been evaluated; a C4 γ-diaminobutyric acid linker was found to be optimal for the hairpin binding (Figure 12.2).

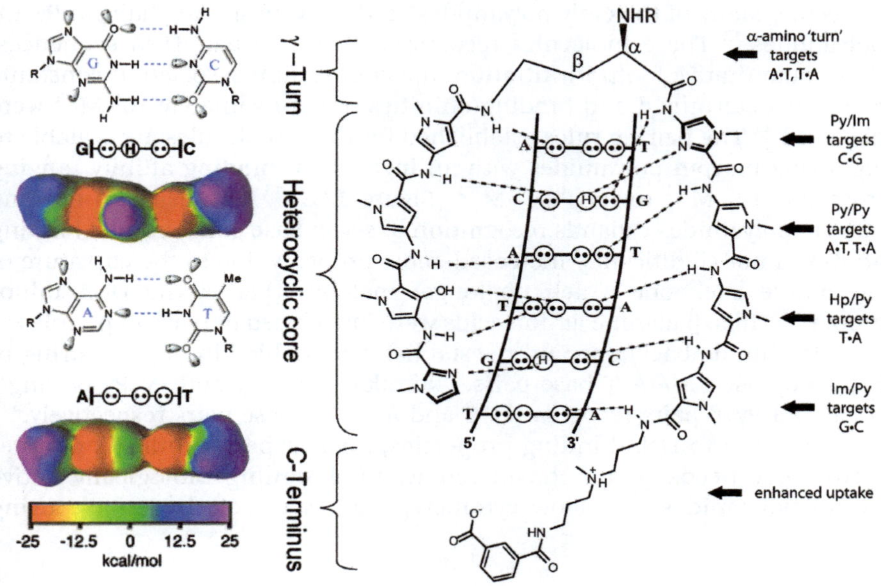


**Figure 12.2** (Left) Chemical structure of Watson–Crick base pairs. Circles with two dots indicate a hydrogen bond acceptor while a circle with an H indicates a hydrogen bond donor. Electronic surfaces of the base pair edges are shown with colours indicating areas of negative and positive electronic charges. (Right) Schematic illustration of hairpin polyamide–DNA recognition through formation of hydrogen bonds with the floor of the DNA minor groove. The γ-turn, heterocycle core and C-terminus modification are indicated by braces and labelled. Pairing rules for the heterocycles are shown to the right.

Covalent linkage of two antiparallel Py–Im polyamide strands was found to result in molecules with ~100-fold higher affinity compared with unlinked dimers.[42] The hairpin architecture has been the prominent motif used in our group due to ease of synthesis and superior biological activity. Each hairpin polyamide consists of the core heterocyclic DNA-recognition domain, a γ-hairpin turn and a C-terminus, each of which can be modified to optimise the properties of a DNA-binding polyamide for a specific application. Several linking approaches were explored beyond the hairpin; H-pin and U-pin motifs, each consisting of linkers of varying alkyl chain lengths.[47–49] In addition, more complex structures such as cycles, tandem hairpin dimers and extended hairpins were characterised, employing techniques such as DNAse1 footprinting and Fe-mediated affinity cleavage to determine binding affinity and specificity.[50–53]

## 12.4 Binding Site Size: β–β, Im–β and Py–β Pairs

The sequence specificity, binding affinity and biological activity of Py–Im polyamides are all affected by many factors, including the size of the molecule. Many of the early polyamides studied were six-ring hairpin Py–Im polyamides.[42] These molecules recognized five base pair DNA sequences. From quantitative footprint titration, the equilibrium association constants could be determined and binding affinities of approximately $10^8$ $M^{-1}$ were observed.[42] The pairing rules established for these molecules are scalable to eight-ring hairpin polyamides with an increase in binding affinity ranging from low $10^8$ $M^{-1}$ to low $10^{10}$ $M^{-1}$ (Figure 12.3).[43] Expansion to ten-ring hairpin polyamides expands recognition to seven base pairs, but the binding affinity is not significantly increased–most probably due to the curvature of the amide backbone which no longer matches that of the DNA minor groove.[45] Thus, β-alanine amino acids were introduced in place of pyrroles to relax the amide backbone and re-establish favourable binding.[40] Pairing of β–β recognises T–A/A–T base pairs.[40] Similar to Im–Py and Py–Py pairings, Im–β and Py–β pairs recognise G–C and A–T/T–A base pairs respectively.[54]

In addition to DNA-binding properties, nuclear permeability is another factor that needs to be considered when designing biologically active Py–Im polyamides.[55–57] Flow cytometry studies of cellular uptake using

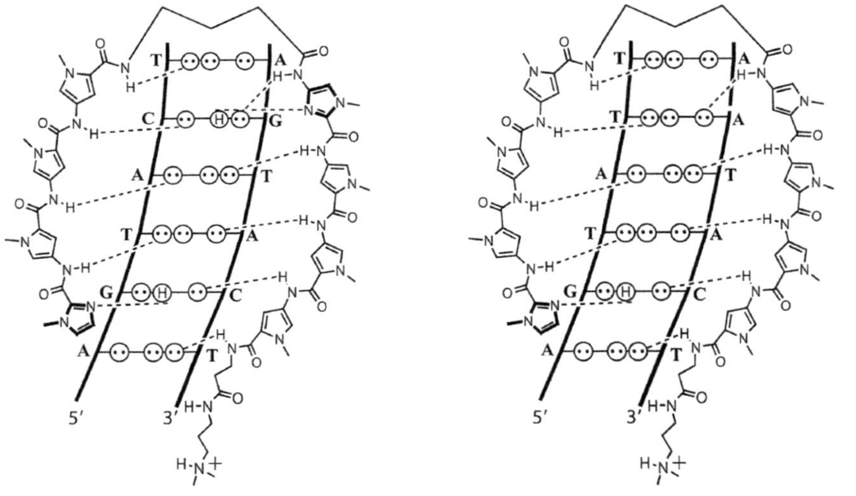

Equilibrium Association Constants ($M^{-1}$)

| Binding Site | ImPyPyPy-γ-ImPyPyPy-β-Dp | ImPyPyPy-γ-PyPyPyPy-β-Dp |
|---|---|---|
| 5'- AGTACT -3' | 3.7 x $10^{10}$ | 5.0 x $10^8$ |
| 5'- AGTATT -3' | 4.1 x $10^8$ | 3.5 x $10^9$ |

Conditions: 10 mM Tris•HCl, 10 mM KCl, 10 mM $MgCl_2$, and 5 mM $CaCl_2$, pH 7.0, 22°C

**Figure 12.3** Eight-ring Py–Im polyamides bind matching sequences with affinity and specificity similar to those of transcription factors. The replacement of a single nitrogen atom with a C–H regulates affinity and specificity by two orders of magnitude.

fluorescently labelled polyamides yielded the estimate that only about 2–20% of the administered dose reaches the nucleus of a cell.[58] Confocal microscopy studies on living cells have shown that nuclear localisation of polyamides is largely influenced by the size of the molecule and generally decreases with increasing molecular size.[55,56] Py–Im polyamides have been synthesised to target from five base pairs up to 16 base pairs; given the trade-off between sequence specificity and nuclear uptake, our laboratory has settled on eight-ring hairpins which recognise six base pairs (Figure 12.4).[43] This results in a binding site size similar to the size of the DNA recognition element for many transcription factors.

## 12.5 The γ-Hairpin Turn and Orientation Preference

The γ-turn of a hairpin polyamide prefers a A–T/T–A base pair for steric reasons; the steric bump of a G–C base pair is energetically unfavourable. A six-ring hairpin was shown to favour an orientation preference of Py–Im polyamides in the minor groove of DNA; antiparallel polyamide subunits align N–C with the walls of the minor groove of DNA in the 5′–3′ direction.[59] Chiral substitutions at the alpha (α) or beta (β) position of the turn can influence the forward (N–C/5′–3′) or reverse (C–N/5′–3′) binding orientation of the molecule.[60,61] Sequence context can also influence the affinity and binding orientation of a polyamide; it is generally accepted that chiral α-substitution and β-substitution turns encourage forward binding in eight-ring hairpins, whereas conformationally flexible hairpin polyamides containing β-alanine sometimes prefer the reverse binding orientation.[62] Modification of the chiral turn also influences the biological activity of polyamides in cell culture as well as in animal models (Figure 12.5).

## 12.6 The C-terminus of the Hairpin

Modifications to the C-terminus of Py–Im polyamides were initially explored because of different methods of solid-phase synthesis. For steric reasons, linkers at the C-terminus generally demand T–A over G–C base pairs.[63] After cleavage from resin, elaboration of the C-terminus usually completes the synthesis. As a result, over 60 C-terminal structures have been tested, and many have been shown to display gene-regulatory activity in the micromolar concentration range. Isophthalic acid conjugated at the C-terminus demonstrated favourable uptake and has been the modification of choice for our gene-modulation studies in cell culture (Figure 12.2).[64] Oxime-linked polyamides have been used in $^{18}$F-labelled PET studies tracing real-time localization of Py–Im polyamides in mice.[65,66] The tolerance toward modifications at the C-terminus of hairpins indicates that further molecular refinements may lead to improved biological activity of polyamides.

| General (5'→3') | Polyamide | $K_a$ ($M^{-1}$) | Sequence context |
|---|---|---|---|
| WW**G**WWWW |  | $3\times10^9$ | 5′-TA**G**TATT-3′ |
| WW**GG**WWW | 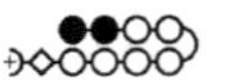 | $5\times10^8$ | 5′-**C**T**GG**TTA-3′ |
| WW**G**W**G**WW | 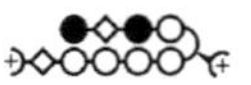 | $4\times10^9$ | 5′-TA**G**T**G**AA-3′ |
| WW**G**WW**G**W |  | $9\times10^9$ | 5′-TA**G**TA**G**T-3′ |
| WW**G**WW**C**W | 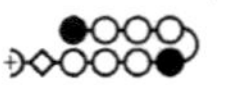 | $3\times10^{10}$ | 5′-TA**G**TA**C**T-3′ |
| WW**G**W**C**WW |  | $2\times10^9$ | 5′-**G**A**G**T**C**TA-3′ |
| WW**GC**WWW | 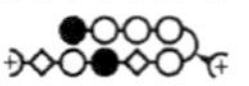 | $5\times10^9$ | 5′-AT**GC**AAA-3′ |
| WW**GGG**WW |  | $3\times10^8$ | 5′-AA**GGG**AA-3′ |
| WW**GG**W**G**W |  | $1\times10^{10}$ | 5′-TA**GG**T**G**T-3′ |
| WW**GG**W**C**W | 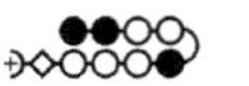 | $1\times10^{10}$ | 5′-AT**GG**T**C**A-3′ |
| WW**GGC**WW |  | $4\times10^8$ | 5′-AA**GGC**AT-3′ |
| WW**G**W**GG**W | 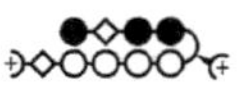 | $4\times10^{10}$ | 5′-TA**G**T**GG**T-3′ |
| WW**G**W**GC**W |  | $2\times10^9$ | 5′-AT**G**A**GC**T-3′ |
| WW**GCG**WW |  | $2\times10^9$ | 5′-AT**GCG**TA-3′ |
| WW**GC**W**G**W |  | $2\times10^9$ | 5′-TA**GC**A**G**T-3′ |
| WW**GC**W**C**W | 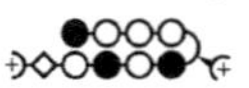 | $9\times10^9$ | 5′-AT**GC**T**C**A-3′ |
| WW**G**W**CG**W |  | $1\times10^{10}$ | 5′-AT**G**A**CG**T-3′ |
| WW**G**W**CC**W |  | $2\times10^9$ | 5′-TA**G**A**CC**A-3′ |
| WW**GCC**WW |  | $7\times10^8$ | 5′-AT**GCC**TA-3′ |
| WW**GGGG**W |  | $2\times10^8$ | 5′-**G**A**GGGG**T-3′ |
| WW**GCGG**W |  | $9\times10^8$ | 5′-AT**GCGG**A-3′ |
| WW**GGCG**W |  | $2\times10^8$ | 5′-**C**A**GGCG**T-3′ |
| WW**GGGC**W |  | $1\times10^8$ | 5′-**C**T**GGGC**A-3′ |
| WW**GCCG**W |  | $2\times10^9$ | 5′-AT**GCCG**T-3′ |
| WW**GGCC**W |  | $9\times10^9$ | 5′-AT**GGCC**A-3′ |
| WW**GCGC**W | 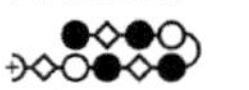 | $3\times10^9$ | 5′-AT**GCGC**A-3′ |
| WW**GCCC**W |  | $1\times10^9$ | 5′-AT**GCCC**A-3′ |

**Figure 12.4** Programmable Py–Im polyamides target a library of six base pair sequences. Im and Py are shown as filled and non-filled circles, respectively; β is shown as a diamond; the Dp at the C-terminus is shown as a half-circle with a plus; the achiral γ-aminobutyric acid turn residue is shown as a semicircle connecting the two subunits; and the chiral diaminobutyric acid turn residue is shown as a semicircle linked to a half-circle with a plus connecting the two subunits.

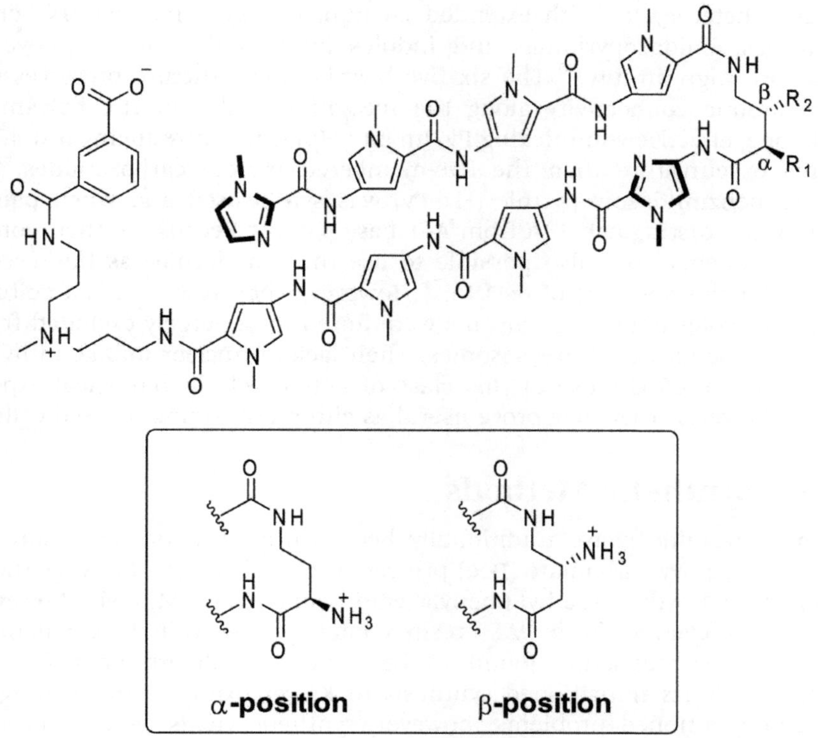


**Figure 12.5** Eight-ring hairpin Py–Im polyamide with chiral substituents on the γ-turn at the α and β positions.

## 12.7 Second Generation Heterocycles for DNA Recognition

The pairing rules for recognition of the four Watson–Crick base pairs by Im–Py and Hp–Py represented a benchmark for the field. That said, Hp-containing hairpins were found to degrade over time and would not be suitable for robust biological experiments. Furthermore, due to sequence-dependent alterations in the shape of the DNA minor groove, some DNA sequences proved difficult to target with the existing repertoire of aromatic amino acids. This motivated efforts to design new heterocycles to distinguish A–T and T–A base pairs based on principles learned from the success of the parent Hp–Py. These led to the development of second-generation aromatic amino acids and fused heterocycle moieties that can engage in sequence-specific DNA recognition in the minor groove. Simple heterocycles screened include thiophenes, furans, oxazoles and pyrazoles.[67–69] The findings from these studies will not be discussed here as they have been summarized in previous reviews.[19,20]

Fused heterocycles with extended $\pi$-conjugated systems such as benzimidazoles, imidazopyridines and indoles bind to the minor groove of DNA with high affinity.[70] The six–five bicyclic-ring structure preserves the same atomic connectivity along the recognition edge of the polyamide backbone, effectively mimicking Py, Im and Hp, while introducing a smaller degree of curvature than the five-membered pyrrole carboxamides.[71–73] Hydroxybenzimidazole–pyrrole (Hz–Py) was synthesised and when paired with Py can distinguish T–A from A–T base pairs.[72] Because of their conjugated $\pi$-systems, it is also possible to use these molecules as fluorescent probes for the presence of dsDNA.[74] However, when tested in cell culture, benzimidazole-containing hairpins were found to be largely excluded from the cell or sequestered in lysosomes. Their lack of nuclear uptake in living cells has limited the use of this class of heterocycle in biological experiments; however, they may prove useful as chromatin stains in fixed cells.

## 12.8 Synthetic Methods

Py–Im polyamides have traditionally been synthesized on solid support using tert-butoxy carbamate (Boc) protection chemistry.[75,76] Early synthetic efforts used 4-hydroxymethyl-phenylacetamidomethyl (PAM) resin; however, polyamides cleaved from PAM resin contain a C-terminal beta-alanine, which was subsequently found to be a negative determinant for cell uptake.[64] Efforts transitioned synthesis to Kaiser oxime resin to mitigate the aforementioned problems; however, synthetic yields were reduced.[76] A typical eight-ring hairpin might take four days to synthesise with a final yield of 1%.[77–79] The use of preassembled trimers and tetramers, synthesized in solution phase, enabled Py–Im polyamide yields of 10% on Kaiser oxime resin.[65]

These methods were later improved with the adaptation of microwave-assisted synthesis which dramatically increased the yield and reduced the time required for each coupling.[80] For comparison, a traditional Py–Py coupling was reduced from 3 hours to 5 minutes with the use of a microwave synthesiser. Using this method, an eight-ring hairpin can by synthesised in nine coupling cycles plus a protecting group exchange and cleavage step in 56% overall yield.[80] If the latter two steps and purification are assumed to be quantitative, this indicates an average yield of 94% per coupling cycle.[80] While other coupling chemistries and protection schemes have been explored in the literature, the availability of Boc-protected pyrrole and imidazole amino acids has made the current methods a mainstay.[81,82] Solid-phase methodologies offer many advantages such as rapid and reliable coupling as well as facile purification due to immobilization of the growing oligomer on a solid support. This makes solid-phase methods ideal for the rapid synthesis of a variety of new oligomers but also intrinsically limits the scale of synthesis.

While significant improvements have been made in solid-phase synthesis of Py–Im polyamides, a solution-phase approach may be more amenable for scale-up. We have demonstrated the feasibility of this approach with the solution-phase synthesis of an eight-ring hairpin polyamide core using Boc-protected dimers, trimers and tetramers of heterocycles.[83] By exploiting differences in the physical solubility properties of starting materials *versus* products, this solution-phase synthesis allowed preparation of multi-gram quantities of a polyamide core without chromatographic purification. Each intermediate is purified by a combination of precipitation, trituration and crystallisation. Once the intermediates have been prepared, the final product is produced in three steps with good overall yield.[83] However, this approach may be more practical for symmetric Py–Im polyamides where identical dimer, trimer or tetramer components can be incorporated in multiple steps.

## 12.9 Disruption of Transcription Factor–DNA Interface

Early X-ray crystallographic studies of antiparallel 2 : 1 binding polyamides revealed that they widen the minor groove by 1–2 Å.[35,36] Similar effects were observed with a hairpin Py–Im polyamide bound to the nucleosome core particle, although these structures were at moderate resolution.[84–86] Such structural perturbations are too modest to explain the many transcription factor–DNA interfaces found to be disrupted by hairpin polyamides.[87–90] Atomic resolution crystal structures of cyclic Py–Im polyamides bound to duplex DNA oligonucleotides, however, indicate that minor groove expansion may be sufficient to perturb protein–DNA interfaces of certain major-groove-binding proteins (Figure 12.6). We observed a minor groove widening of up to 4 Å, with a simultaneous compression of the major groove and bending of the DNA helix towards the major groove by $>15^\circ$ (Figure 12.7).[91,92] Comparison with previously published structures indicated that the changes caused by polyamide binding cause major groove surface geometry to become incompatible with androgen receptor–DNA and glucocorticoid receptor–DNA binding, results that were later confirmed by modelling. In effect, besides the property of programmable sequence-specific DNA recognition, the eight-ring hairpin Py–Im polyamides create local perturbation of the minor and major groove surfaces of the DNA helix. Many human diseases, such as cancer, are caused by overactivity of transcription factors.[93] Sequence-specific disruption of transcription factor–DNA interfaces was identified as a likely mechanism for the activity of Py–Im polyamides affecting gene regulation in cell culture (Figure 12.8).[94]

## 12.10 Inhibition of RNA Polymerase II Elongation

DNA-targeting agents have been shown to inhibit DNA-dependent enzymes, including RNA polymerases, DNA polymerase, topoisomerases and helicases.[95] DNA minor-groove-binders such as distamycin and actinomycin

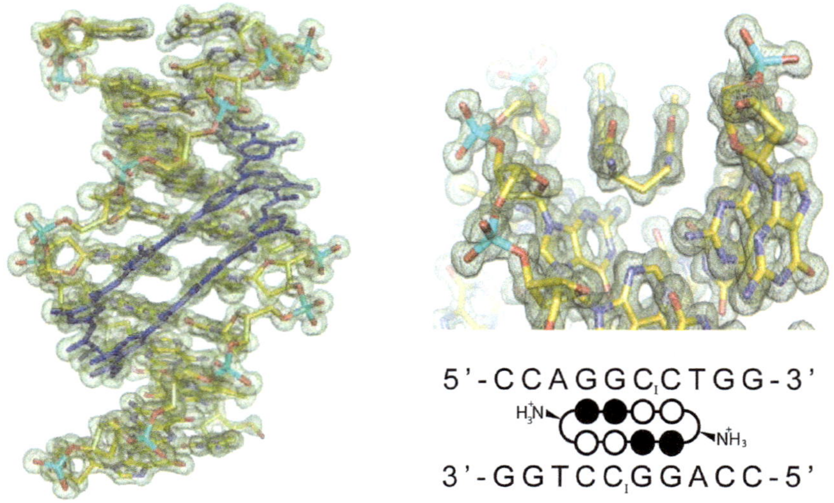


**Figure 12.6** (Left) DNA–cyclic Py–Im polyamide co-crystal structure (PDB 3I5L, 1.18 Å resolution). (Right-Top) View looking down the DNA minor-groove, showing the (R)-α-amine-γ-turn conformation observed in the X-ray crystal structure. The conformation of the α-amino-substituted GABA turn presents the β-methylene down toward the minor-groove floor while orienting the α-ammonium up and out of the minor-groove, relieving possible steric interaction with the sugar–phosphate backbone (minor-groove wall). (Right-bottom) Chemical structure of the cyclic polyamide and DNA sequence. Cyclic polyamide targeting the sequence 5′-WGGCCW-3′ shown with ball-and-stick model superimposed onto the DNA oligonucleotide used for crystallization. Black circles represent imidazoles, open circles represent pyrroles, and ammonium substituted half circles at each end represent the (R)-α-amine-γ-turn.

D have been shown to inhibit RNA polymerase II-mediated transcription in enzymatic studies and in cell culture.[96–98] Similarly, Py–Im polyamides have also been shown to inhibit the elongation reaction catalysed by RNA polymerase II.[99]

Evidence for hairpin polyamides inhibiting RNA polymerase II in cell culture was observed in LNCaP cells.[100] In addition to transcribing mRNA, RNA pol II also functions to detect DNA lesions caused by UV irradiation, cisplatin binding or DNA intercalation.[96–98] In cell culture experiments, transcription blockade has been shown to induce degradation of the RNA pol II large subunit (RPB1), and to function as a signal for p53-mediated apoptosis.[101,102] Interestingly, treatment with α-amanitin also leads to RPB1 degradation and apoptosis, suggesting that inhibition of RNA pol II elongation is sufficient to trigger cell death.[101] Studies in LNCaP cells treated with the eight-ring hairpin Py–Im polyamide ARE-1 targeted to the androgen response element have shown both a dose- and time-dependent degradation of RPB1, each coinciding with the growth inhibition $IC_{50}$ conditions seen for this compound in the same cell line.[100]

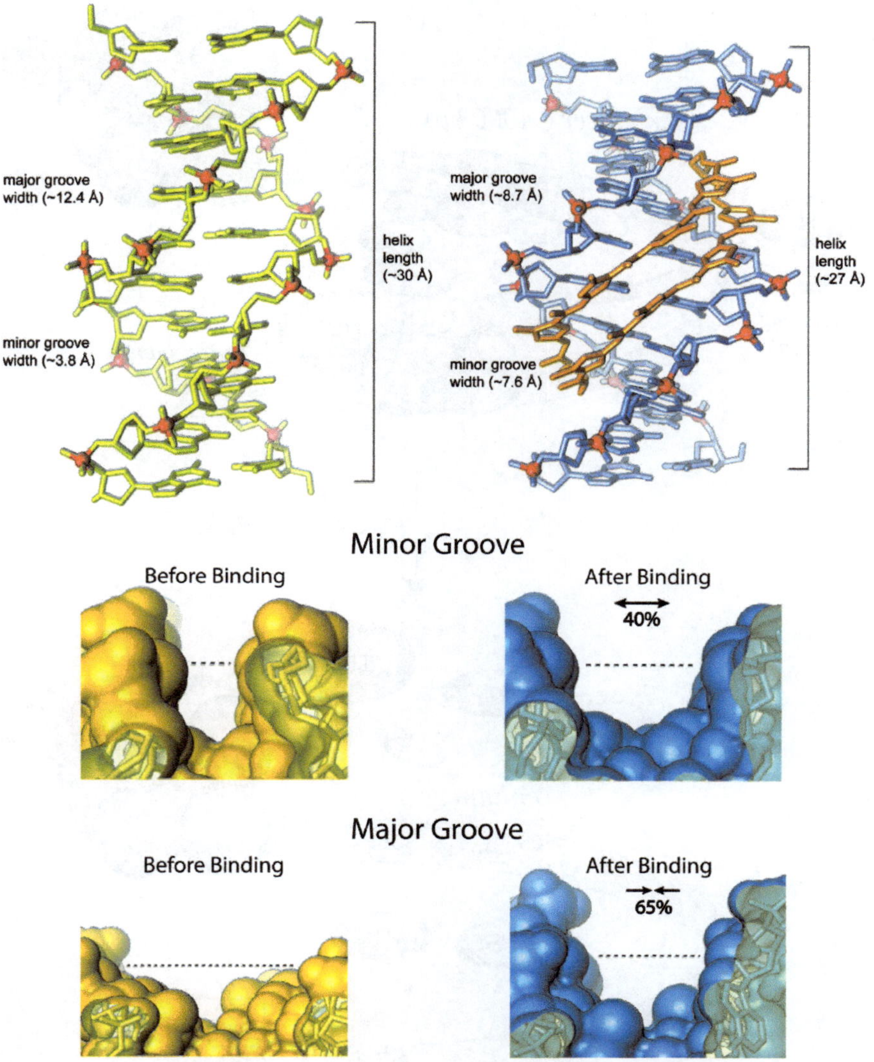


**Figure 12.7** (Top) Native DNA crystal structure (PDB 1D8G, 0.74 Å resolution) and DNA–polyamide co-crystal structure (PDB 3I5L, 1.18 Å resolution) with the major groove and minor groove widths labelled. (Bottom) DNA minor- and major-groove dimensions in the absence (yellow curve and structures) and in the presence (blue curve and structures) of cyclic polyamide. Polyamide has been removed from the blue complex for clarity. Native DNA structure d(5′-CCAGTACTGG-3′)$_2$ solved by Rees and co-workers.

Further evidence of RNA pol II inhibition was gained in a recent biophysical study using biochemical assays, modelling and mutation studies to assess the effect of site-specific inhibition of pol II elongation by Py–Im

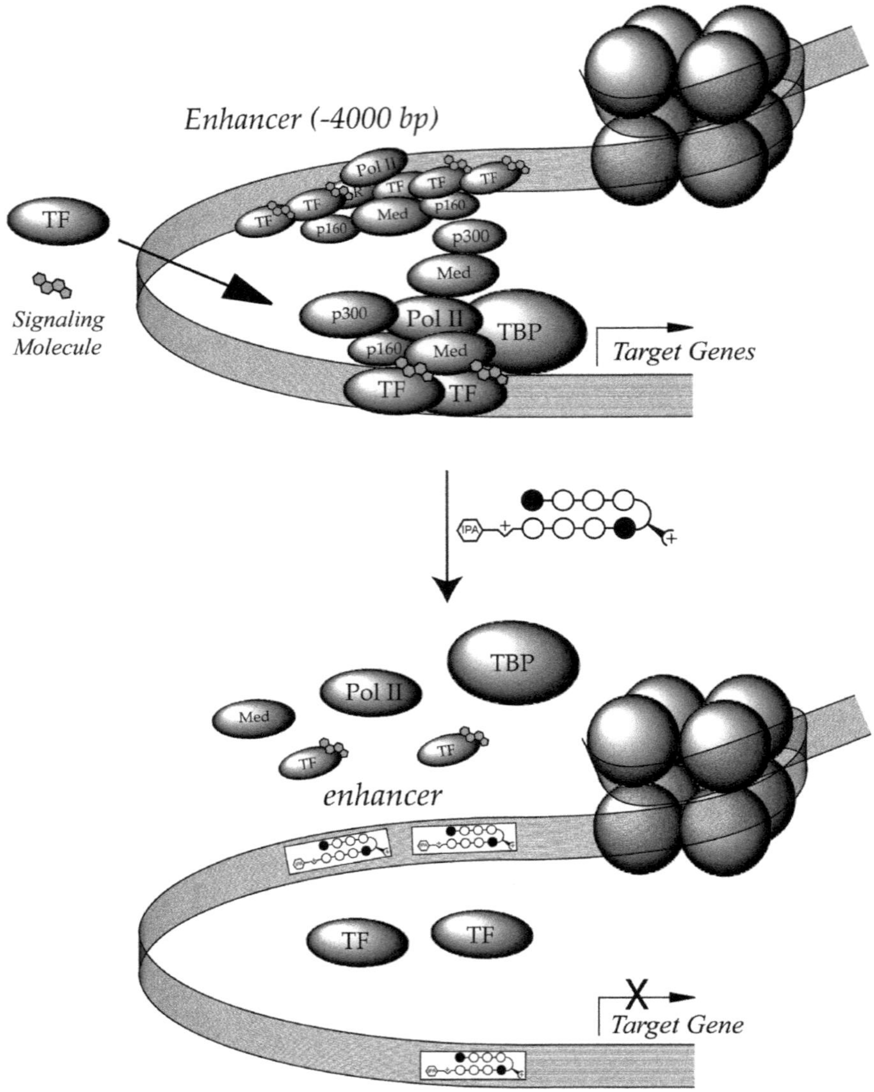


**Figure 12.8** (Top) Model of the transcription complex. Assembly of transcription machinery following transcription factor activation and binding to DNA. (Bottom) Addition of hairpin Py–Im polyamides designed for sequence-specific binding at the promoter and enhancer sites blocks transcription factor (TF) binding, disrupting the assembly of transcriptional machinery and interfering with gene transcription.

polyamides in two binding orientations. In an *in vitro* transcription assay using isolated eukaryotic RNA pol II with a DNA template containing a single polyamide binding site (5′-WGGWCW-3′), D. Wang and coworkers were able to show robust transcription inhibition at nanomolar $IC_{50}$ (Figure 12.9).[103]

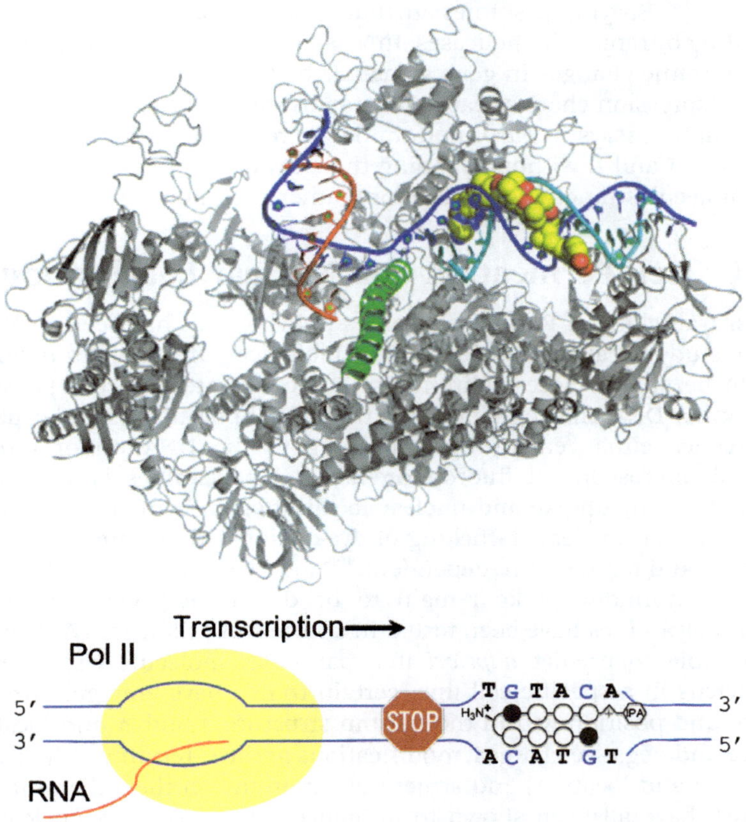


**Figure 12.9** Py–Im polyamides are strong roadblocks for RNA pol II transcription elongation. (Top) Molecular modelling of arrested RNA polymerase II complex in the presence of Py–Im polyamide. Model derived from 10-subunit RNA pol II elongation complex with RPB2 omitted. Protein residues are coloured grey; RNA, template DNA, and non-template DNA are shown in red, blue and cyan, respectively. The Py–Im polyamide is highlighted in yellow. (Bottom) Schematic illustration of RNA pol II transcription blockage by Py–Im polyamide–DNA binding.

The blocked RNA pol II enzyme was trapped on the DNA template for up to 20 hours, and was not rescued by TFIIS, a transcription elongation factor that can rescue arrested pol II and enhance pol II bypass of a variety of translocation barriers.[103] Furthermore, polyamide-induced transcription inhibition was abolished when a single mismatch mutation was introduced into the polyamide binding site, demonstrating the sequence-specificity of the effect.[103]

In addition to transcription inhibition *via* disruption of TF–DNA binding and RNA pol II blockage, studies have shown that Py–Im polyamides also interfere with other DNA-dependent processes. In DU145 prostate cancer cells, a low level of replication stress was observed upon polyamide treatment.[104] Enzymatic studies attribute this response to inhibition of DNA unwinding by replicative helicases. In VCaP prostate cancer cells, profiling of transcriptomic changes in gene signature by LINCS2000 revealed similarity to gene expression changes caused by topoisomerase poisons. Interestingly, in enzymatic assays *in vitro*, polyamides were found to inhibit both topoisomerases I and II without trapping the cleavage complex, suggesting that these molecules function as inhibitors rather than poisons.[105]

## 12.11 Cell Permeation and Nuclear Localisation

Nuclear trafficking of Py–Im polyamides in living cells has been a subject of intense study because in order to exert biological effect, Py–Im polyamides need to be able to access chromatin. Dye-conjugated hairpin polyamides have similar DNA binding kinetics to traditional polyamides and experience fluorescence enhancement upon binding to the DNA minor groove.[106] Confocal microscopy of fluorophore-labelled polyamides is a convenient method to study uptake and nuclear localisation in living cells. Early work has shown that nuclear trafficking of dye-conjugated hairpin polyamides is both ATP- and temperature-dependent.[56] However, the exact mechanisms of hairpin polyamide uptake remain to be determined. Over 100 distinct hairpin polyamides have been tested in a variety of cell lines. While it is still not possible to predict *a priori* if a particular molecule will localise to the nucleus in a specific cell line, certain trends have emerged. Imidazole content and position within the hairpin structure, number and location of positive and negative charges, modifications at the C-terminus (fluorophore, linker size and location) and structural variations on the γ-diaminobutyric acid turn have all been shown to influence cell uptake and nuclear localisation (Figure 12.10).[57,64]

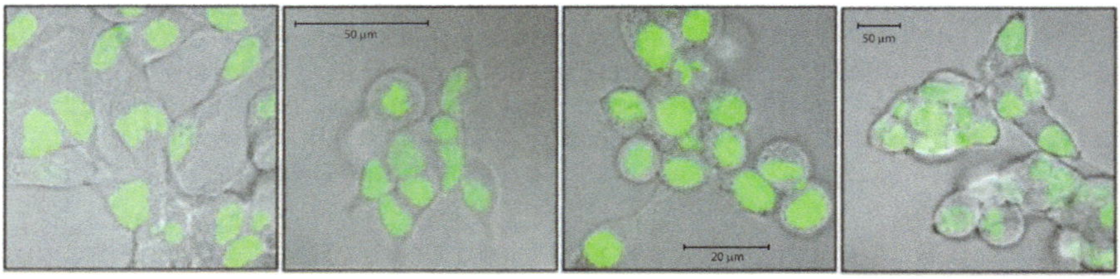


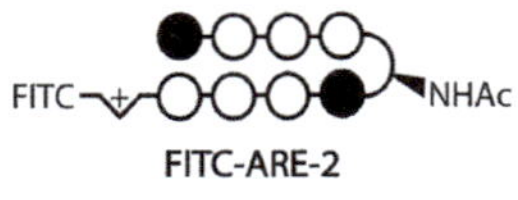


| Cell Line | AR | GR |
|---|---|---|
| LNCaP | + | - |
| LNCaP-AR | +++ | - |
| LREX′ | +++ | + |
| VCaP | +++++ | + |

**Figure 12.10** Cellular localization of Py–Im polyamide–fluorescein conjugate FITC-ARE-2 in live cells.

Modification of the γ-diaminobutyric acid turn may be especially important, particularly when bulky aryl groups are conjugated to compounds containing an amine at the β position of the turn.[107] This effect is probably attributable to the increased lipophilicity of the molecule rather than changes to DNA affinity, since only slight reduction in DNA binding is typically seen with conjugation.[60] Modification of the chiral amine at the α position affords less improvement in biological activity, presumably due to steric clash of bulky substituents with the DNA minor groove wall.

## 12.12 Gene Regulation in Cell Culture

Complex expression networks regulated by transcription factors are essential for normal cellular function, but their dysregulation leads to the development of a myriad of diseases, including cancer. The central regulatory role of transcription factors renders them coveted therapeutic targets in oncology.[93,108,109] Prime examples where transcription factor antagonism has benefited patients include androgen receptor and oestrogen receptor inhibition in the treatment of prostate cancer and breast cancer, respectively.[110–112] However, unlike nuclear hormone receptors which contain ligand-binding domains to effect transcription, most transcription factors lack small-molecule binding pockets and instead rely on extensive weak, non-contiguous contacts to interact with binding partners. The potential of Py–Im polyamides to interfere with protein–DNA interactions was recognised early, and we have since used sequence-specific polyamides to disrupt transcription factor–DNA interfaces so as to inhibit oncogenic signalling in several disease models (Figure 12.11).[86–90,94,113] The ability of Py–Im polyamide conjugates to chemically modify DNA or alter function by recruiting

| Polyamide | Binding Sequence | Target | Disease Model |
|---|---|---|---|
| 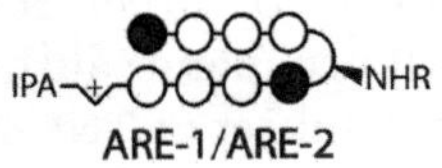 ARE-1/ARE-2 | 5′-WGWWCW-3′ | ARE/GRE (AR, GR, RNA pol II) | Prostate cancer |
|  HRE-1 | 5′-WTWCGT-3′ | HRE (HIF-1) | Solid tumors |
| 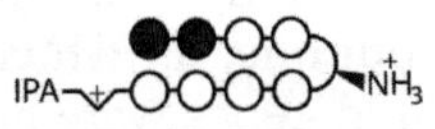 NFκB-1 | 5′-WGGWWW-3′ | NF-κB | Inflammatory disorders |
| 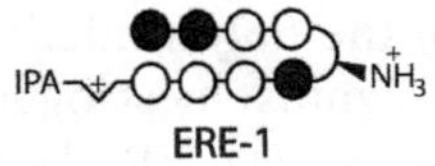 ERE-1 | 5′-WGGWCW-3′ | ERE (ERα, ERRα) | Breast cancer |

**Figure 12.11** Biologically active eight-ring Py–Im hairpin polyamides discussed in this review.

specific proteins to a particular site has also been demonstrated, but has been reviewed elsewhere and is not included here.[19,114–118]

The hypoxia inducible factor (HIF) is a basic helix–loop–helix transcription factor that has been extensively studied for its role in oxygen sensing. In the hypoxic tumour microenvironment, oxygen deprivation leads to stabilisation of HIF-1α. Subsequent heterodimerisation with ARNT allows nuclear translocation and the activation of signalling cascades that ultimately increase tumour vascularisation. Due to the central role of HIF signalling in cancer biology, mechanisms to inhibit HIF signalling have been intensively researched.[119] In one early study, the natural product echinomycin, which binds to 5′-NCGN-3′ sequences in the DNA minor groove,[14] was found to be efficacious against HIF-driven expression of selected genes, demonstrating the feasibility of inhibiting HIF signalling from the TF–DNA interface.[95] To further refine the DNA target, a Py–Im polyamide HRE-1 was designed to target the sequence 5′-WTWCGW-3′, found in the VEGF hypoxia response element (HRE), which has more specificity than echinomycin.[113] Cell culture studies showed that HRE-1 reduced the expression of selected HIF-driven genes, with concomitant reduction of HIF occupancy from gene promoters.[113,120] This molecule was further demonstrated to be efficacious in reducing HIF signalling in xenograft models of glioblastoma and multiple myeloma.[121,122] As with the cell culture data, the HRE-specific polyamide significantly inhibited VEGF protein expression by approximately 50% in treated tumours when compared with controls.[122]

NF-κB is a transcription factor that plays an important role in the inflammatory response, cell proliferation, and survival as well as cancer.[123,124] When induced, NF-κB translocates to the nucleus and binds its target DNA sequence, 5′-GGGRNYYYCC-3′, where R = purine, Y = pyrimidine, N = any base, and activates various gene programmes, depending on the method of induction.[125] Based on the results of chromatin immunoprecipitation–sequencing (ChIP-seq) experiments conducted on the *IL6* and *IL8* promoters, we designed an NF-κB-specific Py–Im polyamide NFκB-1 to target the sequence 5′-WGGWWW-3′. This polyamide reduced NF-κB occupancy at the designated promoters and proved capable of modulating a subset of TNF-α-inducible genes in A549 lung carcinoma cells.[79] Further studies confirmed the effectiveness of NFκB-1 *in vivo*, using both A549-luc-C8 cells and A549-luc-C8 tumour-bearing animals. A representative panel of genes studied by RNA sequencing was further interrogated by quantitative reverse transcription polymerase chain reaction (qRT-PCR); the same trend was observed in cell culture and xenograft models for all genes but one, for which no effect was observed in cell culture. All other interrogated genes were affected more strongly in cell culture than in the xenograft setting, probably due to differences in exposure time and concentration of the polyamide.[126]

Py–Im polyamides have also demonstrated an effect against oestrogen receptor-alpha (ERα)-driven gene expression *in vitro* and *in vivo*. ERα is a ligand-activated nuclear hormone receptor that is active in a majority of breast adenocarcinomas.[127,128] Polyamide ERE-1, targeted to the oestrogen response element (ERE), has been shown to be active against 17β-oestradiol (E2)-induced transcription in ERα-positive T47DKBluc cells, which express luciferase under ERα control. Interestingly, RNA-sequencing analysis showed that ERE-1 had the greatest effect on the genes most strongly affected by E2, and much less effect on the majority of E2-induced genes. *In vivo*, ERE-1 was able to reduce levels of ER-driven luciferase expression in xenograft tumours without significant host toxicity.[129]

Lastly, the nuclear hormone receptors androgen receptor (AR) and glucocorticoid receptor (GR) have both been studied in the context of prostate cancer. Inhibition of AR signalling by anti-androgens is a standard of care for prostate cancer. However, due to nearly identical DNA binding sequences, GR has been shown to compensate for lost AR activity during treatment and to cause a resurgence in the disease.[130] Currently no direct inhibitor of GR exists; however, the dual GR and progesterone receptor inhibitor, mifepristone, is currently being evaluated as a combination therapy with the anti-androgen enzalutamide in patients with metastatic castration-resistant prostate cancer.[131]

Due to the similarities in binding preference of AR and GR, hairpin Py–Im polyamides designed to target the sequence 5′-WGWWCW-3′, can be used to inhibit the signalling of both transcription factors (Figure 12.12).[132] In LNCaP prostate cancer cells, polyamide ARE-1 treatment reduced AR promoter occupancy and expression of selected AR-driven genes, including prostate specific antigen (PSA).[77] This polyamide was able to downregulate PSA by a similar amount to the same concentration of bicalutamide (Casodex), a clinically used anti-androgen. The effects of this molecule on GR-driven gene expression in cell culture experiments were similar.[78] In LREX′ cells, a prostate cancer model that expresses both AR and GR, treatment with the next generation ARE-2 was found to be efficacious towards both AR- and GR-driven gene expression while enzalutamide only attenuated AR signaling.[133]

## 12.13 Global Sequence Analysis of Sequence Specificity

Characterization of Py–Im polyamide binding preferences has traditionally been studied using methidiumpropyl-EDTA-Fe(II) and DNaseI footprinting methods.[13] Methidiumpropyl-EDTA-Fe(II) footprinting is able to identify high-affinity polyamide binding sites at near-nucleotide resolution.[14,15] Affinity cleavage determines the orientation of a bound ligand in the DNA minor groove,[16,17] and quantitative DNaseI footprinting titration can be used to determine equilibrium association constants ($K_a$) at previously identified match and mismatch sites.[43] DNaseI footprinting is, however, limited to interrogating small numbers of binding sites on a DNA fragment (150 bp in size) due to the limited resolving power of a polyacrylamide gel. To comprehensively investigate all binding site possibilities for an eight-ring

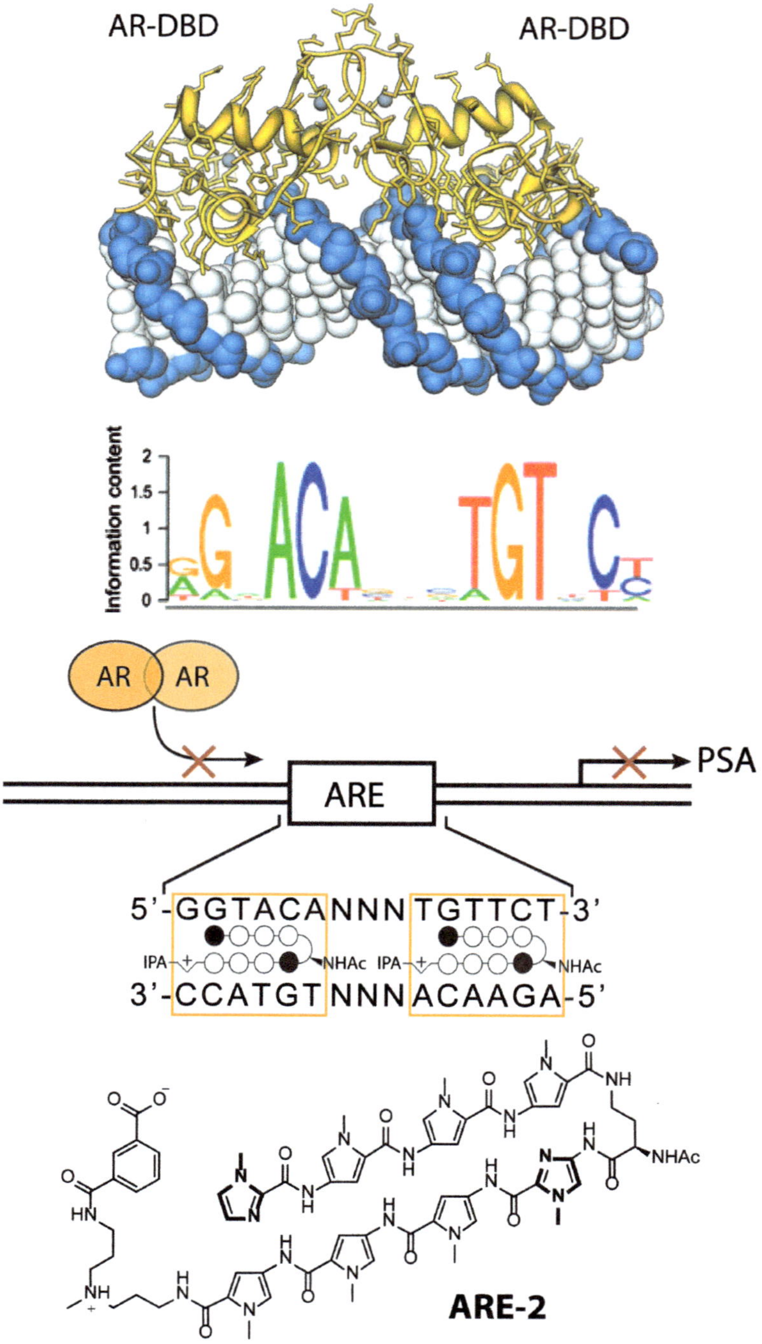
AR-DBD
AR-DBD
Information content
2
1.5
1
0.5
0
AR
AR
PSA
ARE
5'-GGTACANNNTGTTCT-3'
IPA
NHAc
IPA
NHAc
3'-CCATGTNNNACAAGA-5'
ARE-2

hairpin polyamide would require unbiased screening of 2 080 (for 6 bp total) or 32 896 (for 8 bp total) unique binding sites.[17]

The Cognate Site Identifier (CSI) microarray platform is a high-throughput assay used to determine polyamide binding preferences in an unbiased manner. The CSI array presents all 32 896 unique octamer sequences and is scalable to decamers.[134,135] This, and similar, approaches have been used to study transcription factor binding to great effect. Correlating the sequence preference landscape present on the CSI microarray with quantitative DNaseI footprinting enables energetic studies using a vastly expanded repertoire of DNA sequences. This marks a significant step forward for the field of small-molecule–DNA recognition, and enables a better understanding of the variability in the range of affinities for match and mismatch sites caused by the sequence-dependent variability of the DNA minor groove shape (Figure 12.13).

Bind-N-Seq is another high-throughput method that has been used to identify protein–DNA interactions *in vitro*. This method takes advantage of next-generation short-read sequencing technology and represents an advance over the capacity of microarray-based techniques. The large number of copies of each possible decamer ($2.5\times10^7$) ensures sampling of short sequences and provides a context-averaged picture of binding. This platform allows rapid, quantitative identification of high-affinity polyamide binding sites, correlates well with solution-phase and microarray platforms, and can be used to guide the refinement of general polyamide design principles.[136] In addition, some Py–Im polyamides bind narrow purine-rich tracts such as $(GAA)_n$ repeats in a 1 : 1 stoichiometry.[137,138] Bind-N-Seq has proved a powerful method to characterise high-affinity sites 10 bp in size (Figure 12.14).[135]

New mechanistic insights into Py–Im polyamide activity have underlined the importance of mapping polyamide binding to chromatin in living cells. The sequence-specificity of polyamides for DNA has been extensively studied *in vitro* using traditional biochemical and biophysical techniques. However, mapping polyamide binding in the complex cellular environment presents a more daunting problem since DNA in a cell is nucleosome-bound as chromatin and has varying degrees of accessibility. Access by Py–Im polyamides to nucleosomal DNA has nevertheless been demonstrated.[84–86] Recently, techniques were developed to interrogate small-molecule interactions on DNA with living chromatin. The Ansari laboratory has developed

**Figure 12.12** (Top) AR–DNA crystal structure (PDB 1R4I). The DNA corresponds to the androgen direct repeat response element (ARE), and the sequence used is 5′-CCAGAACATCAAGAACAG-3′. Consensus sequence of the ARE. (Bottom) Map of the PSA–ARE site and schematic representation of a Py–Im polyamide targeting the PSA–ARE site. Potential polyamide ARE-2 binding sites (ball-and-stick model) of the consensus ARE site are shown within the yellow boxes. Black circles represent imidazoles and open circles represent pyrroles. Chemical structure of polyamide ARE-2, designed to bind the consensus ARE.

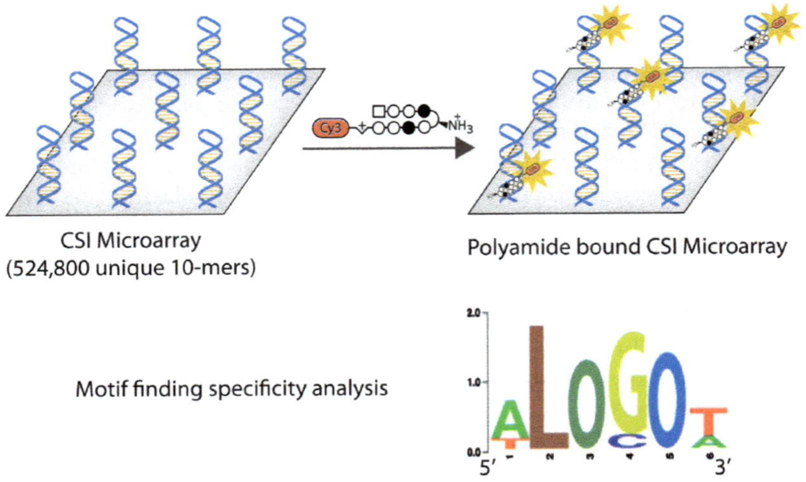


**Figure 12.13** Schematic illustration of a Cognate Site Identifier (CSI) microarray and overview of the experimental approach. (Top left) Each feature on the microarray displays a unique sequence as a DNA hairpin, with all permutations of 10 bp DNA represented on the array. (Top right) A Py–Im polyamide conjugated with a fluorophore (Cy3) is applied to the microarray to obtain a comprehensive ligand-binding profile. (Bottom) The resulting binding profile yields a histogram and position weight matrices (PWM) are generated from the highest intensity data and displayed as a logo.

a genomic-mapping method termed 'crosslinking of small molecules for isolation of chromatin' (COSMIC) with next-generation sequencing (COSMIC-Seq), to map the genome-wide binding events of polyamides.[139,140] We now seek to use this method to address the question of how the genomic architecture and local chromatin landscape affect polyamide occupancy and how polyamide binding perturbs genomic architecture. The genome-wide binding profiles acquired from COSMIC-Seq will allow us to develop Py–Im polyamides with improved specificity.

## 12.14 Animal Studies: Pharmacokinetics and Toxicity

Based on promising cell culture results, an effort was made to evaluate the activity of hairpin polyamides *in vivo*. Using intravenous injection as the route of administration, two ARE-targeted hairpin polyamides ARE-1 and ARE-3 were monitored for plasma bioavailability and tissue distribution, as well as renal and faecal elimination (Figure 12.15). While structurally similar, the two hairpins had quite distinct profiles of distribution. ARE-1 has an R-3,4-diaminobutyric acid linker and was found to have an $AUC_{0-\infty}$ of 77.4 $\mu g\, ml^{-1} \times h$ and a plasma half-life of 5.2 h. ARE-3 has an acetylated R-2,4-diaminobutyric acid linker and was found to have a higher

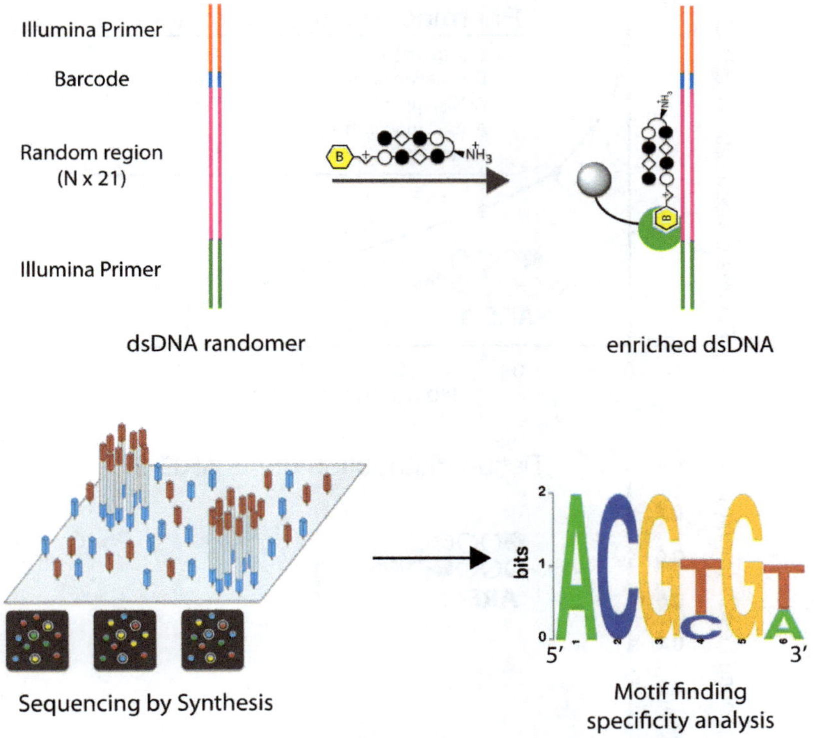


**Figure 12.14** Scheme for Bind-n-Seq analysis of DNA binding Py–Im polyamides. Double-stranded DNA containing a degenerate, 21-bp segment is enriched, purified and analysed *via* high-throughput sequencing and multiplexing by 3-nt barcode. Commonly bound DNA consensus sequences are identified *via* motif searching. Polyamide is shown as a ball-and-stick model with black circles representing imidazoles, open circles representing pyrroles, diamonds representing β-alanines, and hexagons representing a Biotin–$PEG_2$ linker.

$AUC_{0-\infty}$ of 176.5 $\mu g\,ml^{-1} \times h$ despite its shorter plasma half-life of 4.2 h.[141] This may be attributed to decreased water solubility of ARE-3 and increased lipophilicity as implied by the increased tissue distribution.[142,143] Interestingly, polyamide ARE-1 showed only 5.7% renal clearance, which was nearly complete after 4 hours, while ARE-3 showed 46% after 24 hours.[141] Neither compound exhibited significant faecal elimination. In addition to intravenous administration, Py–Im polyamides have also been shown to be bioavailable *via* subcutaneous and intraperitoneal routes of injection.

To further assess the tissue distribution of Py–Im polyamides, $^{14}C$ labelled analogues were employed (Figure 12.15). Analysis of major organ tissues 24 hours after administration identified the liver, kidney, spleen, pancreas and bone marrow as the major tissues of polyamide distribution. Furthermore, long-term studies showed that the molecules are retained in tissues for up

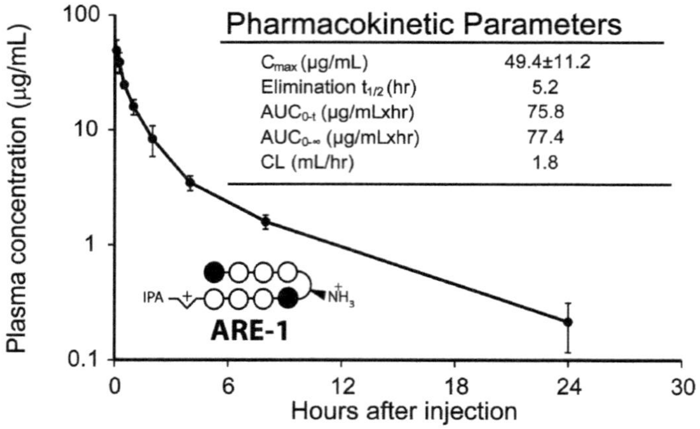


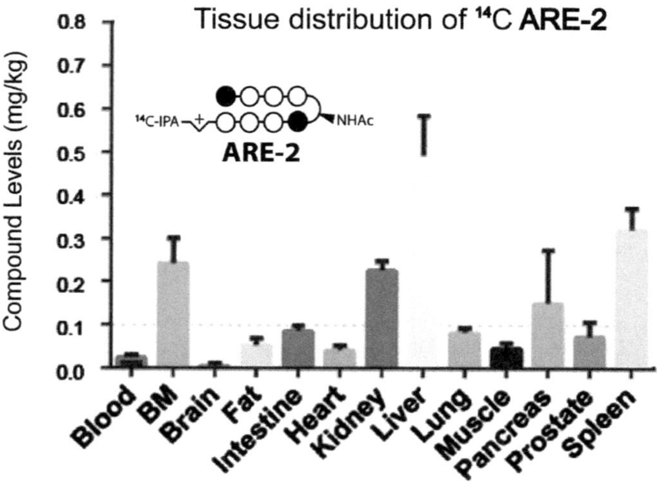


**Figure 12.15** (Top) Plasma levels (plasma concentration *versus* time) and calculated (non-compartmentally) pharmacokinetic parameters of Py–Im polyamide ARE-1. ARE-1 was administered *via* intravenous (IV) tail vein injection at 7.5 mg $kg^{-1}$ (balb/c female mouse). Error bars indicate standard deviation ($n = 3$). (Bottom) Extended tissue distribution analysis of $^{14}C$ Py–Im polyamide ARE-2 in wild-type mice. Injections were performed intraperitoneally at 15 nmol per animal (balb/c male mouse) and organs harvested 24 h following administration. Error bars represent standard deviations ($n = 5$). BM denotes bone marrow. Dashed line indicates the 0.1 mg $kg^{-1}$ threshold.

to a week after single injection and accumulate in the tissues after multiple injections.[144,145]

As new chemical entities, the toxicity profiles of eight-ring hairpin Py–Im polyamides have yet to be fully explored. In the previous pharmacokinetic study of two structurally similar ARE-targeted hairpin polyamides, a distinct difference in animal toxicity in wild type female mice was observed.[138] Further structural studies attributed the increase in toxicity to changing the R-3,4-diaminobutyric acid linker of ARE-1 to an R-2,4-diaminobutyric acid linker in ARE-3. Interestingly, acetylation of the β-amine position increased toxicity while acetylation of the α-amine had the opposite effect (Figure 12.16).[146] Removal of the chiral amine also reduced animal toxicity, but biological activity in cell culture was concomitantly attenuated.

To identify the mechanism of *in vivo* toxicity, wild-type mice received injections of escalating doses of 1, 3, and 10 mg $kg^{-1}$ of Py–Im polyamides and were monitored for signs of adverse effects. Histopathology of major organs was performed at the end of the experiment, or at euthanasia. From these studies, liver damage was identified as the major cause of mortality. This was further confirmed by serum chemistry where aspartate transaminase (AST), alanine transaminase (ALT), and total bilirubin were elevated. Polyamide ARE-2, with the α-acetamide linker, was the only molecule devoid of adverse effects towards the animals at doses of 10 mg $kg^{-1}$.[146] In a follow-up study, increased AST, ALT and total bilirubin were observed at 30 mg $kg^{-1}$ and decline in animal weights was observed at 60 mg $kg^{-1}$.[133]

While ARE-2 was less toxic than ARE-1, this was not due to the decreased systemic exposure. In fact, pharmacokinetic studies showed that ARE-2 has similar plasma distribution parameters to the much more toxic ARE-3 (β acetamide). The nuclear uptake of fluorescein conjugates of polyamides ARE-1, ARE-2 and ARE-3 in mouse livers provided an early indication of why ARE-2 had reduced animal toxicity. Of the three molecules, the fluorescein conjugate of ARE-2 had the lowest amount of liver nuclear uptake.[146] This result was further corroborated in $^{14}C$ studies comparing the tissue distribution of ARE-1 and ARE-2. Polyamide ARE-2 was found to localise to the liver 33% less than ARE-1, suggesting that decreased liver localisation may be the cause for reduced toxicity.[144] However, the exact mechanism for this effect is yet to be understood.

In multiple-dose experiments designed to mimic the treatment regimen administered to animals in xenograft studies, Py–Im polyamides are also well tolerated. In healthy animals, six injections of polyamide at 1 mg $kg^{-1}$ showed no elevation of ALT, AST, total bilirubin, blood urea nitrogen (BUN) or creatinine. Additionally, haematology counts of white blood cells (WBC), red blood cells (RBC), lymphocytes, haemoglobin and neutrophils remained unchanged from pre-treatment counts after six injections.[146]

## 12.15 Xenograft Cancer Models

Intertwined with toxicity studies is the understanding of the efficacy of Py–Im polyamides *in vivo*. The ARE-targeted polyamides have been extensively studied in cell culture, and were among the first to be evaluated in prostate cancer xenograft models. The efficacy seen in LNCaP, VCaP and LREX′ prostate cancer cell lines and xenografts is described below (Figure 12.17).

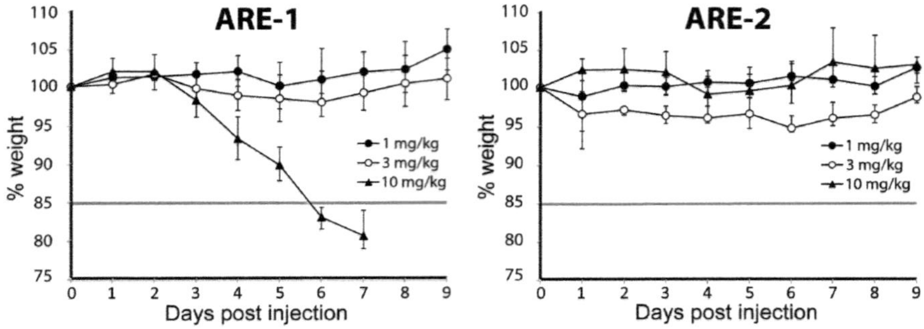


**ARE-2** Hematology

| | Pre- | Post- |
|---|---|---|
| Total WBC (x1,000/μL) | 8±1 | 9±1 |
| Total RBC (x1,000,000/μL) | 10±0.5 | 11±0.5 |
| HGB (g/dL) | 15±1 | 16±1 |
| Neutrophil (% of WBC) | 8±5 | 10±2 |
| Lymphocyte (% of WBC) | 89±5 | 88±1 |

**ARE-2** Serum chemistry analysis

| mg/kg | ALT (U/L) | AST (U/L) | T Bili. (mg/dL) | BUN (mg/dL) | Creat. (mg/dL) |
|---|---|---|---|---|---|
| 1 | 36 ±4.2 | 93 ±41 | 0.2 | 27 ±1 | 0.2 |
| 3 | 29 ±2.8 | 47 ±2.1 | 0.2 | 35 ±4 | 0.3 |
| 10 | 26±2.6 | 47±3.7 | 0.1 | 29 ±2 | 0.2 |
| 1, every 3d, x6 cycles | 134±56 | 80±9 | 0.14±0.05 | 40±5 | 0.1 |

**ARE-2** Histopathology

| mg/kg | cecum | duodenum | heart | ileum | kidney | liver | pancreas | spleen | lung | stomach |
|---|---|---|---|---|---|---|---|---|---|---|
| 1 | - | - | - | - | - | - | - | - | - | - |
| 3 | - | - | - | - | - | - | - | - | - | - |
| 10 | - | - | - | - | - | - | - | - | - | - |

**Figure 12.16** (Top) Weight curves of animals treated with ARE-1 and ARE-2 over nine days. Mice treated with ARE-1 at 10 mg kg$^{-1}$ demonstrated significant weight loss (defined as greater than 15%, orange line), while those treated with ARE-2 at the same dose did not. Error bars indicate maximum and minimum ($n=4$). (Bottom) Treatment with ARE-2 did not significantly affect levels of white blood cells (WBC), red blood cells (RBC), haemoglobin (HGB), neutrophils or lymphocytes. Figures are mean ± standard deviation ($n=5$). Serum analysis of ALT, AST, total bilirubin, BUN and creatinine showed no elevation of these liver and kidney damage markers after treatment at the indicated concentrations. (single dose $n=4$, multiple dose $n=3$). Histopathology analysis of sacrificed animals after a single injection of ARE-2 at the indicated doses. - indicates no damage ($n=2$).

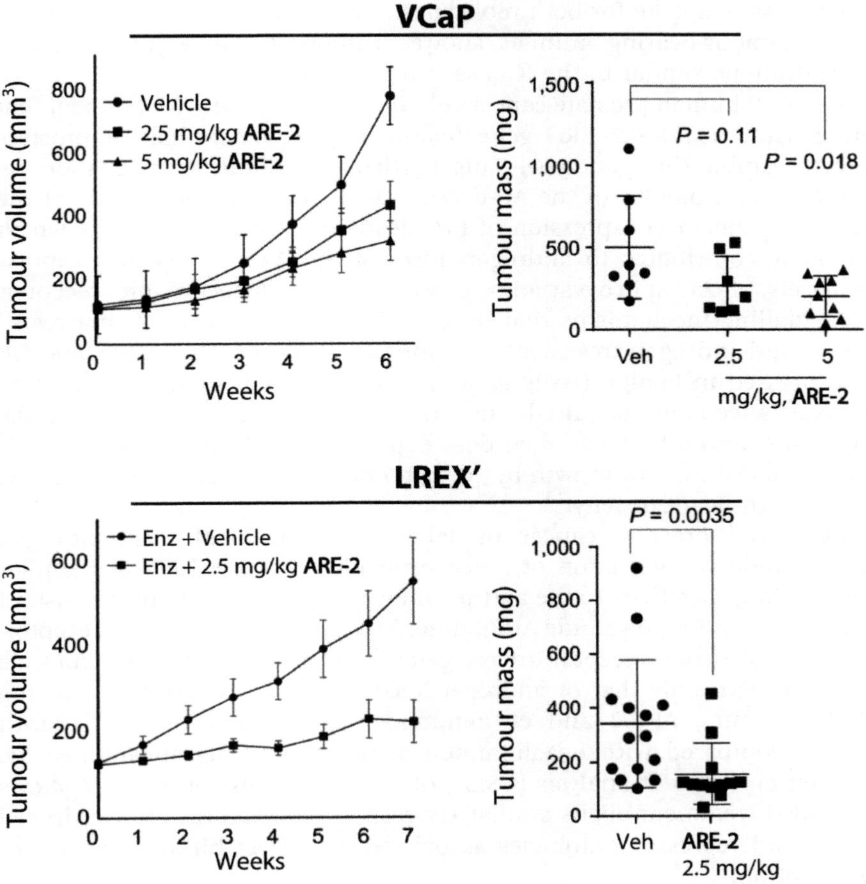


**Figure 12.17** Mice were treated subcutaneously three times per week with ARE-2, in the flank opposite the engrafted tumour. Tumour volumes were measured weekly with callipers; error represents SEM. Whisker plots for tumour mass represent mean and standard deviation. (Top) Tumour volumes and final tumour masses of VCaP xenografts treated with vehicle (20% DMSO/saline, $n=8$) or ARE-2 at 2.5 mg kg$^{-1}$ ($n=7$) or 5 mg kg$^{-1}$ ($n=8$). (Bottom) Tumour volumes and final tumour masses of LREX′ xenografts in castrated animals treated daily with 10 mg kg$^{-1}$ enzalutamide (enz) and vehicle ($n=14$) or enzalutamide and 2.5 mg kg$^{-1}$ ARE-2 ($n=12$).

The LNCaP cell line is a widely used model of hormone-sensitive prostate cancer. In cell culture growth inhibition studies using LNCaP cells, the $IC_{50}$ was found to be 7.0 and 2.1 μM for polyamides ARE-1 and ARE-2 respectively.[100,146] Early studies with ARE-1 in LNCaP xenografts showed tumour reduction after three injections of 1 mg kg$^{-1}$ in tumour-bearing NSG mice.[100] Similar results were seen after six injections of ARE-2.[146] Distribution studies using $^{14}$C radiolabelled analogues of ARE-1 and ARE-2 showed preferential tumour tissue uptake for both molecules. For ARE-2, multiple injections in LNCaP tumour-bearing animals showed tumour tissue accumulation at concentrations similar to the $IC_{50}$ seen in cell culture.[144]

The VCaP human prostate cancer cell line is a commonly used model that contains the TMPRSS2–ERG gene fusion. Approximately half of prostate cancers exhibit this genotype. This particular genomic translocation rearranges the promoter of the AR-driven TMPRSS2 to the gene body of the ERG oncogene. Over-expression of ERG leads to genomic instability which ultimately contributes to androgen insensitivity. VCaP cells also express low levels of AR splice variants as well as the glucocorticoid receptor, two signalling mechanisms that have been identified as conferring resistance to anti-androgen treatment. It is important to note that AR-Vs and GR are expressed in benign tissue as well as malignant neoplasia, suggesting that resistance is not acquired, but rather selected for. In VCaP xenografts, both ARE-1 and ARE-2 exhibited dose-dependent antitumour activity.[105,133] ARE-2 reduced tumour growth by 70% at 5 mg $kg^{-1}$ compared with vehicle, without significant toxicity.[133]

The LREX′ prostate cancer model is resistant to the antiandrogen enzalutamide *via* activation of glucocorticoid receptor, which has similar DNA-binding specificity to the androgen receptor. In enzalutamide-resistant LREX′ cells, Py–Im polyamide ARE-2 interfered with both androgen receptor- and glucocorticoid receptor-driven gene expression, while enzalutamide interfered with only that of androgen receptor. In castrated mice bearing LREX′ tumours, ARE-2 and enzalutamide co-treatment reduced growth by 80% compared with enzalutamide alone without significant toxicity. Gene set enrichment analysis (GSEA) of tumour expression profiles shows that ARE-2 treatment elicits similar UV response signatures as seen in cell culture and represses ontologies associated with DNA binding-dependent transcription.[133]

## 12.16 Formulation

The translational development of Py–Im polyamides into therapeutics necessitates the optimisation of compound formulation to improve solubility and bioavailability in complex organisms. Due to the large molecular weight of eight-ring hairpin Py–Im polyamides (~1 200 MW) and the hydrophobic nature of the pyrrole and imidazole amino acids, solubility in aqueous buffers is limited.[143] To address this issue, polyamide solubility has been screened under several hundred buffer conditions, including many FDA-approved excipients. This study has led to the identification of multiple solvent conditions that improve the solubility and stability of polyamides in solution. The effect of these excipients on bioavailability are being tested in mice and benchmarked against the bioavailability of compounds administered in DMSO–saline. This study will allow us to determine an optimised formulation for polyamide administration.

In addition to formulation improvements, hairpin Py–Im polyamides have also been found to form nanoparticles in solution and to localise preferentially into prostate cancer tumours *in vivo*.[144,145] This observation may be explained by the enhanced permeability and retention (EPR) effect where large macromolecules accumulate in tumour tissue better than in normal tissues.[147–149] This is further supported by the biodistribution profile of a $^{14}C$-labelled polyamide in mice where the major organs of accumulation, the liver, spleen and bone marrow, all possess sinusoidal capillaries with large intercellular gaps to allow macromolecule diffusion.[144,147,150] In addition to tumour retention, the EPR effect also brings the advantages of improving compound stability and providing sustained compound release. These properties of Py–Im polyamides require additional studies using dynamic light scattering and cryo-EM techniques to further understand the characteristics of hairpin Py–Im polyamide nanoparticles. Understanding the parameters of aggregate formation would allow their properties to be fine-tuned and improved with various substrate supports.

## Acknowledgements

We are grateful to our coworkers and collaborators for their splendid efforts in the research described. We thank also the National Institutes of Health GM27681-38, the American Cancer Society, and the Prostate Cancer Foundation for financial support of this research.

## References

1. F. E. Hahn, ed., *Progress in Molecular and Subcelluar Biology*, Springer-Verlag, New York, 1971.
2. *Platinum Coordination Compounds in Cancer Chemotherapy*, ed., T. A. Conners and J. J. Roberts, Springer-Verlag, Heidelberg, 1974.
3. *Mechanism of Action of Antimicrobial and Antitumor Agents*, ed., J. W. Corcoran and F. E. Hahn, Springer-Verlag, New York, 1975.
4. L. H. Hurley, *Nat. Rev. Cancer*, 2002, **2**, 188–200.
5. S. P. Basourakos, L. K. Li, A. M. Aparicio, P. G. Corn, J. Kim and T. C. Thompson, *Curr. Med. Chem.*, 2017, **24**, 1586–1606.
6. F. Arcamone, S. Penco, P. Orezzi, V. Nicolella and A. Pirelli, *Nature*, 1964, **203**, 1064–1065.
7. C. Zimmer and U. Wahnert, *Prog. Biophys. Mol. Biol.*, 1986, **47**, 31–112.
8. M. L. Kopka, C. Yoon, D. Goodsell, P. Pjura and R. E. Dickerson, *Proc. Natl. Acad. Sci. U. S. A.*, 1985, **82**, 1376–1380.
9. M. Coll, C. A. Frederick, A. H. Wang and A. Rich, *Proc. Natl. Acad. Sci. U. S. A.*, 1987, **84**, 8385–8389.
10. J. G. Pelton and D. E. Wemmer, *Proc. Natl. Acad. Sci. U. S. A.*, 1989, **86**, 5723–5727.
11. X. Chen, B. Ramakrishnan, S. T. Rao and M. Sundaralingam, *Nat. Struct. Biol.*, 1994, **1**, 169–175.

12. M. W. Van Dyke, R. P. Hertzberg and P. B. Dervan, *Proc. Natl. Acad. Sci. U. S. A.*, 1982, **79**, 5470–5474.
13. M. W. Van Dyke and P. B. Dervan, *Nucleic Acids Res.*, 1983, **11**, 5555–5567.
14. M. W. Van Dyke and P. B. Dervan, *Science*, 1984, **225**, 1122–1127.
15. P. G. Schultz, J. S. Taylor and P. B. Dervan, *J. Am. Chem. Soc.*, 1982, **104**, 6861–6863.
16. J. S. Taylor, P. G. Schultz and P. B. Dervan, *Tetrahedron*, 1984, **40**, 457–465.
17. P. B. Dervan, *Science*, 1986, **232**, 464–471.
18. P. B. Dervan, *Bioorg. Med. Chem.*, 2001, **9**, 2215–2235.
19. P. B. Dervan and B. S. Edelson, *Curr. Opin. Struct. Biol.*, 2003, **13**, 284–299.
20. P. B. Dervan, A. T. Poulin-Kerstien, E. J. Fechter and B. S. Edelson, *Top. Curr. Chem.*, 2005, **253**, 1–31.
21. W. S. Wade, M. Mrksich and P. B. Dervan, *J. Am. Chem. Soc.*, 1992, **114**, 8783–8794.
22. W. S. Wade, M. Mrksich and P. B. Dervan, *Biochemistry*, 1993, **32**, 11385–11389.
23. M. Mrksich, W. S. Wade, T. J. Dwyer, B. H. Geierstanger, D. E. Wemmer and P. B. Dervan, *Proc. Natl. Acad. Sci. U. S. A.*, 1992, **89**, 7586–7590.
24. B. H. Geierstanger, M. Mrksich, P. B. Dervan and D. E. Wemmer, *Science*, 1994, **266**, 646–650.
25. C. L. Kielkopf, E. E. Baird, P. B. Dervan and D. C. Rees, *Nat. Struct. Biol.*, 1998, **5**, 104–109.
26. D. S. Pilch, N. Poklar, C. A. Gelfand, S. M. Law, K. J. Breslauer, E. E. Baird and P. B. Dervan, *Proc. Natl. Acad. Sci. U. S. A.*, 1996, **93**, 8306–8311.
27. D. S. Pilch, N. Poklar, E. E. Baird, P. B. Dervan and K. J. Breslauer, *Biochemistry*, 1999, **38**, 2143–2151.
28. D. M. Chenoweth, J. L. Meier and P. B. Dervan, *Angew. Chem.*, 2013, **52**, 415–418.
29. M. Mrksich and P. B. Dervan, *J. Am. Chem. Soc.*, 1993, **115**, 2572–2576.
30. M. Mrksich and P. B. Dervan, *J. Am. Chem. Soc.*, 1995, **117**, 3325–3332.
31. S. White, E. E. Baird and P. B. Dervan, *Chem. Biol.*, 1997, **4**, 569–578.
32. S. E. Swalley, E. E. Baird and P. B. Dervan, *J. Am. Chem. Soc.*, 1997, **119**, 6953–6961.
33. C. F. Hsu, J. W. Phillips, J. W. Trauger, M. E. Farkas, J. M. Belitsky, A. Heckel, B. Z. Olenyuk, J. W. Puckett, C. C. Wang and P. B. Dervan, *Tetrahedron*, 2007, **63**, 6146–6151.
34. S. White, J. W. Szewczyk, J. M. Turner, E. E. Baird and P. B. Dervan, *Nature*, 1998, **391**, 468–471.
35. C. L. Kielkopf, S. White, J. W. Szewczyk, J. M. Turner, E. E. Baird, P. B. Dervan and D. C. Rees, *Science*, 1998, **282**, 111–115.
36. C. L. Kielkopf, R. E. Bremer, S. White, J. W. Szewczyk, J. M. Turner, E. E. Baird, P. B. Dervan and D. C. Rees, *J. Mol. Biol.*, 2000, **295**, 557–567.

37. S. White, J. M. Turner, J. W. Szewczyk, E. E. Baird and P. B. Dervan, *J. Am. Chem. Soc.*, 1999, **121**, 260–261.
38. A. R. Urbach, J. W. Szewczyk, S. White, J. M. Turner, E. E. Baird and P. B. Dervan, *J. Am. Chem. Soc.*, 1999, **121**, 11621–11629.
39. J. J. Kelly, E. E. Baird and P. B. Dervan, *Proc. Natl. Acad. Sci. U. S. A.*, 1996, **93**, 6981–6985.
40. J. W. Trauger, E. E. Baird, M. Mrksich and P. B. Dervan, *J. Am. Chem. Soc.*, 1996, **118**, 6160–6166.
41. J. W. Trauger, E. E. Baird and P. B. Dervan, *J. Am. Chem. Soc.*, 1998, **120**, 3534–3535.
42. M. Mrksich, M. E. Parks and P. B. Dervan, *J. Am. Chem. Soc.*, 1994, **116**, 7983–7988.
43. J. W. Trauger, E. E. Baird and P. B. Dervan, *Nature*, 1996, **382**, 559–561.
44. J. W. Trauger, E. E. Baird and P. B. Dervan, *Chem. Biol.*, 1996, **3**, 369–377.
45. J. M. Turner, E. E. Baird and P. B. Dervan, *J. Am. Chem. Soc.*, 1997, **119**, 7636–7644.
46. R. Baliga, E. E. Baird, D. M. Herman, C. Melander, P. B. Dervan and D. M. Crothers, *Biochemistry*, 2001, **40**, 3–8.
47. J. W. Trauger, E. E. Baird and P. B. Dervan, *Angew. Chem., Int. Ed.*, 1998, **37**, 1421–1423.
48. W. A. Greenberg, E. E. Baird and P. B. Dervan, *Chem. Eur. J.*, 1998, **4**, 796–805.
49. B. Olenyuk, C. Jitianu and P. B. Dervan, *J. Am. Chem. Soc.*, 2003, **125**, 4741–4751.
50. A. Heckel and P. B. Dervan, *Chem. - Eur. J.*, 2003, **9**, 3353–3366.
51. J. Y. Cho, M. E. Parks and P. B. Dervan, *Proc. Natl. Acad. Sci. U. S. A.*, 1995, **92**, 10389–10392.
52. P. Weyermann and P. B. Dervan, *J. Am. Chem. Soc.*, 2002, **124**, 6872–6878.
53. D. M. Herman, E. E. Baird and P. B. Dervan, *Chem. - Eur. J.*, 1999, **5**, 975–983.
54. I. Kers and P. B. Dervan, *Bioorg. Med. Chem.*, 2002, **10**, 3339–3349.
55. J. M. Turner, S. E. Swalley, E. E. Baird and P. B. Dervan, *J. Am. Chem. Soc.*, 1998, **120**, 6219–6226.
56. T. P. Best, B. S. Edelson, N. G. Nickols and P. B. Dervan, *Proc. Natl. Acad. Sci. U. S. A.*, 2003, **100**, 12063–12068.
57. B. S. Edelson, T. P. Best, B. Olenyuk, N. G. Nickols, R. M. Doss, S. Foister, A. Heckel and P. B. Dervan, *Nucleic Acids Res.*, 2004, **32**, 2802–2818.
58. C. F. Hsu and P. B. Dervan, *Bioorg. Med. Chem. Lett.*, 2008, **18**, 5851–5855.
59. S. White, E. E. Baird and P. B. Dervan, *J. Am. Chem. Soc.*, 1997, **119**, 8756–8765.
60. D. M. Herman, E. E. Baird and P. B. Dervan, *J. Am. Chem. Soc.*, 1998, **120**, 1382–1391.

61. C. Dose, M. E. Farkas, D. M. Chenoweth and P. B. Dervan, *J. Am. Chem. Soc.*, 2008, **130**, 6859–6866.
62. V. C. Rucker, C. Melander and P. B. Dervan, *Helv. Chim. Acta*, 2003, **86**, 1839–1851.
63. S. E. Swalley, E. E. Baird and P. B. Dervan, *J. Am. Chem. Soc.*, 1999, **121**, 1113.
64. N. G. Nickols, C. S. Jacobs, M. E. Farkas and P. B. Dervan, *Nucleic Acids Res.*, 2007, **35**, 363–370.
65. D. A. Harki, N. Satyamurthy, D. B. Stout, M. E. Phelps and P. B. Dervan, *Proc. Natl. Acad. Sci. U. S. A.*, 2008, **105**, 13039–13044.
66. C. S. Jacobs and P. B. Dervan, *J. Med. Chem.*, 2009, **52**, 7380–7388.
67. M. A. Marques, R. M. Doss, A. R. Urbach and P. B. Dervan, *Helv. Chim. Acta*, 2002, **85**, 4485–4517.
68. S. Foister, M. A. Marques, R. M. Doss and P. B. Dervan, *Bioorg. Med. Chem.*, 2003, **11**, 4333–4340.
69. D. H. Nguyen, J. W. Szewczyk, E. E. Baird and P. B. Dervan, *Bioorg. Med. Chem.*, 2001, **9**, 7–17.
70. T. G. Minehan, K. Gottwald and P. B. Dervan, *Helv. Chim. Acta*, 2000, **83**, 2197–2213.
71. C. A. Briehn, P. Weyermann and P. B. Dervan, *Chem.-Eur. J*, 2003, **9**, 2110–2122.
72. D. Renneberg and P. B. Dervan, *J. Am. Chem. Soc.*, 2003, **125**, 5707–5716.
73. M. A. Marques, R. M. Doss, S. Foister and P. B. Dervan, *J. Am. Chem. Soc.*, 2004, **126**, 10339–10349.
74. D. M. Chenoweth, A. Viger and P. B. Dervan, *J. Am. Chem. Soc.*, 2007, **129**, 2216–2217.
75. E. E. Baird and P. B. Dervan, *J. Am. Chem. Soc.*, 1996, **118**, 6141–6146.
76. J. M. Belitsky, D. H. Nguyen, N. R. Wurtz and P. B. Dervan, *Bioorg. Med. Chem.*, 2002, **10**, 2767–2774.
77. N. G. Nickols and P. B. Dervan, *Proc. Natl. Acad. Sci. U. S. A.*, 2007, **104**, 10418–10423.
78. K. A. Muzikar, N. G. Nickols and P. B. Dervan, *Proc. Natl. Acad. Sci. U. S. A.*, 2009, **106**, 16598–16603.
79. J. A. Raskatov, J. L. Meier, J. W. Puckett, F. Yang, P. Ramakrishnan and P. B. Dervan, *Proc. Natl. Acad. Sci. U. S. A.*, 2012, **109**, 1023–1028.
80. J. W. Puckett, J. T. Green and P. B. Dervan, *Org. Lett.*, 2012, **14**, 2774–2777.
81. N. R. Wurtz, J. M. Turner, E. E. Baird and P. B. Dervan, *Org. Lett.*, 2001, **3**, 1201–1203.
82. B. C. Li, D. C. Montgomery, J. W. Puckett and P. B. Dervan, *J. Org. Chem.*, 2013, **78**, 124–133.
83. D. M. Chenoweth, D. A. Harki and P. B. Dervan, *J. Am. Chem. Soc.*, 2009, **131**, 7175–7181.
84. R. K. Suto, R. S. Edayathumangalam, C. L. White, C. Melander, J. M. Gottesfeld, P. B. Dervan and K. Luger, *J. Mol. Biol.*, 2003, **326**, 371–380.

85. R. S. Edayathumangalam, P. Weyermann, J. M. Gottesfeld, P. B. Dervan and K. Luger, *Proc. Natl. Acad. Sci. U. S. A.*, 2004, **101**, 6864–6869.
86. J. M. Gottesfeld, J. M. Belitsky, C. Melander, P. B. Dervan and K. Luger, *J. Mol. Biol.*, 2002, **321**, 249–263.
87. N. R. Wurtz, J. L. Pomerantz, D. Baltimore and P. B. Dervan, *Biochemistry*, 2002, **41**, 7604–7609.
88. J. A. Ehley, C. Melander, D. Herman, E. E. Baird, H. A. Ferguson, J. A. Goodrich, P. B. Dervan and J. M. Gottesfeld, *Mol. Cell. Biol.*, 2002, **22**, 1723–1733.
89. D. H. Nguyen-Hackley, E. Ramm, C. M. Taylor, J. K. Joung, P. B. Dervan and C. O. Pabo, *Biochemistry*, 2004, **43**, 3880–3890.
90. M. D. Gearhart, L. Dickinson, J. Ehley, C. Melander, P. B. Dervan, P. E. Wright and J. M. Gottesfeld, *Biochemistry*, 2005, **44**, 4196–4203.
91. D. M. Chenoweth and P. B. Dervan, *Proc. Natl. Acad. Sci. U. S. A.*, 2009, **106**, 13175–13179.
92. D. M. Chenoweth and P. B. Dervan, *J. Am. Chem. Soc.*, 2010, **132**, 14521–14529.
93. J. E. Darnell, Jr., *Nat. Rev. Cancer*, 2002, **2**, 740–749.
94. J. M. Gottesfeld, L. Neely, J. W. Trauger, E. E. Baird and P. B. Dervan, *Nature*, 1997, **387**, 202–205.
95. D. H. Kong, E. J. Park, A. G. Stephen, M. Calvani, J. H. Cardellina, A. Monks, R. J. Fisher, R. H. Shoemaker and G. Melillo, *Cancer Res.*, 2005, **65**, 9047–9055.
96. M. Ljungman and F. F. Zhang, *Oncogene*, 1996, **13**, 823–831.
97. J. N. Ratner, B. Balasubramanian, J. Corden, S. L. Warren and D. B. Bregman, *J. Biol. Chem.*, 1998, **273**, 5184–5189.
98. Y. Jung and S. J. Lippard, *J. Biol. Chem.*, 2006, **281**, 1361–1370.
99. L. A. Dickinson, R. J. Gulizia, J. W. Trauger, E. E. Baird, D. E. Mosier, J. M. Gottesfeld and P. B. Dervan, *Proc. Natl. Acad. Sci. U. S. A.*, 1998, **95**, 12890–12895.
100. F. Yang, N. G. Nickols, B. C. Li, G. K. Marinov, J. W. Said and P. B. Dervan, *Proc. Natl. Acad. Sci. U. S. A.*, 2013, **110**, 1863–1868.
101. Y. Arima, M. Nitta, S. Kuninaka, D. W. Zhang, T. Fujiwara, Y. Taya, M. Nakao and H. Saya, *J. Biol. Chem.*, 2005, **280**, 19166–19176.
102. V. T. Nguyen, F. Giannoni, M. F. Dubois, S. J. Seo, M. Vigneron, C. Kedinger and O. Bensaude, *Nucleic Acids Res.*, 1996, **24**, 2924–2929.
103. L. Xu, W. Wang, D. Gotte, F. Yang, A. A. Hare, T. R. Welch, B. C. Li, J. H. Shin, J. Chong, J. N. Strathern, P. B. Dervan and D. Wang, *Proc. Natl. Acad. Sci. U. S. A.*, 2016, **113**, 12426–12431.
104. T. F. Martinez, J. W. Phillips, K. K. Karanja, P. Polaczek, C. M. Wang, B. C. Li, J. L. Campbell and P. B. Dervan, *Nucleic Acids Res.*, 2014, **42**, 11546–11559.
105. A. E. Hargrove, T. F. Martinez, A. A. Hare, A. A. Kurmis, J. W. Phillips, S. Sud, K. J. Pienta and P. B. Dervan, *PLoS One*, 2015, **10**, e0143161.
106. V. C. Rucker, S. Foister, C. Melander and P. B. Dervan, *J. Am. Chem. Soc.*, 2003, **125**, 1195–1202.

107. J. L. Meier, D. C. Montgomery and P. B. Dervan, *Nucleic Acids Res.*, 2012, **40**, 2345–2356.
108. J. E. Yeh, P. A. Toniolo and D. A. Frank, *Curr. Opin. Oncol.*, 2013, **25**, 652–658.
109. A. S. Bhagwat and C. R. Vakoc, *Trends Cancer*, 2015, **1**, 53–65.
110. A. K. Shiau, D. Barstad, P. M. Loria, L. Cheng, P. J. Kushner, D. A. Agard and G. L. Greene, *Cell*, 1998, **95**, 927–937.
111. G. Attard, A. H. Reid, R. A'Hern, C. Parker, N. B. Oommen, E. Folkerd, C. Messiou, L. R. Molife, G. Maier, E. Thompson, D. Olmos, R. Sinha, G. Lee, M. Dowsett, S. B. Kaye, D. Dearnaley, T. Kheoh, A. Molina and J. S. de Bono, *J. Clin. Oncol.*, 2009, **27**, 3742–3748.
112. C. Tran, S. Ouk, N. J. Clegg, Y. Chen, P. A. Watson, V. Arora, J. Wongvipat, P. M. Smith-Jones, D. Yoo, A. Kwon, T. Wasielewska, D. Welsbie, C. D. Chen, C. S. Higano, T. M. Beer, D. T. Hung, H. I. Scher, M. E. Jung and C. L. Sawyers, *Science*, 2009, **324**, 787–790.
113. B. Z. Olenyuk, G. J. Zhang, J. M. Klco, N. G. Nickols, W. G. Kaelin and P. B. Dervan, *Proc. Natl. Acad. Sci. U. S. A.*, 2004, **101**, 16768–16773.
114. A. Z. Ansari, A. K. Mapp, D. H. Nguyen, P. B. Dervan and M. Ptashne, *Chem. Biol.*, 2001, **8**, 583–592.
115. Y. Kwonj, H. D. Arndt, M. Qian, Y. Choi, Y. Kawazoe, P. B. Dervan and M. Uesugi, *J. Am. Chem. Soc.*, 2004, **126**, 15940–15941.
116. R. L. Stafford and P. B. Dervan, *J. Am. Chem. Soc.*, 2007, **129**, 14026–14033.
117. A. K. Mapp, A. Z. Ansari, M. Ptashne and P. B. Dervan, *Proc. Natl. Acad. Sci. U. S. A.*, 2000, **97**, 3930–3935.
118. N. R. Wurtz and P. B. Dervan, *Chem. Biol.*, 2000, **7**, 153–161.
119. A. Giaccia, B. G. Siim and R. S. Johnson, *Nat. Rev. Drug Discovery*, 2003, **2**, 803–811.
120. N. G. Nickols, C. S. Jacobs, M. E. Farkas and P. B. Dervan, *ACS Chem. Biol.*, 2007, **2**, 561–571.
121. V. S. Mysore, J. Szablowski, P. B. Dervan and P. J. Frost, *Mol. Cancer Res.*, 2016, **14**, 253–266.
122. J. O. Szablowski, J. A. Raskatov and P. B. Dervan, *Mol. Cancer Ther.*, 2016, **15**, 608–617.
123. M. Karin, Y. Cao, F. R. Greten and Z. W. Li, *Nat. Rev. Cancer*, 2002, **2**, 301–310.
124. S. I. Grivennikov, F. R. Greten and M. Karin, *Cell*, 2010, **140**, 883–899.
125. T. D. Gilmore, *Oncogene*, 2006, **25**, 6680–6684.
126. J. A. Raskatov, N. G. Nickols, A. E. Hargrove, G. K. Marinov, B. Wold and P. B. Dervan, *Proc. Natl. Acad. Sci. U. S. A.*, 2012, **109**, 16041–16045.
127. A. Manni, B. Arafah and O. H. Pearson, *Cancer*, 1980, **46**, 2838–2841.
128. V. Kumar and P. Chambon, *Cell*, 1988, **55**, 145–156.
129. N. G. Nickols, J. O. Szablowski, A. E. Hargrove, B. C. Li, J. A. Raskatov and P. B. Dervan, *Mol. Cancer Ther.*, 2013, **12**, 675–684.
130. V. K. Arora, E. Schenkein, R. Murali, S. K. Subudhi, J. Wongvipat, M. D. Balbas, N. Shah, L. Cai, E. Efstathiou, C. Logothetis, D. Zheng and C. L. Sawyers, *Cell*, 2013, **155**, 1309–1322.

131. R. Z. Szmulewitz, C. Nabhan, P. H. O'Donnell, J. Kach, T. Karrison, E. Martinez and W. M. Stadler, *J. Clin. Oncol.*, 2016, **34**, TPS5091–TPS5091.
132. D. M. Chenoweth, D. A. Harki, J. W. Phillips, C. Dose and P. B. Dervan, *J. Am. Chem. Soc.*, 2009, **131**, 7182–7188.
133. A. A. Kurmis, F. Yang, T. R. Welch, N. G. Nickols and P. B. Dervan, *Cancer Res.*, 2017, **77**, 2207–2212.
134. C. L. Warren, N. C. Kratochvil, K. E. Hauschild, S. Foister, M. L. Brezinski, P. B. Dervan, G. N. Phillips, Jr. and A. Z. Ansari, *Proc. Natl. Acad. Sci. U. S. A.*, 2006, **103**, 867–872.
135. J. W. Puckett, K. A. Muzikar, J. Tietjen, C. L. Warren, A. Z. Ansari and P. B. Dervan, *J. Am. Chem. Soc.*, 2007, **129**, 12310–12319.
136. J. L. Meier, A. S. Yu, I. Korf, D. J. Segal and P. B. Dervan, *J. Am. Chem. Soc.*, 2012, **134**, 17814–17822.
137. A. R. Urbach and P. B. Dervan, *Proc. Natl. Acad. Sci. U. S. A.*, 2001, **98**, 4343–4348.
138. A. R. Urbach, J. J. Love, S. A. Ross and P. B. Dervan, *J. Mol. Biol.*, 2002, **320**, 55–71.
139. G. S. Erwin, D. Bhimsaria, A. Eguchi and A. Z. Ansari, *Angew. Chem., Int. Ed.*, 2014, **53**, 10124–10128.
140. G. S. Erwin, M. P. Grieshop, D. Bhimsaria, A. Eguchi, J. A. Rodriguez-Martinez and A. Z. Ansari, *J. Vis. Exp.*, 2016.
141. T. W. Synold, B. Xi, J. Wu, Y. Yen, B. C. Li, F. Yang, J. W. Phillips, N. G. Nickols and P. B. Dervan, *Cancer Chemother. Pharmacol.*, 2012, **70**, 617–625.
142. K. A. Muzikar, J. L. Meier, D. A. Gubler, J. A. Raskatov and P. B. Dervan, *Org. Lett.*, 2011, **13**, 5612–5615.
143. A. E. Hargrove, J. A. Raskatov, J. L. Meier, D. C. Montgomery and P. B. Dervan, *J. Med. Chem.*, 2012, **55**, 5425–5432.
144. J. A. Raskatov, J. W. Puckett and P. B. Dervan, *Bioorg. Med. Chem.*, 2014, **22**, 4371–4375.
145. J. A. Raskatov, J. O. Szablowski and P. B. Dervan, *J. Med. Chem.*, 2014, **57**, 8471–8476.
146. F. Yang, N. G. Nickols, B. C. Li, J. O. Szablowski, S. R. Hamilton, J. L. Meier, C. M. Wang and P. B. Dervan, *J. Med. Chem.*, 2013, **56**, 7449–7457.
147. M. I. Setyawati, C. Y. Tay, D. Docter, R. H. Stauber and D. T. Leong, *Chem. Soc. Rev.*, 2015, **44**, 8174–8199.
148. H. Kobayashi, R. Watanabe and P. L. Choyke, *Theranostics*, 2014, **4**, 81–89.
149. J. Fang, H. Nakamura and H. Maeda, *Adv. Drug Delivery Rev.*, 2011, **63**, 136–151.
150. O. Cleaver and D. A. Melton, *Nat. Med.*, 2003, **9**, 661–668.

CHAPTER 13

# *Synthetic Peptides for DNA Recognition Inspired by Transcription Factors*

ABHISHEK IYER,[†] YARA RUIZ GARCIA[†] AND ANNEMIEKE MADDER*

Organic and Biomimetic Chemistry Research Group, Ghent University, 9000 Ghent, Belgium
*Email: annemieke.madder@ugent.be

## 13.1 Transcription Factors as Source of Inspiration for the Design of DNA-binding Peptides

### 13.1.1 Interaction Between dsDNA and Proteins: A Key Factor Regulating Transcription

Proteins are involved in a majority of chemical reactions, play critical roles and are vital for the structure, function and regulation of tissues in the human body. Depending on the functions that proteins perform, they belong to different classes, including antibodies, enzymes and messenger proteins. Special proteins may also provide structure and support for cells and transport cells or store small molecules throughout the body. The subject of interest in this chapter encompasses a class of proteins known as transcription factors (TFs), which are responsible for 'reading' the

[†]These authors contributed equally.

Chemical Biology No. 7
DNA-targeting Molecules as Therapeutic Agents
Edited by Michael J. Waring

Published by the Royal Society of Chemistry, www.rsc.org

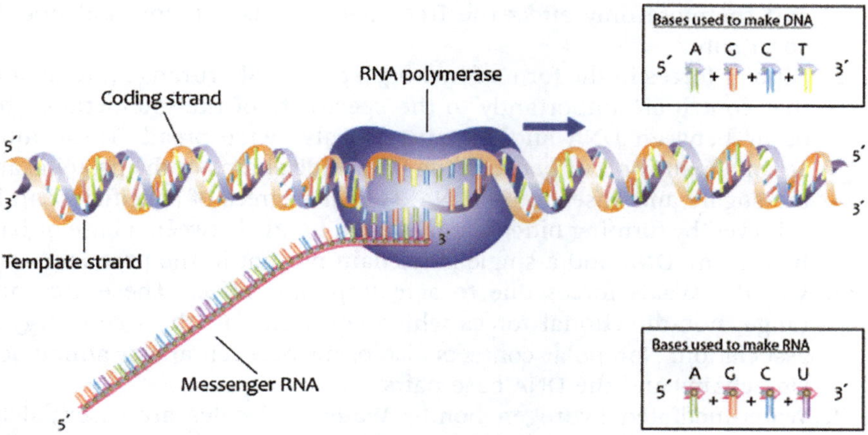


**Figure 13.1** The DNA transcription process.
Figure by Genomics Education Programme (Process of transcription) [CC BY 2.0 (http://creativecommons.org/licenses/by/2.0)], *via* Wikimedia Commons.

information stored in DNA so as to assist further transcription into messenger RNA (mRNA) (Figure 13.1). Thereafter, the information held in mRNA is converted (the process is known as translation) into a 'language' that the body can understand. This language is nothing but a sequence of amino acids, the essential building blocks of proteins.

Transcription factors are key players in this process because they specifically recognise and bind to short regulatory DNA sequences which control the gene. These interactions between DNA and proteins are extremely specific and of very high affinity. However, the lack of any general recognition code for the interaction between a specific amino acid sequence within a protein and its corresponding specific DNA-binding site has been a major challenge in the study of the structure of TFs since this renders the prediction of binding sites virtually impossible. Although a general formulation of these interactions has not been achieved, some general rules can be followed, providing a starting point for elucidating the molecular basis of specific DNA recognition. In the case of TFs, all these interactions occur in the major groove of the DNA, which is wide enough to accommodate the α-helical recognition unit. The four major interaction forces between proteins and nucleic acids, constituting the basis for an efficient design rationale, can be summarised as follows:

1. Electrostatic forces in the form of salt bridges: Long range, electrostatic interactions between groups of opposite charge are quite dominant due to the negatively charged DNA phosphodiester backbone and the positively charged basic side chains of particular amino acids in the protein, such as Lys, Arg and His. Electrostatic forces are non-specific

and therefore only affect the free energy of association and not the specificity.
2. Dipolar forces in the form of hydrogen bonds: Short-range interactions that contribute importantly to the specificity of the interactions that occur between DNA and proteins are also widespread. These interactions mainly occur between the polar side chains in the proteins and the sugars and bases in the DNA. Higher degrees of specificity can be achieved by forming bidentate hydrogen bonds between a base or base pair in the DNA and a single side chain present in the protein.
3. Van der Waals forces due to a hydrophobic effect: These are short range, non-directional forces which contribute to the free energy of association. Non-polar contacts also occur between apolar amino acid side chains and the DNA base pairs.
4. Water-mediated hydrogen bonds: Water molecules are essential for indirect hydrogen bonding between DNA and proteins. They also behave as space fillers in DNA–protein complexes. Hydrogen bond donors and acceptors located especially at the edge of a DNA–protein complex can be influenced by the water content.

Studying these interactions holds the key to synthesis of transcription factor mimics which could find applications in various fields including cancer therapy.

### 13.1.2 Families of Transcription Factors

At present, though there is great interest in the general area of transcriptional regulation, rather limited information is available about transcription factors that regulate human transcription. A list of 1391 sequence-specific dsDNA-binding transcription factors, their functions, genomic organisation and evolutionary conservation has been compiled in an excellent review article.[1] Despite this, a great deal remains unexplored. The detailed study and description of TFs is beyond the scope of this chapter; however, a short summary of the relevant families of TFs will be provided in this section.[2]

The protein structure of TFs has been divided into two broad domains for the purpose of easy understanding:

1. The DNA-binding domain: This part of the protein is responsible for the selective interaction with the DNA control sequence for transcription.
2. The DNA-activation domain: This part is responsible for the interaction with other TFs or with the transcribing RNA polymerase II itself.

For the design of DNA-binding peptides, it is the DNA-binding domains that are most relevant, among which several structurally different forms can be found in eukaryotic TFs. The most frequent, commonly occurring motifs are discussed in more detail in the following sections. While many transcription factors can be classified according to these motifs, others feature recognition elements belonging to different families but are not discussed in detail.

### 13.1.2.1 *The Helix–Turn–Helix (HTH) Motif*

The helix–turn–helix was the first DNA-binding protein motif to be discovered. It was first identified in bacterial proteins, but is now known to be present in hundreds of DNA-binding proteins in both eukaryotes and prokaryotes. It features a very simple secondary structure consisting of two α helices connected by a short, slightly elongated chain of amino acids which constitute the 'turn' (Figure 13.2). The angle between the two helices does not change, mainly due to interactions between the two helices. The C terminal helix is called the recognition helix and fits perfectly into the major groove of DNA. The amino acid side chains of the recognition helix differ from protein to protein, which enables them to recognise the specific DNA sequence to which the protein is required to bind. It is important to note that no high-affinity DNA binding is possible for HTH monomers and that the formation of a dimer (or even higher order species) is required.

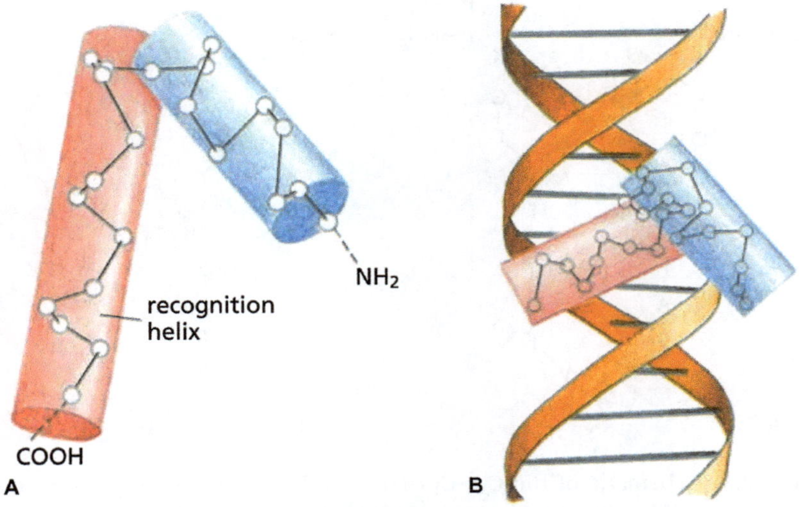


**Figure 13.2** The helix–turn–helix motif is shown in (A). Each white circle denotes the α-carbon atom of an amino acid. The C-terminal α helix, *i.e.* the recognition region, is shown in red. Part (B) of the figure shows how this helix fits into the major groove of DNA, where it is in contact with the edges of the base pairs. It is clear from the figure that the N-terminal α-helix (blue) is present primarily to provide a structural component which helps to position the recognition helix.
Reproduced with permission from DNA-Binding Motifs in Gene Regulatory Proteins Cover of Molecular Biology of the Cell. 4th edition. B. Alberts, A. Johnson, J. Lewis, *et al.*, New York: Garland Science; 2002.

Eukaryotic versions of HTH proteins are called homeodomain transcription factors which, by additional minor groove interactions through the N- or C-terminal arms, have the capacity to bind as monomers.

### 13.1.2.2 *The Zinc-finger Motif*

Zinc-fingers are a second important group of DNA-binding motifs containing one or more zinc ions important for their overall structure. Although we refer to all zinc-coordinated DNA-binding motifs as zinc fingers, in reality this name only refers to the way they appear in schematic drawings that have become familiar ever since they were discovered (Figure 13.3).[3] Structural studies over the years have shown that they can be divided into a number of distinct structural groups. Only the first one discovered is described here. The protein shown in Figure 13.3 belongs to the Cys–Cys–His–His family of zinc finger proteins. It is known to activate the transcription of a eukaryotic ribosomal RNA gene. The structure is simple, consisting of two parts—an α helix and a β sheet—which are held together by a zinc ion (Figure 13.3B). This type of zinc finger is often found in combination with additional zinc fingers which are arranged in succession so that the α helix of each

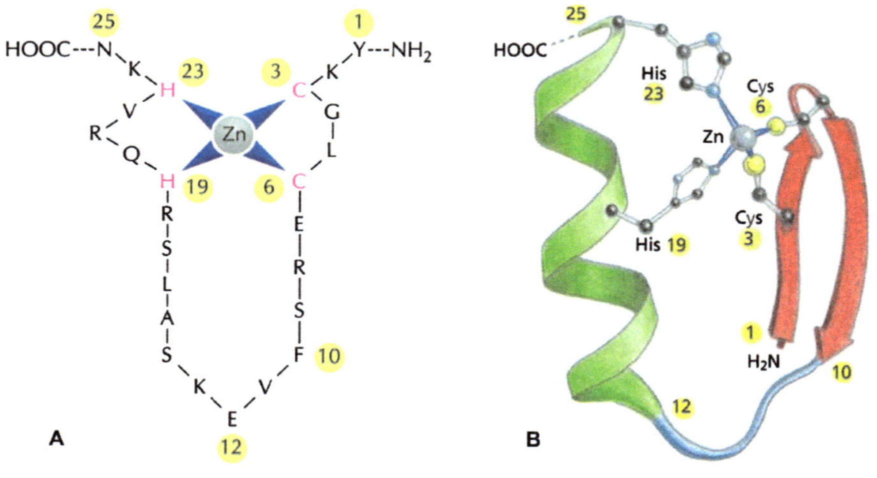


**Figure 13.3** Schematic of the Cys–Cys–His–His family of zinc finger proteins. (A) Representation of a zinc finger showing the amino acid sequence from a frog protein belonging to this class. (B) Schematic drawing showing the three-dimensional structure of this type of zinc finger comprising an antiparallel β sheet followed by an α helix. Amino acids 1 to 10 make up the β sheet and amino acids 12 to 24 make up the α helix. The four amino acids that bind to the zinc atom (Cys 3, Cys 6, His 19, and His 23) hold one end of the α helix tightly connecting it to one end of the β sheet.
Reproduced with permission from DNA-Binding Motifs in Gene Regulatory Proteins Cover of Molecular Biology of the Cell. 4th edition. B. Alberts, A. Johnson, J. Lewis, *et al.*, New York: Garland Science; 2002.

monomeric unit can interact with the major groove of the DNA. The result is a near continuous stretch of α helices along the major groove. Interaction at multiple points ensures that the DNA and protein bind very tightly to each other *via* a repeating basic structural unit (Figure 13.3A). During evolution, the number of zinc finger repeats can be adjusted, thereby tuning the strength and specificity of the DNA–protein interaction. No other TF discussed here has this unique ability to form repeating chains.

#### *13.1.2.3 The Leucine Zipper Motif*

The leucine zipper motif is unique in the sense that, unlike other DNA-binding motifs described previously, it binds to DNA as a dimer instead of a monomer. Usually, the portion of the protein responsible for the dimeric structure is distinct from the portion that is responsible for DNA recognition and binding. This motif, however, combines these two functions in a simple way. The name leucine zipper derives from the way in which the two α helices contributed by each monomer are joined together to form a short coiled-coil. The helices are held together by strong hydrophobic interactions between the amino acid side chains particularly of leucines extending from one side of either helix. Almost directly beyond the dimerization interface, the two α helices diverge, forming a Y-shaped structure that allows the amino acid side chains to interact with the major groove of DNA, gripping the DNA double helix like a pair of tweezers (Figure 13.4).

#### *13.1.2.4 The Helix–Loop–Helix motif*

The final important DNA-binding motif discussed here, and related to the leucine zipper motif, is the helix–loop–helix (HLH) motif. It must be emphasised that the HLH should not be confused with the HTH discussed earlier. An HLH motif consists of a short α helix connected to a second, longer α helix by a loop. The loop provides flexibility allowing one helix to fold back and squeeze against the other. As shown in Figure 13.5, this two-helix structure binds to DNA while at the same time binding to the HLH motif of a second HLH protein. As with leucine zipper proteins, the second HLH protein can be the same, resulting in a homodimer, or different, resulting in a heterodimer. In either case, the two α helices extending from the dimerization interface interact specifically with the DNA.

### 13.1.3 Detailed Analysis of the Primary and Secondary Structures of the GCN4 TF Bound to Its Target DNA as a Basis for the Design of Synthetic DNA-binding Mimics

The GCN4 transcription factor (TF) is arguably one of the most well-investigated transcription factors. Although it is a yeast amino acid

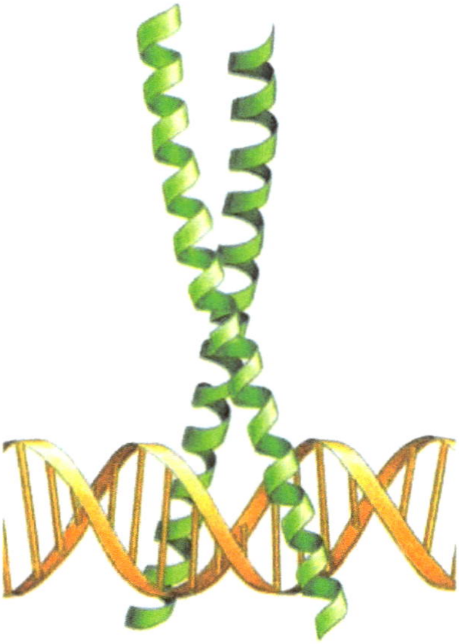

**Figure 13.4** A leucine zipper dimer bound to DNA. The bottom parts of the two α-helical monomers are involved in DNA-binding whereas the top, α-helical, leucine zipper regions allow for dimerization. Each of the two α helices binds to one-half of a symmetrical DNA sequence. The structure shown is that of the yeast GCN4 protein which regulates transcription in response to the availability of amino acids in the environment.[4] Reproduced with permission from DNA-Binding Motifs in Gene Regulatory Proteins Cover of Molecular Biology of the Cell. 4th edition. B. Alberts, A. Johnson, J. Lewis, *et al.* New York: Garland Science; 2002.

metabolism regulatory transcription factor[6] controlling the activation of several genes in response to amino acid starvation, it is of particular importance, especially to synthetic chemists, because of its well-defined dimerization domain and DNA recognition region. This makes it one of the easiest TFs to study and mimic, which, in turn, allows chemists to design and synthesise a diverse range of synthetic mimics of GCN4 TF—a strategy which could eventually be used in cancer therapy by targeting TFs in humans.[7]

The GCN4 TF is a fairly large protein comprising 281 amino acids with a molecular weight slightly over 31 kDa. The full-length sequence, although known, is of little importance. The amino acids forming the two important domains—the dimerization domain and the basic recognition region (Figure 13.6A) are, however, of much interest. The very first crystal structure of GCN4 TF bound to DNA was solved at a resolution of 2.9 Å by Ellenberger in 1992.[4] Subsequently, Keller improved it further, achieving a resolution of 2.2 Å in 1995 for the same TF bound to DNA and containing the ATF/CREB recognition sequence. Crystallographic data is freely available from the Protein Data Bank (pdb: 1YSA). A complete set of interactions is described in Figure 13.6.

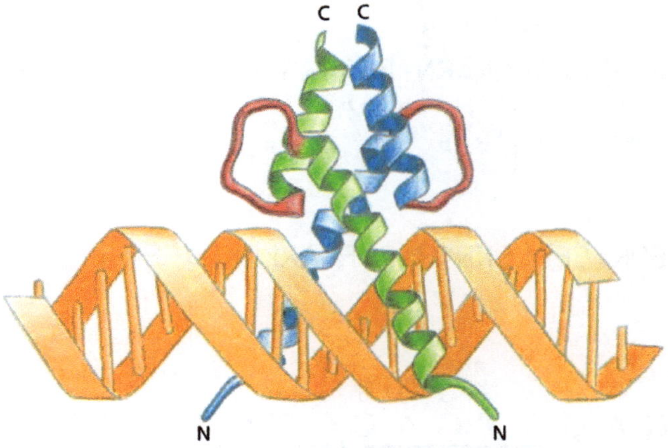


**Figure 13.5** Schematic showing a helix–loop–helix dimer bound to DNA. The two monomeric units are held together in a four-helix bundle. Two α helices arise from each unit and are connected by a flexible loop of protein (red). The HLH is capable of recognising a specific DNA sequence due to the two α helices that protrude out from the four-helix bundle.[5] Adapted with permission from DNA-Binding Motifs in Gene Regulatory Proteins Cover of Molecular Biology of the Cell. 4th edition. B. Alberts, A. Johnson, J. Lewis J, *et al.* New York: Garland Science; 2002.

The secondary structure of the GCN4 TF is quite well defined and is a perfect example of a homodimer. The bZIP dimer is a pair of continuous α helices that form a parallel coiled-coil over the C-termini of the recognition regions. The recognition region encompasses around 30 residues and gradually diverges towards the N-termini while simultaneously passing through the major groove of the DNA-binding site. The interactions between the bases and phosphate groups present in DNA and the basic residues present in the protein constitute the main contacts of this region between the DNA and the protein. The leucine zipper region of the GCN4 TF is stabilised by hydrophobic and non-polar interactions within the monomer forming an extensive van der Waals contact surface. Inter- and intra-helical salt bridges are also present, and although these are fewer they also contribute to the formation of the dimer. The basic region of each monomer adopts the optimal conformation to bind to its respective half-site of the target DNA sequence. Although there are no sharp bends or kinks in either GCN4 monomer, the rigid coiled coil imposes a gentle bend of the α-helix towards the axis of the DNA. Consequently, it is able to grip the major groove like a pair of tweezers and interact specifically with the base pairs located there. The contacts of this region with the major groove are based on the positively charged polar side chains of the arginines and lysines, which establish hydrogen bonds with the oxygen atoms of the phosphate backbone and the base pairs. The N-terminal end of the basic region continues as a helix beyond the region of DNA contact and is the most flexible part of the protein.

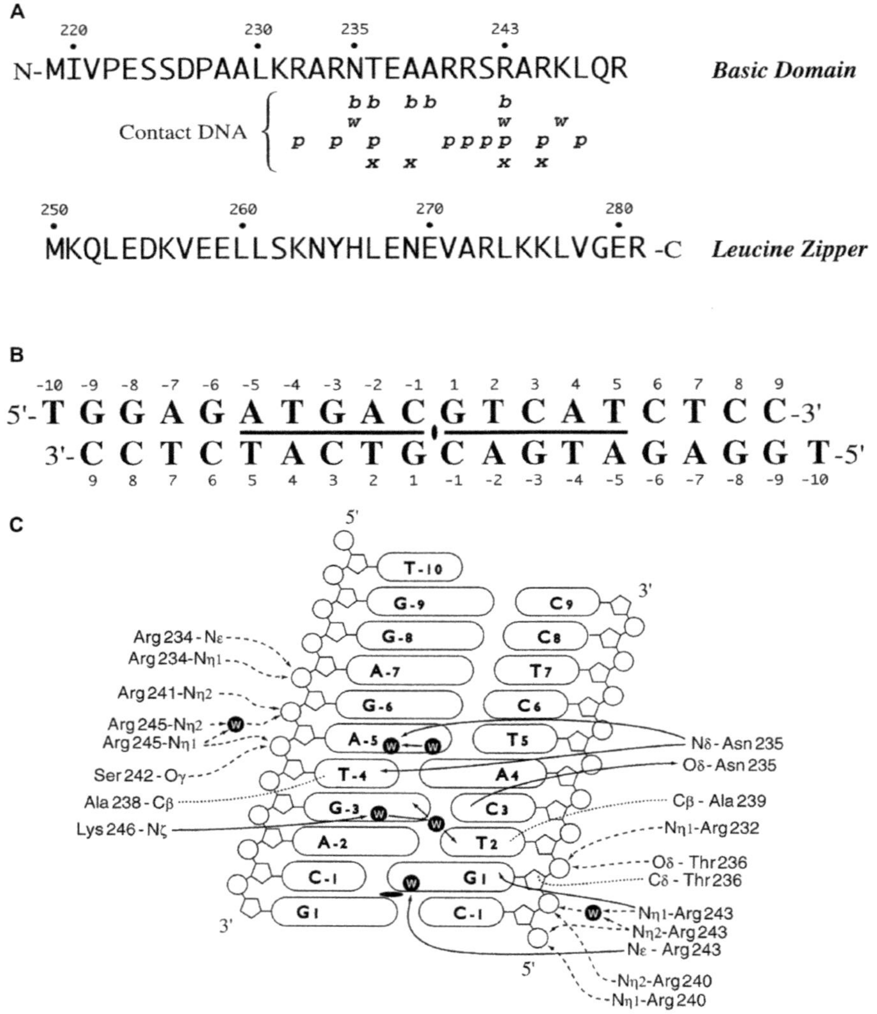


**Figure 13.6** The protein and DNA sequences of the crystallized GCN4–DNA complex and complete interaction summary. Reproduced from *Journal of Molecular Biology*, **254**, W. Kellera, P. König and T. J. Richmond, 657–667, 1995, Crystal Structure of a bZIP/DNA Complex at 2.2 Å: Determinants of DNA Specific Recognition. Copyright © 1995 Academic Press, with permission from Elsevier. (A) GCN4-bZIP domain numbered according to the full-length sequence. Residues that make one or more DNA contacts are indicated: b, direct to base; w, *via* water to base; p, direct to phosphate; x, *via* water to phosphate. (B) Self-complementary 18 bp palindromic ATF/CREB DNA with terminal thymidine. A molecular and crystallographic twofold axis passes through the C1–G1 base step. The specific recognition sequence is marked with a bar. (C) Protein–DNA interaction summary for one basic domain and DNA half-site. The location of the twofold axis is indicated. Direct and water (W) mediated interactions with bases are shown as continuous lines, interactions with the phosphate backbone as broken lines and hydrophobic interactions as dotted lines. Arrows indicate the direction of hydrogen bonds.

## 13.2 Design and Synthesis of TF Mimics as DNA Binding Peptides

### 13.2.1 Design of TF Mimics: Replacement of the Dimerization Domain by Non-peptide Scaffolds

Next we describe pioneering experiments carried out in earlier years, where the work of some of the main players in the area will be discussed alongside our own contributions to the field. Part of this work has been described in greater detail in earlier reviews.[8–10]

#### *13.2.1.1 Early TF Mimics*

A first successful step towards simplification of the GCN4 TF dates from 1990 thanks to the group of Peter S. Kim, who delivered the first proof-of-concept for a miniaturised version of the GCN4 TF selectively recognising its ATF/CREB sequence. Replacement of the leucine zipper by a disulphide bond as a dimerization link joining the two basic regions resulted in a construct with the correct properties to grip the DNA major groove as a tweezer. After several synthetic attempts, the length of the peptide sequence of the basic region within the artificial construct could be optimised. The required fragment consisted of the 23-residue N-terminal part of GCN4 -DPAALKRARNTEAARRSRARKLQ- complemented by a GGC linker containing the cysteine residue to connect the binding region with its counterpart *via* a disulphide bond.[11] An example from Goddard's group[12] further built on these ideas with a design consisting of three fragments (each corresponding to the basic region of the v-Jun transcription factor) linked by disulphide bonds through Cys residues at the C and/or N terminus of the recognition helices. Gel retardation experiments showed that the trimeric peptide with these basic regions was able to recognise a 16 base-pair ds DNA site.

The success of Kim's reductionist approach was inspirational for the work of Alanna Schepartz, who reported on the use of metal complexes as alternative dimerization units in 1993. These constructs were made to assess the effect of geometry on the affinity and sequence selectivity[13] of dsDNA recognition (Figure 13.7). A later study by the group of A. Peacock confirmed the need of proper geometry for selective dsDNA recognition from the major groove.[14]

#### *13.2.1.2 From Covalent to Non-covalent Dimerization*

In line with the previously discussed strategies to substitute the leucine zipper domain by a synthetic moiety, T. Morii *et al.*, in their early contributions to the field, described the constrained positioning of DNA-binding regions by using enantiomeric and $C_2$-symmetric templates derived from 9,10-dihydrophenanthrene-9,10-diol.[15–17]

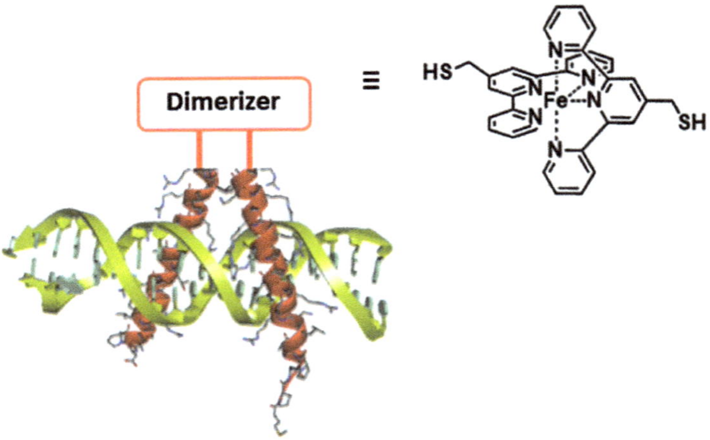


**Figure 13.7** A Fe-complex as an artificial dimerization domain for TF miniaturisation.

Subsequently they were the first to successfully connect two binding region (BR) peptides by a non-covalent interaction. This non-covalent approach was designed to enhance binding selectivity and increase responsiveness of the systems to changing protein concentrations. Furthermore it could help in understanding the role of non-covalent protein dimer formation in affinity modulation and protein cooperativity for DNA binding. Morii *et al.* showed that the inclusion of an adamantane (Ad)-derivatised BR (G-peptide) within a β-cyclodextrin (β-CD)-bearing BR counterpart was sufficient to form a stable dimer with high selectivity towards the ATF/CREB sequence (Figure 13.8).

This non-covalent strategy has been further exploited to design heterodimeric peptides to recognise non-palindromic DNA tracts by combining the DNA-binding sequences of two different basic leucine zipper domains: the yeast transcriptional activator GCN4 and an enhancer binding protein C/EBP.[19] Further extension of the β-CD/Ad inclusion principle allowed a method to be developed for cooperative DNA binding by peptide homo-oligomers constructed from the basic region of GCN4.[20–22] By this means G–Ad–CD peptides containing a β-CD as well as an Ad moiety were shown to selectively and cooperatively bind to double- and triple-direct repeat sequences while the affinity of the monomeric peptide for the specific monomeric half-site was reduced by formation of an intramolecular host–guest complex of the peptide in the absence of DNA which substantially reduces the affinity of the monomeric peptide.

### *13.2.1.3 Towards Dynamic Recognition with Switchable Systems*

The group of J. L. Mascareñas and M. E. Vázquez has played a major role in the development of TF-inspired DNA binding peptides, contributing a rich

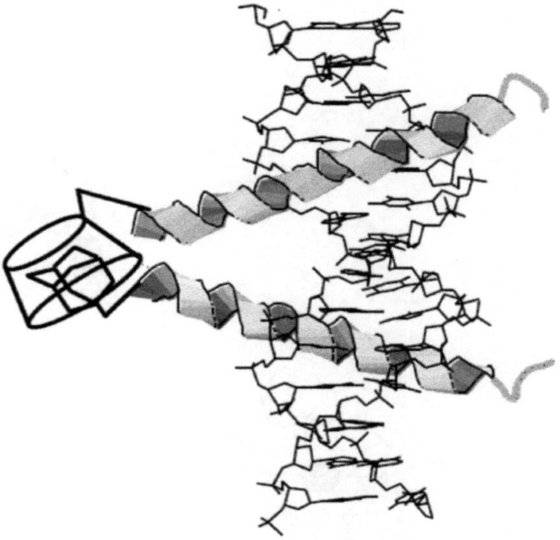

**Figure 13.8** Schematic representation showing the G–Ad/G–CD dimer bound at the AP1 sequence.[10] Grey ribbons represent the basic region peptides in the helical conformation. The β-cyclodextrin and the adamantane group are shown at the C-terminus of the helices. Coordinates for the basic region peptides and DNA are adopted from the GCN4–AP1 complex.[18] Reproduced from Sato *et al.*, Chemical Approaches Untangling Sequence-Specific DNA Binding by Proteins, *Chemistry – A European Journal*, John Wiley and Sons, © 2002 WILEY-VCH Verlag GmbH & Co. KGaA, Weinheim.

repertoire of original designs in which the use of different external triggers to modulate DNA binding is a recurrent theme.

*Light-induced DNA binding*: Starting from their first design in 2000,[23] a photo switchable azobenzene scaffold was employed as a dimerization unit that can undergo geometrical changes in response to irradiation (Figure 13.9A). Thus, the two appended basic regions can adopt a *cis* or *trans* conformation depending on the wavelength of the irradiation. The geometrical distance between the benzylic carbons and the orientation of the peptides resulting from the *cis* conformation were shown to favour major groove DNA binding. However, azobenzene-based photoswitches are known to relax quickly back to the *trans* state, resulting in poor generation of the desired *cis* state under classical illumination conditions.[24] Additionally, the required wavelength for generation of the *cis* isomer, typically below 400 nm, implies a low tissue penetration and easy disturbance by other biomolecules present. Therefore, ruthenium bipyridyl complexes bearing photolabile amino groups were studied as DNA recognition units, resulting in high affinity DNA binding bZIP derivatives with a $K_d$ of 12 nM at 20 °C. DNA interaction can be switched off by irradiation with long-wavelength visible light. This constitutes one of the first examples of a system allowing

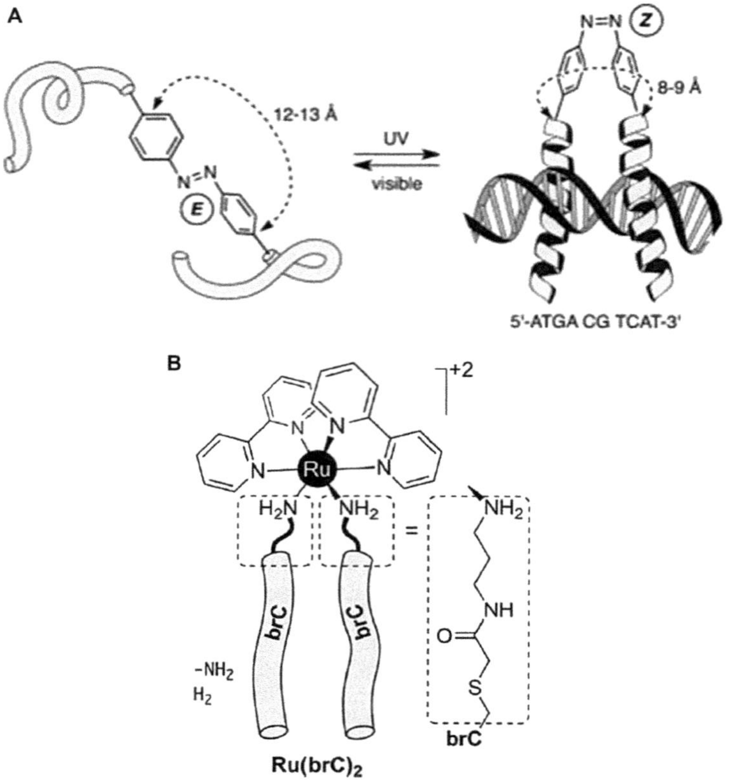


**Figure 13.9** Mascareñas' light-switchable systems: (A) The original azobenzene construct. Reproduced from Caamaño *et al.*, A Light-Modulated Sequence-Specific DNA-Binding Peptide, *Angewandte Chemie, International Edition*, John Wiley and Sons, © 2000 WILEY-VCH Verlag GmbH, Weinheim, Fed. Rep. of Germany. (B) The visible-light-responsive ruthenium bispyridyl complex. Reproduced from ref. 25 with permission from the Royal Society of Chemistry.

conditional deactivation, which should be exploitable in other cases where dimerization is a prerequisite for DNA binding.[25]

An alternative approach towards light-controlled DNA binding involved the conjugation of negatively charged oligo-Glu sequences to the N-termini of the basic region peptides connected through a photolabile 3-amino-3-(2-nitrophenyl)propionic acid (ANP)-linker. Electrostatic repulsive forces originating from the appended negative stretch of peptide interfere with DNA binding. Simple UV-irradiation results in cleavage and thus restoration of the DNA binding ($K_d = 1\ \mu M$).[26]

*Metal-ion-induced DNA binding:* Given the wealth of metal-binding motifs found in proteins, metal-ion regulation of DNA binding was a next logical step in the design of systems that could be influenced by changes in the dsDNA local environment. A first report on this principle, albeit on a protein rather than peptide scale, was published by the Futaki group[27] who developed metal-responsive bZIP proteins by introducing metal-coordinating iminodiacetic acid side chains into the leucine zipper region of the GCN4 DNA-binding domain. DNA-binding properties were preserved for the modified protein but the DNA-binding capability could be counteracted by the addition of $Co^{II}$, causing a decrease in helical content.

Building on this principle and elegantly combining it with inspiration from the previously discussed Cys-dimerization approach, Mascareñas and his team designed a system where the dsDNA binding selectivity could be modulated by two different external stimuli, allowing the system to slide from one dsDNA recognition box to another depending on the exact environmental conditions (Figure 13.10). Exploiting advances in the development of terpyridine-based $Ni^{II}$ coordination sites, and with precedent in the successes in the field of GCN4 TF mimics, they introduced a disulphide bond at the N-termini and a metal–terpyridine complex at the C-termini.[28] The monomers with the specific functionalities at their termini were mixed with two ds-oligonucleotides [containing the activating transcription factor/cAMP response element binding protein (ATF/CRE) and the inverted sequence respectively]. The disulphide bond was formed after addition of Ellman's reagent [(5,5′-dithiobis-(2-nitrobenzoic acid) (DTNB)]. A complex within the major groove of the dsDNA containing the inverted ATF/CRE site was generated. After addition of TCEP to the dsDNA-peptide complex, the interaction was disrupted by reduction of the disulphide bond. Therefore, both peptide monomers were free to create a new dimer at the C-terminus when $Ni(ClO_4)_2$ was added to the mixture. The new binding site corresponded to the ATF/CRE sequence. Complete turn-over could be achieved by adding the chelating agent DTNB, regenerating the binding affinity towards the inverted sequence. This study represented the first proof-of-concept regarding the possibility of synthesising molecules capable of achieving binding at two different DNA sites with $K_d$ values of 299 nM and 375 nM (for the SS system) respectively.

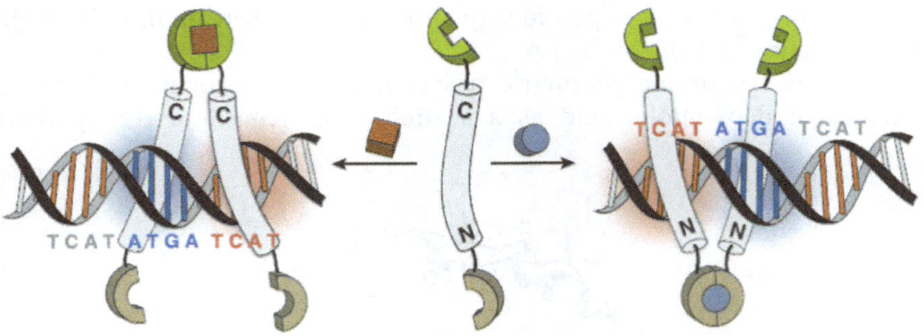


**Figure 13.10** Mascareñas' dual responsive synthetic bZIP peptides. Reprinted by permission from Macmillan Publishers Ltd: *Nature Communications* (ref. 28) Copyright 2013.

### 13.2.1.4 Alternative Dimerization Scaffolds for Enhanced Therapeutic Potential

*Steroid-based scaffolds:* Recently, we have reported on our efforts to develop a dipodal peptidosteroid mimic of the GCN4 TF.[29–31] The idea of a steroidal scaffold as a dimerization moiety for the appendage of GCN4br type peptides was inspired by the significant attention that steroidal structures have gained in fields such as supramolecular and bioorganic chemistry, pharmacology and nanotechnology. The steroid nucleus is a tetracyclic hydrocarbon perhydrocyclopentanophenanthrene unit composed of three six-membered rings (A, B and C) and a five-membered ring (D). It further contains two angular methyl groups at positions C18 and C19. The A and B rings are in a *cis* configuration which allows the nucleus to adopt a curved shape. It must be noted that the *trans* configuration is also possible, but not of interest for the current chapter. Due to the specific arrangement of the hydroxyl groups on the core structure, bile acids possess an amphiphilic character with a convex hydrophobic upper and a concave hydrophilic lower side (Figure 13.11).

Moreover, the specific choice of the steroid scaffold as a dimerizing unit for TF models was based on several observations such as its ability to enhance the proteolytic stability of the attached peptides,[32–34] its amphiphilicity,[35] its conformational properties, ensuring correct positioning of the two appended chains,[36] and its potential to increase cellular uptake and bioavailability.[33] In the case of the GCN4 TF, this moiety substituting the leucine zipper domain may be expected to promote helicity of the appended basic region peptides and maintain a correct distance between the peptides for major groove recognition. From the different types of steroids, we chose deoxycholic acid as a starting template for the attachment of the DNA-binding peptides as it had the right properties to replace the leucine zipper.[37] In view of its rigid framework and the presence of two hydroxyl groups at the C3 and C12 positions in an α-stereochemistry, deoxycholic acid has the ability to orientate the peptides in a parallel arrangement. In addition, the distance between C3 and C12 is 7–8 Å,[38] which ensures sufficient distance for the anchoring of peptides to allow further insertion into the major groove (Figure 13.11).

Beside its favourable geometric properties, we also considered the advantages of deoxycholic acid as a scaffold with regard to the synthetic

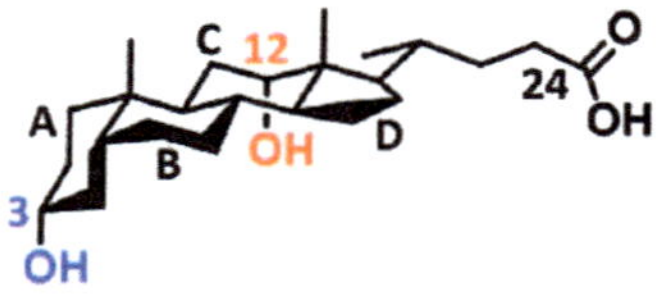


**Figure 13.11** Deoxycholic acid serves as a template for GCN4br dimerization by peptide attachment to C3 and C12.

strategy. While earlier efforts were centred around a linear solid-phase strategy,[29] we later adopted a solution-phase strategy involving a copper-catalysed azide–alkyne cycloaddition (CuAAC) reaction. The geometric, steric and electronic properties of the 1,2,3-triazole resemble a trans-amide bond while also affording resistance to enzymatic degradation,[39–42] hydrolysis and oxidation, making it an attractive heterocyclic moiety to replace unstable linkers under physiological conditions in biologically active compounds. In addition, successful replacement of two amino acids in α-helical peptides by a triazole unit has been shown not to perturb significantly the secondary peptide structure.[39] We had previously demonstrated the need for spacers between the steroid nucleus and the peptides for effective DNA recognition.[43] On account of these considerations deoxycholic acid was modified at the alcohol positions by attachment of azide and alkyne linkers. The GCN4br peptides on the other hand were functionalised at the C-terminus with either alkyne or azide-modified amino acids, Fmoc–propargylglycine–OH (Fmoc–Pra–OH) or Fmoc–azidohomoalanine–OH (Fmoc–Aha–OH) (Figure 13.12) and CuAAC reactions were performed in solution.[31]

The DNA binding affinity of the mimics was evaluated using electrophoretic mobility shift assay (EMSA). Constructs possessing an aliphatic spacer showed DNA affinity, while models with aromatic linkers did not. From this study, we concluded that flexibility is mandatory for achieving DNA recognition in the major groove. Kd calculations from competition experiments with 0.5–2 mM competitor DNA revealed values between 40 and 300 nM for the different constructs.

We further studied cell uptake in RAW264.7 cells. Results indicated that the deoxycholic acid-coupled peptides are mainly internalised *via* active transport at 37 °C. The most likely explanation was that the peptides followed an endocytotic pathway.

*Cyclodextrin scaffolds:* Encouraged by the outcome of our peptidosteroid conjugates with regard to the synthesis and influence of the linker on the DNA recognition ability, we chose to explore other structural features such as the distance between the C-termini of both basic regions in artificial TF models. Cyclodextrins, like cholic acid, also possess a rigid structure and therefore their dimensions are fixed and well-characterised. Cyclodextrins are known to find ample application in supramolecular chemistry[44] and as therapeutic delivery vehicles.[45] Conversion of the primary alcohols at the rim to azides allows the attachment of basic region peptides by CuAAC. Based on the diameter of the primary rim of the α, β and γ-cyclodextrins which are 5.7, 7.8 and 9.5 Å respectively,[46] three different peptide-cyclodextrin conjugates were synthesized, keeping in mind that the width of the major groove of DNA is 11.2 Å. By conjugating the GCN4 basic region peptides to α, β and γ-cyclodextrin derivatives, the influence of the dimerization distance upon DNA recognition could be investigated. From a synthetic point of view, their chemical reactivity allowed the selective functionalisation of hydroxyl groups in two opposite positions at the primary rim for the dimerization (Figure 13.13).

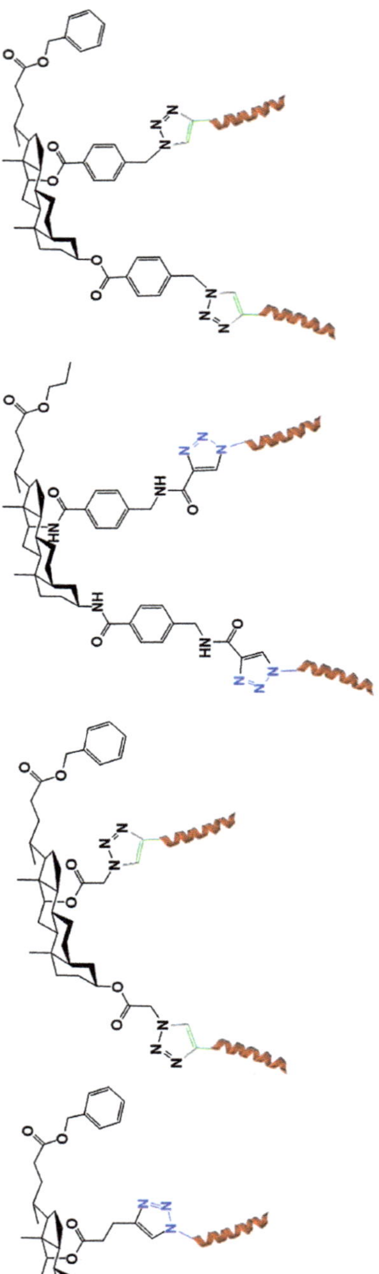

**Figure 13.12** Chemical structures of the peptidosteroid conjugates for ATF/CREB recognition.

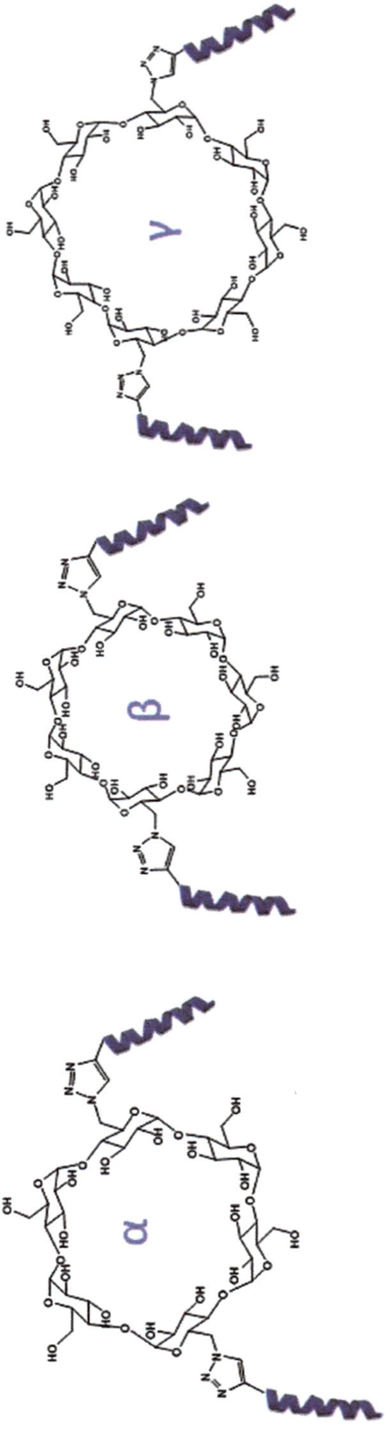


**Figure 13.13** Chemical structures of the peptide–cyclodextrin conjugates for ATF/CREB recognition.

From electrophoresis experiments, we calculated $K_d$ values of $50 \pm 20$, $30 \pm 20$ and $100 \pm 60$ nM respectively, with the β derivative showing the highest affinity for DNA. Unlike the peptidosteroid conjugates, with which we saw that the linker had the potential to completely obstruct sequence-selective DNA recognition, variation of the distance between the C-termini did not influence DNA binding to any great extent. It should be noted that these results are specific for the basic region GCN4 peptide sequence as well as the given spacers. If these features are varied, different conclusions can be expected.[47]

### 13.2.1.5 *Beyond GCN4 Mimics: Towards More Challenging TFs*

Using the expertise acquired in GCN4 miniaturisation, we decided to attempt the miniaturisation of the cMyc/Max bHLH transcription factor, whose role in the development of tumours renders it an interesting target for potential gene therapy.[48] cMyc/Max binds to DNA in a sequence-specific manner and its function involves transcription from promoters containing the E-box palindromic sequence CACGTG. The biological functions of the cMyc-Max bHLH ZIP transcription factor as an oncoprotein have not been extensively investigated. One of the reasons could be the lack of good synthetic routes or ligation strategies, which might provide sufficient material to investigate the functions of the protein in greater detail. Due to the relevance of this oncoprotein to cancer development, Kent *et al.*[49] developed a first approach in the form of the total synthesis of cMyc-Max protein. Unlike the natural protein, Kent synthesised a covalently-linked dimer mimic by convergent chemical ligation of unprotected peptide segments (Figure 13.14).

Notwithstanding the lack of synthetic peptide models for the cMyc/Max TF, a successful mimic of the MyoD–MyoD bHLH type TF was reported by the Morii group.[50] This encouraged us to modify our successful peptidosteroid conjugation strategy to synthesise a heterodimeric counterpart. Previously, we attempted to synthesise an artificial mimic of this oncoprotein, which unfortunately lacked DNA-binding capabilities.[43] It was suspected that a spacer between the steroid-like dimerizer moiety and the basic region of the peptides might facilitate correct positioning of the peptides within the major groove. This was attempted *via* the addition of one helical turn (approximately four amino acids) at the C-termini to the 17-residue cMyc/Max peptides cMyc (NVKRRTHNVLERQRRNELKRGC) and Max (ADKRAHHNALERKRRDHIKDGAha), together with the incorporation of an aliphatic linker on the final peptidosteroid conjugate to increase flexibility and enable the correct conformation for successful binding in the DNA major groove. We chose 4-maleimido-butyric acid and pentynoic acid as linkers to allow selective conjugation of the two peptide strands onto the heterodipodal scaffold. The first conjugation at the C12 position was performed *via* a thiol-ene conjugation through a Michael nucleophilic addition to a cysteine-contained peptide resulting in a succinimidyl thioether

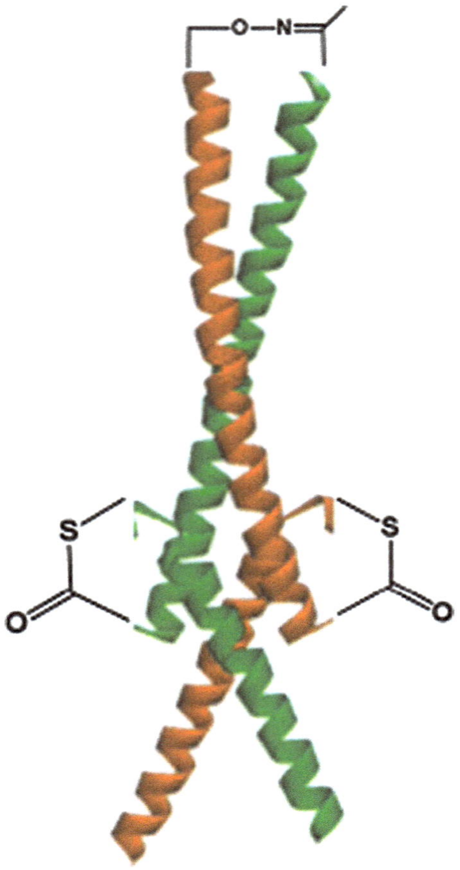


**Figure 13.14** Structure of Myc/Max bHLH protein mimic synthesised by Kent. The green ribbon represents the residues that constitute the Max peptide, while the orange one refers to the Myc peptide. Oxime ligation and thioester formation enable connection of the two different fragments.

moiety. The second conjugation was achieved *via* CuAAC at the C3 alkyne-functionalised scaffold. Both succinimidyl thioether and triazole moieties are generally accepted as stable linkages under physiological conditions (Figure 13.15).

EMSA studies showed that binding was less strong than for our homodimer GCN4 mimics. On the other hand, adding competitor DNA influenced the binding pattern, and the construct was shown to retain specificity for the E-box sequence to some extent. Peptide binding was still observed even when a onefold to fourfold excess of competitor ssDNA was used. We concluded that the peptide was specific for the target sequence, but only at a high excess concentration of the peptide (ratio 1 : 400).[51]

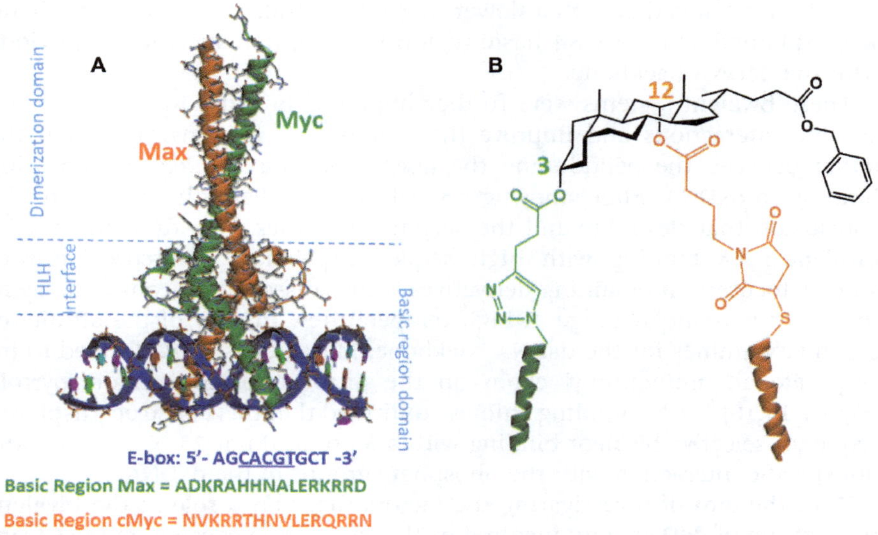


**Figure 13.15** (A) Crystal structure of the cMyc–Max protein and (B) miniature model synthesised by Ruiz *et al.*[51]

### 13.2.2 Moving Away From Dimer-based Major Groove Binding: Miscellaneous Peptide Conjugates for Combined Major and Minor Groove Recognition

Driven by the desire to achieve even higher sequence specificity with artificial dsDNA binders, Mascareñas and his team also investigated systems allowing the recognition of both grooves.[52] As was demonstrated in 1995 by Stanojevic and Verdine,[53] a 24-mer GCN4 basic region monomeric peptide can bind in the major groove selectively if it is covalently connected to the dsDNA in the appropriate position. This discovery formed the inspiration for a system where the GCN4 basic region was linked to a construct with affinity for the minor groove thus allowing the basic region monomer to bind selectively to the adjacent major groove bearing the CRE sequence. As a result of the affinity of tripyrroles for binding selectively to dsDNA at A–T rich sites, distamycin A was chosen as a unit for minor groove recognition. The linker between the basic region peptide and the oligopyrrole was chosen based on X-ray structure analysis in order to optimise the distance between the two counterparts and to ensure adequate passage across the phosphate backbone of the dsDNA. The interaction between the artificial construct and dsDNA bearing the CRE and the A–T rich sites was studied by circular dichroism. An increase in magnitude of the signals at 222 and 330 nm reflects the increased helicity of the GCN4 basic region when binding to the major groove and the insertion of distamycin A into the minor groove respectively ($K_d = 3 \pm 0.2$ nM at 4 °C and 50 nM at 23 °C). Using gel mobility shift assays, a shifted band corresponding to both minor and major groove interactions was observed, together with a slower-migrating band. This less intense band was explained as the GCN4 basic region forming non-specific interactions with the dsDNA backbone.[52]

These bivalent systems were further improved in order to suppress non-specific interactions and improve the affinity.[54] It was envisaged that the tether between the peptide and the polypyrrole was the key to improving binding to dsDNA. After studying several models, the authors came to the conclusion that flexibility and the secondary amines were fundamental for bivalent DNA binding with high sequence-specificity. Moreover, further studies focused on obtaining derivatives whose dissociation constants lay in the lower nanomolar range at high temperatures. In view of the affinity of secondary amines for the dsDNA backbone, Mascareñas *et al.* decided to incorporate an aminopropyl chain on the nitrogen of the middle pyrrole (Figure 13.16).[55] DNA binding studies confirmed that this construct displayed sequence-selective bivalent binding with a $K_d$ of 5 nM at 23 °C, without any nonspecific interaction with the phosphate groups of the dsDNA.

With the aim of investigating the factors that play a role in the bivalent recognition of dsDNA, and inspired by the earlier non-covalent studies from Morii *et al.*,[56] the adamantyl–cyclodextrin encapsulation system was used as a non-covalent linkage between both counterparts. A polypyrrole chain bearing a β-cyclodextrin was combined with an adamantyl-derivatised GCN4 peptide and showed dsDNA binding with an affinity of 63 ± 7 nM. Further EMSA studies were focused on elucidating the sequence-specificity of the system where it was revealed that a dsDNA sequence mutated at the BR binding site did not induce any nonspecific binding, in contrast to previous models.[57]

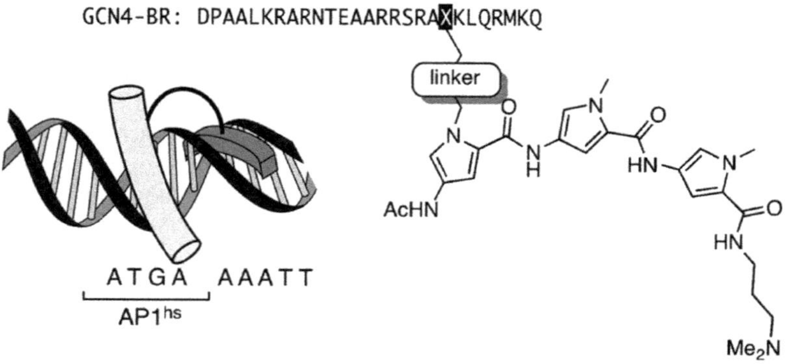


**Figure 13.16** Left: cartoon representing the simultaneous interactions of the basic region–tripyrrole conjugate with a DNA composite site that results from the recognition of the major groove by the peptide helix (grey cylinder), and of the minor groove by the tripyrrole unit (dark grey). Right: general structure of the hybrids.
Reproduced from Pazos *et al.*, DNA Recognition by Synthetic Constructs, *ChemBioChem*, John Wiley and Sons, Copyright © 2011 WILEY-VCH Verlag GmbH & Co. KGaA, Weinheim.

From a synthetic point of view, oligopyrroles are not easily accessible in the lab, as the monomer synthesis takes up to seven steps and involves inconvenient handling of unstable intermediates. More easily accessible constructs were sought, so propamidine and aza-bisbenzamidine derivatives were proposed as candidates for minor groove recognition units for covalent linkage to the GCN4 BR. Although the propamidine and aza-bisbenzamidine-based mimics showed lower sequence-specificity compared with the tripyrrole derivatives, the latter one allowed the study of DNA interaction by fluorescent spectroscopy methods. When evaluating the DNA binding profile of the peptide–bisbenzamidine conjugate, non-specific interactions with the backbone of the dsDNA were observed instead of specific electrostatic interactions within the major groove. Upon increasing the dsDNA concentration, a 1:1 binding mode was observed, consistent with the results of the EMSA study.[58]

Very recently, the principles of metal-mediated dsDNA binding and bivalent recognition were elegantly combined in a heterodimeric complex consisting of the GCN4 basic region (Asp226–Arg249) containing two histidines, connected to a bis(benzamidine) minor groove binding unit with a bipyridine ligand. Both components were brought together by $Ni^{II}$ at their coordination sites to construct the heterodimer which bound ds DNA with a $K_d$ of 0.45 μM at 4 °C (Figure 13.17). Reversibility of the system was demonstrated by the addition of external nickel chelators that sequestered the cation.[59] A key factor in this successful design is the dual role of the metal as an alpha-helix-nucleating factor and as a heterodimerization staple. This supramolecular system combines some of the key characteristics of naturally

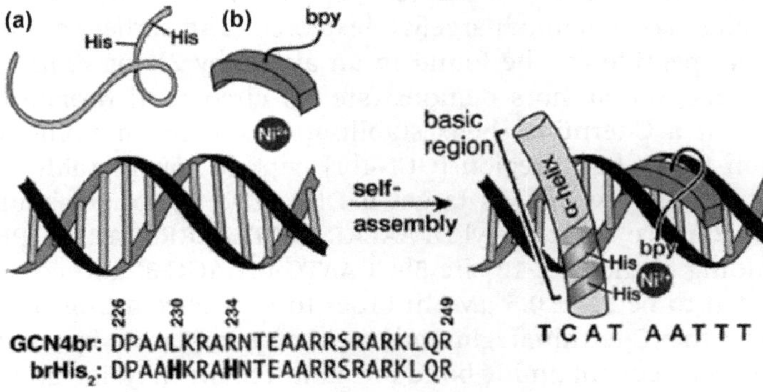


**Figure 13.17** Outline of the cooperative assembly strategy. (a) Unfolded brHis2, which will fold into an alpha-helix in the complex (right). (b) Bis(benzamidine) minor-groove binder. brHis2: Aba-DPAAHKRAHNTEAARRS-RARKLQR-NH2, where Aba: 4 acetamidobenzoic acid; bpy: bipyridine chelator tethered to the bis(benzamidine).
Reproduced from Sánchez *et al.*, Reversible Supramolecular Assembly at Specific DNA Sites: Nickel-Promoted Bivalent DNA Binding with Designed Peptide and Bipyridyl–Bis(benzamidine) Components, *Angewandte Chemie, International Edition*, © 2014 WILEY-VCH Verlag GmbH & Co. KGaA, Weinheim.

occurring DNA-binding proteins, namely bivalence, selectivity, responsiveness to external agents, and reversibility.

### 13.2.3 Further Structure Minimisation: Monomeric Stapled Peptides as GCN4 TF Mimics

It has been shown (*vide supra*) that dimerization of basic region peptides through non-peptide scaffolds can enable DNA binding.[29] However, construction of smaller dimeric mimics of this TF, although feasible, has proved to be synthetically challenging.[15,60] Previous attempts at more thorough structural minimisation and reduction of complexity using the monomeric GCN4 peptide have shown that DNA binding is greatly reduced due to the loss of secondary structure.[61] Indeed, since the basic region of the GCN4 transcription factor cannot adopt a helical fold in solution by itself nor bind to DNA[12] for entropic reasons,[62] an external factor forcing the peptide into a helical conformation is needed. For this reason, unlike the previously discussed synthetic bZip models in which DNA binding is induced *via* dimerization, our group embarked on trying to achieve DNA binding by stabilising a single α helix *via* peptide stapling, reasoning that enhancing the helicity within the monomer should sufficiently stabilise the conformation to allow DNA binding.[63] Stapled peptides have been used extensively for improving helicity,[64] increasing cell-penetration,[65,66] enhancing proteolytic stability and strengthening peptide–protein interactions (PPIs).[67] In all the abovementioned cases, the benefits of stapled peptides have been demonstrated.[68] However, the use of stapled peptides in the miniaturisation of zipper proteins has remained largely unexplored. The earliest example of a stapled GCN4 peptide can be found in an article by Zhang *et al.*[61] In this published work, the authors demonstrate *via* circular dichroism (CD) experiments that a C-terminal helix stabilisation on the monomeric GCN4 transcription factor basic region (GCN4br) peptide considerably improves its binding affinity towards its target dsDNA. The $K_d$ for the monomeric GCN4br sequence Nα-Acetyl-DPAALKRARNTEAARRSRARKLQRMKQLE-$CONH_2$ binding to the CRE-12 site 5′-GCAATGACGAGC-3′ and complement was calculated to be $3.9 \pm 0.5$ μM. In order to introduce a C-terminal helix stabilisation, the C-terminal glutamic acid was replaced by an aspartic acid residue. Subsequent amide bond formation between lysine and aspartic acid resulted in the stapled peptide sequence Nα-Acetyl-(cyclo-Lys–Asp)-DPAALKRARNTEAAR15RSRARKLQRKAAAD-$CONH_2$ which has $K_d = 0.65 \pm 0.09$ μM, six times lower than that of the native GCN4br sequence towards the same dsDNA. Alternatively, we examined DNA binding induced by stapling using the same GCN4br model peptide, by comparing the i, i+4 and i, i+7 stapling methods and varying the positions of the staple along the helix.[63] Additionally, cellular uptake of the constructs was investigated using fluorescently labelled versions of the peptide. Initial studies were dedicated to selecting the most suitable method for peptide stapling among the ones reported in the literature.[65,69–72] To ensure easy and cost-effective modification, the use of unnatural amino acids was avoided. While hydrocarbon stapling is generally considered the best stapling methodology for improving helicity of peptides for subsequent enhancement of peptide–protein interactions, the labour-intensive preparation of the required modified amino acid building blocks, coupling of the non-natural amino acids as well as peptide folding on resin is a cause for concern when stapling a peptide by this method.[73] Therefore, in our study, we opted for cysteine-mediated cross-linking,[68] in view of the commercial availability and/or easy synthesis of cross-linking moieties, mild reaction conditions and easy scalability due to synthesis in solution.

### *13.2.3.1 Design and Synthesis of Suitable Stapled Peptides*

A detailed analysis of the essential contacts for binding of the GNC4 protein to its cognate DNA sequence, as derived from the reported crystal structure, was described earlier by Ellenberger and colleagues (pdb file: 1YSA).[4] Various dimeric peptides based on the D226–Q248 basic region of the GCN4 TF have been shown to retain their DNA binding properties.[28,74]

Based on these studies, amino acids within this sequence, indicated as not being involved in DNA contacts,[75] were identified and systematically replaced by Cys according to an i, i+4 or i, i+7 format (Figure 13.18). These positions were specifically chosen as they stabilise a peptide helix. Furthermore, molecular modelling visualisation based on the pdb file 1YSA[4] was used to ensure that the introduced linkers pointed away from the DNA and not towards it, thereby avoiding any steric repulsion which might arise due to peptide stapling. In this way, three different peptides, comprising the D226–Q248 sequence from the DNA binding basic region of GCN4, containing a double Cys substitution (at positions 237/244 for one, 229/233

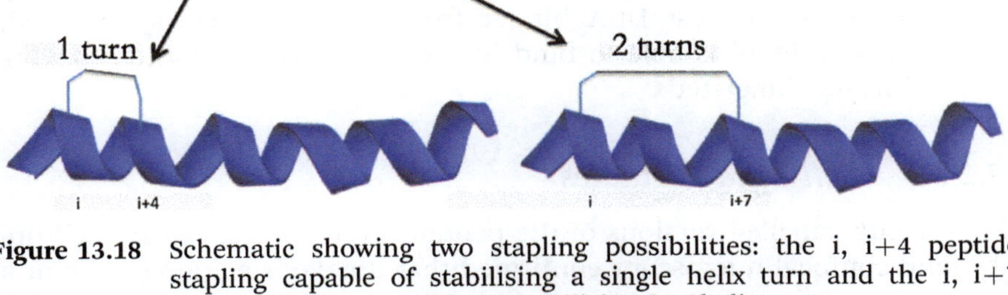


**Figure 13.18** Schematic showing two stapling possibilities: the i, i+4 peptide stapling capable of stabilising a single helix turn and the i, i+7 peptide stapling capable of stabilising two helix turns.

for two and 233/237 for three) were synthesised on solid support, cleaved and subsequently treated with various linkers yielding a series of five stapled peptides as shown in Figure 13.19.

Cross-linkers a, c and d are commercially available. The stapling moiety b has not been used thus far for peptide stapling and was designed and synthesised as a more polar alternative to the biphenyl and bipyridine cross-linkers.

#### *13.2.3.2 DNA Binding Studies*

DNA binding capacity of all peptides was evaluated through EMSA titration of various peptide concentrations into the DNA sequence 5′-CGG ATG ACG TCA TTT TTT TTC-3′ containing the cognate monomeric GCN4 binding site GTCAT. The wild-type peptide 4 does not bind to DNA under the chosen EMSA conditions. For all the synthetic constructs 1a–c, 2d and 3d, enhanced DNA binding as compared with peptide 4 could be observed. Earlier, DNA binding with monomeric peptides was observed through the use of a grafting strategy, whereby the crucial contact residues of the GCN4 binding region were specifically positioned on an avian pancreatic polypeptide.[76] Apparently, our simple and straightforward stapling strategy also succeeds in constraining the peptide into a suitable conformation for DNA binding.

From experiments described in the literature[61,77] and from our own data, a general conclusion can be drawn regarding monomeric DNA-binding stapled peptides. The major challenge in the case of these peptides is that increasing the helicity needs to be complemented with a degree of flexibility in order to accommodate the conformational change which occurs when the peptide binds to DNA. Since this conformational change is more significant in the case of DNA-binding peptides compared with PPIs, a too-tight locking of the peptide into a helical conformation may well result in inability to accomplish DNA binding. Indeed, our data show that peptide stapling by providing an N- or C-terminal helix stabilisation, rather than centrally in the sequence, gives better results in terms of DNA binding. Although in the absence of DNA a low helical content can be observed, apparently the conformation can be adjusted into a structure whereby the contacts between the positively charged side chains, mainly involving Lys and Arg residues, and the negatively charged backbone of the DNA can be maximised without a high entropic penalty. We believe peptide 2d fits these criteria satisfactorily and hence it is the best DNA binder from the constructs synthesised. Moreover, peptide 2d is able to bind in a dimeric fashion without having been artificially dimerized.

#### *13.2.3.3 Cell Uptake Studies*

Fluorescently labelled versions of the peptides were also tested for cell uptake, and confocal microscopy confirmed that uptake is achieved even at a low concentration of 0.25 μM for all the peptides at 37 °C. The observed cell

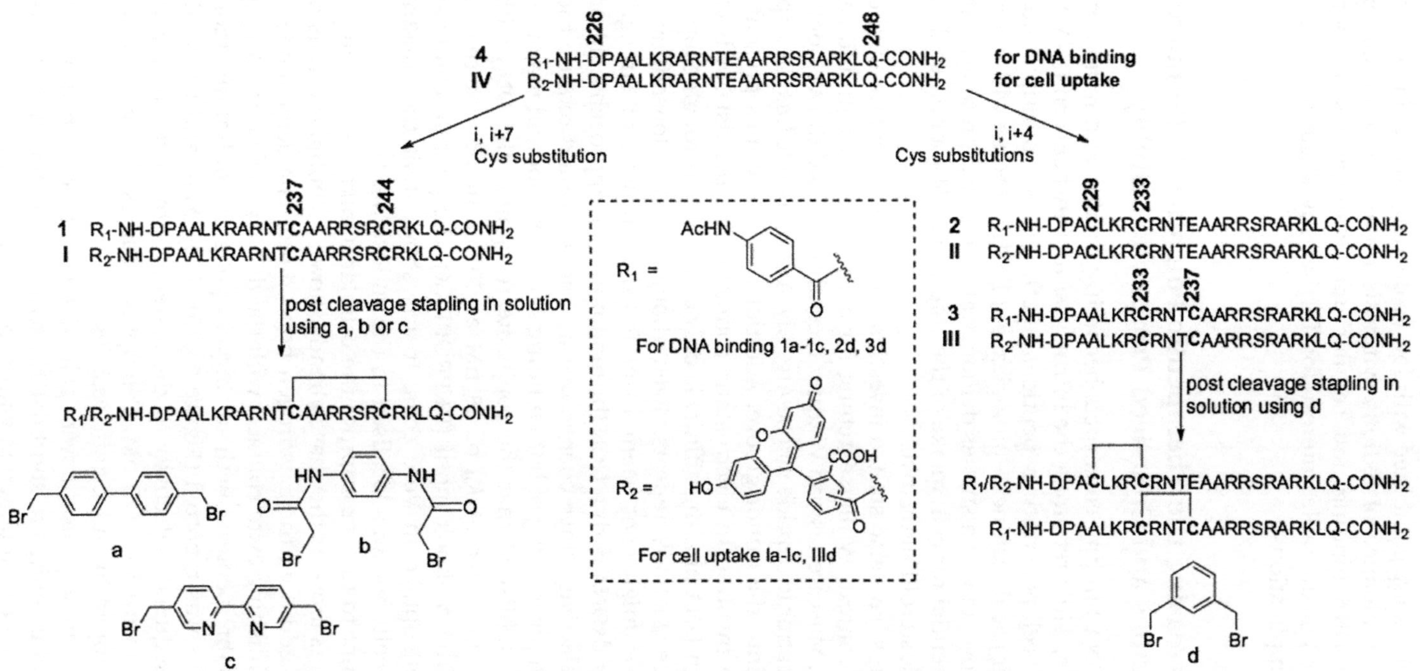


**Figure 13.19** Synthetic stapled peptides 1a–c, 2d and 3d for DNA binding and unmodified basic region peptide 4. The peptides are N-terminally capped with a p-acetylamino benzoic acid (ABA) moiety to permit UV-based detection and analysis in the case of DNA binding. For cell uptake studies, the ABA moiety was replaced by fluorescein.

uptake properties are combined with low cytotoxicity and enhanced peptide stability. The avoidance of tedious synthesis routes, in combination with the observed DNA binding and cellular uptake properties, render these constructs of considerable interest for further investigation of potential therapeutic applications.

### 13.2.4 Increasing the Therapeutic Potential: Cell-uptake Studies and Enhanced Proteolytic Stability

With the dsDNA binding constructs becoming smaller and more synthetically accessible, while retaining the intrinsic similarity between DNA-binding regions and cell-penetrating peptides—both feature multiple positively charged amino acid residues—biological applications come within reach. As discussed above, our own studies demonstrate increased proteolytic stability as well as potential for cell uptake of the stapled GCN4 monomers,[63] as well as the peptidosteroid derivatives.[31]

Other groups have also started investigating the potential for applications in a cellular context. While searching for even more efficient and selective DNA binders, Mascareñas and Vázquez recently exploited the AT-hook motif, a short oligocationic peptide whose Arg–Gly–Arg sequence has the capability of inserting into the minor groove. Additionally, the lysines present in the sequence are involved in electrostatic contacts with the dsDNA backbone. However, high DNA binding affinity is only achievable through cooperativity between three AT-hook repeats. Even though such a low-affinity minor groove anchor might not seem ideal for the purpose of strengthening interaction of dsDNA with a GCN4br monomer, the approach held promise for greater selectivity since cooperative interaction would require the exact proximity and positioning of both binding sites in minor and major grooves. It could be established that the conjugate recognises the DNA with high affinity, with an apparent $K_d$ of 28 nM at 20 °C. In view of the cationic character of the AT-hook motif (RKPRGRPKK), the peptide conjugate was tested for cell uptake in Vero cells. It was shown that the construct internalises in cells and was localised in the endosomes.[78]

In an attempt to further improve the cellular uptake, interest in guanidinium oligomers arose as they have already shown potential for translocation into cells. In addition, their hydrogen bond donor character renders them ideal for establishing extra contacts with the dsDNA backbone. Conjugation of the GCN4br peptide with a bicyclic oligoguanidinium moiety at the C-terminus was reported (Figure 13.20). The EMSA and CD studies revealed high sequence-selectivity of the system, both in the major and the minor grooves with $K_d = 170$ nM. Cell-uptake studies showed that the conjugate gets trapped in endocytotic vesicles.[79]

Although promising results were gained with the developed models, recognition remains limited to short dsDNA sequences (9 bp). In order to target long DNA sequences, interaction with two consecutive major grooves could

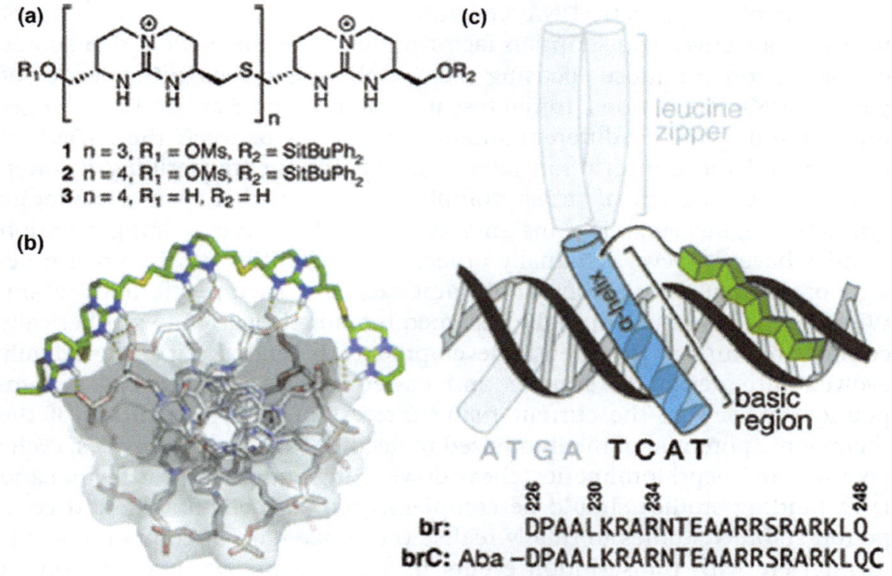


**Figure 13.20** (a) Oligoguanidines used in the study. (b) Proposed structure of the interaction between an oligoguanidine and the DNA; a view along the DNA axis showing potential hydrogen bonds. (c) Cartoon representation of a designed hybrid bound to the DNA. The sequences of the GCN4 basic region (br) and the target peptide (brC) are also shown. Reproduced from ref. 79 with permission from The Royal Society of Chemistry.

be achieved by joining constituent TF binding modules. By conjugating different modules *via* a minor-groove-binding moiety with a cationic character, selectivity and cell uptake could be boosted. This type of binding, involving major–minor–major groove recognition, could be achieved by conjugating a GCN4 BR peptide, an AT-hook motif and the zinc finger peptide region of GAGA.[58,80] This trifunctional model showed excellent selectivity in EMSA studies and bound dsDNA with an affinity of 28 nM at 20 °C. Despite the large size of the conjugate, cell internalisation tests were carried out in Vero cells and unfortunately it localised in the endosomes.[81]

To sum up, the conjugates discussed in this section combine synthetic accessibility with nanomolar affinity and selectivity, further complemented by good cell-penetration properties, allowing us to anticipate potential applications in the modulation of gene transcription processes.

## 13.3 Conclusions and Considerations for Future Design

Given the routine deciphering of the human genome and the growing body of knowledge it has proffered on the genetic origins of disease, the development of synthetic DNA-recognition entities has opened up a major research objective. Transcription factor proteins have become a rich source of inspiration for those focusing on deciphering a recognition code for peptide–DNA interactions. Initial research mainly aimed at gaining a deeper understanding of the different interaction modes between the dsDNA of interest and the transcription factor responsible for transcribing it, often leading to the design of rather complex constructs that involve elaborate synthetic engineering. The insights generated from these first-generation mimics have, however, gradually unlocked possibilities for the precise design of compounds that are more amenable for therapeutic application. Whereas the constructs initially designed remained clearly too synthetically complex for further therapeutic development, the endeavour has gradually evolved into design of simpler and easier-to-access peptide-based compounds. In view of the current rising interest in, and recognition of the therapeutic potential of medium-sized molecules such as macrocycles, cyclic peptides and peptidomimetics, these downsizing efforts are worthwhile and DNA binding studies should be complemented with cell-uptake and compound stability studies to finally realise the conception of compounds able to interfere with transcription events in a therapeutically relevant way. It becomes evident from the examples discussed in the last section that the future development of such compounds as genetic tools will require focusing on avoiding endosomal entrapment, potentially by incorporating endosomal escape signals and nuclear localisation signals that further direct the required translocation into the nucleus.

## Acknowledgements

We acknowledge funding from the Marie Curie Early Stage Research Training Fellowship of the European Community's Seventh Framework Programme under contract number PITN-GA-2010-238679. The FWO-Vlaanderen and BOF-UGent are also acknowledged for financial support. We additionally thank all coworkers and collaborators involved in the research described.

## References

1. J. M. Vaquerizas, S. K. Kummerfeld, S. A. Teichmann and N. M. Luscombe, A census of human transcription factors: function, expression and evolution, *Nat. Rev. Genet.*, 2009, **10**, 252–263.
2. V. W. Yang, Issues and Opinions in Nutrition Eukaryotic Transcription Factors: Identification, Characterization, *J. Nutr.*, 1998, 2045–2051.
3. M. S. Lee, G. P. Gippert, K. V. Soman, D. A. Case and P. E. Wright, Three-dimensional solution structure of a single zinc finger DNA-binding domain, *Science*, 1989, **245**, 635–637.
4. T. E. Ellenberger, C. J. Brandl, K. Struhl and S. C. Harrison, The GCN4 basic region leucine zipper binds DNA as a dimer of uninterrupted α

helices: crystal structure of the protein-DNA complex, *Cell*, 1992, **71**, 1223–1237.
5. A. R. Ferré-D'Amaré, G. C. Prendergast, E. B. Ziff and S. K. Burley, Recognition by Max of its cognate DNA through a dimeric b/HLH/Z domain, *Nature*, 1993, **363**, 38–45.
6. D. Pathak and P. B. Sigler, Updating structure-function relationships in the bZip family of transcription factors, *Curr. Opin. Struct. Biol.*, 1992, **2**, 116–123.
7. J. E. Darnell, Transcription Factors as Targets for Cancer Therapy, *Nature*, 2002, **2**, 740–749.
8. E. Pazos, J. Mosquera, M. E. Vázquez and J. L. Mascareñas, DNA recognition by synthetic constructs, *ChemBioChem*, 2011, **12**, 1958–1973.
9. M. E. Vázquez, A. M. Caamaño, J. L. Mascareñas, Y. Mori, K. Imoto and K. Makino, From transcription factors to designed sequence-specific DNA-binding peptides, *Chem. Soc. Rev.*, 2003, **32**, 338–349.
10. S. Sato, M. Hagihara, K. Sugimoto and T. Morii, Chemical Approaches Untangling Sequence-Specific DNA Binding by Proteins, *Chem. - Eur. J.*, 2002, **8**, 5066–5071.
11. R. V. Talanian, C. J. McKnight and P. S. Kim, Sequence-specific DNA binding by a short peptide dimer, *Science*, 1990, **249**, 769–771.
12. C. Park, J. L. Campbell and W. A. Goddard, Can the Monomer of the Leucine Zipper Proteins Recognize the Dimer Binding Site without Dimerization?, *J. Am. Chem. Soc.*, 1996, **118**, 4892–4896.
13. B. Cuenoud and A. Schepartz, Altered specificity of DNA-binding proteins with transition metal dimerization domains, *Science*, 1993, **259**, 510–513.
14. E. Oheix and A. F. A. Peacock, Metal-ion-regulated miniature DNA-binding proteins based on GCN4 and non-native regulation sites, *Chem. -Eur. J.*, 2014, **20**, 2829–2839.
15. T. Morii, M. Simomura and S. Morimoto, Sequence-Specific DNA Binding by a Geometrically Constrained Peptide Dimer, *J. Am. Chem. Soc.*, 1993, **32**, 1150–1151.
16. M. Okagami, M. Ueno, K. Makino, M. Shimomura, I. Saito, T. Morii and Y. Sugiura, Sequence-specific DNA binding by covalently constrained peptide dimers of the basic leucine zipper protein GCN4, *Bioorg. Med. Chem.*, 1995, **3**, 777–784.
17. T. Morii, Y. Saimei, M. Okagami, K. Makino and Y. Sugiura, Factors Governing the Sequence-Selective DNA Binding of Geometrically Constrained Peptide Dimers, *J. Am. Chem. Soc.*, 1997, **119**, 3649–3655.
18. J. S. Kim and C. O. Pabo, Getting a handhold on DNA: design of poly-zinc finger proteins with femtomolar dissociation constants, *Proc. Natl. Acad. Sci. U. S. A.*, 1998, **95**, 2812–2817.
19. M. Ueno, M. Sawada, K. Makino and T. Morii, Recognition of non-palindromic DNA sequence by a peptide heterodimer with artificial dimerization module, *J. Am. Chem. Soc.*, 1994, **116**, 11137–11138.

20. T. Morii, J. Yamane, Y. Aizawa, A. Keisuke Makino and Y. Sugiura, Cooperative Oligomerization Enhances Sequence-Selective DNA Binding by a Short Peptide, *J. Am. Chem. Soc.*, 1996, **118**, 10011–10017.
21. Y. Aizawa, Y. Sugiura, M. Ueno, Y. Mori, K. Imoto, K. Makino and T. Morii, Stability of the dimerization domain effects the cooperative DNA binding of short peptides, *Biochemistry*, 1999, **38**, 4008–4017.
22. T. Morii, T. Tanaka, S. Sato, M. Hagihara, Y. Aizawa and K. Makino, A general strategy to determine a target DNA sequence of a short peptide: application to a d-peptide, *J. Am. Chem. Soc.*, 2002, **124**, 180–181.
23. A. M. Caamano, M. E. Vázquez, J. Martinez-Costas, L. Castedo and J. L. Mascareñas, A light-modulated sequence-specific DNA-binding peptide, *Angew. Chem., Int. Ed.*, 2000, **39**, 3104–3107.
24. O. Sadovski, A. A. Beharry, F. Zhang and G. A. Woolley, Spectral Tuning of Azobenzene Photoswitches for Biological Applications, *Angew. Chem., Int. Ed.*, 2009, **48**, 1484–1486.
25. J. Mosquera, M. I. Sánchez, M. E. Vázquez, J. L. Mascareñas, C. Y. Cheung, S. Yue and V. L. Singer, Ruthenium bipyridyl complexes as photocleavable dimerizers: deactivation of DNA-binding peptides using visible light, *Chem. Commun.*, 2014, **50**, 10975.
26. A. Jiménez-Balsa, E. Pazos, B. Martínez-Albardonedo, J. L. Mascareñas and M. E. Vázquez, Temporary Electrostatic Impairment of DNA Recognition: Light-Driven DNA Binding of Peptide Dimers, *Angew. Chem., Int. Ed.*, 2012, **51**, 8825–8829.
27. Y. Azuma, M. Imanishi, T. Yoshimura, T. Kawabata and S. Futaki, Cobalt(II)-Responsive DNA Binding of a GCN4-bZIP Protein Containing Cysteine Residues Functionalized with Iminodiacetic Acid, *Angew. Chem.*, 2009, **121**, 6985–6988.
28. J. Mosquera, A. Jiménez-Balsa, V. I. Dodero, M. E. Vázquez and J. L. Mascareñas, Stimuli-responsive selection of target DNA sequences by synthetic bZIP peptides, *Nat. Commun.*, 1874, **2013**, 4.
29. L. L. G. Carrette, T. Morii and A. Madder, Peptidosteroid Tweezers Revisited: DNA Binding Through an Optimised Design, *Eur. J. Org. Chem.*, 2014, 2883–2891.
30. D. Verzele and A. Madder, Synthetic Progress in cMyc-Max Oncoprotein Miniaturization: Semi-Online Monitoring Gives Solid-Phase Access to Hydrophobic b(-HLH-)ZIP Peptidosteroid Tweezers, *Eur. J. Org. Chem*, 2013, **2013**, 673–687.
31. Y. Ruiz García, A. Iyer, D. Van Lysebetten, Y. Vladimir Pabon, B. Louage, M. Honcharenko, B. G. De Geest, C. I. Edvard Smith, R. Strömberg and A. Madder, Sequence-selective DNA recognition and enhanced cellular up-take by peptide–steroid conjugates, *Chem. Commun.*, 2015, 17552–17555.
32. W. Kramer, G. Wess, A. Enhsen, E. Falk, A. Hoffmann, G. Neckermann, G. Schubert and M. Urmann, Modified bile acids as carriers for peptides and drugs, *J. Controlled Release*, 1997, **46**, 17–30.

33. C. A. Bodé, T. Bechet, E. Prodhomme, K. Gheysen, P. Gregoir, J. C. Martins, C. P. Muller and A. Madder, Towards the conformational mimicry of the measles virus HNE loop: design, synthesis and biological evaluation of a cyclic bile acid-peptide conjugate, *Org. Biomol. Chem.*, 2009, **7**, 3391–3399.
34. D. B. Salunke, B. G. Hazra and V. S. Pore, Steroidal conjugates and their pharmacological applications, *Curr. Med. Chem.*, 2006, **13**, 813–847.
35. C. A. Bode, C. P. Muller and A. Madder, Validation of a solid-phase-bound steroid scaffold for the synthesis of novel cyclic peptidosteroids, *J. Pept. Sci.*, 2007, **13**, 702–708.
36. A. Madder, L. Li, M. H. De, N. Farcy, H. D. Van, F. Fant, G. Vanhoenacker, P. Sandra, A. P. Davis and C. P. J. De, Evaluation of a two-stage screening procedure in the combinatorial search for serine protease-like activity, *J. Comb. Chem.*, 2002, **4**, 552–562.
37. H. Li and L.-X. Wang, Cholic acid as template for multivalent peptide assembly, *Org. Biomol. Chem.*, 2003, **1**, 3507–3513.
38. L. J. D. Souza and U. Maitra, Design, Synthesis, and Evaluation of Bile Acid-Based Molecular Tweezers, *J. Org. Chem.*, 1996, **3263**, 9494–9502.
39. W. S. Horne, M. K. Yadav, C. D. Stout and M. R. Ghadiri, Heterocyclic Peptide Backbone Modifications in an α-Helical Coiled Coil, *J. Am. Chem. Soc.*, 2004, **126**, 15366–15367.
40. H. C. Kolb and K. B. Sharpless, The growing impact of click chemistry on drug discovery, *Drug Discovery Today*, 2003, **8**, 1128–1137.
41. C. W. Tornoe, C. Christensen and M. Meldal, Peptidotriazoles on Solid Phase: [1,2,3]-Triazoles by Regiospecific Copper(I)-Catalyzed 1,3-Dipolar Cycloadditions of Terminal Alkynes to Azides, *J. Org. Chem.*, 2002, **67**, 3057–3064.
42. I. E. Valverde, A. Bauman, C. A. Kluba, S. Vomstein, M. A. Walter and T. L. Mindt, 1,2,3-Triazoles as amide bond mimics: triazole scan yields protease-resistant peptidomimetics for tumor targeting, *Angew. Chem., Int. Ed. Engl.*, 2013, **52**, 8957–8960.
43. D. Verzele and A. Madder, Synthetic Progress in cMyc-Max Oncoprotein Miniaturization: Semi-Online Monitoring Gives Solid-Phase Access to Hydrophobic b(-HLH-)ZIP Peptidosteroid Tweezers, *Eur. J. Org. Chem.*, 2013, 673–687.
44. R. Mejia-Ariza, L. Graña-Suárez, W. Verboom, J. Huskens, P. Dhanaraj, N. Mallik, F. J. Ahmad, Y. Yen, S. J. Forman and A. Raubitschek, *et al.*, Cyclodextrin-based supramolecular nanoparticles for biomedical applications, *J. Mater. Chem. B*, 2017, **5**, 36–52.
45. H. Arima, K. Motoyama and T. Higashi, Potential therapeutic application of dendrimer/cyclodextrin conjugates with targeting ligands as advanced carriers for gene and oligonucleotide drugs, *Ther. Delivery*, 2017, **8**, 215–232.
46. M. E. Brewster and T. Loftsson, Cyclodextrins as pharmaceutical solubilizers, *Adv. Drug Delivery Rev.*, 2007, **59**, 645–666.

47. Y. Ruiz García, J. Zelenka, V. Pabon, A. Iyer, M. Buděšínský, T. Kraus, C. I. E. Smith and A. Madder, Cyclodextrin – peptide conjugates for sequence specific DNA binding, *Org. Biomol. Chem.*, 2015.
48. S. K. Nair and S. K. Burley, X-ray structures of Myc-Max and Mad-Max recognizing DNA: Molecular bases of regulation by proto-oncogenic transcription factors, *Cell*, 2003, **112**, 193–205.
49. L. E. Canne, A. R. Ferre-D'Amare, S. K. Burley and S. B. H. Kent, Total Chemical Synthesis of a Unique Transcription Factor-Related Protein: cMyc-Max, *J. Am. Chem. Soc.*, 1995, **117**, 2998–3007.
50. T. Morii, M. Simomura, S. Morimoto, I. Saito and D. N. A. Sequence-Specific, Binding by a Geometrically Constrained Peptide Dimer, *J. Am. Chem. Soc.*, 1993, **32**, 1150–1151.
51. Y. Ruiz Garcia, V. Pabon-Martinez, C. I. E. Smith and A. Madder, Specific dsDNA recognition by a mimic of the DNA binding domain of the c-Myc/Max transcription factor, *Chem. Commun.*, 2017, **53**, 6653–6656.
52. M. E. Vázquez, A. M. Caamaño, J. Martinez-Costas, L. Castedo and J. L. Mascareñas, Design and Synthesis of a Peptide That Binds Specific DNA Sequences through Simultaneous Interaction in the Major and in the Minor Groove, *Angew. Chem., Int. Ed.*, 2001, **40**, 4723–4725.
53. D. Stanojevic and G. L. Verdine, Deconstruction of GCN4/GCRE into a monomeric peptide-DNA complex, *Nat. Struct. Mol. Biol.*, 1995, **2**, 450–457.
54. J. B. Blanco, M. E. Va, L. Castedo, L. Mascareñas and U. Asociada, A Synthetic Miniprotein that Binds Specific DNA Sequences by Contacting Both the Major and the Minor Groove, *Chem. Biol.*, 2003, **10**, 713–722.
55. J. B. Blanco, O. Vázquez, J. Martínez-costas, L. Castedo and J. L. Mascareñas, High Affinity, Sequence Specific DNA Binding by Synthetic Tripyrrole – Peptide Conjugates, *Chem. – Eur. J.*, 2005, 4171–4178.
56. M. Ueno, A. Murakami, K. Makino and T. Morii, Arranging quaternary structure of peptides by cyclodextrin-guest inclusion complex: sequence-specific DNA binding by a peptide dimer with artificial dimerization module, *J. Am. Chem. Soc.*, 1993, **115**, 12575–12576.
57. J. B. Blanco, V. I. Dodero, M. E. Vázquez, M. Mosquera, L. Castedo and J. L. Mascareñas, Sequence-Specific DNA Binding by Noncovalent Peptide–Tripyrrole Conjugates, *Angew. Chem., Int. Ed.*, 2006, **45**, 8210–8214.
58. M. I. Sánchez, O. Vázquez, M. E. Vázquez and J. L. Mascareñas, Sequence-Selective DNA Recognition with Peptide-Bisbenzamidine Conjugates, *Chem. – Eur. J.*, 2013, **19**, 9923–9929.
59. M. I. Sánchez, J. Mosquera, M. E. Vázquez and J. L. Mascareñas, Reversible supramolecular assembly at specific DNA Sites: Nickel-promoted bivalent DNA binding with designed peptide and bipyridyl-bis(benzamidine) components, *Angew. Chem., Int. Ed.*, 2014, **53**, 9917–9921.
60. M. Vázquez, A. M. Eugenio and J. L. Mascareñas, A Light-Modulated Sequence-Specific, *Angew. Chem., Int. Ed. Engl.*, 2000, **112**, 3234–3237.
61. M. I. N. Zhang, B. Wu, H. Zhao and J. W. Taylor, The Effect of C-Terminal Helix Stabilization on Specific DNA Binding by Monomeric GCN4 Peptides, *J. Pept. Sci.*, 2002, **136**, 125–136.

62. L. K. Henchey, A. L. Jochim and P. S. Arora, Contemporary strategies for the stabilization of peptides in the a -helical conformation, *Curr. Opin. Chem. Biol.*, 2008, **12**, 22–26.
63. A. Iyer, D. Van Lysebetten, Y. Ruiz García, B. Louage, B. G. De Geest and A. Madder, Stapling monomeric GCN4 peptides allows for DNA binding and enhanced cellular uptake, *Org. Biomol. Chem.*, 2015, **13**, 3856–3862.
64. K. Estieu-gionnet and G. Guichard, Stabilized helical peptides: overview of the technologies and therapeutic promises, *Drug Discovery*, 2011, **6**, 937–963.
65. M. M. Madden, A. Muppidi, Z. Li, X. Li, J. Chen and Q. Lin, Bioorganic & Medicinal Chemistry Letters Synthesis of cell-permeable stapled peptide dual inhibitors of the p53-Mdm2/Mdmx interactions via photoinduced cycloaddition, *Bioorg. Med. Chem. Lett.*, 2011, **21**, 1472–1475.
66. H. Inhibitor, H. Zhang, Q. Zhao, S. Bhattacharya, A. A. Waheed, X. Tong, A. Hong, S. Heck, F. Curreli and M. Goger, *et al.*, A Cell-penetrating Helical Peptide as a Potential, *J. Mol. Biol.*, 2008, **378**, 565–580.
67. C. J. Brown, S. T. Quah, J. Jong, A. M. Goh, P. C. Chiam, K. H. Khoo, M. L. Choong, M. A. Lee, L. Yurlova and K. Zolghadr, *et al.*, Stapled peptides with improved potency and specificity that activate p53, *ACS Chem. Biol.*, 2013, **8**, 506–512.
68. H. Jo, N. Meinhardt, Y. Wu, S. Kulkarni, X. Hu, K. E. Low, P. L. Davies, W. F. DeGrado and D. C. Greenbaum, Development of α-helical calpain probes by mimicking a natural protein-protein interaction, *J. Am. Chem. Soc.*, 2012, **134**, 17704–17713.
69. Y.-W. Kim, T. N. Grossmann and G. L. Verdine, Synthesis of all-hydrocarbon stapled α-helical peptides by ring-closing olefin metathesis, *Nat. Protoc.*, 2011, **6**, 761–771.
70. Ø. Jacobsen, H. Maekawa, N. Ge, C. H. Görbitz and P. Rongved, Stapling of a 3 10 -Helix with Click Chemistry, *J. Org. Chem.*, 2011, **76**, 1228–1238.
71. W. S. Horne, Promoting peptide α-helix formation with dynamic covalent oxime, *Chem. Commun.*, 2011, **47**, 1–4.
72. Y. Demizu, N. Yamagata, S. Nagoya, Y. Sato and M. Doi, Enantioselective epoxidation of a , b -unsaturated ketones catalyzed by stapled helical L -Leu-based peptides, *Tetrahedron*, 2011, **67**, 6155–6165.
73. L. D. Walensky and G. H. Bird, Hydrocarbon-Stapled Peptides: Principles, Practice, and Progress, *J. Med. Chem.*, 2014, **57**, 6275–6288.
74. A. Jimønez-balsa, E. Pazos, B. Martínez-albardonedo, J. L. Mascareñas and M. E. Vázquez, Temporary Electrostatic Impairment of DNA Recognition: Light- Driven DNA Binding of Peptide Dimers, *Angew. Chem., Int. Ed.*, 2012, 8825–8829.
75. W. Keller, P. Ko and T. J. Richmond, Crystal Structure of a bZIP/DNA Complex at 2. 2 Å: Determinants of DNA Specific Recognition, *J. Mol. Biol.*, 1995, **254**, 657–667.
76. N. J. Zondlo and A. Schepartz, Highly Specific DNA Recognition by a Designed Miniature Protein, *J. Am. Chem. Soc.*, 1999, **121**, 6938–6939.

77. L. Guerrero, O. S. Smart, G. A. Woolley and R. K. Allemann, Photocontrol of DNA Binding Specificity of a Miniature Engrailed Homeodomain, *J. Am. Chem. Soc.*, 2005, **127**, 15624–15629.
78. J. Rodríguez, J. Mosquera, J. R. Couceiro, M. E. Vázquez and L. José, The AT-Hook motif as a versatile minor groove anchor for promoting DNA binding of transcription factor fragments, *Chem. Sci.*, 2015, **6**, 4767–4771.
79. J. Mosquera, M. I. Sánchez, J. Valero, J. De Mendoza, M. E. Vázquez and J. L. Mascareñas, Sequence-selective DNA binding with cell-permeable oligoguanidinium-peptide conjugates, *Chem. Commun.*, 2015, **51**, 4811–4814.
80. O. Vázquez, M. E. Vázquez, J. B. Blanco, L. Castedo and J. L. Mascareñas, Specific DNA Recognition by a Synthetic, Monomeric Cys2His2 Zinc-Finger Peptide Conjugated to a Minor-Groove Binder, *Angew. Chem., Int. Ed.*, 2007, **46**, 6886–6890.
81. J. Rodríguez, J. Mosquera, R. García-Fandiño, M. E. Vázquez and J. L. Mascareñas, A designed DNA binding motif that recognizes extended sites and spans two adjacent major grooves, *Chem. Sci.*, 2016, **7**, 3298–3303.

CHAPTER 14

# *Targeting DNA Mismatches with Coordination Complexes*

JACQUELINE K. BARTON,* ADAM N. BOYNTON AND KELSEY M. BOYLE

Division of Chemistry and Chemical Engineering, California Institute of Technology, Pasadena, CA 91125, USA
*Email: jkbarton@caltech.edu

## 14.1 Introduction—Transition Metal Complexes as Non-covalent Probes for Nucleic Acids

Coordination complexes offer distinct advantages as probes for nucleic acid structure. Transition metals possess an array of unique photophysical, spectroscopic and reactive properties, which can be finely tuned by varying the identity of the metal center and surrounding ligands. Furthermore, the relative ease with which different ligands can be exchanged for one another allows the researcher to explore a range of DNA recognition properties. Owing to this modular nature, one can generate families of closely related complexes to assist in structure–function studies.

This chapter focuses on transition metal complexes that target DNA through non-covalent interactions, with emphasis on rhodium and ruthenium complexes capable of selectively binding to DNA base pair mismatches. Typically, these positively charged, water-soluble compounds are coordinatively saturated and substitutionally inert. Hence, in stark contrast to agents such as cisplatin, there is no direct coordination of the metal

Chemical Biology No. 7
DNA-targeting Molecules as Therapeutic Agents
Edited by Michael J. Waring

Published by the Royal Society of Chemistry, www.rsc.org

center to the DNA or aqueous solvent. Additionally, these are rigid, three-dimensional systems with well-defined stereochemistry, all highly valuable properties in the context of non-covalent DNA recognition.

Perhaps the simplest non-covalent interaction between a cationic transition metal complex and DNA is electrostatic association with the duplex (Figure 14.1a). Also referred to as ionic site binding, the driving force behind this mode is the electrostatic attraction between the positively charged transition metal complex and the negatively charged DNA polyanion.[1]

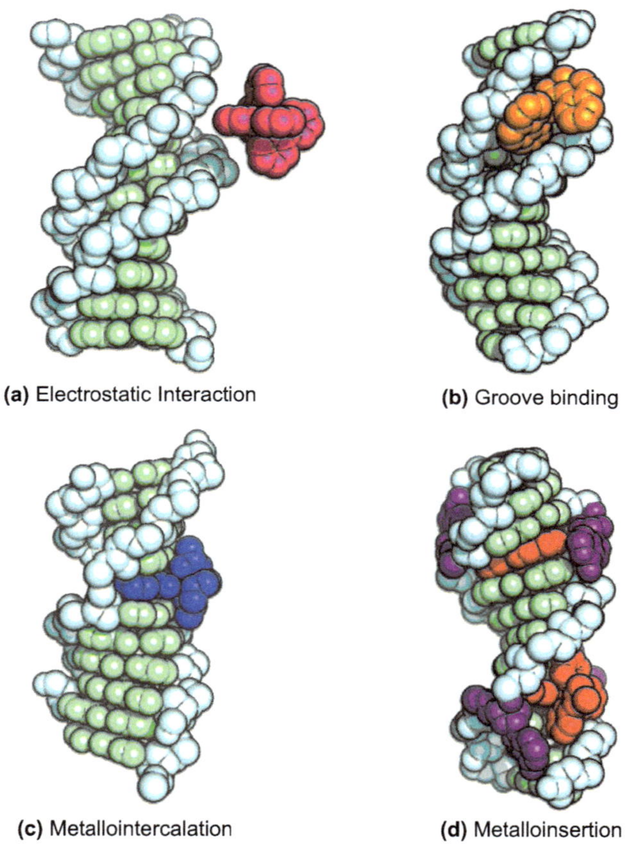


**Figure 14.1** Four modes of non-covalent interactions between transition metal complexes and DNA: (a) electrostatic interactions between $[Ru(bpy)_3]^{2+}$ and the negatively charged DNA, (b) groove binding of $[Cu(phen)_2]^{2+}$ in the minor groove of DNA, (c) intercalation of $[Rh(R,R)\text{-}Me_2trien)(phi)]^{3+}$ between DNA base pairs and (d) metalloinsertion of $[Rh(bpy)_2(chrysi)]^{3+}$ into two A–C mismatched sites. In all images, the DNA sugar–phosphate backbone is shown in blue, well-matched DNA base pairs are shown in green, and the mismatched bases in part (d) are highlighted in purple. Figure created from PDB structures 4C64, 1VTJ, 3GSK, 454D and CCDC structures ABAFAN and LUQGEL.

However, this binding mode offers no site-specific molecular recognition of the DNA. Electrostatic binding is embodied by the classic ruthenium polypyridyl complex $[Ru(bpy)_3]^{2+}$ (bpy = 2,2′-bipyridine).[1–4] Although the electrostatic interactions may have some stabilizing effects, this attraction results in very weak binding affinities towards the duplex, on the order of $10^2$ $M^{-1}$.[2]

Groove binding is another type of non-covalent interaction that a transition metal complex may undergo with DNA (Figure 14.1b). This mode involves hydrophobic association in the minor groove of the helix. An example of a minor groove-specific binder is $Cu(phen)_2^{2+}$ (phen = 1,10-phenanthroline), which was thoroughly studied by Sigman *et al.* and functions as a DNA cleaving agent in the presence of hydrogen peroxide.[5–7] Another complex that is known to associate with the DNA helix *via* groove binding is $[Ru(phen)_3]^{2+}$, whose interactions with DNA have been characterized extensively by Barton *et al.*[8] It is critical to note, however, that $[Ru(phen)_3]^{2+}$ is a chiral molecule; owing to its octahedral coordination geometry and three bidentate phen ligands, it possesses a left-handed enantiomer, designated as Λ, and a right-handed enantiomer, or Δ. Early work led to the deduction that the Λ isomer exhibits preferential binding with B-DNA through association in the minor groove.[2,4]

In contrast to Λ-$[Ru(phen)_3]^{2+}$, the Δ isomer preferentially binds to DNA through a third non-covalent interaction called intercalation. In this binding mode, a planar, aromatic, heterocyclic ligand stacks between the DNA base pairs, typically from the major groove, causing slight unwinding of the helix as well as a small increase in rise (Figure 14.1c).[9] Lippard *et al.* pioneered the early development of metallointercalators in the 1970s with their square planar platinum(II) complexes.[10,11] In the case of $[Ru(phen)_3]^{2+}$, the phen ligands possess a sufficiently extended π-system to intercalate, albeit partially, between the base pairs of the helix. This intercalation raises the binding affinity towards the DNA relative to groove binding or electrostatic associations. However, the binding affinity of $[Ru(phen)_3]^{2+}$ towards DNA is still relatively meager, on the order of $10^3$ $M^{-1}$. Increasing the intercalative binding affinity for a transition metal probe can be accomplished by increasing the surface area of the intercalating ligand, so as to achieve greater π-stacking between the base pairs of the helix.

Two of the best-known examples of strongly intercalating ligands are phi (9,10-phenanthrequinone diimine) and dppz (dipyrido[3,2-*a*:2′,3′-*c*]phenazine). Ruthenium(II) complexes bearing the dppz ligand, such as $[Ru(bpy)_2(dppz)]^{2+}$, favorably intercalate into B-form DNA with an affinity on the order of $>10^6$ $M^{-1}$.[12] While $[Ru(bpy)_2(dppz)]^{2+}$ binds somewhat non-specifically, it does exhibit a preference towards AT-rich regions in the duplex.[13] $[Ru(bpy)_2(dppz)]^{2+}$ and related complexes are described in greater detail later in this chapter.

Phi is another well-studied intercalating ligand, primarily in the context of rhodium(III) complexes. A classic intercalating complex is $[Rh(phen)_2(phi)]^{3+}$, which binds double helical DNA with high affinity ($\geq 10^7$ $M^{-1}$) owing to the ability of the phi ligand to overlap significantly with the base pairs.[14] Phi complexes of rhodium are capable of cleaving DNA at their intercalative site upon irradiation with UV light, a property with tremendous utility for determining site-specific DNA binding. For example, $[Rh(phen)_2(phi)]^{3+}$ exhibits some shape-selective recognition by preferentially photocleaving DNA at 5′-Py–Py–Pu-3′ sites. This sequence in particular generates a greater opening of the major groove, which alleviates steric interactions between the DNA and the ancillary phen ligands, allowing the complex to intercalate more deeply. Metallointercalators have also been developed as sequence-specific recognition agents, exemplified by Δ-α-[Rh[(*R*,*R*)-$Me_2$trien]phi]$^{3+}$((*R*,*R*)-$Me_2$trien = 2*R*,9*R*-diamino-4,7-diazadecane), which photocleaves specifically at the sequence 5′-TGCA-3′.[15] A high-resolution crystal structure of this metallointercalator bound to DNA revealed that the DNA slightly unwinds at the binding site to permit deep intercalation of the rhodium complex, which occurs from the major groove (Figure 14.2).[16] The sequence-specificity

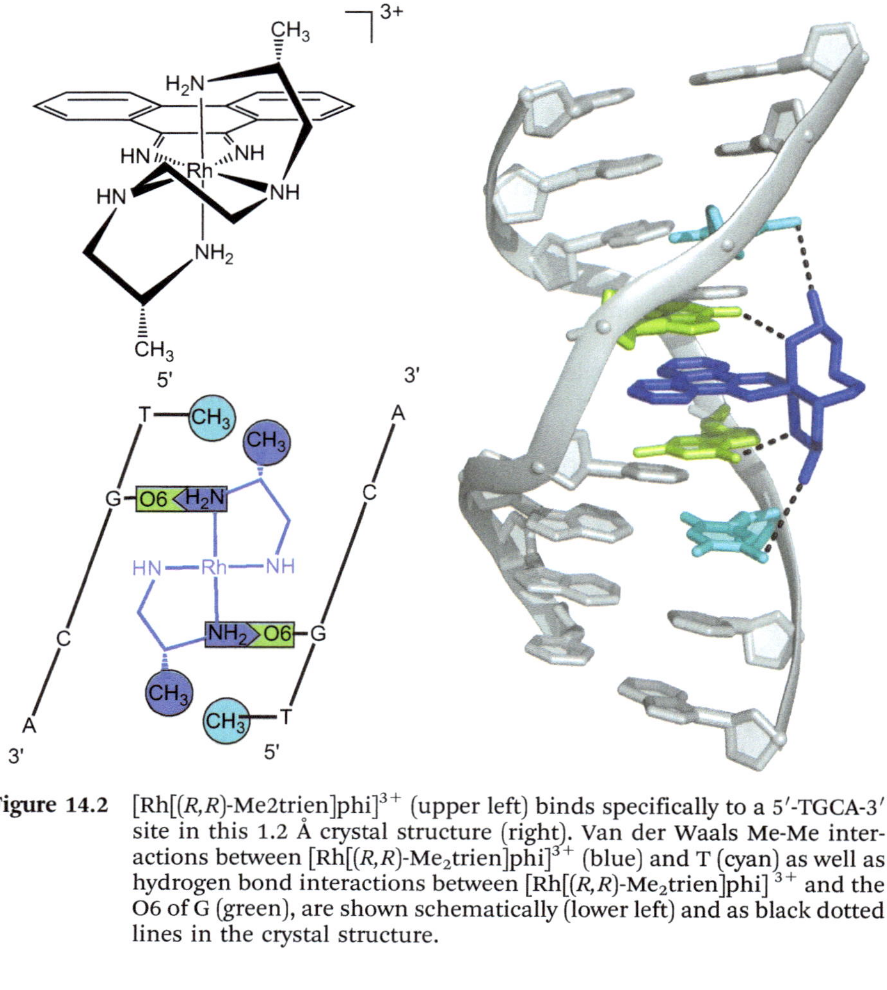


**Figure 14.2** [Rh[(*R*,*R*)-Me2trien]phi]$^{3+}$ (upper left) binds specifically to a 5′-TGCA-3′ site in this 1.2 Å crystal structure (right). Van der Waals Me-Me interactions between [Rh[(*R*,*R*)-$Me_2$trien]phi]$^{3+}$ (blue) and T (cyan) as well as hydrogen bond interactions between [Rh[(*R*,*R*)-$Me_2$trien]phi]$^{3+}$ and the O6 of G (green), are shown schematically (lower left) and as black dotted lines in the crystal structure.

results from an ensemble of non-covalent interactions between functionalities on the ancillary $Me_2$trien ligand posed into the major groove by rigid intercalation of the octahedral complex. The intercalation results in a doubling of the rise at the binding site, and thus, the phi ligand behaves simply as if it were another base pair in the helix.

The molecules described thus far represent several approaches to target DNA with coordination complexes. Moving forward, however, it becomes interesting not only to target DNA, but to target DNA in a manner that is therapeutically relevant. Furthermore, this targeting should be selective towards a biomarker of cancer, so as to avoid off-target damage that is often seen with traditional DNA-binding therapeutics, such as cisplatin.[17] Even the most selective complex described here, Δ-α-[Rh[(*R*,*R*)-$Me_2$trien]phi]$^{3+}$, serves merely as a proof-of-concept molecule since its target, 5′-TGCA-3′ DNA, is not specifically implicated in any disease. With this considered, it becomes the task of the researcher not only to target DNA, but to identify and target DNA biomarkers that are unique to disease.

## 14.2 Rhodium Metalloinsertors: Probes for DNA Mismatches

We have considered that potential targets for the design of new complexes might include sites of DNA damage or DNA lesions, which can be caused by both endogenous and exogenous sources such as replication errors or interaction with reactive oxygen species.[18,19] Since the genomic integrity of DNA is essential to mutation-free replication, cells usually have several mechanisms to repair these types of DNA insults.[20–22] One such repair mechanism, mismatch repair (MMR), is responsible for the correction of DNA base pair mismatches, which, if left uncorrected, propagate into point mutations upon cellular replication.[23,24] Deficiencies in this repair pathway are associated with many types of cancer, including tumorigenesis in up to 75% of individuals with Lynch syndrome (a genetic condition involving MMR deficiencies) and up to 20% of all solid tumors.[25,26] Additionally, cancers featuring microsatellite instabilities (MSI) are associated with resistance to many first-line cancer treatments, such as cisplatin and alkylating agents.[27,28] As such, MMR deficiencies and their associated increase in uncorrected DNA base pair mismatches represent a promising target for therapeutic and diagnostic design.

The first example of a small transition metal complex capable of selectively targeting a DNA base pair mismatch came in 1997 with the development of a unique rhodium complex, $[Rh(bpy)_2(chrysi)]^{3+}$ (chrysi = 5,6-chrysenequinone diimine) (Figure 14.3).[29] $[Rh(bpy)_2(chrysi)]^{3+}$ was inspired by complexes such as $[Rh(phen)_2(phi)]^{3+}$ and $[Rh(phi)_2(bpy)]^{3+}$, which exploit inert, redox-capable metal centers that allow them to cleave DNA upon irradiation with UV-light.[30,31] To impart mismatch selectivity upon this family of complexes, the phi ligand framework was expanded upon with an additional phenyl ring

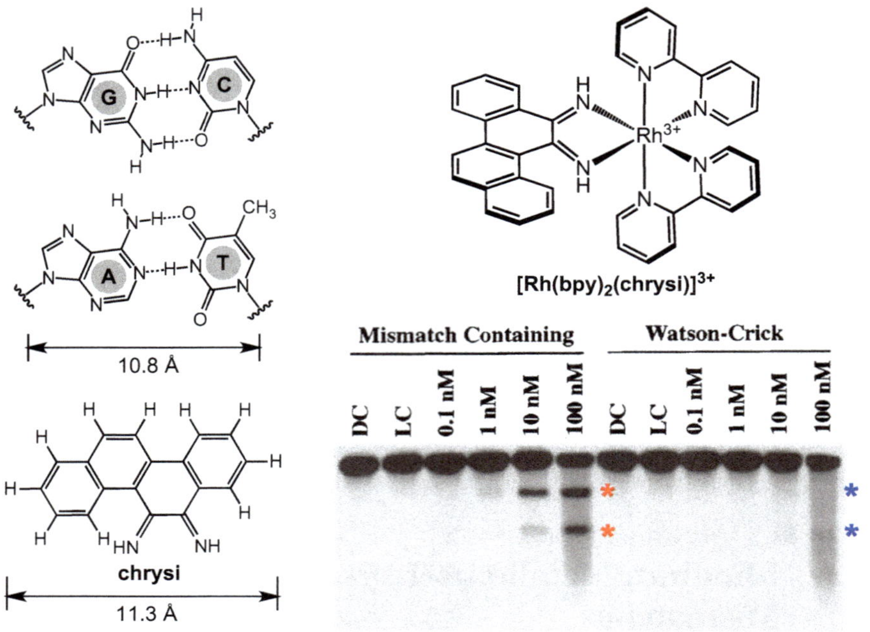


**Figure 14.3** (Left) The chrysi ligand is 0.5 Å wider than a well matched base pair, making intercalation unfavorable. (Upper right) The structure of $[Rh(bpy)_2(chrysi)]^{3+}$, which is capable of site-selective photocleavage at a single CC mismatch in a 2725-base pair linearized plasmid (Lower Right), resulting in two photocleavage products (red *). The equivalent well-matched plasmid is not cleaved by the metal complex (blue *).

to generate the chrysi ligand. At 11.3 Å, the chrysi ligand is about 2.1 Å wider than phi and about 0.5 Å wider than a well-matched base pair, making intercalation a less favorable binding possibility due to steric clashing between the ligand and the sugar–phosphate backbone (Figure 14.3). Instead, it was proposed that rhodium complexes containing this bulky ligand would preferentially interact with DNA at thermodynamically destabilized sites, such as mismatches and abasic sites, which may be better able to accommodate the expansive ligand.

As predicted, $[Rh(bpy)_2(chrysi)]^{3+}$ was shown to photocleave DNA in a surprisingly selective and robust manner. Initial experiments revealed the rhodium complex could photocleave mismatches of all identities, and that the extent of photocleavage (and therefore the level of detection) was dependent primarily on the thermodynamic destabilization of the targeted mismatch.[32] Overall, it was found that 80% of mismatches in all possible sequence contexts (as well as thermodynamically destabilized abasic sites and single-base bulges) were detected by $[Rh(bpy)_2(chrysi)]^{3+}$.[33,34] Guanine-containing mismatches are frequently undetected by $[Rh(bpy)_2(chrysi)]^{3+}$ due to their low destabilization; they are in fact quite similar to well-matched base pairs. Remarkably, in binding a highly destabilized site, such as a C–C mismatch, high selectivity of the Rh complex is maintained even in a large 2725 base pair linearized plasmid containing only a single C–C mismatch: incubation of $[Rh(bpy)_2(chrysi)]^{3+}$ with this plasmid produced discrete cleavage products that corresponded exclusively with cleavage at the mismatched site (Figure 14.3).[33] In comparison, no photocleavage was observed in an analogous well-matched plasmid.

These early studies indicated that $[Rh(bpy)_2(chrysi)]^{3+}$ was a promising answer to DNA mismatch targeting, but it was not fully understood *how* $[Rh(bpy)_2(chrysi)]^{3+}$ targeted mismatches until $\Delta$-$[Rh(bpy)_2(chrysi)]^{3+}$ was co-crystallized with DNA containing an AC mismatch in 2006 (Figure 14.1d).[35] The crystal structure revealed that the rhodium complex did not bind DNA mismatches *via* intercalation, but instead bound DNA through insertion. This distinct binding mode was initially proposed by L. S. Lerman in 1961, but the binding mode had not been directly observed until this structure was identified.[36] Accordingly, $[Rh(bpy)_2(chrysi)]^{3+}$ and future complexes became known as rhodium metalloinsertors and their binding mode as metalloinsertion.

In metalloinsertion, the chrysi ligand of the $\Delta$-enantiomer *inserts* into the DNA helix from the minor groove at a mismatch site by displacing and ejecting the mismatched bases. This insertion allows the chrysi ligand essentially to replace the destabilized DNA mispair and $\pi$-stack with the neighboring well-matched bases. Unlike intercalation, there is little unwinding of the DNA nor a rise in the DNA helix. This is a highly stable binding interaction, with most metalloinsertors having binding affinities on the order of $10^6$ to $10^7$ $M^{-1}$. Moreover, for most metalloinsertors, the binding is highly enantioselective; only the $\Delta$-isomer can fit into the right-handed minor groove. The features of this unique binding mode also clarified the thermodynamic dependence of metalloinsertion on mismatch stability—proper hydrogen bond contacts between well-matched base pairs are too stabilizing to permit insertion of the chrysi ligand. This binding mode was later verified with additional crystal structures using A–A-mismatched DNA as well as an NMR-solution structure using C–C-mismatched DNA.[37,38] In particular, the NMR structure confirmed that the crystallographically observed metalloinsertion was not simply an artifact of crystal packing forces.

It is worthwhile to note that, since the discovery of $[Rh(bpy)_2(chrysi)]^{3+}$, several other mismatch-targeting small molecules, both organic and inorganic, have been developed by various research groups.[39–44] These molecules use diverse strategies to target mismatches, from exploitation of the thermodynamic destabilization (similar to metalloinsertors) to matching the hydrogen bond patterns of specific base-pair mismatches.

## 14.3 Rhodium Metalloinsertors in the Cell

After nearly a decade of *in vitro* work elucidating the binding properties and selectivity of rhodium metalloinsertors, it became clear that these complexes could find even greater significance as potential therapeutics. As discussed previously, deficiencies in MMR are associated with some cancers and can lead to an increase in persistent DNA base-pair mismatches.[45] However, due to the high fidelity and proofreading abilities of DNA polymerases, even MMR-deficient cells have extremely few mismatches. Since there are so few mismatch targets, it is necessary that a therapeutically relevant small molecule should have a generic affinity for all mismatches, not just certain mispairs. As such, rhodium metalloinsertors, which have high selectivity and bind an abundance of mismatches in various sequence contexts, are an ideal therapeutic scaffold for MMR-deficient cancers.

To test if *in vitro* results would translate to *ex vivo* cellular models, rhodium metalloinsertors were tested *via* cell proliferation and cell viability assays in isogenic MMR-proficient and MMR-deficient cell lines. Remarkably, $[Rh(bpy)_2(chrysi)]^{3+}$ preferentially targets MMR-deficient cells over their MMR-proficient counterparts.[46] In fact, all metalloinsertors, regardless of ancillary ligand identity, exhibit selectivity (defined here as preferential cytotoxicity) towards MMR-deficient cells or, in some cases, no selectivity, but *never* selectivity towards MMR-proficient cells (Figure 14.4).[47,48] This result is in stark contrast to the majority of DNA-targeting therapeutics, including cisplatin (a covalent DNA binder) and etoposide (a DNA intercalator), among others, which preferentially target the MMR-*proficient* cell line (Figure 14.4).[49–51] Unsurprisingly, many of these therapeutics build up drug resistance in MMR-deficient cancers, revealing an even greater need for an effective and selective MMR-deficiency-targeting agent.[27]

Since the discovery of this unique biological selectivity, several generations of rhodium metalloinsertors have helped us better understand some elements behind this biological activity, though the detailed mechanism of action is still largely unknown. Some of the earliest metalloinsertor studies

**Figure 14.4** Anti-proliferative activity of (a) a generic rhodium metalloinsertor, (b) the alkylating agent, MNNG, and (c) the ubiquitous therapeutic, cisplatin, as measured *via* bromodeoxyuridine (BrdU)-incorporation ELISA. It is of interest to note that both the alkylating agent and cisplatin preferentially kill the MMR proficient (MMR+, green) cell line over the MMR deficient (MMR−, red) cell line. This result is reflective of the resistance many MMR-deficient cancers exhibit towards these classes of therapeutics.

in cells revealed that only the Δ-enantiomer of $[Rh(bpy)_2(chrysi)]^{3+}$ inhibited cell growth, with the Λ-enantiomer having little cellular effect; this high enantioselectivity would be consistent with metalloinsertion into the right-handed B-DNA helix.[46] This enantioselectivity also indicated that the metal complexes did not undergo decomposition once inside the cell, and that any biological activity was likely to be due to the metalloinsertor as a whole, not free ligands. Similarly, early studies showed that the binding affinity of five metalloinsertors correlated well with their selectivity towards the MMR-deficient cell lines—that is, smaller complexes with larger binding affinities, such as $[Rh(NH_3)_4(chrysi)]^{3+}$, have greater biological selectivity, whereas bulky complexes with smaller binding affinities, such as $[Rh(DIP)_2(chrysi)]^{3+}$ (DIP = 4,7-diphenyl-1,10-phenanthroline), have minimal biological selectivity.[47] This study provided the first piece of *ex vivo* evidence to support the hypothesis that metalloinsertors targeted mismatched nucleic acids within the cell. The mechanism of cell death caused by metalloinsertors has also been explored and was found to occur *via* a necrotic pathway following both flow cytometry and inhibition experiments.[52] Recent experiments indicate that a necrotic cell death pathway may be significant to cellular selectivity: conjugates combining a rhodium metalloinsertor with a cisplatin moiety have similar mismatch binding affinities to monomeric metalloinsertors and similar ligand environments, yet they exhibit no biological selectivity.[53,54] Unlike classic rhodium metalloinsertors, these conjugates were found to trigger apoptotic cell death similar to that seen with cisplatin. Though causation cannot be invoked at this stage, this correlation may indicate that the mode of cell death could be important to the selectivity of these complexes.

A greater understanding of the biological activity of these complexes was revealed in a comprehensive study of metalloinsertors. A large family of metalloinsertors differing primarily in the lipophilicity of their ancillary ligands was synthesized and examined for biological activity, including nuclear and mitochondrial localization.[48] Despite similar binding affinities and uptake characteristics in whole cells, the most hydrophilic of these complexes produced significant selectivity between MMR-proficient and MMR-deficient cell lines whereas the most lipophilic of these complexes produced no selectivity. By isolating the nuclear and mitochondrial components and examining their rhodium content by inductively coupled plasma-mass spectrometry (ICP-MS), it was revealed that organelle localization profile was perhaps the best indicator of biological selectivity of rhodium metalloinsertors. To understand this activity, it is perhaps best to compare two of the tested complexes, $[Rh(DPAE)_2(chrysi)]^{3+}$ (DPAE = 2-(di(pyridine-2-yl)amino)ethanol) and $[Rh(PrDPA)_2(chrysi)]^{3+}$ (PrDPA = N-propyl-N-(pyridin-2-yl)pyridin-2-amine), which differ only slightly in their ancillary ligands by the presence of an ethanol or propyl group, respectively (Figure 14.5).[48,55] $[Rh(DPAE)_2(chrysi)]^{3+}$ and $[Rh(PrDPA)_2(chrysi)]^{3+}$ have similar binding affinities of $K_B = 6.8 \times 10^6\ M^{-1}$ and $2.5 \times 10^6\ M^{-1}$, respectively, as well as similar whole-cell uptake patterns. The complexes differ primarily in their lipophilicity, with Log P = −1.5 and

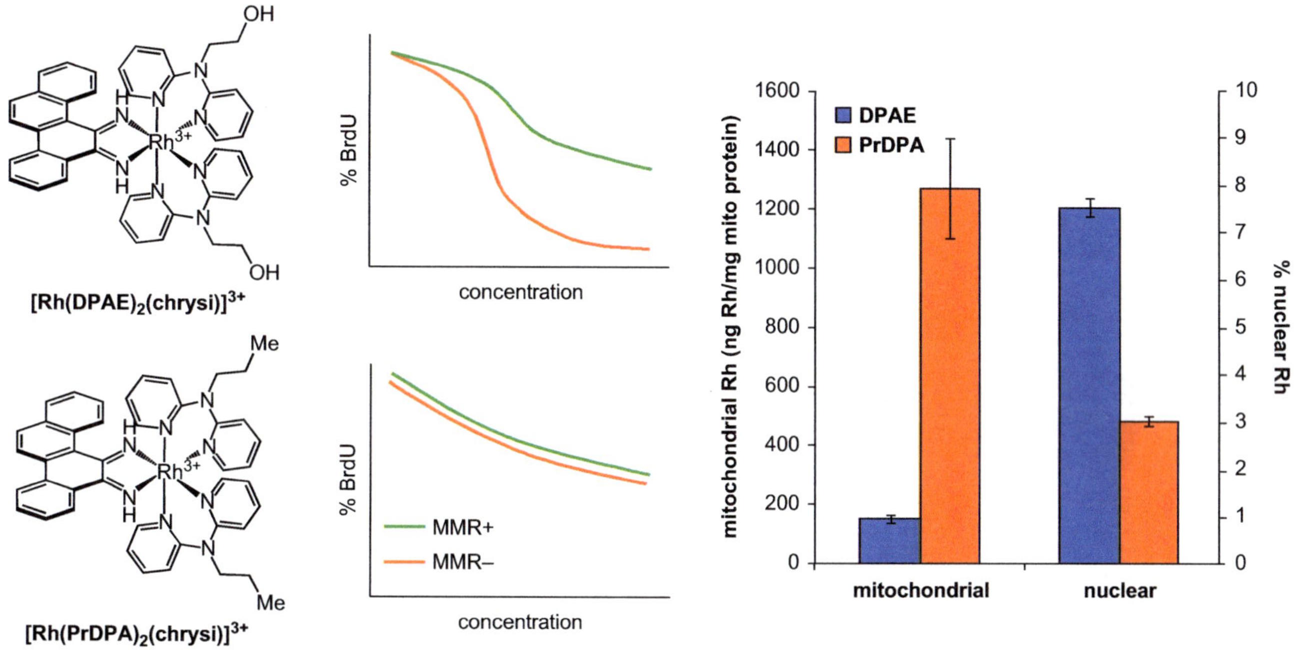


**Figure 14.5** Cell proliferation observations (as measured *via* BrdU-incorporation ELISA) for two similar metalloinsertors, [Rh(DPAE)$_2$(chrysi)]$^{3+}$ and [Rh(PrDPA)$_2$(chrysi)]$^{3+}$ (Left), which contain hydrophilic or lipophilic ligands, respectively. (Center) Despite having similar DNA binding affinities and similar whole-cell uptake, [Rh(DPAE)$_2$(chrysi)]$^{3+}$ shows selective behavior toward MMR-deficient (MMR−) cells over the MMR proficient (MMR+) cells, whereas [Rh(PrDPA)$_2$(chrysi)]$^{3+}$ exhibits no selectivity. (Right) This is explained by the subcellular localization profiles of each complex. While both complexes enter the nucleus at high enough concentrations for DNA binding, [Rh(PrDPA)$_2$(chrysi)]$^{3+}$ has significant localization into the mitochondria, leading to off target and non-selective cytotoxicity.

−1.0 for $[Rh(DPAE)_2(chrysi)]^{3+}$ and $[Rh(PrDPA)_2(chrysi)]^{3+}$, respectively. Both complexes are found in the nucleus at high enough concentrations to bind mismatches, indicating that both complexes *should* be capable of cell selectivity. However, the more lipophilic metalloinsertor, $[Rh(PrDPA)_2(chrysi)]^{3+}$ (as well as other metalloinsertors bearing lipophilic ligands), had significant rhodium localization in the mitochondria, and this high mitochondrial localization was correlated with the absence of preferential inhibition in the matched MMR-proficient/deficient cell lines (Figure 14.5). Conversely, the more hydrophilic metalloinsertor, $[Rh(DPAE)_2(chrysi)]^{3+}$ (as well as other metalloinsertors bearing hydrophilic ligands), exhibited comparatively low mitochondrial localization and a higher percentage accumulation in the nucleus, which correlated with selective activity towards the MMR deficient cells (Figure 14.5). This localization difference is not necessarily unexpected, since it is understood that lipophilic cations often accumulate in mitochondria due to large membrane potentials, yet the significant biological implication for metalloinsertors is surprising.[56] As with the study correlating binding affinity to biological selectivity, these studies strongly indicate that the biological selectivity of these complexes does not arise from targeting of mitochondrial DNA, and is instead observed when the metalloinsertors accumulate to a greater extent in the nucleus and, by extension, interact with *nuclear* DNA. This study also shows the importance of ancillary ligand in designing metalloinsertors as viable therapeutic agents.

In this study, one metalloinsertor exhibited particularly interesting and promising *ex vivo* results. In addition to showing high selectivity towards the MMR-deficient cell line (with almost no inhibition of the MMR-proficient cell line), $[Rh(phen)(chrysi)(DPE)]^{2+}$ (DPE = 1,1-di(pyridin-2-yl)ethanol) was significantly more potent than $[Rh(HDPA)_2(chrysi)]^{3+}$ (HDPA = 2,2′-dipyridylamine), the previously most potent metalloinsertor, despite the two complexes having similar mismatch binding affinities.[48] Additionally, complexes related to $[Rh(phen)(chrysi)(DPE)]^{2+}$, in which the DPE ligand is derivatized to have its pyridyl group replaced by a phenyl, methyl or n-hexyl group, all exhibit high selectivity and even greater potency.[49] One of these derivatives, $[Rh(phen)(chrysi)(PPO)]^{2+}$ (PPO = 1-methyl-1-(pyrid-2-yl)ethanol), shows a potency of 320 nM, making this a promising candidate for therapeutic use *in vivo* (Figure 14.6). In addition to this interesting biological activity, these complexes all exhibit an unusual ligand coordination, in which the hydroxyl group of the DPE-like ligand is bound directly to the Rh center, creating a Rh–O bond and decreasing the overall charge of these complexes from 3+ to 2+. This structural change was associated with additional interesting properties: specifically, both the Δ and Λ enantiomers of $[Rh(phen)(chrysi)(DPE)]^{2+}$ were shown to bind DNA *in vitro* and selectively target MMR deficiencies *ex vivo*, in both cases at slightly different concentrations, suggesting that there was no racemization of the complexes.[49]

Though it is not clear how this racemic binding occurs, the pKa of these new complexes may provide some insight. Previous generations of

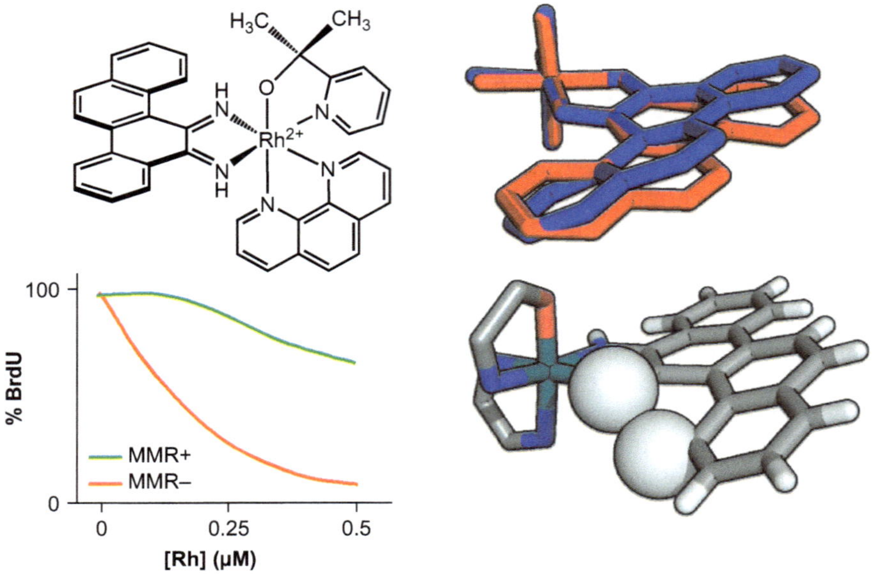


**Figure 14.6** The structure of $[Rh(phen)(chrysi)(PPO)]^{2+}$ (upper left) and its antiproliferative activity (lower left) as measured *via* BrdU-incorporation ELISA. $[Rh(phen)(chrysi)(PPO)]^{2+}$ preferentially targets MMR-deficient (MMR−) cells over MMR-proficient (MMR+) cells. $[Rh(phen)(chrysi)(PPO)]^{2+}$ is currently the most selective and potent metalloinsertor. When the inserting chrysi ligand of $[Rh(phen)(chrysi)(DPE)]^{2+}$ (blue) is overlain with the chrysi ligand of $[Rh(bpy)_2(chrysi)]^{3+}$, (red) there are is a noticeable buckling of the $[Rh(phen)(chrysi)(DPE)]^{2+}$ chrysi ligand (upper right) due to steric clashing with an imine proton with high pKa value (lower right).

metalloinsertors, such as $[Rh(bpy)_2(chrysi)]^{3+}$ and $[Rh(HDPA)_2(chrysi)]^{3+}$, have pKa values of 5.6 and 7.0 pH units, respectively, meaning these complexes deprotonate at physiological pH. Deprotonation occurs at the imine proton of the chrysi ligand, which otherwise clashes with one of the aromatic chrysi protons, and allows the chrysi ligand to lay planar relative to the rhodium center (Figure 14.6).[49] Rh–O-containing complexes have pKa values far above physiological pH at 8.3–8.9 pH units. With such high pKa values, the imine proton of the chrysi ligand in these complexes is unable to deprotonate, forcing the chrysi ligand to buckle relative to the rhodium center (Figure 14.6). This significant distortion of the inserting ligand can be observed crystallographically and undoubtedly perturbs the metalloinsertion binding mode of these complexes with DNA. A difference in DNA binding would help to explain how both enantiomers are biologically active, and may also be at the root of the increased potency and selectivity of these complexes.

Although rhodium metalloinsertors are arguably the best characterized and most well-studied probes for DNA base-pair mismatches, very little is known about their biological mechanisms. Determining the structure of the potent lesion formed by the Δ and Λ enantiomers of the newest generation of Rh–O metalloinsertors could give valuable insight into why these new complexes have such unique activity. For instance, it is possible that $[Rh(phen)(chrysi)(PPO)]^{2+}$ forms a more stable interaction with DNA mispairs, which may interrupt cellular processes, such as replication and transcription, if the lesion cannot be resolved. Similarly, it is possible that these metalloinsertors create a greater distortion in the DNA structure upon mismatch binding, leading to easier recognition of the lesion by DNA-processing proteins. In any case, understanding how these complexes bind DNA mismatches will allow us to improve the design of metalloinsertors and create newer generations of plausible drug candidates. In addition to better understanding the structure of the DNA lesion, it is essential to learn more about the proteins or pathways involved in processing the metalloinsertor–DNA complex. At this point, we still do not know which proteins are responsible for recognizing the DNA lesion or initiating pathways that result in cell death. Furthermore, even the role of MMR proteins is unclear—is biological selectivity a result of metalloinsertors binding more mismatches in MMR-deficient cells, or is it that cell death is avoided because MMR-proficient cells can process and remove these lesions? To answer these questions is essential, and experiments are currently underway to better understand how and why these metalloinsertors work and their potential value as therapeutics.

## 14.4 Luminescent Ruthenium Complexes as Probes for DNA Mismatches

In addition to developing octahedral rhodium complexes as potential therapeutics targeted to mismatches, a substantial amount of research has focused on designing luminescent transition metal complexes to serve as spectroscopic probes of nucleic acids. In particular, ruthenium(II) polypyridyl complexes possess an intense metal-to-ligand charge transfer (MLCT) transition which, upon excitation, yields a bright, long-lived emission in the visible region. This characteristic contrasts with the rhodium intercalators and metalloinsertors, which are non-luminescent upon excitation. However, like the rhodium complexes discussed in this chapter, ruthenium(II) polypyridyl compounds are coordinatively saturated, making them inert to ligand substitution and stable in aqueous solutions.

With respect to investigating luminescent probes for nucleic acids, one of the most extensively studied compounds is the aforementioned $[Ru(bpy)_2(dppz)]^{2+}$, the well-known molecular 'light switch' for DNA (Figure 14.7).[12,57,58] In organic, aprotic solvents such as acetonitrile, this complex luminesces brightly from an MLCT excited state, analogously to other ruthenium(II) polypyridyl complexes such as the archetypal $[Ru(bpy)_3]^{2+}$.[59,60] However, in aqueous solution, the excited state emission of $[Ru(bpy)_2(dppz)]^{2+}$ is entirely quenched due to hydrogen bonding interactions between the solvent water

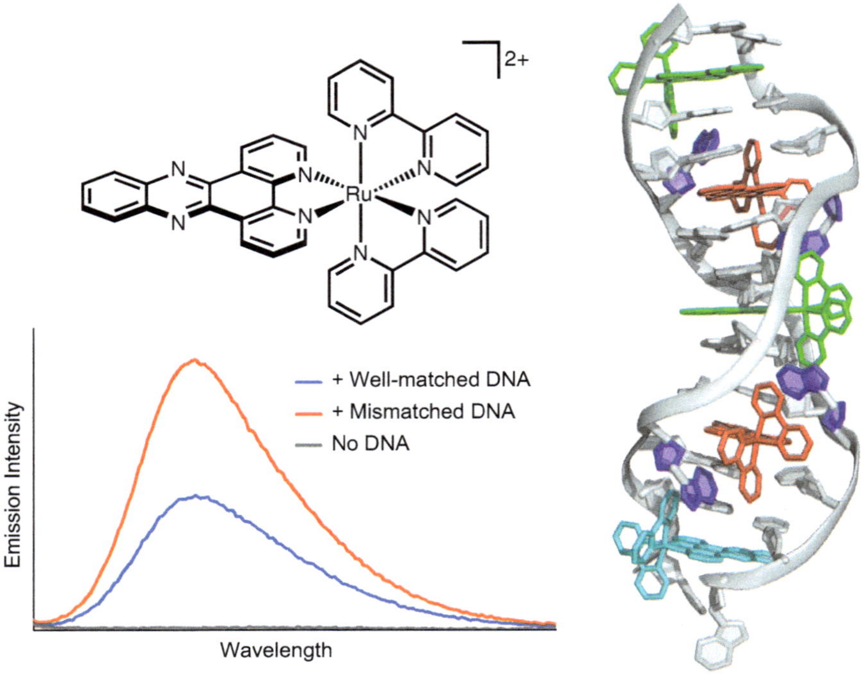


**Figure 14.7** (Left) $[Ru(bpy)_2(dppz)]^{2+}$ is a molecular 'light switch' for duplex DNA in aqueous solution, and luminesces more brightly in the presence of DNA containing a mismatch. (Right) $\Delta$-$[Ru(bpy)_2(dppz)]^{2+}$ was co-crystallized with a short DNA duplex containing two AA mismatches (mismatched base pairs are highlighted in purple), and three binding modes of the complex were observed: metalloinsertion (red) at the mismatched sites, intercalation (green) at well-matched sites, and end-capping between two duplexes (cyan).

molecules and phenazine nitrogen atoms of the dppz ligand. However, when duplex DNA is present in aqueous solution, $[Ru(bpy)_2(dppz)]^{2+}$ exhibits an intense emission as if it were in an organic solvent like acetonitrile. The dppz ligand intercalates into duplex DNA quite readily, owing to its large, planar, aromatic surface area. By intercalating between the base pairs of the duplex, the dppz nitrogen atoms are shielded from the bulk aqueous solvent. This shielding eliminates the quenching observed in pure water, hence the DNA light switch effect.

While $[Ru(bpy)_2(dppz)]^{2+}$ intercalates non-specifically into duplex DNA, the complex luminesces more brightly in the presence of a DNA mismatch or abasic site relative to completely well-matched DNA (Figure 14.7).[61] A crystal structure of $\Delta$-$[Ru(bpy)_2(dppz)]^{2+}$ bound to a 12-mer DNA oligonucleotide containing two AA mismatches was obtained at 0.92 Å resolution (Figure 14.7).[62] The asymmetric unit revealed five ruthenium complexes bound to the DNA duplex. In this structure, the complex was bound to the

DNA in three distinct binding modes: (i) intercalation at well-matched sites, (ii) metalloinsertion at the mismatched sites, and (iii) end-capping between two duplexes. At the mismatched sites, the complex binds from the minor groove and inserts deeply into the base stack, completely extruding the mismatched adenosines. Interestingly, at both mismatched sites, each adenosine is displaced back towards the minor groove and stacks with the bpy ancillary ligands of the ruthenium complex. Furthermore, because the dppz ligand is relatively narrow and also symmetric, a minimal widening of the DNA minor groove is observed. This structural change contrasts with the metalloinsertion of $[Rh(bpy)_2(chrysi)]^{3+}$, in which a more substantial widening of the minor groove occurs to accommodate the large, asymmetric chrysi ligand. The crystal structure of $[Ru(bpy)_2(dppz)]^{2+}$ bound to DNA also reveals two complexes bound *via* intercalation at well-matched sites. Here, both complexes are intercalated in the minor groove; this observation differs from what was seen in solution-phase experiments, namely that $[Ru(bpy)_2(dppz)]^{2+}$ intercalates at well-matched sites *via* the major groove.[13,61,63,64] However, in this structure, there are clear stacking interactions between the bpy ligands of the intercalated complexes and the neighboring mismatched adenosines, interactions that ostensibly stabilize the complexes in the minor groove at well-matched sites. Thus, these results indicate that there is likely to be a relatively small energetic difference when comparing intercalation from the minor and major groove for $[Ru(bpy)_2(dppz)]^{2+}$. Importantly, this crystal structure revealed that metalloinsertion at destabilized sites in the DNA duplex is still the preferred binding mode despite the narrow dppz ligand compared with the sterically cumbersome chrysi ligand.

Despite exhibiting a 1.5- to 2-fold luminescence enhancement in the presence of a DNA defect compared with fully well-matched DNA, $[Ru(bpy)_2(dppz)]^{2+}$ is not mismatch-specific. A mismatch-specific luminescent compound could be particularly valuable in the context of developing diagnostic probes for detecting deficiencies in MMR. Numerous efforts have been undertaken to design a mismatch-specific ruthenium 'light switch' complex. This effort has included investigations on ruthenium(II) complexes bearing chrysi and related ligands naphthi, phzi and acri (Figure 14.8).[65] However, such derivatives are found to show no luminescence at ambient temperatures, presumably due to hydrogen bonding between water and the ligand imine protons, which leads to vibrational relaxation. Other studies included adding functional groups to dppz to increase its steric bulk, as illustrated by the ligands dppae and tactp (Figure 14.8). However, no improvement in the luminescence differential between mismatched and well-matched DNA was observed for these derivatives compared with the parent $[Ru(bpy)_2(dppz)]^{2+}$ complex. Additionally, ligands possessing extended aromatic rings such as pyrenephen and tactp have a propensity to aggregate in aqueous solution.

Other ligands that were studied in attempts to achieve mismatch-specificity include pqp and tpqp, developed as even bulkier versions of chrysi; however, unlike the diimine-containing ligands, pqp and tpqp were

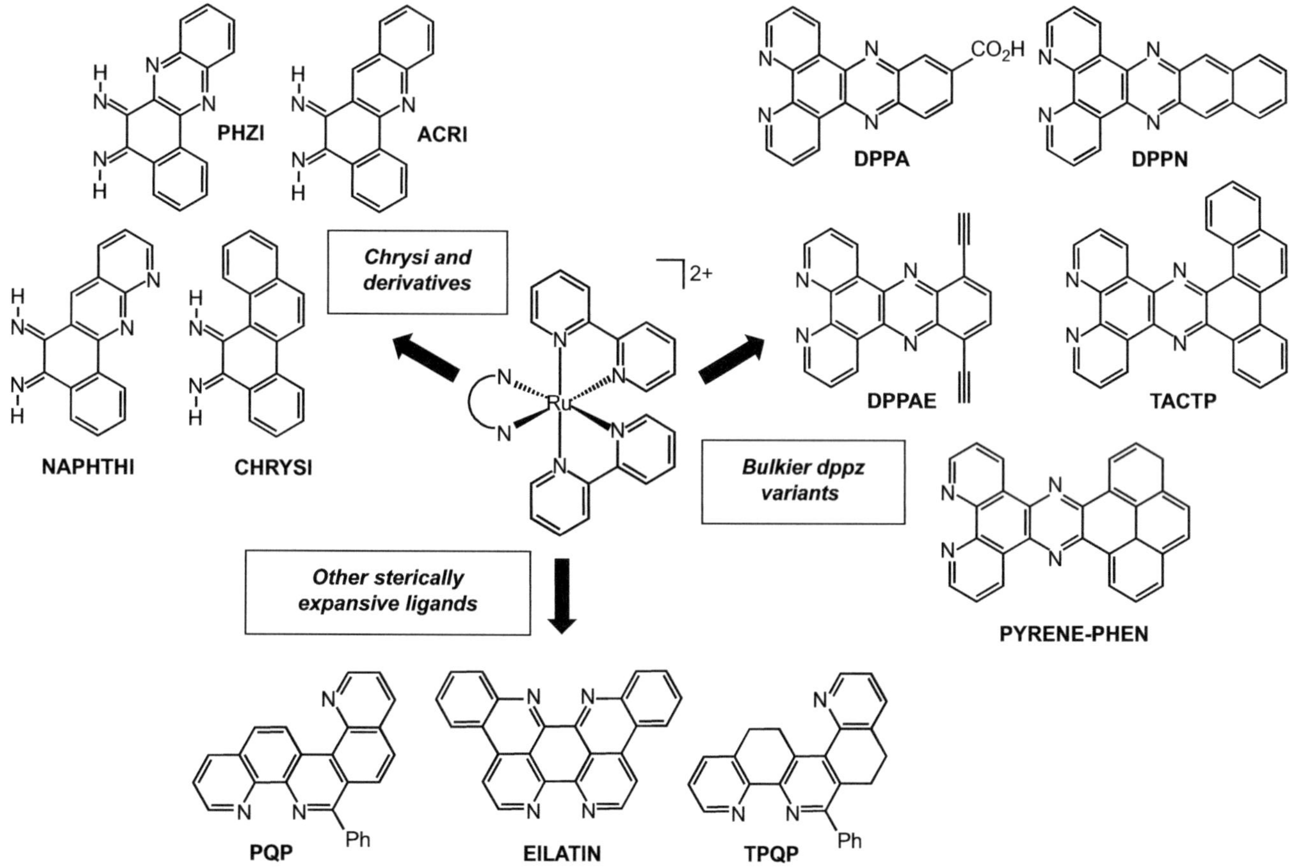


**Figure 14.8** An array of $[Ru(bpy)_2(\mathbf{L})]^{2+}$ complexes bearing bulky ligands (**L**) have been investigated over the years in efforts to obtain a mismatch-specific luminescent ruthenium(II) complex.

designed to coordinate directly to the ruthenium center *via* the heterocyclic nitrogen atoms.[66] $[Ru(bpy)_2(pqp)]^{2+}$ shows no detectable luminescence, while $[Ru(bpy)_2(tpqp)]^{2+}$ emits weakly in aqueous solution. However, $[Ru(bpy)_2(tpqp)]^{2+}$ shows no increase in emission intensity upon the addition of either well-matched or mismatched DNA. That being said, luminescence polarization and $[Fe(CN)_6]^{4-}$ quenching experiments revealed that $[Ru(bpy)_2(tpqp)]^{2+}$ binds more tightly at a C–C mismatch site compared with well-matched base pairs in the duplex. This preferential binding to the mismatch was confirmed by a $^1O_2$ sensitization assay, which revealed preferential damage to the guanine adjacent to the C–C mismatch compared with other guanine sites in the DNA. Another study explored the binding of $[Ru(bpy)_2(eilatin)]^{2+}$ (Figure 14.8) to well-matched and mismatched DNA duplexes.[67] Eilatin is a symmetric, aromatic ligand that is notably expansive. While competitive DNA photocleavage experiments with the metalloinsertor $[Rh(bpy)_2(chrysi)]^{3+}$ revealed some preferential binding of $[Ru(bpy)_2(eilatin)]^{2+}$ to a C–C mismatch, competition experiments with the non-specific metallointercalator $[Rh(bpy)_2(phi)]^{3+}$ showed that the ruthenium complex is still readily capable of binding to well-matched sites in the duplex. These results emphasized the notion that while a planar, aromatic, bulky ligand such as chrysi leads to high mismatch-specificity, simply increasing the steric size does not necessarily correlate with an increase in mismatch specificity.

Another avenue for developing a diagnostic probe for DNA mismatches that has been explored involves fluorescently tagged rhodium metalloinsertors. In one design, a negatively charged fluorophore, Oregon Green 514, was tethered to a positively charged rhodium metalloinsertor to generate the conjugate RhOG (Figure 14.9).[68] The motivation for this design was that the fluorescence of the dye would be quenched in free solution, or in the presence of well-matched DNA, due to ion-pairing with the rhodium moiety. However, if a mismatch were present, the rhodium would undergo metalloinsertion, causing the negatively charged dye to be repelled by the sugar–phosphate DNA backbone, resulting in 'turn-on' fluorescence. As expected, the fluorescence of the RhOG conjugate is considerably quenched relative to free Oregon Green, and compared with an equimolar solution of free Oregon Green and free metalloinsertor. Importantly, in fluorescence titrations of the conjugate with well-matched and C–C-mismatched 17-mer oligonucleotides, the conjugate exhibits a 3.2-fold brighter emission intensity in the presence of the sequence containing the mismatch. Additionally, polyacrylamide gel electrophoresis (PAGE) experiments utilizing $^{32}P$-labeled oligonucleotides revealed specific photocleavage at the C–C mismatch. However, one drawback of this RhOG conjugate is that even in the presence of the mismatched sequence, the maximum fluorescence reached by the fluorophore was only 6% of that of free Oregon Green; rhodium quenching, even in the extended form, could not be eliminated.

More recently, another rhodium–fluorophore conjugate has been studied, in which a positively charged cyanine dye was tethered to a metalloinsertor scaffold *via* a PEG linker.[69] In this design, the cyanine dye was chosen

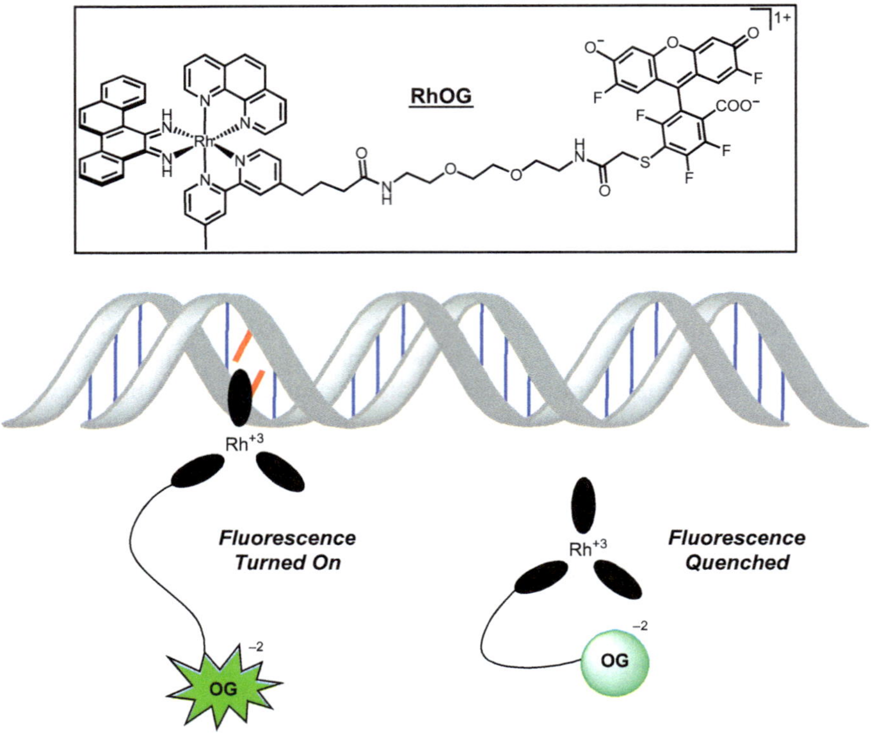


**Figure 14.9** Illustration of the design behind the rhodium metalloinsertor-Oregon green conjugate (RhOG).

because it weakly fluoresces in aqueous solution, but it emits quite brightly upon intercalation into DNA. The concept behind this conjugate design is that in the presence of well-matched DNA, the rhodium metalloinsertor will have a low affinity for the duplex, and thus only a weak emission from the conjugate will be observed in solution. Conversely, the metalloinsertor would readily bind to a mismatch site, thus permitting the cyanine dye to intercalate between the base pairs of the duplex, yielding a significant increase in fluorescence. Indeed, upon the addition of well-matched DNA, no change in the conjugate's cyanine fluorescence is seen; however, upon the addition of the duplex containing the C–C mismatch, a substantial increase in fluorescence intensity is observed. The complex serves as a sensitive luminescent probe for mismatches in genomic DNA.

While conjugated organic fluorophores certainly play an important role in the context of DNA detection, the use of a singular transition metal center that is inherently luminescent is still highly attractive owing to ease of synthesis and the ability to study longer-lived emission lifetimes *via* time-resolved spectroscopy. Thus far, the discussion regarding the development of a mismatch-specific luminescent transition metal complex has relied on making modifications to the steric bulk of the intercalating ligand; clearly, this approach has seen minimal success with respect to achieving a mismatch-specific luminescent response. Thus, it is fair to ask the following question: what if no changes are made to the intercalating ligand, and attention is instead paid to the ancillary ligands of the transition metal scaffold? Perhaps by appending functional groups to the ancillary ligands, one could induce intercalative discrimination at well-matched sites due to steric clashing with the sugar–phosphate backbone, yet encourage insertion at a destabilized mismatch site. Recently, this approach has proved very effective for luminescent platinum and ruthenium-based complexes.[70,71] Fung *et al.* synthesized a series of $[Pt^{II}(C'N'N)(NHC)]^{+}$ complexes (HC′N′N = 6-phenyl-2,2′-bipyridine; NHC = *N*-heterocyclic carbine), the planar C′N′N ligand serving as the intercalating moiety. These compounds weakly luminesce in aqueous solution, but emit much more brightly upon binding to DNA. A range of functional groups, different alkyl chains, aromatic rings and alcohols, were appended to the NHC ligand, which is out-of-plane to the intercalating ligand. Indeed, as the steric bulk of the added functionalities increases, the binding to well-matched DNA decreases, as evidenced by lower emission intensities. While the various derivatives displayed varied luminescence responses to mismatched DNA, one derivative exhibited a 15-fold luminescence increase in the presence of a C–C mismatch relative to well-matched DNA for one particular sequence.[70]

Utilizing an analogous approach, a variant of the ruthenium light switch complex $[Ru(phen)_2(dppz)]^{2+}$ has been generated that also exhibits very high selectivity in binding mismatched DNA (Figure 14.10).[71] This derivative, $[Ru(Me_4phen)_2(dppz)]^{2+}$, demonstrates that by simply introducing methyl groups into the phen ancillary ligands, one can achieve significant luminescence discrimination between mismatched and well-matched DNA. Titrations with well-matched and mismatched DNA duplexes revealed that the complex possesses nearly a 30-fold higher binding affinity for a C–C mismatch compared with well-matched base pairs. Furthermore, excited state lifetime measurements showed that when bound to the mismatch, the complex possesses a lifetime of ~160 ns, whereas at a well-matched site, the lifetime is much shorter, nearly 30 ns. These lifetime data indicate that when bound to the mismatch, the complex sits more deeply within the duplex and is less accessible to quenching by the aqueous solvent. This observation supports the notion of metalloinsertion occurring at the mismatch. Additionally, the luminescence intensity of the complex in the presence of different types of DNA mismatches correlates with the thermodynamic destabilization associated with the mismatches, an observation consistent with metalloinsertion. Minor-groove-specific quenching by $Cu(phen)_2{}^{2+}$ also implicates binding of $[Ru(Me_4phen)_2(dppz)]^{2+}$ *via* the minor groove at the mismatch. Δ-$[Ru(Me_4phen)_2(dppz)]^{2+}$ was modeled into a crystal structure of the Rh complex bound to a DNA mismatch (Figure 14.10) to show that metalloinsertion from the minor groove at a mismatch can accommodate the bulky $Me_4$phen ligands. At well-matched sites, due to the observed short lifetime, the complex is likely to be predominantly bound in a ‘side-on’

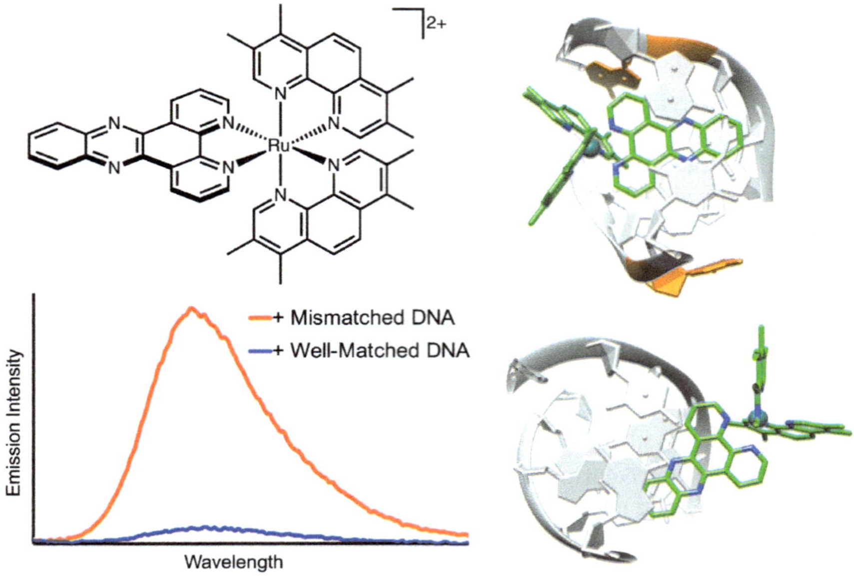


**Figure 14.10** (Left) By introducing methyl groups into the phen ancillary ligands, $[Ru(Me_4phen)_2(dppz)]^{2+}$ exhibits high luminescence discrimination between well-matched and mismatched DNA. (Right) Views down the helix axis of $\Delta$-$[Ru(Me_4phen)_2dppz]^{2+}$ modeled into the crystal structures of DNA duplexes. At the mismatch site (top, mismatched bases highlighted in orange) the complex is capable of undergoing metalloinsertion *via* the minor groove. At a well-matched site (bottom), the compound probably intercalates side-on from the major groove.

orientation, leaving one of the dppz phenazine nitrogen atoms exposed to hydrogen bond interactions with the bulk solvent. Certainly both this work and the work of Fung *et al.* demonstrate that ancillary ligand modification offers a promising approach in the design of mismatch-specific transition metal complexes.

## 14.5 Conclusions and Future Directions

As described, DNA has been a fruitful target for non-covalent binding by transition metal complexes based upon their shapes and symmetries. DNA base-pair mismatches, furthermore, provide a particularly interesting target due to their strong association with several types of cancer. Our group has focused on this molecular target in designing non-covalent DNA probes that selectively bind to the thermodynamically destabilized mismatched sites. Rhodium metalloinsertors, such as $[Rh(phen)(chrysi)(PPO)]^{2+}$, bear an expansive chrysi ligand that facilitates binding to DNA mismatched sites with remarkably high selectivity. Moreover, this *in vitro* selectivity translates into a biological selectivity, where the rhodium complexes show a high level of selective toxicity in mismatch-repair-deficient *versus* -proficient cell lines.

While much still needs to be determined regarding the mechanism behind this high potency and what proteins and pathways are being selectively activated, it is clear from the biological activity that these metalloinsertors provide a promising strategy for new therapeutic design. In parallel, ruthenium 'light switch' complexes, such as $[Ru(Me_4phen)_2(dppz)]^{2+}$, have been successfully utilized as metalloinsertors that function as 'light switches' for DNA mismatches. These luminescent probes, in binding mismatches through metalloinsertion, produce a selective, low-background, signal-on method to detect mismatches *in vitro*. The goal now is to develop these ruthenium complexes as sensitive tools for early detection of cancers deficient in mismatch repair. Overall, these mismatch-targeting rhodium and ruthenium complexes represent a new strategy for therapeutic and diagnostic design, one based upon the application of octahedral metal complexes targeted to thermodynamically destabilized DNA mismatches through non-covalent insertion. Through these designs, a new generation of potent metal diagnostics and therapeutics may emerge.

## Acknowledgements

We are grateful to our many coworkers and collaborators for their efforts in carrying out the research described. We thank also the NIH, Moore Foundation, and Amgen for their financial support of this research.

## References

1. N. J. Turro, J. K. Barton and D. A. Tomalia, *Acc. Chem. Res.*, 1991, **24**, 332–340.
2. A. M. Pyle, J. P. Rehmann, R. Meshoyrer, C. V. Kumar, N. J. Turro and J. K. Barton, *J. Am. Chem. Soc.*, 1989, **111**, 3051–3058.
3. J. K. Barton, J. M. Goldberg, C. V. Kumar and N. J. Turro, *J. Am. Chem. Soc.*, 1986, **108**, 2081–2088.
4. C. V. Kumar, J. K. Barton and N. J. Turro, *J. Am. Chem. Soc.*, 1985, **107**, 5518–5523.
5. T. B. Thederahn, M. D. Kuwabara, T. A. Larsen and D. S. Sigman, *J. Am. Chem. Soc.*, 1989, **111**, 4941–4946.
6. D. S. Sigman, A. Mazumder and D. M. Perrin, *Curr. Opin. Chem. Biol.*, 1993, **93**, 2295–2316.
7. D. S. Sigman and C. B. Chen, *Annu. Rev. Biochem.*, 1990, **59**, 207–236.
8. J. K. Barton, *Science*, 1986, **233**, 727–735.
9. K. E. Erkkila, D. T. Odom and J. K. Barton, *Chem. Rev.*, 1999, **99**, 2777–2796.
10. K. W. Jennette, S. J. Lippard, G. A. Vassiliades and W. R. Bauer, *Proc. Natl. Acad. Sci.*, 1974, **71**, 3839–3843.
11. P. J. Bond, R. Langridge, K. W. Jennette and S. J. Lippard, *J. Proc. Natl. Acad. Sci.*, 1975, **72**, 4825–4829.
12. A. E. Friedman, J.-C. Chambron, J.-P. Sauvage, N. J. Turro and J. K. Barton, *J. Am. Chem. Soc.*, 1990, **112**, 4960–4962.

13. R. E. Holmlin, E. D. A. Stemp and J. K. Barton, *Inorg. Chem.*, 1998, **37**, 29–34.
14. A. Sitlani, E. C. Long, A. M. Pyle and J. K. Barton, *J. Am. Chem. Soc.*, 1992, **114**, 2303–2312.
15. A. H. Krotz, B. P. Hudson and J. K. Barton, *J. Am. Chem. Soc.*, 1993, **115**, 12577–12578.
16. C. L. Kielkopf, K. E. Erkkila, B. P. Hudson, J. K. Barton and D. C. Rees, *Nat. Struct. Biol.*, 2000, **7**, 117–121.
17. M. C. Shoshan and S. Linder, *Expert Opin. Drug Metab. Toxicol.*, 2008, **4**, 273–280.
18. E. C. Friedberg, L. D. McDaniel and R. A. Schultz, *Curr. Opin. Genet. Dev.*, 2004, **14**, 5–10.
19. A. L. Jackson and L. A. Loeb, *Mutat. Res.*, 2001, **477**, 7–21.
20. A. Sancar, L. A. Lindsey-Boltz, K. Unsal-Kacmaz and S. Linn, *Annu. Rev. Biochem.*, 2004, **73**, 39–85.
21. S. Jackson and J. Bartek, *Nature*, 2009, **461**, 1071–1078.
22. T. Helleday, E. Petermann, C. Lundin, B. Hodgson and R. A. Sharma, *Nat. Rev. Cancer*, 2008, **8**, 193–204.
23. W. E. Glaab and K. R. Tindall, *Carcinogenesis*, 1997, **18**, 1–8.
24. R. Parsons, G. M. Li, M. J. Longley, W. H. Fang, N. Papadopoulos, J. Jen, A. Delachapelle, K. W. Kinzler, B. Vogelstein and P. Modrich, *Cell*, 1993, **75**, 1227–1236.
25. E. Barrow, J. Hill and D. Gareth Evans, *Fam. Cancer*, 2013, **12**, 229–240.
26. I. I. Arzimanoglou, F. Gilbert and H. R. Barber, *Cancer*, 1998, **82**, 1808–1820.
27. D. Fink, S. Aebi and S. B. Howell, *Clin. Cancer Res.*, 1998, **4**, 1–6.
28. L. P. Martin, T. C. Hamilton and R. J. Schilder, *Clin. Cancer Res.*, 2008, **14**, 1291–1295.
29. B. A. Jackson and J. K. Barton, *J. Am. Chem. Soc.*, 1997, **119**, 12986–12987.
30. K. Uchida, A. M. Pyle, T. Morii and J. K. Barton, *Nucleic Acids Res.*, 1989, **17**, 10259–10279.
31. A. M. Pyle, M. Y. Chiang and J. K. Barton, *Inorg. Chem.*, 1990, **29**, 4487–4495.
32. B. A. Jackson and J. K. Barton, *Biochemistry*, 2000, **39**, 6176–6182.
33. B. A. Jackson, V. Y. Alekseyev and J. K. Barton, *Biochemistry*, 1999, **38**, 4655–4662.
34. B. M. Zeglis, J. A. Boland and J. K. Barton, *J. Am. Chem. Soc.*, 2008, **130**, 7530–7531.
35. V. C. Pierre, J. T. Kaiser and J. K. Barton, *Proc. Natl. Acad. Sci.*, 2007, **104**, 429–434.
36. L. S. S. Lerman, *J. Mol. Biol.*, 1961, **3**, 18–30.
37. B. M. Zeglis, V. C. Pierre, J. T. Kaiser and J. K. Barton, *Biochemistry*, 2009, **48**, 4247–4253.
38. C. Cordier, V. C. Pierre and J. K. Barton, *J. Am. Chem. Soc.*, 2007, **129**, 12287–12295.

39. A. David, N. Bleimling, C. Beuck, J. M. Lehn, E. Weinhold and M. P. Teulade-Fichou, *ChemBioChem*, 2003, **4**, 1326–1331.
40. A. Granzhan and M. P. Teulade-Fichou, *Chem. – Eur. J.*, 2009, **15**, 1314–1318.
41. T. Takada, A. Ashida, M. Nakamura and K. Yamana, *Bioorg. Med. Chem.*, 2013, **21**, 6011–6014.
42. K. Nakatani, S. Sando and I. Saito, *Nat. Biotechnol.*, 2001, **19**, 51–55.
43. K. Nakatani, S. Sando, H. Kumasawa, J. Kikuchi and I. Saito, *J. Am. Chem. Soc.*, 2001, **123**, 12650–12657.
44. H. Chen, C. Dou, Y. Wu, H. Li, X. Xi and P. Yang, *J. Inorg. Biochem.*, 2009, **103**, 827–832.
45. T. A. Kunkel and D. A. Erie, *Annu. Rev. Genet.*, 2015, **49**, 291–313.
46. J. R. Hart, O. Glebov, R. J. Ernst, I. R. Kirsch and J. K. Barton, *Proc. Natl. Acad. Sci. U. S. A.*, 2006, **103**, 15359–15363.
47. R. J. Ernst, H. Song and J. K. Barton, *J. Am. Chem. Soc.*, 2009, **131**, 2359–2366.
48. A. C. Komor, C. J. Schneider, A. G. Weidmann and J. K. Barton, *J. Am. Chem. Soc.*, 2012, **134**, 19223–19233.
49. A. C. Komor and J. K. Barton, *J. Am. Chem. Soc.*, 2014, **136**, 14160–14172.
50. M. M. de las Alas, S. Aebi, D. Fink, S. B. Howell and G. Los, *J. Natl. Cancer Inst.*, 1997, **89**, 1537–1541.
51. A. Fedier, V. A. Schwarz, H. Walt, R. D. Carpini, U. Haller and D. Fink, *Int. J. Cancer*, 2001, **93**, 571–576.
52. R. J. Ernst, A. C. Komor and J. K. Barton, *Biochemistry*, 2011, **50**, 10919–10928.
53. A. G. Weidmann and J. K. Barton, *Inorg. Chem.*, 2014, **53**, 7812–7814.
54. A. G. Weidmann and J. K. Barton, *Inorg. Chem.*, 2015, **54**, 9626–9636.
55. A. G. Weidmann, A. C. Komor and J. K. Barton, *Philis. Trans. R. Soc.*, 2013, **371**, 20120117.
56. M. P. Murphy, *Biochim. Biophys. Acta*, 2008, **1777**, 1028–1031.
57. Y. Jenkins, A. E. Friedman, N. J. Turro and J. K. Barton, *Biochemistry*, 1992, **31**, 10809–10816.
58. R. M. Hartshorn and J. K. Barton, *J. Am. Chem. Soc.*, 1992, **114**, 5919–5925.
59. E. J. C. Olson, D. Hu, A. Ho, A. M. Jonkman, M. R. Arkin, E. D. A. Stemp, J. K. Barton and P. F. Barbara, *J. Am. Chem. Soc.*, 1997, **119**, 11458–11467.
60. M. K. Brennaman, J. H. Alstrum-Acevedo, C. N. Fleming, P. Jang, T. J. Meyer and J. M. Papanikolas, *J. Am. Chem. Soc.*, 2002, **124**, 15094–15098.
61. M. H. Lim, H. Song, E. D. Olmon, E. E. Dervan and J. K. Barton, *Inorg. Chem.*, 2009, **48**, 5392–5397.
62. H. Song, J. T. Kaiser and J. K. Barton, *Nat. Chem.*, 2012, **4**, 615–620.
63. C. M. Dupureur and J. K. Barton, *J. Am. Chem. Soc.*, 1994, **116**, 10286–10287.
64. C. M. Dupureur and J. K. Barton, *Inorg. Chem.*, 1997, **36**, 33–43.

65. A. J. McConnell, M. H. Lim, E. D. Olmon, H. Song, E. E. Dervan and J. K. Barton, *Inorg. Chem.*, 2012, **51**, 12511–12520.
66. E. Ruba, J. R. Hart and J. K. Barton, *Inorg. Chem.*, 2004, **43**, 4570–4578.
67. B. M. Zeglis and J. K. Barton, *Inorg. Chem.*, 2008, **47**, 6452–6457.
68. B. M. Zeglis and J. K. Barton, *J. Am. Chem. Soc.*, 2006, **128**, 5654–5655.
69. A. Nano, A. N. Boynton and J. K. Barton, *J. Am. Soc.*, 2017, **139**, 17301–17304.
70. S. K. Fung, T. Zou, B. Cao, T. Chen, W.-P. To, C. Yang, C.-N. Lok and C.-M. Che, *Nat. Commun.*, 2016, **7**, 10655.
71. A. N. Boynton, L. Marcélis and J. K. Barton, *J. Am. Chem. Soc.*, 2016, **138**, 5020–5023.

CHAPTER 15

# *CRISPR Highlights and Transition of Cas9 into a Genome Editing Tool*

THOMAS BENTIN

Department for Cellular and Molecular Medicine, University of Copenhagen, Blegdamsvej 3c, 2200 Copenhagen N, Denmark
Email: bentin@sund.ku.dk

## 15.1 Introduction

In a book devoted to DNA recognition, it is impossible not to mention clustered, regularly interspaced, short palindromic repeats (CRISPR)–CRISPR associated genes (*cas*). CRISPR-*cas* systems have enabled routine human cell genome editing in research laboratories, rapidly advancing our understanding of the human genome and raising hopes for the development of tools to address outstanding clinical challenges. The history of the discovery of CRISPR-*cas*,[1–3] its transition into a genome editing tool[4] and its possible transfer into clinical use[5] have been recently reviewed. Here, are highlighted key experiments that underpinned the discovery and transition of CRISPR-*cas* into a human genome editing tool.

CRISPR-*cas* systems provide immunity in archaea and bacteria towards bacteriophages (phages, viruses that infect bacteria) and invasive plasmids. Immunity occurs in three steps: (i) adaptation, during which foreign DNA segments are incorporated into the CRISPR locus providing genetic memory, (ii) expression, where short CRISPR RNAs (crRNAs) are generated

Chemical Biology No. 7
DNA-targeting Molecules as Therapeutic Agents
Edited by Michael J. Waring

Published by the Royal Society of Chemistry, www.rsc.org

and (iii) interference, where the invading genome is targeted *via* crRNA-directed cleavage.

The majority of sequenced archaea genomes and half of the bacteria genomes include CRISPR–*cas* systems. Current classification divides CRISPR–*cas* systems into classes, types and subtypes according to *cas* gene composition and architecture.[6] Class 1 CRISPR–*cas* systems, encompassing types I, III and IV, involve multiprotein crRNA–effector complexes. Class 2 CRISPR–*cas* systems, including types II and V, involve single-protein crRNA–effector modules.

## 15.2 The Discovery of CRISPR

The first credited observation of what came to be known as CRISPR loci was made during studies of the *iap* gene in the Gram-negative bacterium *Escherichia coli*. The authors observed repeated sequences spaced apart 3′ to the *iap* gene, the biological function of which was unknown.[7] Subsequent studies revealed CRISPR loci in other microbial organisms including Gram-positive bacteria[8] and archaea,[9,10] leading to the recognition that CRISPRs constitute a widespread family of repeats[11] containing a repeat-spacer structure where the spacer sequence varies. Further *in silico* analyses revealed CRISPR-associated (*cas*) genes in CRISPR-containing prokaryotes, which were absent from CRISPR-negative prokaryotes[12] (Figure 15.1).

## 15.3 CRISPRs Contain Foreign DNA Elements, Suggesting a Role in Immunity

The hypothesis that CRISPRs could form microbial adaptive immune systems came in 2005 from several independent sources using *in silico* analyses.[13–15] In one work, Mojica and coworkers reasoned that the widespread nature of CRISPRs within prokaryotes and their structural conservation predicted biological function and that identifying the source of CRISPR spacers could provide clues to determine this function.[14] By comparing thousands of spacers with known prokaryotic genomes fewer than one hundred sequences were found to have a known homologue. The majority of these homologues were of foreign origin, including sequences from phages and conjugative plasmids (plasmids capable of lateral transfer among bacteria through tubes called pili), both of which are transmissible genetic elements. Importantly, CRISPR spacer-carrier strains were protected against infection with the corresponding pathogen, strongly suggesting a role for CRISPR in prokaryotic immunity.[14] In a separate and almost simultaneous work, Pourcel and co-workers focusing on species of the genus *Yersinia* (including the plague bacterium *Yersinia pestis*) arrived at similar conclusions after finding strong similarity among *Yersinia* spacers and phage-derived sequences.[15] Bolotin and co-workers, focusing on *Streptococcus* species similarly found spacers with homology to phages

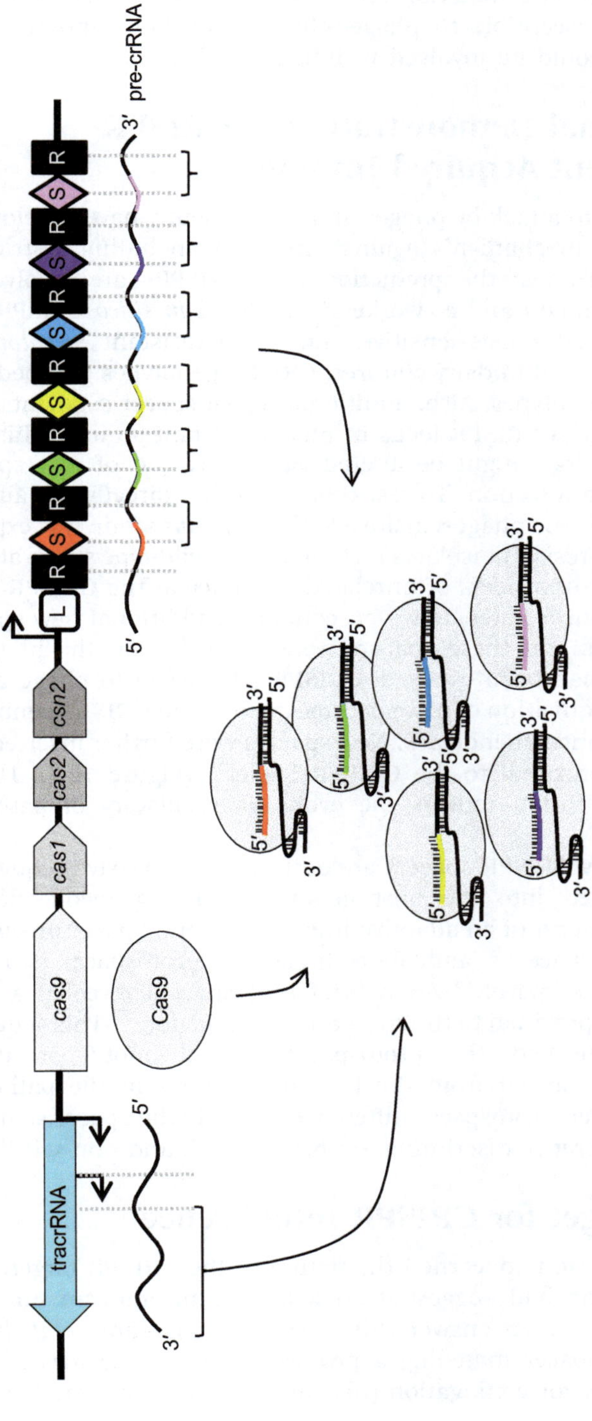


**Figure 15.1** Schematic showing a type II CRISPR–*cas* locus from *Streptococcus pyogenes.* Top: genomic structure of the tracrRNA gene (left), *cas* genes (centre), and CRISPR array (right). Below: expression of tracrRNA, SpCas9 (grey ovals) and pre-crRNA enabling SpCas9–crRNA–tracrRNA complex formation. RNase III (not shown)-mediated RNA processing sites are indicated with vertical broken grey lines. The *cas9* gene product is involved in adaptation, expression and interference; the *cas1* and *cas2* gene products are involved in adaptation, the *csn2* gene product is an ancillary protein (ref. 6). Mature crRNA (39–42 nt); tracrRNA (~75 nt). R = repeat, S = spacer, L = leader, from where CRISPR transcription initiates.

and other foreign genetic material, and observed that strains with more spacers were less susceptible to phage-infection, further supporting the idea that CRISPRs could be involved in immunity.[13]

## 15.4 Functional Demonstration of CRISPR-dependent Acquired Immunity

Bacteria are subject to attack by phages and consequently have developed a number of defence mechanisms against infection, including restriction/modification loci. To test the prediction that CRISPRs are involved in immunity,[13–15] Barrangou and co-workers[16] analysed *in silico* CRISPR spacers of closely related phage-sensitive and phage-resistant *Streptococcus thermophilus* strains used in dairy cultures. CRISPR genotypes grouped with phage-resistance phenotypes. Also, additional repeat-spacer segments were occasionally found in a CRISPR locus of phage-resistant strains. This suggested that CRISPR loci might be altered *via* acquisition of new spacers during adaptation to infection. To test this prediction directly the authors used genetics. A wild-type phage-sensitive *S. thermophilus* strain was exposed to phage, and phage-resistant isolates recovered. The wild-type strain already contained a number of spacers of unrelated sequence at the CRISPR locus investigated. Resistant isolates, however, contained additional new spacers at the CRISPR locus and these spacers were derived from the phage to which the bacteria had been exposed. CRISPR adaptation to phage attack therefore involves acquisition of new spacer sequence and CRISPR genotypes predict phage-resistance phenotypes. New spacers were further inserted in a polarised manner proximal to the CRISPR leader[16] (Figure 15.1). Hence, CRISPR-array organization reflects the evolutionary history of pathogen attack.[16]

In summary, many CRISPR spacers are derived from invasive sequences, which were inscribed into the host genome, enabling recollection of pathogen attack as a part of an adaptive immune system. Spacer precursors are known as protospacers[17] and, more broadly, a protospacer is a DNA target that matches a spacer.[18] Analysis of protospacers revealed a short conserved sequence proximal to the spacer-identical region.[13] This sequence element is now termed the protospacer-adjacent motif or PAM[17] (Figure 15.2) and is absent from the host but present in the pathogen; one function of spacer–protospacer differences outside the spacer sequence is to facilitate interference discrimination between self and non-self.[19]

## 15.5 The Target for CRISPR Interference

An outstanding question concerned the nature of the CRISPR target, DNA or RNA. Early reports had suggested an RNAi-like mechanism but little direct evidence existed. To answer this question Marraffini *et al.* (2008) selected a CRISPR spacer matching a protospacer sequence located in a gene that is required for conjugation (plasmid transfer from one bacterial

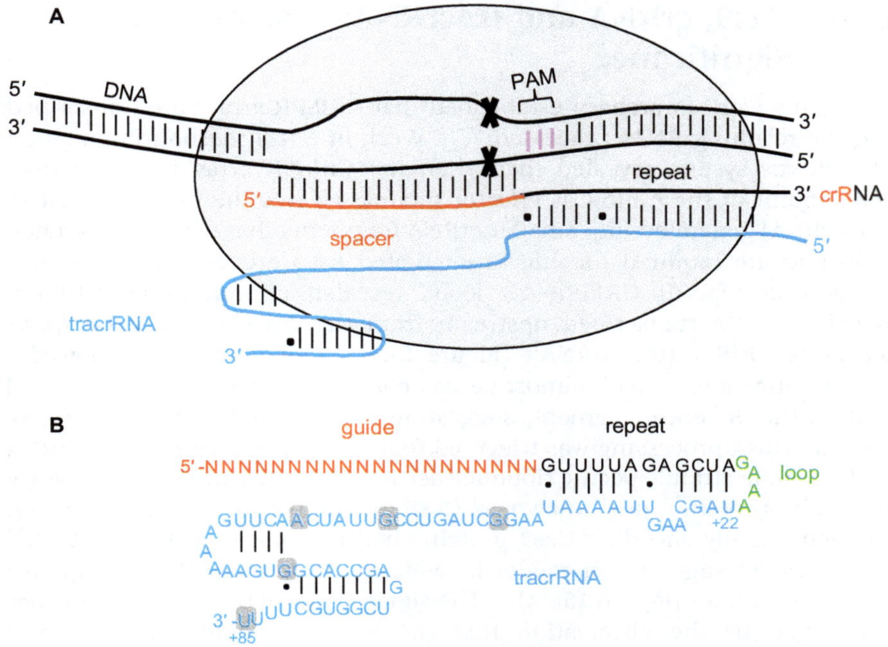


**Figure 15.2** Schematic showing dual-RNA-guided SpCas9 DNA targeting (A) and sgRNA design (B). The X's indicate positions of hydrolytic DNA cleavage. The SpCas9 protospacer is 20 bp; the PAM sequence, indicated in magenta, is NGG for SpCas9. Grey oval: SpCas9. Black dots and vertical lines indicate G–U and Watson–Crick base pairs, respectively. Spacer and repeat signifies the origin of these segments in the CRISPR locus (see Figure 15.1). In B, nucleotide positions indicated with a grey background indicate tracrRNA lengths, which were tested in early sgRNA designs (ref. 28, 30, 32, 41).

strain to another) and which is present in nearly all conjugative plasmids of Staphylococci. To distinguish whether the spacer target was DNA or mRNA, a self-splicing intron was introduced into the protospacer thereby destroying the DNA target while preserving the mRNA target, which correctly formed following transcription and splicing.[20] The plasmid was then conjugated into a recipient strain harbouring a CRISPR spacer homologous to the protospacer within the gene in question. The results were clear. No DNA transfer was observed when using the plasmid containing an intact protospacer, reflecting interference. In contrast, DNA transfer increased dramatically when using the plasmid containing the self-splicing intron, indicating abrogation of interference. Hence, CRISPR interference required an intact DNA target, and altering the DNA target overcame CRISPR defences. Besides DNA interference, CRISPR–*cas* systems that additionally[21] or exclusively[22] show RNA interference have been described but these are not further addressed here.

## 15.6 Cas9, crRNA and tracrRNA: Discovery and Significance

Pioneering work in archaea established that CRISPR arrays are transcribed and the resulting RNAs processed.[23–25] Work in *E. coli* harbouring a type I CRISPR–*cas* system revealed that when the CRISPR array is transcribed, one subunit of the multi-Cas effector module cleaves the precursor crRNA (pre-crRNA) generated into smaller crRNA fragments that remain associated with- and are required for effector-mediated interference.[26] Analysis of a *S. pyogenes* type II CRISPR–*cas* locus revealed an additional CRISPR-associated RNA transcribed upstream from the *cas* genes, coined *trans*-activating CRISPR (tracr) RNA[18] (Figure 15.1). Notably, tracrRNA showed a 25-nucleotide stretch with almost perfect complementarity to crRNA located to the CRISPR repeat segment, suggesting hybridization between the two. Correct crRNA processing was triggered following hybridization to tracrRNA by the double-strand-specific ribonuclease III (RNase III) and in the presence of a large protein[18] now known as Cas9 or SpCas9 for *S. pyogenes* Cas9. The gene family encoding Cas9 proteins had previously been identified.[13] These results suggested a model in which crRNA, tracrRNA and SpCas9 formed a complex (Figure 15.2A).[18] The significance of tracrRNA was further underlined by the observation that the gene encoding tracrRNA (and genes encoding crRNA, SpCas9 and RNase III) were required to confer phage immunity and because tracrRNA is conserved among different bacteria.[18]

## 15.7 Biochemistry of type II CRISPR–Cas-mediated DNA Cleavage

The *in vitro* requirements for DNA cleavage by CRISPR–Cas type II systems were resolved in two almost simultaneous landmark papers in the summer of 2012.[27,28] These studies formed the basis for the subsequent genome editing revolution. In these reports, SpCas9[28] and *Streptococcus thermophilus* Cas9 (StCas9),[27] were shown to function as RNA-guided sequence-directed endonucleases, hence defining biochemically the minimal and sufficient requirements for type II CRISPR interference. In the work by Jinek *et al.*,[28] recombinant SpCas9 purified from *E. coli* was complexed with crRNA and tracrRNA. The team showed that SpCas9–crRNA–tracrRNA complex cleaves plasmid DNA and oligodeoxyribonucleotide duplex harbouring a complementary protospacer, and that SpCas9 uses separate nuclease domains to cleave each of the two DNA strands three nucleotides upstream of the PAM in a magnesium-dependent reaction (Figure 15.2). Using deletions, the team further defined the minimal crRNA and tracrRNA structures required for DNA binding and cleavage. Using DNA targets containing single point mutations, it was shown that 5′ proximal crRNA mismatches are tolerated whereas 3′ proximal mismatches abrogate SpCas9-mediated DNA cleavage.

Using targets with contiguous mismatches, it was shown that at least 13 base-pairs of complementarity between crRNA and protospacer target adjacent to the PAM are required to elicit SpCas9-mediated cleavage. The work further established that both the tracrRNA and the PAM are essential for recruitment of SpCas9. Importantly, genetic fusion of crRNA with tracrRNA produced single guide RNAs (sgRNAs) that endowed the resulting SpCas9–sgRNA complexes with DNA target-specific cleavage activity (Figure 15.2). By changing the sgRNA 5′ proximal 20 nucleotides (guide) complementary to the protospacer target, the resulting SpCas9–sgRNA complexes could be programmed to cleave pre-determined targets.[28] Structural studies using SpCas9–sgRNA complexes[29] subsequently confirmed the predicted RNA hydrogen bonding pattern.[28]

Rather than using purified Cas9 subsequently activated with crRNA–tracrRNA,[28] Gasiunas *et al.* purified a StCas9–crRNA–tracrRNA complex following recombinant expression of an engineered type II *S. thermophilus* CRISPR–*cas* locus in *E. coli*.[27] Affinity purification of StCas9 co-purified crRNA with a size of 42 nucleotides. The purified complex cleaved DNA protospacer target strands three nucleotides upstream of the PAM sequence and each strand was cut by separate nuclease domains. Also, both protospacer and PAM were required for StCas9 recruitment to the DNA protospacer target. Importantly, by changing the spacer of the crRNA, DNA target specificity could be reprogrammed to multiple targets as exemplified *in vitro* by double plasmid digestion using StCas9 activated with different crRNAs.

The authors of these works suggested that "RNA-programmed Cas9 could offer considerable potential for gene targeting and genome editing applications"[28] and that "these findings pave the way for development of unique molecular tools for cellular DNA surgery".[27] Many researchers were similarly convinced and courageously embarked into uncharted territory although a number of unknowns remained, including whether Cas9, an enormous protein, could be functionally expressed in human cells.

## 15.8 First Human Cell Genome Editing Using CRISPR–Cas9

The ease with which Cas9 could be programmed to target formation of double-strand DNA breaks (DSB) immediately encouraged researchers to try out these new molecular scissors in human cells[30–32] including human pluripotent stem cells[33] as well as in other systems including bacteria,[34] yeast,[35] zebrafish,[36,37] *Drosophila*[38] and mice.[39] Cells are naturally hardwired to repair DSBs rapidly and this feature of cell physiology is exploited in genome engineering. DSBs are repaired either by non-homologous end joining (NHEJ), an efficient but error-prone process useful for gene knockout, or *via* homology-directed repair (HDR), an inefficient but accurate repair process that requires the presence of a repair template, which under physiological conditions in somatic cells is provided by the sister chromatid.

In genome editing, the repair template or donor is commonly provided as a synthetic DNA fragment containing homology arms to the target.

To investigate the possibility of using CRISPR–*cas* for human cell genome editing, Cong *et al.* (2013) employed a *S. pyogenes* type II CRISPR–*cas* system.[30] The authors designed a plasmid expressing human-codon-optimized SpCas9 fused to green fluorescent protein (GFP) enabling its visualisation and nuclear localisation signals (NLS) to direct the recombinant protein to the nucleus. The construct was expressed in human cells along with tracrRNA and a pre-crRNA containing a single spacer flanked by repeats. When targeting a protospacer in the *EMEX1* gene, site-specific indel formation (insertions or deletions) ranging from a few percent up to 27% was observed. Such indels are due to repair *via* NHEJ. Interestingly, the engineered pre-crRNA was processed to active crRNA by the host without a requirement for microbial RNase III. As predicted by the *in vitro* results of Jinek *et al.*[28] (*vide supra*), crRNAs containing single mismatches targeting the 5′ region of the protospacer target had little effect on indel formation whereas crRNAs harbouring single mismatches targeted towards the 3′ region of the protospacer abolished mutagenesis. Expression of a CRISPR array containing two spacers targeted to protospacers in different genes promoted indel formation at the designated targets. Targeting the two spacers to different protospacers located 118 bp apart within the same gene promoted deletion of the intervening sequence. A 'nickase' variant of SpCas9 containing one functional nuclease domain and one nuclease domain inactivated by mutation, reduced NHEJ to below the limit of detection while preserving low levels of HDR. Finally, after co-transfection of a repair donor homologous to the *EMEX1* gene harbouring unique restriction sites, restriction fragment polymorphism was observed in 0.5–0.7% of cells, indicating repair *via* HDR.

Simultaneously, another research team investigating SpCas9 confirmed and expanded these results thereby underpinning the generality of the utility of the systems for human cell genome editing. Mali *et al.* (2013) engineered human-codon-optimized SpCas9 containing a single C-terminal NLS and co-expressed this protein with sgRNA targeting the *AAVS1* locus in several human cell types.[32] Analysis of genomic DNA from these cells by deep sequencing revealed indel formation *via* NHEJ in 2–37% of targets. Targeting the SpCas9–sgRNA complex in combination with a repair donor to a protospacer in a chromosomal *gfp* reporter gene inactivated by mutation, restored GFP fluorescence in 7% of cells. Simultaneous use of two sgRNAs targeted 19 bp apart in the same gene led to deletion of the intervening sequence. Using nickase SpCas9 increased the HDR to NHEJ repair ratio. Finally, the SpCas9–sgRNA complex, when combined with a kilobase pair repair donor, promoted integration of the donor into a native target locus.

Together these results established that SpCas9 can be harnessed for programmable genome editing using simple three-component (SpCas9–crRNA–tracrRNA) or two-component (SpCas9–sgRNA) systems. It also revealed that multiplexing (simultaneous targeting of multiple loci) is possible and suggested that means exist to modulate the HDR *vs.* NHEJ repair ratio. No prior technology for genome editing had ever been that accessible.

Of course it was obvious to attempt to use Cas9 for genome editing. Success, however, was in no way certain. For example, Cas9 could have simply misfolded or for other reasons failed to reach the nucleus. Even if reaching the nuclear compartment, Cas9 could have been unable to access chromatin or, alternatively, cleaved genomic DNA indiscriminately rather than seeking out unique targets guided by sgRNA. Cas9 could have been incompatible with human cell physiology for a variety of other reasons. Fortunately, none of these reservations proved insurmountable.

## 15.9 DNA Target Specificity of Cas9

While target editing efficacy is of importance, high target specificity is likewise a critical aspect of genome editing applications. Following the initial reports on SpCas9-mediated genome editing,[30,32] more comprehensive off-target editing analyses were conducted.[40–42] Early studies had revealed that sgRNA–target mismatches were tolerated in the 5′ guide region both *in vitro*[28] and *in vivo*[30] consistent with a model in which target specificity is governed by a PAM proximal seed sequence.[43] Systematic mutagenesis of the guide–target interface involving single or multiple nucleotide mismatches confirmed this general model, highlighting the importance of the PAM in licensing protospacer cleavage,[40,41] but also revealing further complexities in target discrimination.[40–42] For example, mismatched guide–target hybrids can affect *in vivo* specificity depending on position (as indicated by the seed model) but also depending on target sequence and mismatch identity.[40,41] Furthermore, off-target cellular editing *in vivo* was observed with some sgRNAs carrying multiple mismatches,[40–42] in some cases surpassing the editing frequency at the fully matched target.[40] Editing was even observed in some cases with multiple PAM distal mismatches[40,41] and with a non-standard PAM sequence (NAG rather than NGG).[41]

Early human-cell genome engineering with SpCas9 was performed using sgRNA designs with different 3′ tracrRNA tail lengths yielding either efficient genome editing when using a long 3′ tail[32] or inefficient genome editing when using a short 3′ tail.[30] Comparative testing of the importance of sgRNA tail length revealed both amplified sgRNA transcript abundance as well as elevated genome editing with increased 3′ tail length[41] (Figure 15.2). However, while longer sgRNA designs showed improved genome editing, shorter ones displayed increased specificity, revealing a trade-off between activity and specificity.[42] Furthermore, SpCas9–sgRNA titration altered the absolute activity but also the specificity of indel formation, revealing that maximal specificity may be achieved using minimal SpCas9–sgRNA amounts,[41,42] though not in all cases.[40] On a separate note, one of the studies also revealed that highly-methylated CpG chromatin at the human *SERPIN5B* locus is a substrate for SpCas9–sgRNA[41] suggesting, more generally, that heterochromatin could be in reach of Cas9-mediated editing. Additional evidence supporting the finding that Cas9 can access heterochromatin targets has accumulated but the topic is not further developed here.

The observations described above enabled establishment of guidelines for target selection including (i) a genuine NGG PAM should be present, (ii) there should be minimal similarity to other genomic sequences, (iii) off-targets should contain multiple PAM proximal mismatches and (iv) SpCas9–sgRNA concentrations employed should be kept minimal.[41] Computational sgRNA design tools were developed[41] [see review by Chira *et al.*[4]]. Nevertheless, the dominant off-target editing activities observed with wild-type SpCas9 provided a sobering reminder that much needed to be ironed out.

If target specificity was strictly governed by the PAM proximal 12, 14 or 16 bp, a given sgRNA would theoretically yield 190, 12 and $<1$ fully matched targets, respectively, in a random sequence the size of a human haploid genome. Given that PAM distal mismatches in particular seem to be tolerated with some regularity, this clearly limits the potential of SpCas9 as a specific genome editing tool. However, because SpCas9–sgRNA binding to a target does in fact involve 20 bp corresponding to a chance find of only 0.003 fully matched targets per human haploid genome, perhaps a hidden target-discrimination potential could be unearthed from this remarkable enzyme.

## 15.10 High-fidelity CRISPRs

The off-target editing proclivity of wild-type SpCas9 promoted a search for solutions to the specificity challenge. A general solution to off-target genome editing was pioneered by two groups using structure-guided protein engineering.[44,45] Both groups based their work on the same general idea previously proposed[46] that the energy of SpCas9–sgRNA binding to the cognate target surpasses that required for DNA cleavage. If correct, then reducing that energy might selectively reduce off-target editing while preserving the intended on-target editing.

When a SpCas9–sgRNA complex binds to its designated target, the DNA strands are separated and the negatively charged non-complementary DNA strand is held in a positively charged groove within SpCas9, promoting strand cleavage (Figure 15.3). If the energy of the SpCas9–sgRNA–ncDNA interaction at on-target sites surpasses that required for strand separation, then reducing the number of positively charged amino acids within the non-complementary DNA binding-groove might prevent DNA unwinding and cleavage at off-targets. Indeed, by singly mutagenizing each of 31 basic amino acids to alanine followed by combination of promising mutants, one team of researchers created enhanced (e) SpCas9 variants carrying three amino acid substitutions with preserved on-target editing and reduced off-target editing.[45]

In a separate study, the target DNA-guide binding-site of SpCas9 (Figure 15.3) was targeted for mutagenesis under a similar assumption that the energy of SpCas9–sgRNA bound to a fully matched target exceeds that required for strand cleavage, and so reducing that energy might

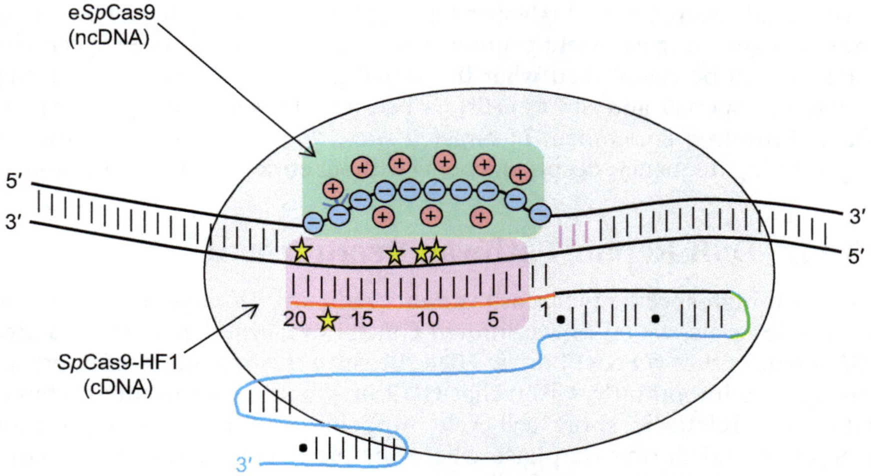


**Figure 15.3** Schematic showing principles employed to generate high-specificity SpCas9. Improved specificity SpCas9 variants were attained by disrupting electrostatic interactions between SpCas9 and the non-complementary DNA strand: eSpCas9-vs.1 (K810A, K1003A, R1060A), eSpCas9-vs.1-1 (K848A, K1003A, R1060A). Green rectangle: non-complementary DNA binding-site. Basic amino acids (+), DNA backbone negative charge (−). High fidelity SpCas9 was achieved by disrupting SpCas9 hydrogen bonds or salt bridges to the complementary DNA strand: SpCas9-HF1 (N497A, R661A, Q695A, Q926A). Purple rectangle: target DNA-guide binding-site. Position of hydrogen bonds or salt bridges to the complementary DNA strand and guide RNA (yellow stars). Other symbols are as in Figure 15.2.

preserve on-target editing while disrupting off-target binding and editing.[44] Four amino acids forming DNA sequence neutral hydrogen bonds and salt bridges were substituted with alanines. Combination of the single mutations into all possible combinations of double, triple and quadruple mutants yielded SpCas9 variants, each and every one showing high activity at the correct target. Importantly, quadruply mutated SpCas9, named SpCas9-HF1 for high fidelity variant 1, showed considerably reduced cleavage at most predicted off-target sites.[†]

Performance enhancing protein-engineering could be applied to other Cas9 or Cas9-like proteins for which ternary protein–sgRNA–DNA structures are available including *Staphyloccocus aureus* Cas9,[47] and *Acidaminococcus* sp. CpfI,[48] both of which are functional when transplanted into human cells.[49,50]

[†]Contrary to the hypothesis that engineered high-specificity Cas9 variants owe their performance enhancing characteristics to reduced DNA binding affinity,[44,45] it was recently shown that DNA cleavage specificity is governed by proofreading. Proofreading entails Cas9 internal domain movements. At off-targets, eSpCas9 and SpCas9-HF1 are trapped in a catalytically inactive conformation preventing DNA cleavage licensing.[56]

The cited examples of high-specificity SpCas9 variants show how clever protein engineering can yield genome-editing tools with improved specificity. It remains to be established what the actual genome-wide off-target editing activity of eSpCas9 and SpCas9-HF1 is because the sequencing approaches adopted involved enrichment of targeted sites.[44,45] To establish the true off-target editing frequency, deep sequencing without enrichment will be required.

## 15.11 DSB Repair Pathway Recruitment

Even with high-specificity SpCas9 variants reducing off-target editing below detectable levels, we still have limited control over whether Cas9-mediated DSB repair occurs *via* NHEJ or *via* HDR. Because the outcomes differ greatly this matters importantly. Given that HDR in somatic cells occurs preferentially in a relatively short cell-cycle time interval following replication in S-phase and during G2-phase where the sister chromatids are in close apposition, NHEJ is the dominant mechanism in gene editing. Also, as mentioned above, genome editing of somatic cells *via* HDR generally requires a synthetic repair donor. Nevertheless, NHEJ is disruptive by nature and resultant indels can lead to frameshifts in protein-coding genes, which is not compatible with correction, albeit useful for gene knockout. NHEJ can be suppressed, for example by drugs targeting DNA ligase IV, a key NHEJ enzyme.[51,52] Alternatively, recent advances in germ cell editing involving human preimplantation embryos successfully harnessed a germ-line-specific HDR mechanism promoting a high-level correction of a pathogenic human *MYPBC3* gene mutation causing myocardial disease. Following injection of SpCas9–sgRNA ribonucleoprotein complex targeting the mutation along with heterozygotic carrier-sperm into normal M-phase human oocytes, 72.4% of the resulting zygotes contained homozygote wild type alleles as compared with 47.2% in the control (50% expected). Interestingly, HDR involved the maternal wild-type homologue and no synthetic repair donor was required. HDR occurred without evidence of off-target editing or genome instability, as investigated using genome sequencing without enrichment and karyotyping.[53] This result established that, in at least some cell systems and cell cycle stages, efficient gene correction *via* HDR can be attained using SpCas9. However, since these results are germ-line-specific, they may not be transferable to somatic cells.

## 15.12 Therapeutics

Cas9-mediated genome editing continues to yield new information on human gene function at an unprecedented rate. A multitude of pre-clinical genome editing studies aimed at correcting or knocking out disease-associated genes have been conducted. Several clinical trials involving Cas9 are either ongoing or scheduled at www.clinicaltrials.gov. These trials are predominantly *ex-vivo* trials involving experimental treatments of diseases ranging from HIV infection to various cancers including virally derived cancers, blood cancers and solid tumours (Table 15.1). Interestingly,

**Table 15.1** Scheduled and ongoing clinical trials involving CRISPR–Cas9 at the clinicaltrials.gov registry. HPV (human papilloma virus), CCR5 (cell-surface receptor used for HIV entry), CAR T-cells (chimeric antigen receptor T-cells), PD1 (programmed cell death 1 gene), KO (knock-out), EBV (Epstein Barr Virus), CTL (cytotoxic T-lymphocytes).

| Indication | Target | Administration | Status | Reference |
|---|---|---|---|---|
| HPV related malignancy | HPV viral genes | Topical | Not yet recruiting | NCT03057912 |
| HIV infection | CCR5 CD34+ cells | *Ex-vivo* | Recruiting | NCT03164135 |
| Leukaemia/lymphoma | CAR T-cells | *Ex-vivo* | Recruiting | NCT03166878 |
| Oesophageal cancer | PD1 KO T-cells | *Ex-vivo* | Recruiting | NCT03081715 |
| Bladder cancer stage IV | PD1 KO T-cells | *Ex-vivo* | Not yet recruiting | NCT02863913 |
| Prostate cancer | PD1 KO T-cells | *Ex-vivo* | Not yet recruiting | NCT02867345 |
| Metastatic renal cell carcinoma | PD1 KO T-cells | *Ex-vivo* | Not yet recruiting | NCT02867332 |
| Metastatic non-small cell lung cancer | PD1 KO T-cells | *Ex-vivo* | Recruiting | NCT02793856 |
| EBV-associated cancers | PD1 KO EBV-CTLs | *Ex-vivo* | Recruiting | NCT03044743 |

all planned trials will be conducted in China. No CRISPR–Cas9 trials are currently listed for Europe or the USA at clinicaltrials.gov.

Given the potential catastrophic consequences of off-target editing involving cancer-critical genes, paired with the observation of extensive off-target editing using SpCas9, it seems prudent to consider preferential use of high-specificity Cas9 variants. In addition to using highly precise editing enzymes, safe and efficient delivery methods should be used [reviewed in ref. 54] and ideally these should enable only transient Cas9–sgRNA expression, for example *via* delivery of purified Cas9–sgRNA complex, to limit the potential for continued DSB induction following repair. As far as possible, unbiased genome-wide sequence quality control following editing should be conducted. Whether delivering Cas9 encoded on a plasmid (typically >10 kb), as an RNA (~5 kb) or as a ribonucleoprotein in complex with sgRNA (~160 kDa), these are very large structures which are challenging to transport into cells. Delivery involving DNA (viruses in particular) also entails a potential for integration, oncogene activation, continued Cas9 expression and long-term chromosomal instability. For comparison, antisense oligonucleotides, aptamers and siRNAs are further into clinical development, with six drugs approved by the FDA as of 2017, showing proof of concept for oligonucleotide-analogue-based therapy.[55] Whether CRISPR–Cas9 will similarly enable therapeutic benefits with acceptable side effects remains to be established. Given the rapid progress since the first applications of SpCas9 in human genome editing, however, there are now grounds for a cautious hope that further understanding of its workings within human cells, coupled with additional developments, could transform CRISPR–*cas* systems into fully-fledged genome surgical tools for treating a variety of human diseases and for other purposes. In this respect, it is worth keeping in mind that there are many additional Cas9 proteins besides SpCas9 with exciting properties, and that type II systems constitute only a minority of CRISPR-*cas* systems[6] in Nature's microbial treasure trove.

## References

1. R. Barrangou and P. Horvath, A decade of discovery: CRISPR functions and applications, *Nat. Microbiol.*, 2017, **2**, 17092.
2. E. S. Lander, The Heroes of CRISPR, *Cell*, 2016, **164**, 18–28.
3. F. J. Mojica and F. Rodriguez-Valera, The discovery of CRISPR in archaea and bacteria, *FEBS J.*, 2016, **283**, 3162–3169.
4. S. Chira, D. Gulei, A. Hajitou, A. A. Zimta, P. Cordelier and I. Berindan-Neagoe, CRISPR/Cas9: Transcending the Reality of Genome Editing, *Mol. Ther. -Nucleic Acids*, 2017, **7**, 211–222.
5. T. I. Cornu, C. Mussolino and T. Cathomen Refining strategies to-translate genome editing to the clinic. *Nat. Med.*, 2017, **23**, 415–423.
6. K. S. Makarova, Y. I. Wolf, O. S. Alkhnbashi, F. Costa, S. A. Shah, S. J. Saunders, R. Barrangou, S. J. Brouns, E. Charpentier and D. H. Haft, *et al.*, An updated evolutionary classification of CRISPR-Cas systems, *Nat. Rev. Microbiol.*, 2015, **13**, 722–736.
7. Y. Ishino, H. Shinagawa, K. Makino, M. Amemura and A. Nakata, Nucleotide sequence of the iap gene, responsible for alkaline phosphatase isozyme conversion in Escherichia coli, and identification of the gene product, *J. Bacteriol.*, 1987, **169**, 5429–5433.
8. P. W. Hermans, S. D. van, E. M. Bik, P. E. de Haas, J. W. Dale and J. D. van Embden, Insertion element IS987 from Mycobacterium bovis BCG is located in a hot-spot integration region for insertion elements in Mycobacterium tuberculosis complex strains, *Infect. Immun.*, 1991, **59**, 2695–2705.
9. F. J. Mojica, G. Juez and F. Rodriguez-Valera, Transcription at different salinities of Haloferax mediterranei sequences adjacent to partially modified PstI sites, *Mol. Microbiol.*, 1993, **9**, 613–621.
10. F. J. Mojica, C. Ferrer, G. Juez and F. Rodriguez-Valera, Long stretches of short tandem repeats are present in the largest replicons of the Archaea Haloferax mediterranei and Haloferax volcanii and could be involved in replicon partitioning, *Mol. Microbiol.*, 1995, **17**, 85–93.
11. F. J. Mojica, C. Diez-Villasenor, E. Soria and G. Juez, Biological significance of a family of regularly spaced repeats in the genomes of Archaea, Bacteria and mitochondria, *Mol. Microbiol.*, 2000, **36**, 244–246.
12. R. Jansen, J. D. Embden, W. Gaastra and L. M. Schouls, Identification of genes that are associated with DNA repeats in prokaryotes, *Mol. Microbiol.*, 2002, **43**, 1565–1575.
13. A. Bolotin, B. Quinquis, A. Sorokin and S. D. Ehrlich, Clustered regularly interspaced short palindrome repeats (CRISPRs) have spacers of extrachromosomal origin, *Microbiology*, 2005, **151**, 2551–2561.

14. F. J. Mojica, C. Diez-Villasenor, J. Garcia-Martinez and E. Soria, Intervening sequences of regularly spaced prokaryotic repeats derive from foreign genetic elements, *J. Mol. Evol.*, 2005, **60**, 174–182.
15. C. Pourcel, G. Salvignol and G. Vergnaud, CRISPR elements in Yersinia pestis acquire new repeats by preferential uptake of bacteriophage DNA, and provide additional tools for evolutionary studies, *Microbiology*, 2005, **151**, 653–663.
16. R. Barrangou, C. Fremaux, H. Deveau, M. Richards, P. Boyaval, S. Moineau, D. A. Romero and P. Horvath, CRISPR provides acquired resistance against viruses in prokaryotes, *Science*, 2007, **315**, 1709–1712.
17. F. J. Mojica, C. Diez-Villasenor, J. Garcia-Martinez and C. Almendros, Short motif sequences determine the targets of the prokaryotic CRISPR defence system, *Microbiology*, 2009, **155**, 733–740.
18. E. Deltcheva, K. Chylinski, C. M. Sharma, K. Gonzales, Y. Chao, Z. A. Pirzada, M. R. Eckert, J. Vogel and E. Charpentier, CRISPR RNA maturation by trans-encoded small RNA and host factor RNase III, *Nature*, 2011, **471**, 602–607.
19. L. A. Marraffini and E. J. Sontheimer, Self versus non-self discrimination during CRISPR RNA-directed immunity, *Nature*, 2010, **463**, 568–571.
20. L. A. Marraffini and E. J. Sontheimer, CRISPR interference limits horizontal gene transfer in staphylococci by targeting DNA, *Science*, 2008, **322**, 1843–1845.
21. C. R. Hale, P. Zhao, S. Olson, M. O. Duff, B. R. Graveley, L. Wells, R. M. Terns and M. P. Terns, RNA-guided RNA cleavage by a CRISPR RNA-Cas protein complex, *Cell*, 2009, **139**, 945–956.
22. O. O. Abudayyeh, J. S. Gootenberg, S. Konermann, J. Joung, I. M. Slaymaker, D. B. Cox, S. Shmakov, K. S. Makarova, E. Semenova and L. Minakhin, *et al.*, C2c2 is a single-component programmable RNA-guided RNA-targeting CRISPR effector, *Science*, 2016, **353**, aaf5573.
23. R. K. Lillestol, P. Redder, R. A. Garrett and K. Brugger, A putative viral defence mechanism in archaeal cells, *Archaea*, 2006, **2**, 59–72.
24. T. H. Tang, J. P. Bachellerie, T. Rozhdestvensky, M. L. Bortolin, H. Huber, M. Drungowski, T. Elge, J. Brosius and A. Huttenhofer, Identification of 86 candidates for small non-messenger RNAs from the archaeon Archaeoglobus fulgidus, *Proc. Natl. Acad. Sci. U. S. A.*, 2002, **99**, 7536–7541.
25. T. H. Tang, N. Polacek, M. Zywicki, H. Huber, K. Brugger, R. Garrett, J. P. Bachellerie and A. Huttenhofer, Identification of novel non-coding RNAs as potential antisense regulators in the archaeon Sulfolobus solfataricus, *Mol. Microbiol.*, 2005, **55**, 469–481.
26. S. J. Brouns, M. M. Jore, M. Lundgren, E. R. Westra, R. J. Slijkhuis, A. P. Snijders, M. J. Dickman, K. S. Makarova, E. V. Koonin and J. van der Oost, Small CRISPR RNAs guide antiviral defense in prokaryotes, *Science*, 2008, **321**, 960–964.
27. G. Gasiunas, R. Barrangou, P. Horvath and V. Siksnys, Cas9-crRNA ribonucleoprotein complex mediates specific DNA cleavage for adaptive immunity in bacteria, *Proc. Natl. Acad. Sci. U. S. A.*, 2012, **109**, E2579–E2586.

28. M. Jinek, K. Chylinski, I. Fonfara, M. Hauer, J. A. Doudna and E. Charpentier, A programmable dual-RNA-guided DNA endonuclease in adaptive bacterial immunity, *Science*, 2012, **337**, 816–821.
29. H. Nishimasu, F. A. Ran, P. D. Hsu, S. Konermann, S. I. Shehata, N. Dohmae, R. Ishitani, F. Zhang and O. Nureki, Crystal structure of Cas9 in complex with guide RNA and target DNA, *Cell*, 2014, **156**, 935–949.
30. L. Cong, F. A. Ran, D. Cox, S. Lin, R. Barretto, N. Habib, P. D. Hsu, X. Wu, W. Jiang and L. A. Marraffini, *et al.*, Multiplex genome engineering using CRISPR/Cas systems, *Science*, 2013, **339**, 819–823.
31. M. Jinek, A. East, A. Cheng, S. Lin, E. Ma and J. Doudna, RNA-programmed genome editing in human cells, *eLife*, 2013, **2**, e00471.
32. P. Mali, L. Yang, K. M. Esvelt, J. Aach, M. Guell, J. E. DiCarlo, J. E. Norville and G. M. Church, RNA-guided human genome engineering via Cas9, *Science*, 2013, **339**, 823–826.
33. Q. Ding, S. N. Regan, Y. Xia, L. A. Oostrom, C. A. Cowan and K. Musunuru, Enhanced efficiency of human pluripotent stem cell genome editing through replacing TALENs with CRISPRs, *Cell Stem Cell*, 2013, **12**, 393–394.
34. W. Jiang, D. Bikard, D. Cox, F. Zhang and L. A. Marraffini, RNA-guided editing of bacterial genomes using CRISPR-Cas systems, *Nat. Biotechnol.*, 2013, **31**, 233–239.
35. J. E. DiCarlo, J. E. Norville, P. Mali, X. Rios, J. Aach and G. M. Church, Genome engineering in Saccharomyces cerevisiae using CRISPR-Cas systems, *Nucleic Acids Res.*, 2013, **41**, 4336–4343.
36. N. Chang, C. Sun, L. Gao, D. Zhu, X. Xu, X. Zhu, J. W. Xiong and J. J. Xi, Genome editing with RNA-guided Cas9 nuclease in zebrafish embryos, *Cell Res.*, 2013, **23**, 465–472.
37. W. Y. Hwang, Y. Fu, D. Reyon, M. L. Maeder, S. Q. Tsai, J. D. Sander, R. T. Peterson, J. R. Yeh and J. K. Joung, Efficient genome editing in zebrafish using a CRISPR-Cas system, *Nat. Biotechnol.*, 2013, **31**, 227–229.
38. S. J. Gratz, A. M. Cummings, J. N. Nguyen, D. C. Hamm, L. K. Donohue, M. M. Harrison, J. Wildonger and K. M. O'Connor-Giles, Genome engineering of Drosophila with the CRISPR RNA-guided Cas9 nuclease, *Genetics*, 2013, **194**, 1029–1035.
39. B. Shen, J. Zhang, H. Wu, J. Wang, K. Ma, Z. Li, X. Zhang, P. Zhang and X. Huang, Generation of gene-modified mice via Cas9/RNA-mediated gene targeting, *Cell Res.*, 2013, **23**, 720–723.
40. Y. Fu, J. A. Foden, C. Khayter, M. L. Maeder, D. Reyon, J. K. Joung and J. D. Sander, High-frequency off-target mutagenesis induced by CRISPR-Cas nucleases in human cells, *Nat. Biotechnol.*, 2013, **31**, 822–826.
41. P. D. Hsu, D. A. Scott, J. A. Weinstein, F. A. Ran, S. Konermann, V. Agarwala, Y. Li, E. J. Fine, X. Wu and O. Shalem, *et al.*, DNA targeting specificity of RNA-guided Cas9 nucleases, *Nat. Biotechnol.*, 2013, **31**, 827–832.
42. V. Pattanayak, S. Lin, J. P. Guilinger, E. Ma, J. A. Doudna and D. R. Liu, High-throughput profiling of off-target DNA cleavage reveals RNA-programmed Cas9 nuclease specificity, *Nat. Biotechnol.*, 2013, **31**, 839–843.

43. E. Semenova, M. M. Jore, K. A. Datsenko, A. Semenova, E. R. Westra, B. Wanner, J. van der Oost, S. J. Brouns and K. Severinov, Interference by clustered regularly interspaced short palindromic repeat (CRISPR) RNA is governed by a seed sequence, *Proc. Natl. Acad. Sci. U. S. A.*, 2011, **108**, 10098–10103.
44. B. P. Kleinstiver, V. Pattanayak, M. S. Prew, S. Q. Tsai, N. T. Nguyen, Z. Zheng and J. K. Joung, High-fidelity CRISPR-Cas9 nucleases with no detectable genome-wide off-target effects, *Nature*, 2016, **529**, 490–495.
45. I. M. Slaymaker, L. Gao, B. Zetsche, D. A. Scott, W. X. Yan and F. Zhang, Rationally engineered Cas9 nucleases with improved specificity, *Science*, 2016, **351**, 84–88.
46. Y. Fu, J. D. Sander, D. Reyon, V. M. Cascio and J. K. Joung, Improving CRISPR-Cas nuclease specificity using truncated guide RNAs, *Nat. Biotechnol.*, 2014, **32**, 279–284.
47. H. Nishimasu, L. Cong, W. X. Yan, F. A. Ran, B. Zetsche, Y. Li, A. Kurabayashi, R. Ishitani, F. Zhang and O. Nureki, Crystal Structure of *Staphylococcus aureus* Cas9, *Cell*, 2015, **162**, 1113–1126.
48. T. Yamano, H. Nishimasu, B. Zetsche, H. Hirano, I. M. Slaymaker, Y. Li, I. Fedorova, T. Nakane, K. S. Makarova and E. V. Koonin, *et al.*, Crystal Structure of Cpf1 in Complex with Guide RNA and Target DNA, *Cell*, 2016, **165**, 949–962.
49. F. A. Ran, L. Cong, W. X. Yan, D. A. Scott, J. S. Gootenberg, A. J. Kriz, B. Zetsche, O. Shalem, X. Wu and K. S. Makarova, *et al.*, In vivo genome editing using *Staphylococcus aureus* Cas9, *Nature*, 2015, **520**, 186–191.
50. B. Zetsche, J. S. Gootenberg, O. O. Abudayyeh, I. M. Slaymaker, K. S. Makarova, P. Essletzbichler, S. E. Volz, J. Joung, J. van der Oost and A. Regev, *et al.*, Cpf1 is a single RNA-guided endonuclease of a class 2 CRISPR-Cas system, *Cell*, 2015, **163**, 759–771.
51. V. T. Chu, T. Weber, B. Wefers, W. Wurst, S. Sander, K. Rajewsky and R. Kuhn, Increasing the efficiency of homology-directed repair for CRISPR-Cas9-induced precise gene editing in mammalian cells, *Nat. Biotechnol.*, 2015, **33**, 543–548.
52. T. Maruyama, S. K. Dougan, M. C. Truttmann, A. M. Bilate, J. R. Ingram and H. L. Ploegh, Increasing the efficiency of precise genome editing with CRISPR-Cas9 by inhibition of nonhomologous end joining, *Nat. Biotechnol.*, 2015, **33**, 538–542.
53. H. Ma, N. Marti-Gutierrez, S. W. Park, J. Wu, Y. Lee, K. Suzuki, A. Koski, D. Ji, T. Hayama and R. Ahmed, *et al.*, Correction of a pathogenic gene mutation in human embryos, *Nature*, 2017, **548**, 413–419.
54. H. Yin, K. J. Kauffman and D. G. Anderson, Delivery technologies for genome editing, *Nat. Rev. Drug Discovery*, 2017, **16**, 387–399.
55. C. A. Stein and D. Castanotto, FDA-Approved Oligonucleotide Therapies in 2017, *Mol. Ther.*, 2017, **25**, 1069–1075.
56. J. S. Chen, Y. S. Dagdas, B. P. Kleinstiver, M. M. Welch, A. A. Sousa, L. B. Harrington, S. H. Sternberg, J. K. Joung, A. Yildiz and J. A. Doudna, Enhanced proofreading governs CRISPR-Cas9 targeting accuracy, *Nature*, 2017, **550**, 407–410.

# Subject Index